Lecture Notes in Artificial Intelligence 7243

Subseries of Lecture Notes in Computer Science

Helena Caseli Aline Villavicencio
António Teixeira Fernando Perdigão (Eds.)

Computational Processing of the Portuguese Language

10th International Conference, PROPOR 2012
Coimbra, Portugal, April 17-20, 2012
Proceedings

Springer

Series Editors

Randy Goebel, University of Alberta, Edmonton, Canada
Jörg Siekmann, University of Saarland, Saarbrücken, Germany
Wolfgang Wahlster, DFKI and University of Saarland, Saarbrücken, Germany

Volume Editors

Helena Caseli
UFSCAR, São Carlos, SP, Brazil
E-mail: helenacaseli@dc.ufscar.br

Aline Villavicencio
UFRGS, Porto Alegre, RS, Brazil
E-mail: alinev@gmail.com

António Teixeira
Universidade de Aveiro, Aveiro, Portugal
E-mail: ajst@ua.pt

Fernando Perdigão
Universidade de Coimbra, Coimbra, Portugal
E-mail: fp@deec.uc.pt

ISSN 0302-9743 e-ISSN 1611-3349
ISBN 978-3-642-28884-5 e-ISBN 978-3-642-28885-2
DOI 10.1007/978-3-642-28885-2
Springer Heidelberg Dordrecht London New York

Library of Congress Control Number: 2012933122

CR Subject Classification (1998): I.2, H.3, H.4, I.4, I.5, H.2.8

LNCS Sublibrary: SL 7 – Artificial Intelligence

Typesetting: Camera-ready by author, data conversion by Scientific Publishing Services, Chennai, India

Printed on acid-free paper

Springer is part of Springer Science+Business Media (www.springer.com)

Preface

The International Conference on Computational Processing of Portuguese – PROPOR – is the main event in the area of natural language processing that is focused on Portuguese and the theoretical and technological issues related to this language. It welcomes contributions for both written and spoken language processing.

The event is hosted in Brazil and in Portugal. The meetings have been held in Lisbon/Portugal (1993), Curitiba/Brazil (1996), Porto Alegre/Brazil (1998), Évora/Portugal (1999), Atibaia/Brazil (2000), Faro/Portugal (2003), Itatiaia/Brazil (2006), Aveiro/Portugal (2008), and Porto Alegre/Brazil (2010).

This meeting has been a highly productive forum for the progress of this area and to foster the cooperation among the researchers working on the automated processing of the Portuguese language. PROPOR brings together research groups promoting the development of methodologies, resources, and projects that can be shared among all researchers and practitioners in the field.

The tenth edition of this event was held at the University of Coimbra, Coimbra, Portugal. It had two main tracks: one for language processing and another for speech processing. This event hosted a special Demonstration Session and a satellite event named "Págico," consisting in an evaluation contest for non-trivial information seeking in Portuguese using Wikipedia as target. This edition of PROPOR featured two invited talks by internationally renowned researchers as well as tutorials on symbolic and statistical approaches to natural language processing and analysis and visual feedback of the singing voice.

A total of 86 submissions were received, 61 for the language track and 25 for the speech track, by authors in worldwide institutions from countries like Brazil, China, Germany, Portugal, and Spain. Each submission was evaluated by at least three members from a multidisciplinary and international scientific committee.

This volume gathers a selection of the 47 best papers accepted to be presented at this meeting, of which 24 are full papers, corresponding to an acceptance rate of 27%. These papers cover the areas related to automatic acquisition of information, linguistic description and processing, language resources, language applications and speech production, speech processing and applications.

We would like to express our thanks to everyone involved in the organization of the event, to the scientific committee members for their excellent work, to the researchers who kindly accepted to contribute to the event by delivering tutorials and invited talks, and to the institutions, organizations, and funding agencies which allowed the realization of this event, namely, University of Coimbra, IT (Instituto de Telecomunicações), FCT (Portuguese National Founding Agency), ISCA (International Speech Communication Association), SIG-IL (the ISCA Special Group on Iberian Languages), CEPLN (the Special Interest Group on Natural Language Processing of the Brazilian Computer Society), and ACL (Association for Computational Linguistics).

April 2012

Helena de Medeiros Caseli
Aline Villavicencio
António Teixeira
Fernando Perdigão

Organization

General Chair

Fernando Perdigão	Universidade de Coimbra/IT, Portugal

Program Chairs

Aline Villavicencio	UFRGS, Brazil – Language
António Teixeira	Universidade de Aveiro/IEETA, Portugal – Speech

Editorial Chair

Helena de Medeiros Caseli	UFSCAR, Brazil

Demo Session Chair

Alberto Abad	L2F INESC-ID, Portugal

PhD and MSc/MA Dissertation Contest Chair

Jorge Baptista	Universidade do Algarve, Portugal

Local Organizing Committee

Luís Sá	UC/IT, Portugal
Sara Candeias	IT, Portugal
Ana R. Luís	UC/CELGA, Portugal
Carla Lopes	IPLEI/IT, Portugal
Hugo Gonçalo Oliveira	UC/CISUC, Portugal

Program Committee

Alberto Abad	INESC-ID, Portugal
Alberto Simões	UM, Portugal
Aldebaro Klautau	UFPA, Brazil
Alexandre Agustini	PUC-RS, Brazil
Aline Villavicencio	UFRGS, Brazil
Amália Andrade	UL, Portugal
Amália Mendes	UL, Portugal

Ana Bocorny	PUC-RS, Brazil
Ana Luís	UC, Portugal
Andre Adami	UCS, Brazil
Andreia Rauber	UCPEL, Brazil
Antonio Bonafonte	UPC, Spain
António Branco	UL, Portugal
Antonio Rubio	UG, Spain
António Serralheiro	INESC-ID, Portugal
António Teixeira	Universidade de Aveiro, Portugal
Ariadne Carvalho	Unicamp, Brazil
Ariani Di Felippo	UFSCAR, Brazil
Belinda Maia	UP, Portugal
Bento da Silva	UNESP, Brazil
Berthold Crysmann	CNRS Paris-Diderot, France
Carlos Prolo	PUC-RS, Brazil
Carlos Teixeira	UL, Portugal
Carmen García Mateo	UV, Spain
Caroline Gasperin	TouchType, UK
Caroline Hagège	Xerox Research Centre, France
Catarina Oliveira	Universidade de Aveiro, Portugal
Ciro Martins	Universidade de Aveiro, Portugal
Cristiane Killian	UFRGS, Brazil
Daniela Braga	Microsoft, China
Dante Barone	UFRGS, Brazil
Diana Santos	University of Oslo, Norway
Doroteo Torre Toledano	UAM, Spain
Eduardo Lleida	UZ, Spain
Eric Laporte	Université Paris Est, France
Eva Navas	UBC, Spain
Fabio Kepler	USP, Brazil
Fábio Violaro	Unicamp, Brazil
Fernando Resende	UFRJ, Brazil
Gaël Harry Dias	UBI, Portugal
Gladis Almeida	UFSCAR, Brazil
Helena de Medeiros Caseli	UFSCAR, Brazil
Irene Rodrigues	UE, Portugal
Isabel Falé	Universidade Aberta, Portugal
Isabel Trancoso	INESC-ID/IST, Portugal
Ivandré Paraboni	USP, Brazil
Jean-Luc Minel	Université de Paris X, France
João Balsa	UL, Portugal
João Luís Rosa	USP-SC, Brazil
João Paulo Neto	INESC-ID/IST, Portugal
João Veloso	UP, Portugal
Joaquim Ferreira da Silva	UNL, Portugal
Joaquim Llisterri	UAB, Spain

Jorge Baptista	Universidade do Algarve, Portugal
José Gabriel Lopes	UNL, Portugal
José João Almeida	UM, Portugal
Julia Hirschberg	Columbia University, USA
Laura Alonso Alemany	University National of Cordoba, Argentina
Leandro Oliveira	Embrapa, Brazil
Leandro Wives	UFRGS, Brazil
Lúcia Rino	UFSCAR, Brazil
Lucia Specia	University of Wolverhampton, UK
Luís Oliveira	INESC-ID, Portugal
Luís Sá	UC, Portugal
Luísa Coheur	INESC-ID/IST, Portugal
Luiz Pizzato	University of Sydney, Australia
Magali Duran	USP-SC, Brazil
Mara Abel	UFRGS, Brazil
Marcelo Finger	USP, Brazil
Marco Gonzalez	PUC-RS, Brazil
Maria das Graças Volpe Nunes	USP-SC, Brazil
Maria Helena Mira Mateus	ILTEC, Portugal
Maria José Finatto	UFRGS, Brazil
Mário Silva	INESC-ID/IST, Portugal
Michel Gagnon	Ecole Polytechnique, Canada
Miguel Sales Dias	Microsoft-MLDC, Portugal
Nuno Cavalheiro Marques	UNL, Portugal
Nuno Mamede	INESC-ID/IST, Portugal
Pablo Gamallo	University of Santiago de Compostela, Spain
Palmira Marrafa	UL, Portugal
Paulo Gomes	UC, Portugal
Paulo Quaresma	UE, Portugal
Plínio Barbosa	Unicamp, Brazil
Ranniery Maia	Toshiba, UK
Renata Vieira	PUC-RS, Brazil
Ricardo Ribeiro	INESC-ID/ISCTE-IUL, Portugal
Robert Dale	Macquarie University, Australia
Ronaldo Martins	Univas, Brazil
Rove Chishman	Unisinos, Brazil
Rubén San-Segundo	UPM, Spain
Ruy Luiz Milidiú	PUC-Rio, Brazil
Sandra Aluisio	USP-SC, Brazil
Sergio Freitas	UnB, Brazil
Solange Rezende	USP-SC, Brazil
Stanley Loh	UCPEL, Brazil
Steven Bird	University of Melbourne, Australia
Thiago Pardo	USP-SC, Brazil
Tracy Holloway King	Microsoft, USA

Valéria Feltrim	UEM, Brazil
Vera Strube de Lima	PUC-RS, Brazil
Violeta Quental	PUC-Rio, Brazil
Vitor Rocio	Universidade Aberta, Portugal
Viviane Moreira	UFRGS, Brazil

Steering Committee

António Teixeira	Universidade de Aveiro, Portugal (Chair)
Fernando Perdigão	Universidade de Coimbra, Portugal
Jorge Baptista	Universidade do Algarve, Portugal
Renata Vieira	PUC-RS, Brazil
Violeta Quental	PUC-RJ, Brazil

Table of Contents

Phonology, Morphology and POS-Tagging

Acquisition

Language Resources

Linguistic Description, Syntax and Parsing

Semantics

Opinion Analysis

Natural Language Processing Applications

Speech Production and Phonetics

Speech Resources

Speech Processing and Applications

Verb Analysis in a Highly Inflective Language with an MFF Algorithm

António Branco and Filipe Nunes

University of Lisbon
{Antonio.Branco,Filipe.Nunes}@di.fc.ul.pt

Abstract. We introduce the MFF algorithm for the task of verbal inflection analysis. This algorithm follows an heuristics that decide for the most frequent inflection feature bundle given the set of admissible feature bundles for a verb input form. This algorithm achieves a significantly better level of accuracy than the ones offered by current stochastic tagging technology commonly used for the same task.

Keywords: ambiguity resolution, verbal inflection, morphological analysis, tagging, tense, aspect, mood, Romance languages, Portuguese.

1 Introduction

In highly inflective languages, the morphological information associated with each token plays an important role in the processing of these languages. For instance, inflection features such as *case*, *number* or *gender* contribute to resolve syntactic ambiguity and may help to partly determine the underlying argument structure (e.g. *case* in free word-order languages). Some other features, e.g. conveying morphological information on *tense*, *person* or *polarity*, may even be the only sources on the basis of which some pieces of semantic information may be determined.

Being an important processing phase, morphological analysis of inflection turns out to be also a challenging one as it has to cope with non trivial ambiguity resolution.

From a broad viewpoint, inflectional ambiguity originates at two interdependent layers: on the one hand, for a given word form, different substrings may happen to qualify, in alternative, as admissible affixes. On the other hand, a given affix may happen to convey more than one admissible feature value. In order to decide what feature values happen to be actually conveyed by an occurrence of a given word, information on the context of that word has to be used to help determining, from its admissible feature values, the ones that are instantiated in that specific occurrence.

A family resemblance emerges between this task and the task of part-of-speech (POS) tagging. Accordingly, it has been common wisdom to approach the task of morphological analysis of inflection as a tagging task (Chanod and Tapanainen, 1995; Hajič and Hladká, 1998; Ezeiza *et al.*, 1998; Hakkani-Tür *et al.*, 2000; Tufiş, 1999; Cucerzan and Yarowsky, 2002; Trushkina and Hinrichs, 2004).

H. Caseli et al. (Eds.): PROPOR 2012, LNAI 7243, pp. 1–11, 2012.

In the present paper, our goal is to present results showing that, at least for some natural languages, there may be advantages to depart from this view in what concerns the analysis of verbal inflection involving time related morphological information.

In Section 2, we present the results of experiments where the task of verbal inflection is approached as a tagging task. The language used in our experiments is a Romance language, Portuguese, a head-initial, highly inflective language in the verbal domain.

In Section 3, we examine in more detail the problem space for verbal inflection and discuss possible alternative ways to conceptualize the task at stake. In Section 4, we present the improvement obtained when tackling this task by using a quite straightforward heuristics to approach to it.

Finally, in Section 5, we discuss the results obtained, and in Section 6, we present concluding remarks.

2 Verb Inflection Analysis as Tagging

As recurrently reported in the literature on POS tagging, inflective languages may raise a problem for stochastic approaches. Besides the typical POS tags, in these languages tokens need to be tagged with a plethora of additional tags representing values for inflection features. This typically requires a much larger tagset, with the consequent worsening of the data sparseness effect.

Nevertheless, this negative impact may to a large extent be compensated by the fact that viewing inflection analysis as a task similar to (or an extension of) POS tagging permits to take advantage of the results accumulated in this domain, whose state of the art accuracy with the best scoring methods is in the range 97%-98%.

Hence, in order to set up a verbal analyzer, we resorted to a state-of-the-art approach embodied in one of the best performing implementations. We used TnT (Brants, 2000), that implements a HMM approach with back off and suffix analysis.

For the training data, we used a corpus of a moderate size (261 385 tokens), with the portion of the CINTIL corpus (Barreto *et al.*, 2006) containing excerpts from news (3/5) and novels (2/5). This is a corpus manually annotated with a large tagset, including a subset with 80 tags (bundles of feature values) for verbal inflection in Portuguese. The models were trained over 90% of this corpus, and the remaining 10% held out for evaluation.

In order to get a perception of how the verbal analyzer may compare to a POS-only tagger trained under the same settings, we used TnT to produce a POS tagger on the basis of the annotation present in the same corpus (tagset of size 69), and with the same evaluation procedure. The resulting tagger scored 96.87% of accuracy.

Next, we developed several verbal inflection disambiguators.

2.1 Experiment T1

In a first experiment, the training data was prepared so that the hidden states are word tokens concatenated with their POS tags (accurately hand annotated), and the symbols to be emitted are the verb inflection tags, for verbal tokens, and a designated null symbol, for the remaining tokens.

The evaluation of this verbal featurizer showed a score of 93.34% for accuracy.

2.2 Experiment T2

As in real applications the POS tags are assigned automatically and henceforth are not all correct, it is relevant to study the outcome of the verbal analyzer when it works in a more realistic setting, viz. over the output of the POS tagger initially referred to above.

When running the verbal analyzer under these circumstances, a score of 92.22% for accuracy was obtained. The noise introduced in POS assignment by the tagger has thus a detrimental effect on the morphological analysis of verbal inflection of over 1% point.

2.3 Experiment T3

Searching for alternatives to possibly alleviate this negative impact of the tagger on the performance of the featurizer, another experiment was yet carried out: POS tagger and verbal featurizer were trained as a single classifier, in the more usual setting of using a larger tagset whose tags were extended with morphological information. In the training data for this encompassing, single pass tagger, a plain word is taken as a hidden state, and the emitted symbol is the tag resulting from the concatenation of the POS tag with the inflectional tag for that word.

No improvement was obtained, as this solution evaluated even slightly worst, to 92.06% of accuracy.

The sequence of results obtained, with decreasing scores from Experiment 1 to Experiment 3, is quite as expected:

Table 1. Accuracy of HMM-based classifiers used for verb inflection analysis

Input	accurate POS	automatic POS	raw text
Output	Infl tags	Infl tags	POS+Infl
Accuracy	93.34%	92.22%	92.06%

By using a POS tagger with only around 97% accuracy in Experiment 2, it is expected that some of the POS-tags corresponding to verbs will be misplaced, and that the verbal analyzer trained over tokens ending in that tag ends up by having a poorer accuracy than in Experiment 1, where it is run over data manually POS annotated. As for the drop from Experiment 2 to Experiment 3, apparently, with a training data with the size of the working corpus we used, the benefits of a slightly larger tagset (with size 148), more than doubling the size of the initial POS only tagset (size 69), were canceled by the sparseness of the data available to significantly estimate the relevant parameters.

Against this background, what turns out to be interesting is the comparison between the accuracy of the POS-only tagger (96.87%) and the accuracy of the best verbal inflection analyzer (93.34%). This drop of more than 3.5% points is unlikely to be due to the mere extending of the tagset size from 69 to 80, even more so that no strong correlation exists between the two inventories of tags (Elworthy, 1995). This decrease is thus rather more likely to be found in the different scattering of the occurrence of the different tags in the tagsets along the training corpus: As there are 27 823 verbal tokens in the training corpus for the verbal analyzer in Experiment 1, one of the 80 tags, viz. the null tag associated with tokens that are not verbs, decorates as much as 88.18% of the tokens in that corpus. This is a much more uneven scattering of tags than the usual one observed in a training corpus for a POS only tagger.

3 How to Improve?

In order to look for improvement, and leaving aside the costly solution of constructing a larger corpus, it is worth to further ponder about the possible reasons underlying the results above.

3.1 A Look into the Problem Space

From the 30 976 verb tokens occurring in our working corpus, around 1/2 are lexically ambiguous, i.e. their inflection suffixes are in correspondence with more than one bundle of feature values. And considering the vocabulary of size 21 814 with types of conjugated verb forms collected from the corpus, one finds an ambiguity rate of 1.42.

It turns out that every feature involved in the bundle of features of verbal inflection may display ambiguity.

It may emerge in terms of *Mood* values (e.g. verb *dar*, to give): *dê* :
<Conjuntivo, Presente, 3rd, Singular> or
<Imperativo, 2nd Courtesy, Singular>

In terms of *Tense* values: *deram* :
<Indicativo, Pretérito Perfeito, 3rd, Plural> or
<Indicativo, Pretérito-Mais-Que-Perfeito, 3rd, Plural>

In terms of *Polarity* values: *dêmos* :
<Imperativo, Afirmativo, 1st, Plural> or
<Imperativo, Negativo, 1st, Plural>

In terms of *Person* values: *dava* :
<Indicativo, Pretérito Imperfeito, 1st, Singular> or
<Indicativo, Pretérito Imperfeito, 3rd, Singular>

In terms of *Number* values (e.g. verb *partir*, to leave): *parti* :
<Indicativo, Pretérito Perfeito, 1st, Singular> or
<Imperativo, 2nd, Plural>

Or in terms of *Gender* values (e.g. verb *assentar*, to lay): *assente* :
<Particípio Passado, 3rd, Singular, Masculine> or
<Particípio Passado, 3rd, Singular, Feminine>

Taking into account the information sources possibly relevant to resolve each one of the above feature value ambiguities, it is compelling to separate the features listed above into two groups. One group includes the first two features in the list above, Mood and Tense, in as much as for their values to be determined, "non local" information needs to be accessed. For instance, the sentential context preceding a verb (typically including the Subject, among other constituents) can hardly be seen a imposing any sensible constraint on its tense value, and the same can actually be observed with respect to any other set of words occurring in a "local" context window. Instead, as the information possibly relevant for disambiguation here is mostly discourse-based, the relevant information sources turn out to be found "non locally", outside any reasonably sized text window.

The other group contains the other four features, Polarity, Person, Number and Gender. Due to opposite circumstances, for their values to be determined, "local" information tend to be helpful for the resolution of ambiguity. As an example, a negative word in the clause containing the verb will help to decide on the polarity of the Imperative verb forms.

Nevertheless, this division into groups of features according to their need of "local" vs. "non local" sources of information for ambiguity resolution may turn out not to correspond to a distinction "easy" vs. "hard" cases for tagging approaches. This may be especially true when null subject languages, like Portuguese, are taken into account. In this case, the verb affixes with feature values for Person, Gender and Number more often than not turn out to be the only place where the information on the person, number and gender values of the Subject is expressed in the text, and no source of information other than the verb form is available to resolve their possible ambiguity. With these considerations in place, we can turn now to possible options to improve the performance of the verbal inflection analysis task.

3.2 Previous Work

A few strategies have been tried to alleviate the detrimental effects that highly inflective languages bring about for the performance of stochastic tagging technology. Past approaches include the tagging by inflection groups (Hakkani-Tür *et al.*, 2000), the tiered tagging (Tufiş, 1999), or a "second pass" with contextual-agreement models to tackle non adjacent dependencies for features like Gender (Cucerzan and Yarowsky, 2002).

In the tagging by inflection groups, the complex, longer tags that include information on POS and different bundles of feature values are broken down into their components. Each such subtag is then envisaged as being dependent on the previous subtags, either in the same or on the previous word tokens.

In the tiered tagging scheme, by means of trial and error, a subset of the complete tagset is used to train the tagger. That reduced tagset contains tags from which it is possible to "map back onto the appropriate tag in the large tagset in more than 90% of the cases" (Tufiş, 1999:29).

Finally, in the second pass approach, Gender agreement is modeled via a window-weighted global feature consensus displaying the best results for a window of size ±3.

These can be seen as different attempts to explore the old *divide et impera* approach to improve tagging technology, typically by dividing the whole tagging task possibly into a sequence of "easier" and more circumscribed tagging subtasks.

In any case, however, the issues raised in the previous subsection are not substantially addressed by these attempts: As the latter keep resorting to "local" sources of information for the optimization of their possible decomposition and interleaving, they hardly bring better chances to an improved account of disambiguation either when access to "non-local" information sources are needed (e.g. in the case of features such as Mood or Tense expressing time related information) or when one is faced with the absence of information sources extracted via shallow processing procedures (e.g. in the case of Person, Gender or Number features in null subject contexts).

3.3 Another Perspective

Against this background, it may be worth taking a step back and envisage the current task under a broader, linguistically informed perspective. Like any other predicator, a verb form may have different senses, that is may be ambiguous between the conveying of different relations between entities of the world (e.g. Portuguese ambiguous *fui* translates into English as *was* or *went*). Furthermore, while denoting temporally anchored state of affaires, a verb may be also ambiguous among the conveying of different time-aspect relations between temporal entities, including at least the utterance time, reference time and event time (Reichenbach, 1947). Accordingly, different admissible feature values of Tense or Mood for a verb form can be taken as different "senses" of that verb form, and to a large extent, the admissible inflection feature bundles for each verb form can be seen as their identifiers.

Under this perspective, to analyze a verb in a given context is thus to pick the (inflection) sense (made out of the semantic information conveyed by its inflectional suffixes) under which it occurs in that context. This suggests that there may be as much ground to conceptualize morphological analysis as a tagging task as there is for it to be conceptualized as a word sense disambiguation (WSD) task.

Following Pedersen and Mihalcea's (2005) overview on WSD methods, supervised learning methods forms a major group for WSD. Repeated experimenting and comparison has shown that, for this class of methods, features entered into the feature vector representations "tend to differentiate among methods more than the learning algorithms" of the methods. Good sets of features used for WSD typically include keywords, collocations, bigrams, POS or verb-subject and verb-object relations in the context window around the word to be disambiguated. In connection with the discussion above, a preliminary reflection on the possible impact of the above "local" features to discriminate among different inflectional verb feature bundles leads one to accept that the accuracy values of the top performing WSD systems — 70-73% for English lexical sample task in Senseval-3 (Mihalcea and Edmonds, 2004) — may not be within easy reach for an analyzer based on the tuning of feature vectors appropriate for verbal featurization.

Knowledge-based methods form another major group of methods for WSD. They include algorithms based on Machine readable dictionaries, Selectional restrictions, and Measures of semantic similarity based on ontologies. From these, the method based on Selectional restrictions is more targeted at verbs, but it does not seem to be the case that the semantic classes of the actual complements of a verb token may be significant to determine its inflectional feature values.

Heuristic-based methods are yet another subclass of knowledge-based methods for WSD, including methods based on Lexical chains, Most frequent sense, One sense per discourse, and One sense per collocation. From these, the one with best reported results is the heuristic that simply picks the most frequent sense, with a quite surprisingly good accuracy, not that far from the figures obtained with other, much more sophisticated WSD methods (Chklovski and Mihalcea, 2003).

As this heuristics appears to be a simple and yet promising WSD method to approach verbal inflection analysis, it inspired a resolution algorithm that led to surprisingly good results, as reported in the next Section.

4 MFF Algorithm

In the experiments reported below we used the Most Frequent Feature Bundle (MFF) algorithm to verbal inflection analysis, with the following outline:

Let TD be the training data, ST the set of verbal inflection tags occurring in the training data, VF the target verb form, and AT the set of its admissible inflection tags:

1. If VF was observed in TD, from the tags $T_1..T_n$ in AT, pick T_k such that VF_T_k is more frequent in TD than any other VF_T_n;

2. Else if at least one tag $T_1..T_n$ in AT was observed in TD, pick T_k such that T_k is more frequent in TD than any other T_n;

3. Else pick a tag at random from AT.

To turn this algorithm into an analyzer we resorted to the same training corpus used in the experiments described so far to obtain the relevant frequency scores for the verb forms. We also developed and used a fully accurate rule-based verbal lemmatization tool that for each input verb form delivers its set of admissible inflection tags, described in (Branco *et al.*, 2007).

4.1 Experiment D1

In line with the settings of Experiment T1, in a first experiment, the analyzer was run over an input with POS accurately hand annotated. It scored 96.92% in accuracy.

This is a result ca. 3% points better than the one obtained with the HMM analyzer of Experiment T1.

4.2 Experiment D2

In line with the settings of Experiment T2, in a second experiment, the analyzer was run over a more realistic setting, i.e. over an input whose POS tags were automatically assigned by the POS only tagger. This version of the analyzer scored 94.73% accuracy.

As expected, the noise of incorrect POS tags affecting the quality of the input led to a drop in accuracy. It can be observed that the over 3% points of tagging errors introduced by the tagger led to a drop of over 2% points in the accuracy of the analyzer. Again, the analyzer shows a better result than the one obtained by the HMM-based analyzer under the same circumstances, in Experiment T2, faring over 2.5% points better.

4.3 Experiment D3a

Given the expected tail of rarely observed items that may lead to low confidence decisions and an impoverished accuracy of the analyzer, we have experimented with alternative versions of the basic MFF algorithm described above. These versions result from adjusting the condition of applicability of Step 1 so that observed items are ignored when their frequency in the training data is below a certain threshold, and the procedure skips then to Step 2. When the more frequent sequence verb form-tag is too rare in the training data, it is ignored: The chosen tag is then the admissible one for that verb form with the highest frequency in the training data.

The analyzer was successively run over accurately POS tagged evaluation corpus with the threshold ranging from 0 to 4:

Table 2. Accuracy of verbal inflection analysis over input text with accurate POS tags

Threshold	0	**1**	2	3	4
Accuracy	96.92	**96.98**	96.98	96.88	96.82

These data indicate that, with respect to the first version of the algorithm (threshold = 0), improvement in accuracy is obtained when items with frequency 1 are discarded in Step1, while discarding also items with higher frequencies lead instead to poorer accuracy scores. The best result scored thus as much as 96.98%, over 3.6% points better than the result from the HMM analyzer in Experiment 1.

4.4 Experiment D3b

A similar testing was performed by running the analyzer over an evaluation corpus whose POS tags were automatically assigned by the POS-only tagger. The results obtained are summarized in the table below:

Table 3. Accuracy of verbal inflection analysis over input text automatically POS tagged

Threshold	0	**1**	2
Accuracy	94.73%	**96.51%**	95.62%

These data indicate that, with respect to the first version of the algorithm (threshold = 0), improvement in accuracy is obtained also when discarding items with frequency 1 in Step1.

Interestingly while in Experiment D3a a mere 0.02% points of improvement is obtained, in Experiment D3b the improvement raises to almost 2.22% points, reaching a score (96.51%) that is clearly better than the second best score (95.62%), and represents an improvement of as much as 4.3% points over the HMM analyzer in Experiment 2.

5 Discussion and Future Work

The results obtained with the MFF are surprisingly good:

Table 4. Comparison of best results

approach \ input	accurate POS	automatic POS
HMM-based	93.34%	92.22%
MFF	96.98%	96.51%

The MFF analyzer scores almost 4.3% points better than the best HMM analyzer when the morphological analysis is performed over data automatically POS tagged. To the best of our knowledge, this is reportedly the best score for this task, at least in what concerns a language from the Romance family.[1]

Also when compared to related but different tasks, the results are very interesting. POS taggers accuracy with state of the art performance lies in the range of 97%-98%. The POS tagger developed here approaches this range, with a 96.87% accuracy score. This implies that, when trained and evaluated over the same language and data set, for its task, the MFF analyzer attains an accuracy level of 96.98%, which is at least as high as the accuracy of a state of the art stochastic POS tagger.

Also when compared to the typical performance of WSD systems, this score of the MFF approach to verbal inflection analysis is quite surprising. According to the overview in (Pedersen and Mihalcea, 2005), the upper bound for WSD may be set to 97%-99%, which is deemed to be the best human performance with few and clearly distinct senses, and in Senseval-3, the WSD systems coping with the English Lexical Sample task scored in the range of 70%-73%.

[1] The best result reported by Cucerzan and Yarowsky, 2002, obtained over a small 1k token evaluation corpus, scores below 94.7%.

5.1 Error Analysis

When looking in detail to the errors produced by the MFF analyzer, we see that most errors (71.42%) are produced when it has to handle unknown words not observed in the training data.

Also interesting to note is that close to 4/5 (79.31%)[2] result from an incorrect decision on a specific pair of tags for infinitive verb forms — the `3rdPerson` tag vs. the `NonInflected` tag —, corresponding to the decision whether or not the form is an instance of inflected infinitive (e.g. *dar*, to give).

The second largest group of errors, covering over 1/10, (12.06%) results from an incorrect option between the values 1st vs. 3rd for Person, a widespread source of ambiguity in Portuguese as in this language, for each regular verb, 7 out of the 11 tenses inflecting for Person have identical forms (e.g. *dava*, *dera*, *daria*, *dê*, *desse*, *der*, *dar* from verb *dar*, to give). And the third largest group of errors (6.89%) involves again the inflected Infinitive, in a decision between this tense and the Subjunctive Future (e.g. *amar*, to love). The remaining errors scatter by minor groups related to tense distinctions.

Accordingly, the overwhelming majority of the errors (over 90%) result from an incorrect decision for *Person* value (or its absence). Hence a major challenge to future work is to try to complement the MFF algorithm with a procedure, partly building on agreement relations, that helps to improve the decision on verb forms ambiguous between 3rd vs. 1st or null *Person* values.

6 Conclusions

In this paper, we discussed an alternative to current POS-tagging inspired methods to the verbal inflection disambiguation task, and experimented with a WSD inspired approach. As a first step, we considered the simple and yet reasonably successful approach based on the Most Frequent Sense heuristics. Accordingly, and for the inflection analysis task at stake, we designed the Most Frequent Feature Bundle (MFF) algorithm, which follows a few quite straightforward procedures that decide for the most frequent feature bundle given the set of admissible inflection-driven feature bundles for the input verb form.

Experimentation revealed that this algorithm permits to obtain significantly better levels of accuracy than the ones offered by current stochastic tagging technology commonly used for the same task. In particular, it permitted to develop a verbal inflection analyzer that, to the best of our knowledge, attains the best level of accuracy (ca. 97%) for this task in what concerns languages from the Romance language family.

---

[2] In this subsection, the specific values, between brackets, are taken from the results obtained with threshold set to 0.

References

1. Branco, A., Costa, F., Nunes, F.: The Processing of Verbal Inflection Ambiguity: Characterization of the Problem Space. In: Actas do XXI Encontro Anual da Associação Portuguesa de Linguística (2007)
2. Barreto, F., Branco, A., Ferreira, E., Mendes, A., Nascimento, M.F., Nunes, F., Silva, J.: Open Resources and Tools for the Shallow Processing of Portuguese: The TagShare Project. In: Proceedings of the 5th International Conference on Language Resources and Evaluation, LREC 2006 (2006)
3. Brants, T.: TnT - A Statistical Part-of-Speech Tagger. In: Proceedings of the 6th Applied Natural Language Processing Conference (ANLP 2000), pp. 224–231 (2000)
4. Chanod, J.-P., Tapanainen, P.: Tagging French – Comparing a Statistical and a Constraint-based Method. In: Proceedings of the 7th Conference of the European Chapter of the Association for Computational Linguistics (EACL 1995), pp. 149–156 (1995)
5. Chklovski, T., Mihalcea, R.: Exploiting Agreement and Disagreement of Human Annotators for Word Sense Disambiguation. In: Proceedings of the Conference on Recent Advances on Natural Language Processing, RANLP 2003 (2003)
6. Cucerzan, S., Yarowsky, D.: Bootstrapping a Multilingual Part-of-speech Tagger in One Person-day. In: Proceedings of The Sixth Conference on Natural Language Learning (CoNLL 2002), pp. 132–138 (2002)
7. Elworthy, D.: Tagset Design and Inflected Languages. In: Procedings of EACL SIGDAT Workshop From Texts to Tags: Issues in Multilingual Language Analysis, pp. 1–10 (1995)
8. Ezeiza, N., Aduriz, I., Alegria, I., Arriola, J., Urizar, R.: Combining Stochastic and Rule-Based Methods for Disambiguation in Agglutinative Languages. In: Proceedings of the 36th Annual Meeting of the Association for Computational Linguistics and the 17th Intemational Conference on Computational Linguistic (ACL-COLING 1998), pp. 380–384 (1998)
9. Hajič, J., Hladká, B.: Tagging Inflective Languages: Prediction of Morphological Categories for a Rich, Structured Tagset. In: Proceedings of the 36th Annual Meeting of the Association for Computational Linguistics and the 17th Intemational Conference on Computational Linguistic (ACL-COLING 1998), pp. 483–490 (1998)
10. Hakkani-Tür, D., Oflazer, K., Tür, G.: Statistical Morphological Disambiguation for Agglutinative Languages. In: Proceedings of the 18th International Conference on Computational Linguistics (COLING 2000), pp. 285–291 (2000)
11. Mihalcea, R., Edmonds, P. (eds.): Proceedings of Senseval-3: The Third International Workshop on the Evaluation of Systems for the Semantic Analysis of Text. Association for Computational Linguistics (2004)
12. Pedersen, T., Mihalcea, R.: Advances in Word Sense Disambiguation, Annual Conference of the Association for Computational Linguistcis, ACL 2005, Tutorial notes (2005)
13. Reichenbach, H.: Elements of Symbolic Logic. Macmillan, New York (1947)
14. Trushkina, J., Hinrichs, E.: A Hybrid Model for Morphosyntactic Annotation of German with a Large Tagset. In: Proceedings of Empirical Methods in Natural Language Processing (EMNLP 2004), pp. 238–246 (2004)
15. Tufis, D.: Tiered Tagging and Combined Language Models Classifiers. In: Matoušek, V., Mautner, P., Ocelíková, J., Sojka, P. (eds.) TSD 1999. LNCS (LNAI), vol. 1692, pp. 28–33. Springer, Heidelberg (1999)

Automatic Analysis of Portuguese Verb Morphology
Solving Ambiguities Caused by Thematic Vowel Allomorphs

Vera Vasilévski[1], Leonor Scliar-Cabral[1], and Márcio José Araújo[2]

[1] Universidade Federal de Santa Catarina (UFSC, CAPES)
sereiad@hotmail.com, lsc@th.com.br
[2] Universidade Tecnológica Federal do Paraná (UTFPR)
marciomjapr@gmail.com

Abstract. We present an automatic morphological analyzer of Portuguese verbs, and discuss its performance in relation to its efficiency and the quality of the results, in dealing with the thematic vowel. We also show the principles that guided its development and how we translated most of the grammatical rules we turned into algorithms. The discussion focuses on the ambiguities caused by thematic vowel allomorphs, by the phenomenon called vowel harmony, and on the resolution of them. The creation of an automatic verb lexicon file was helpful at solving these problems.

Keywords: Portuguese verbs, morphology, automatic analysis.

1 Introduction

This paper discusses some of the problems, and their solutions, while building an automatic morphological analyzer of Portuguese verbs. The phase presented here deals with the automatic analysis of the Portuguese verbs thematic vowels. This study highlights the ambiguities caused by the thematic vowel behavior, and reports their disambiguation.

This tool was developed as part of the project called Portuguese Automatic Morphological Analysis [1], whose main goal is producing the grammar for automatic morphological analysis of Brazilian Portuguese corpora. The file pau003.cha is the corpus we use to validate the computational resource presented here, that is, to test the effectiveness of our algorithms. This file has 10,688 utterances, that consist of six hours of conversation among five people – four adults and one child (the target), the participants – and is available for download [2]. Such corpus comprises adult speech, when they talk among themselves, the speech they address to the child, and the child's speech addressed to adults. We work in partnership with the Childes Project [3].

Partial analyses of this work have been presented [4], [5], [6], [7], [8], [9], [10], [11], [12] and the one we describe here refers specifically to the automatic morphological analysis of regular verbs, but we aim to cover the irregular verbs as well. In order to support project development, we created a computer program that hosts several tools that establish interface with other software. Such program is called Laça-palavras [9], and the morphological analyzer identified here is one of its tools.

H. Caseli et al. (Eds.): PROPOR 2012, LNAI 7243, pp. 12–23, 2012.

2 The System of Regular Verbs of Portuguese

The computational tool we present here is based on grammatical rules, i.e., we did not use machine learning based on a training dictionary. Grammatical rules were converted into algorithms and tested within the corpus. A profound and exhaustive study of the grammatical rules that govern the Portuguese verbal system preceded the design of the tool, consulting the literature on the subject [13], [14], [15], [16], [17], [18], [19, [20], [21]. Portuguese verbal system is considered quite predictable [13], and this encourages the creation of a computational tool based on its rules. The program was designed for the written system.

There are three verb conjugations in Portuguese, and the thematic vowel (TV) reveals them. The three thematic vowels are: "a" for the 1st conjugation (C), "e" for the 2nd conjugation, and "i" for the 3rd conjugation. Portuguese verbs are subject to a phenomenon called vowel harmony [22], [23], which refers to the influence that the sound of a vowel in a word has on neighboring or nearby sounds, and that makes them similar and symmetrical [23]. In the Portuguese verb system, such influence is of the underlying thematic vowel of the 1st person singular of the Indicative Present tense on the preceding vowel, i. e., the last stem vowel, if in the Infinitive, this stem vowel is /e/ or /o/. For instance, the verb "levar" (to carry) belongs to the 1st conjugation, so its thematic vowel is /a/, i.e., [+low] and the last stem vowel is /e/, i.e., [-high], [-low]. In the 1st singular person of the Indicative Present tense, the stem vowel copies the feature [+low]. The verb "sofrer" (to suffer) belongs to the 2nd conjugation, so its thematic vowel is /e/, i.e., [-high], [-low], and the last stem vowel is /o/, i.e., also [-high], [-low]. In the 1st person singular of the Indicative Present tense, the stem vowel remains the same. The verb "dormir" (to sleep) belongs to the 3rd conjugation, so its thematic vowel is /i/, i.e., [+high], and the last stem vowel is /o/, i.e., [-high], [-low]. In the 1st person singular of the Indicative Present tense, the stem vowel copies the feature [+high] and becomes /u/, i.e., /'ew 'durmu/. All the derived forms (Present Subjunctive, Negative Imperative and Affirmative Imperative (except 2nd persons)) from de 1st person singular of the Indicative Present tense maintain the copied feature. Since the Portuguese written system does not signal the difference between [-high], [-low], i.e., /e/, /o/, and their [+low] vowel counterparts in the verbal system (their reading being ambiguous), only the verbal harmony acting in the 3rd conjugation is contemplated by the rules in our program.

Verbs are regular when themes of the main parts are similar. The main parts of the Portuguese verbal system are: 1st person singular of the Indicative Present tense; 2nd person of the Indicative Present tense; 2nd person singular of the Indicative Past Perfect tense and Infinitive (the Past Participle does not have any derived tense). Verbs are irregular when themes of the main parts are not similar. There may be other minor irregularities, although irregular verbs are not irregular at all tenses and persons, namely in the perfective subsystem (except 1st and 3rd singular persons). Most Portuguese verbs (types) are regular, but the majority of occurrences (tokens) belong to irregular verbs.

The Infinitive is the recurrent form of regular verbs. Given a regular verb in its Infinitive form, it is possible to conjugate it easily, for three single and plural grammatical persons. On the other hand, it is more difficult to extract the morphemes from a conjugated verb, i.e., dismembering its person-number and tense-mood suffixes.

In Portuguese, there are three moods: Indicative (6 simple tenses), Subjunctive or Conjunctive (3 simple tenses), and Imperative, which can be Affirmative or Negative. There are also the nominal forms: Infinitive, Gerund and Participle. The personal Infinitive is an idiosyncratic Portuguese construction, exclusive to this language. The symbols relative to all tenses and moods are listed in section 4.

A verb, in Portuguese, is the inflectional word par excellence, thanks to its complexity and multiplicity of inflections. The notions of Mood, Tense, Person, and Number coalesce into two suffixes: TM and PN are attached to the theme in a rigorous order. The theme is constituted by the stem followed by the thematic vowel of the corresponding conjugation. In the general pattern, the stem is invariant and carries the lexical meaning of the verb. The broad formula of Portuguese verb structure is [13]: Theme(Stem + TV) + Suffixes(TM + PN). Taking into account the allomorphs of the suffixes, it is possible for one or both of them to be zero – unless the TV is zero (Ø), in which case only one can; that formula represents the general verb morphology structure in Portuguese. In principle, there are 13 mood-tense morphemes, in which there are only sporadically allomorphs. Moreover, there are six person-number suffixes, and also five allomorphs, indicating speakers, listeners, and referents we talk about [13].

We developed a specific tool the algorithm of which contains the morphological rules of the three verb conjugations, respecting mood/tense and person/number [6]. We then observed and registered the ambiguities, in order to correct them, as well as creating an automatic verbal lexicon for the working corpus.

3 The Software Laça-palavras

Before moving on to the tool, it is important to summarize the working processes and resources of Laça-palavras (LP), which is the environment in which this analyzer runs. Laça-palavras functioning principles [9] are described [8], as well as some of its tools [5], [7], [17] and some results from its implementation [10], [11].

Laça-palavras reads files created by the software Clan [3], but also provides its own tools. We detail here LP's main resources: 1) searching words in the corpus, indicating the kind of discourse – adult talking to child (chad), child's talking to adult (ch), and adult talking to adult (adad) –, filtering specific words to make it possible to work with the parts of speech for statistical reports; 2) inserting the line %pho in the corpus for automatic phonological transcription, signaling stress, through the interface with the software Nhenhém [23]. LP enables the refining of the broad transcription into phonetic one; 3) inserting a line for verb morphological automatic analysis called %mor, the algorithm of which is the focus of this study.

4 Morphological Analysis Patterns and Conventions

The correct functioning of the program depends on the methodology employed, namely on the compatibility of symbols for the features that the program will control [24], [25], on the exhaustive classification of verb forms, and on the delimitation of the tasks the program will accomplish. It was considered relevant for the research to label the verbs directly in the corpus as @v for regular verbs, i.e., default; @vi for irregular verbs and @va for auxiliary verbs, to facilitate Laça-palavras search, the automatic morphological analysis, and, consequently, the %mor line outputs. However, to make corpus reading easier, after finishing the search, these symbols – and also others – may be erased. It is worth remembering that all auxiliary verbs are irregular, but we decided to mark them separately, bearing in mind the subsequent computations of the verbal phrases and compound tenses (afterwards) [7].

Labels for verbal categories in %mor line outputs are the following: PI → Present Indicative; PII → Past Imperfect Indicative; PPI → Past Perfect Indicative; PMI → Past Plusperfect Indicative; FPI → Present Future Indicative; FPPI → Past Future Indicative; PS → Present Subjunctive; PIS → Past Imperfect Subjunctive; FS → Future Subjunctive; IMA → Affirmative Imperative; IMN → Negative Imperative; INF → Infinitive; GER → Gerund; PAR → Participle [12].

Grammatical persons are designated by 1, 2, 3 singular (S), and plural (P), that is: 1S (I), 2S (you), 3S (he/she/it), 1P (we), 2P (you), 3P (they). Ambiguities caused by tense-mood morphemes behavior have been addressed [12].

5 Rules for Automatic Morphological Analysis

Formalization of the whole set of allomorphic rules was made to cover exhaustively all possibilities of occurrences of regular verbs, in this way they are characterized by a strong power of predictability and generalization.

5.1 Grammar Rules

The first sets of rules we developed comprehend the thematic vowels, followed by the rules of tense-mood morphemes and, finally, the personal-number morphemes set of rules, for regular verbs. First, the allomorphic rules were formalized, for subsequent inclusion in the program. The figures below show the formalizations of the TV rules for the three Portuguese verb conjugations. The symbol # means "end of the word".

The first set of allomorphic rules refers to the thematic vowel "a" (Fig.1), characterizing the 1st conjugation. It may be read as follows:

- morpheme |a| becomes null (∅) before the morpheme of 1st person singular of the Present tense (ŏ), therefore, in final word position, and before the tense-mood morpheme of the Present Subjunctive ("e") and similar Imperative forms;
- morpheme |a| becomes nasalized, codified as "ã", when stressed, before the written codification of the 3rd person plural "o", therefore, in final word position;
- morpheme |a| becomes "e", before the written representation of the semivowel /j/, which is "i" (1st person singular of the Past Perfect Indicative), copying the feature [-posterior] and raising the feature [+low] into [-high], [-low];
- morpheme |a| becomes "o", before the written representation of the semivowel /w/ which is written "u" (3rd person singular of the Past Perfect Indicative), copying the feature [+round] and raising the feature [+low] into [-high], [-low];
- morpheme |a| remains the same in the other contexts.

<table>
<tr><th>TV</th><th>is written</th><th>as</th><th>in this context</th><th>Examples</th></tr>
<tr><td>|-a-|</td><td>→</td><td>Ø</td><td>__ŏ#</td><td>cantØo (I sing)</td></tr>
<tr><td></td><td></td><td></td><td>__e</td><td>cantØe, cantØes</td></tr>
<tr><td></td><td></td><td>ã</td><td>'__o#</td><td>estão (they are)</td></tr>
<tr><td></td><td></td><td>e</td><td>__ i#</td><td>cantei</td></tr>
<tr><td></td><td></td><td>o</td><td>__u#</td><td>cantou</td></tr>
<tr><td></td><td></td><td>a</td><td>...</td><td>cantamos</td></tr>
</table>

Fig. 1. Allomorphic rules of the thematic vowel "a" of the 1st C

The second set of allomorphic rules refers to the thematic vowel "e" (Fig.2), characterizing the 2nd conjugation. It may be read as follows:

- morpheme |e| becomes null (∅) before the morpheme of 1st person singular of the Present tense (ŏ), therefore, in final word position, before the tense-mood morpheme of the Imperfect Past Indicative ("ia"), before the tense-mood morpheme of the Present Subjunctive ("a") and similar Imperative forms, and before decodification of the 1st person singular of the Past Perfect Indicative "i";
- morpheme |e| becomes "i", before morphemes "t" or "d", representing Participle, therefore neutralizing with the 3rd conjugation thematic vowel;
- morpheme |e| remains the same in the other contexts.

<table>
<tr><th>TV</th><th>is written</th><th>as</th><th>in this context</th><th>Examples</th></tr>
<tr><td>|-e-|</td><td>→</td><td>∅</td><td>__ŏ#</td><td>comØo (I eat)</td></tr>
<tr><td></td><td></td><td></td><td>__ia</td><td>comØia</td></tr>
<tr><td></td><td></td><td></td><td>__a</td><td>comØa</td></tr>
<tr><td></td><td></td><td></td><td>__i#</td><td>comØi</td></tr>
<tr><td></td><td></td><td>i</td><td>'__t-</td><td>escrito (written)</td></tr>
<tr><td></td><td></td><td></td><td>__d-</td><td>comido (eaten)</td></tr>
<tr><td></td><td></td><td>e</td><td>...</td><td>comemos</td></tr>
</table>

Fig. 2. Allomorphic rules of the thematic vowel "e" of the 2nd C

The third set of allomorphic rules refers to the thematic vowel "i" (Fig.3), characterizing the 3rd conjugation. It may be read as follows:

- morpheme |i| becomes null (∅) before the morpheme of 1st person singular of the Present tense (ŏ), therefore, in final word position, before the tense-mood

morpheme of the Imperfect Past Indicative ("ia"), before the tense-mood morpheme of the Present Subjunctive ("a") and before decodification of the 1st person singular of the Past Perfect Indicative "i";
- morpheme |i| remains the same in the other contexts.

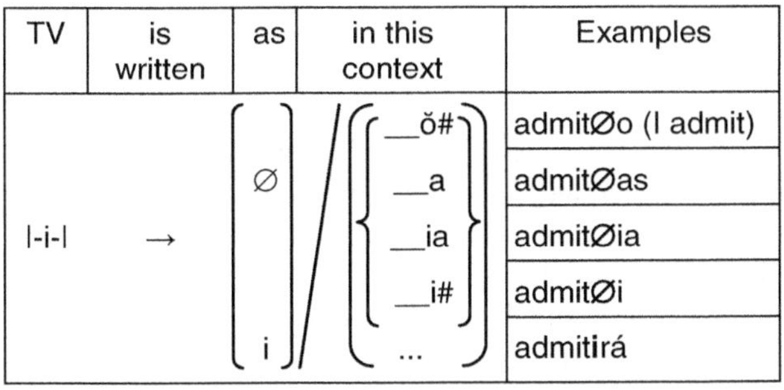

Fig. 3. Allomorphic rules of the thematic vowel "i" of the 3rd C

Using such rules and taking support from the literature [13-22], we built a picture of the morphological behavior of the Portuguese verbs thematic vowels (Tab.1). We put the situation where their allomorphs occur in parentheses.

Table 1. Allomorphs of Portuguese verbs thematic vowels

Conj.	TV	Allomorphs				Observations
1	a	**Ø** (1S-PI, PS, IMN, 3S/3P-IMA)	**e** (1S-PPI)	**o** (3S-PPI)	**ã** (3P-PI)	
2	e	**Ø** (1S-PI, PS, IMN, 3S/3P-IMA, PII, 1S-PPI)	**i** (PA)			opposition e/i suppressed in 2S-IMA and 2S, 3S, and 3P-PI
3	i	**Ø** (1S-PI, PS, IMN, 3S/3P-IMA, PII, 1S-PPI)				opposition e/i suppressed in 2S-IMA and 2S, 3S, and 3P-PI

The next step was to convert those grammatical rules into algorithms, with specific rules demanded by the computational environment, in such a way that they could communicate with the sets of rules commanding the tense-mood and person-number suffixes. Some rules related to the thematic vowel are commented below.

As we can see in Figs.1, 2 and 3, the TV of the three conjugations disappears before "o", which represents the 1S person-number Present Indicative suffix. At this point, a problem emerges: if these forms have lost their TV, how is it possible to rescue their Infinitive forms? The best way we found to rescue the correct Infinitive form was building a program which compared each of the alternatives with the forms registered in the verbal lexicon until one of them matched the only possibility.

5.2 Verbal Lexicon

The creation of the verbs lexicon was one of requirements for preparing the program and also codifying the verbs occurrences into @v for regular verbs; @vi for irregular verbs and @va for auxiliary verbs in the corpus pau003.cha, as has already been mentioned. These decisions were meant to facilitate the automatic grammar to be used by other researchers. We instructed the morphological analyzer to create a file which listed each Infinitive generated from the verbs tokens of the corpus, so that they became available for the user to choose. The program excludes the non-matching ones and saves the file. In this way the verb lexicon of the corpus is created. It is up to the user to create the lexicon gradually or all at once. When analyzing a verb form, the program compares the responses generated with the verb lexicon file content, and only shows the alternative that matches it. We will be back to this point in section 6.

The creation of the lexicon step by step will be described in a forthcoming paper. It is worth remembering that the regular verb lexicon contains only the Infinitive forms of the verbs, nothing else. All morphological rules concerning thematic vowels, tense-mood suffixes, and person-number suffixes and their allomorphs are algorithms belonging to the morphological analyzer.

6 The Automatic Morphological Analyzer

6.1 Functioning

The morphological analyzer checks the verbs contained in a corpus previously prepared, and are loaded into the system that hosts it, that is, Laça-palavras. The labeled verb forms are automatically recognized, read and analyzed, and the result is displayed as the output in the %mor line. Internally, when the morphological analyzer identifies a verb form, the program compares it with its internal sets of rules, and breaks it down into morphemes. The program outputs are the identification of the Infinitive form of the analyzed regular verb form and the respective labeled verbal categories, according to the Portuguese verb structure general formula (Fig.4).

A piece of the analyzer main function code is shown below, in which the subfunction Verbar is responsible for applying verb morphological rules:

```
...
//checks each word in the line selected
foreach (string s in plvs){
    cVerbo cv = new cVerbo(s);
    //if the word is a verb, runs verb morph rules
    if (!(cv.Verbo == null)){
      ...
```

```
  //if Verbar returns true, then, verb morph rules were
  //applied successfully
  if (cv.verbar()){
    //checks possible results, shows the correct one
    foreach (cVerb fcv in cv.LVerb){
      //search the verb in the lexical corpus
      cVerbo VrbCprs = LxcCpr.Find(fcv.Verb);
      //displays the verb found in lexical corpus
      if(VrbCprs!=null){
        ListViewItem lvi = new ListViewItem();
        lvi.Name = VrbCprs.Verb;
        lvi.Text = VrbCprs.Verb;
        lvi.SubItems.Add(VrbCprs.Theme);
        lvi.SubItems.Add(VrbCprs.TM);
        lvi.SubItems.Add(VrbCprs.PN);
        lvi.SubItems.Add(VrbCprs.PAR);
        lvi.SubItems.Add(VrbCprs.GER);
        lvi.SubItems.Add(VrbCprs.Theme + "&" +
        VrbCprs.TenseAndMood + "&" + VrbCprs.Person);
        lvVisMor.Items.Insert(contV, lvi);
        break;
... [final curly braces removed]
```

Fig.4 shows the main screen of the tool, applied to the morphological analysis of the verb form "botou" (put, 3rd singular person, Past Perfect), which occurs in line 2166 of the corpus, said by the participant ISI (child's father).

The field Participantes (participants) allows the user to choose the participants whose statements he wants to check. When the corpus is loaded, the program fills this field with the participants automatically, so the user can make a selection. After this choice, the number of utterances found for the selected participant(s) appears in the field Enunciados (utterances), and their statements are displayed, in Clan's format, right below. To check the verbs of a statement, the user just has to click on it, and the field Análise (analysis) gives the morphological analysis of the verb that appears in the selected statement. The morphological analyzer can be used for testing any isolated Portuguese verb token, and is not restricted to Clan format files. Any text file (txt) is readable by the program.

As the cursor goes down through the statements, each identified verb is parsed automatically by the program. When the program finds more than one verb in the same line, as happens in line 2155, it analyzes the second verb right after the previous one. In Fig.4, the field Análise shows that the verb form "botou" belongs to the 1st C, because its Infinitive form (Inf) is "botar"; the thematic vowel ("a") gave place to an allomorph, so it became "o"; there is no overt tense-mood morpheme, so, it becomes Ø; and its person-number suffix is "u". These pieces of information together show that the verb form "botou" is conjugated as 3S, in the Past Perfect Indicative. The resulting automatically analyzed formula is: boto&PPI&3S.

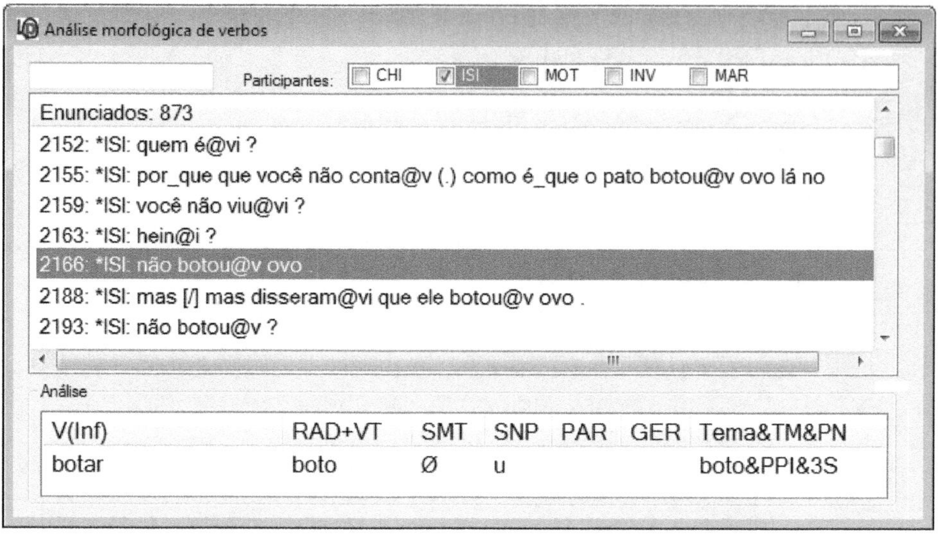

Fig. 4. Main screen of the Automatic Morphological Analyzer

6.2 Dealing with Ambiguities of the Portuguese System of Verbs

The program reproduces the cases where morphological rules are ambiguous [26], as well as cases where the thematic vowel is suppressed, which also cause ambiguities, since the lack of TV prevents conjugational recognition. So, if there is no TV, the verb can belong to any of the three conjugations, because it is not possible to distinguish which is the conjugation. Reproducing the ambiguities of the Portuguese verb system by the program proves that its algorithm corresponds to this system.

We will discuss here only the ambiguities of the Portuguese verb system related to the thematic vowel. We have seen that, in Present Indicative, 1S, the TV of the three conjugations disappears, and the personal-number morpheme "o" is added directly to the stem: this renders an automatic grammatical rule-based program unable to rescue the Infinitive form of a verb under these conditions. The program correctly breaks down verb forms such as "acabo" (I finish), "temo" (I fear), and "admito" (I admit), 1st, 2nd, and 3rd C, respectively (Tab.2), however, the field V(Inf) generated three possibilities for each entry, caused by the loss of the thematic vowel: "acabar", *"acaber", and *"acabir"; *"temar", "temer", and *"temir"; and *"admitar", *"admiter", and "admitir".

Table 2. Program response to the entries "acabo", "temo", and "admito"

V(Inf)	(RAD+VT)		SMT	SNP	Tema&TM&PN
acabar	acab	Ø	Ø	o	acabØ&PI&1S
temer	tem	Ø	Ø	o	comØ&PI&1S
admitir	admit	Ø	Ø	o	admitØ&PI&1S

The possibilities of mistakes multiply when another vowel comes right before the thematic vowel, as in “averiguar” (to check), “atear” (to light), “moer” (to grind), “esvair” (to faint). The entry “ateava” generated the following outputs:

Análise

V(Inf)	RAD+VT	SMT	SNP	PAR	GER	Tema&TM&PN
atear	atea	va	Ø			atea&PII&1S/3S
ateavar	ateava	Ø	Ø			ateava&PI&3S
ateaver	ateavØ	a	Ø			ateavØ&PS&1S/3S
ateavir	ateavØ	a	Ø			ateavØ&PS&1S/3S
ateavar	ateava	Ø	Ø			ateava&IMA&2S
ateaver	ateavØ	a	Ø			ateavØ&IMA&3S
ateavir	ateavØ	a	Ø			ateavØ&IMA&3S

Fig. 5. Laça-palavras first response to the entry “ateava”

In this case, again, only the Portuguese verb lexicon is able to disambiguate the program responses: lexicons are not learned by rules, it is necessary to acquire them. For the verb lexicon works properly, it is crucial that one of those outputs be the correct one, and this always happens. After the creation of the lexicon as described, the only form that belongs to the lexicon file is the first. So, the program excludes the others, before presenting the analysis to the user. The sequence of events the morphological analyzer follows in this situation is displayed in the flowchart below (Fig.6):

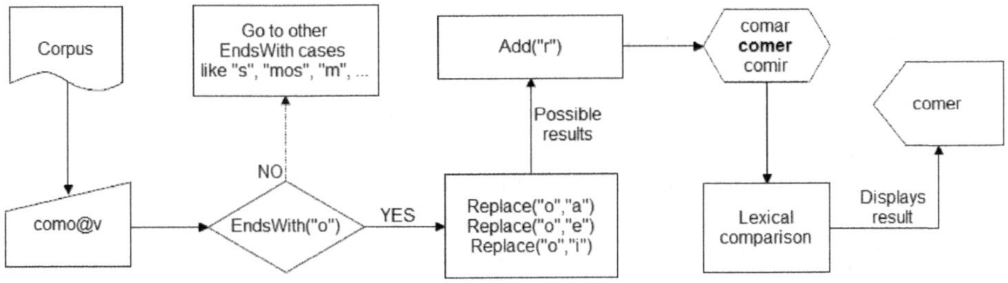

Fig. 6. Deduction of verb Infinitive form, from the entry “como” (I eat)

Another situation covered by the automatic lexicon, in association with computational instructions, concerns the rescuing of regular stems that undergo transformations required by the grapheme values, according to which the phoneme /g/, when followed by the phonemes /e/ or /i/, is coded into the grapheme “gu”, as happens, for instance, to the verbs “ligar” → “liguei” /li'gej/ (to call → I called); the same is true with the phoneme /k/, before the phonemes /e/ and /i/, which is coded into the grapheme “qu”, for instance, in the verbs “ficar” → “fiquei” (to stay → I stayed); and with the phoneme /s/, coded into “ç”, whenever a stem ending with the grapheme “c” occurs before the [+posterior] phonemes in the verbal system, i. e., /o/ or /a/, as happens to "esquecer" → "esqueço", "esqueça" (to forget → I forgot, forget; Subjunctive and Imperative moods). As Fig.7 makes clear, the only correct response is the first.

Análise

V(Inf)	RAD+VT	SMT	SNP	PAR	GER	Tema&TM&PN
ligar	ligue	Ø	i			ligue&PPI&1S
liguar	ligue	Ø	i			ligue&PPI&1S
ligueer	ligueØ	Ø	i			ligueØ&PPI&1S
ligueir	ligueØ	Ø	i			ligueØ&PPI&1S
liguer	ligue	Ø	i			ligue&IMA&2P
ligueir	ligueØ	Ø	i			ligueØ&IMA&2P

Fig. 7. Laça-palavras first response to the entry "liguei"

Finally, the comparison between program generated responses and the lexicon covers 3^{rd}C vowel harmony. When the analyzer generates the infinite forms *"durmar", *"durmer", and *"durmir" for the entry "durmo", while comparing them to the lexicon, it finds "dormir" and chooses it, because it has an instruction that informs that "o" can be replaced by "u", under the conditions we have mentioned. Preliminary tests indicate that it is possible to develop specific automatic rules for vowel harmony recognition, without using the lexicon file. Nevertheless, the lexicon of the verbs is the resource that solves this question for now.

7 Conclusion and Outlooks

So far, the work with rules for automatic analysis of inflected regular verb forms concerning the thematic vowels as well as their allomorphs has been successfully finished. The ambiguities caused by the morphological behavior of the thematic vowel were mainly solved by the creation of the automatic verb lexicon, and also by the creation of specific computational rules. Of course, as the work progresses, we can find better solutions. The next step is to conclude the disambiguation of cases where different persons and tenses keep the same verb form. A great part of this work has been done [12]. We have also been working with person-number suffixes, and the results will be forthcoming soon.

The morphological analyzer was tested with the file pau003.cha, and it has proved to be efficient, giving adequate responses: empirical testing found at least a 90% success rate on the program performance, but it is not yet finished. It seems obvious that the lexicon will improve the results; anyway, we have presented the creation of a specific lexicon, extracted automatically from the working corpus, therefore facilitating and reducing the user's job, without disregarding his/her indispensable decisions. Although many issues could not be addressed here, we have dealt with the important ones, and we expect to handle the other matters in further papers shortly.

References

1. Scliar-Cabral, L.: Codificação da morfologia do PB e análise da fala dirigida à criança: expansão. Projeto de pesquisa aprovado pelo CNPq (2010-2015)
2. http://childes.psy.cmu.edu/data/Romance/Portuguese/Florianopolis.zip
3. MacWhinney, B.: The CHILDES Project: Tools for Analyzing Talk, 3rd edn. Lawrence Erlbaum Associates, Mahwah (2000)
4. Vasilévski, V.: Divisão silábica automática de texto escrito baseada em princípios fonológicos. In: III Enpole, Cd-Rom, São Cristóvão (2010)
5. Vasilévski, V.: O hífen na separação silábica automática. Revista do Simpósio de Estudos Lingüísticos e Literários – SELL 1(3), 657–676 (2011)
6. Vasilévski, V.: Programa para processamento automático das unidades verbais do PB. In: VII Congresso Internacional da Abralin, Curitiba (2011)
7. Vasilévski, V.: An automatic system for verb morphological analysis of BP. In: VIII ENAL, Apresentação Oral, Juiz de Fora (2011)
8. Scliar-Cabral, L., Vasilévski, V.: Descrição do português com auxílio de programa computacional de interface. In: II JDP, Cuiabá (2011)
9. Vasilévski, V., Araújo, M. J.: Laça-palavras: sistema eletrônico para descrição do PB. LAPLE-UFSC, v.2010-2012, Florianópolis, https://sites.google.com/site/sisnhenhem/
10. Costa, R.F.S., Scliar-Cabral, L.: Regularização do sistema verbal pela criança. In: SILC: Homenagem aos 40 anos da PGL, PGLit e PGI da UFSC, UFSC, Florianópolis (2011)
11. Vasilévski, V.: Diferenças entre input e intake: evidências na aquisição de pronomes interrogativos. In: SILC: Homenagem..., UFSC, Florianópolis (2011)
12. Vasilévski, V., Araújo, M.J.: Tratamento dos sufixos modo-temporais na depreensão automática da morfologia dos verbos do português. Linguamática 3(2), 107–118 (2011)
13. Câmara Jr., J.M.: Estrutura da língua portuguesa. Vozes, 26th edn., Petrópolis (1986)
14. Câmara Jr., J.M.: História e estrutura da língua portuguesa. Padrão, 2nd edn., Rio de Janeiro (1976)
15. Scliar-Cabral, L.: Emergência gradual das categorias verbais no Português brasileiro. Alfa 51(1), 223–234 (2007)
16. Scliar-Cabral, L.: Codificação da Morfologia do PB e Análise da Fala Dirigida à Criança. Fórum Lingüístico 5(2), 69–82 (2008)
17. Scliar-Cabral, L.: Análise Automática da Morfologia do PB (Plataforma CHILDES): aquisição da morfologia verbal. In: VII Congresso Internacional da Abralin, Curitiba (2011)
18. Scliar-Cabral, L.: Princípios do sistema alfabético do português do Brasil. Contexto, São Paulo (2003)
19. Basílio, M.: Estruturas lexicais do português. Vozes, Petrópolis (1980)
20. Assis Rocha, L.C.: Estruturas morfológicas do português. UFMG, Belo Horizonte (1999)
21. Gonçalves, C.A.: Flexão & Derivação em português. UFRJ, Rio de Janeiro (2005)
22. Imanishi, E.C.: O processo de metafonia nos verbos. Dissertação de mestrado, PUC, Campinas (1975)
23. Vasilévski, V.: Construção de um programa computacional para suporte à pesquisa em fonologia do português do Brasil. Tese de doutorado, UFSC, Florianópolis (2008)
24. Beal, J., Corrigan, K., Moisl, L.: Creating and Digitizing Language Corpora: Synchronic Databases, vol. 1. Palgrave-Macmillan, Houndmills (2007)
25. Hauser, R.: Principles of computational morphology. Computational Linguistics 47 (1990)
26. Parisse, C., Le Normand, M.T.: Automatic disambiguation of the morphosyntax in spoken language corpora. Behavioral Res. Methods, Instr. and Computers 32, 468–481 (2000)

Coordination of *-mente* Ending Adverbs in Portuguese: An Integrated Solution

Jorge Baptista[1,4], Lucas Nunes Vieira[1,2], Cláudio Diniz[3,4], and Nuno Mamede[3,4]

[1] Universidade do Algarve / Faro, Portugal
[2] Université de Franche-Comté / Besançon, France
[3] Instituto Superior Técnico, Universidade Técnica de Lisboa / Lisboa, Portugal
[4] Spoken Language Lab., INESC-ID Lisboa / Lisboa, Portugal
jbaptis@ualg.pt, {lucasnvieira,cfpdiniz}@gmail.com,
Nuno.Mamede@inesc-id.pt

Abstract. Portuguese *-mente* ending adverbs constitute a large, morphologically homogenous, but syntactically and semantically diverse lexical set. When coordinated, the first adverb loses the adverbial suffix and takes the shape of the base adjective, in the feminine-singular form. This raises the issue of its part-of-speech (POS) classification (adverb or adjective?), but especially its adequate parsing, since it may then be incorrectly analyzed as a modifier of a preceding noun. However, the POS tagging can not be adequately performed prior to some minimal syntactic analysis. The size of the lexicon involved (more than 7,000 adverbs) and the scarcity of instances, even in large corpora, make it ineffective to leave only for the POS tagger the task of solving this adjective/reduced adverbial form ambiguity. This paper proposes an integrated solution, where a rule-base disambiguating module and a POS statistical tagger combine to produce more accurate tagging and better parsing results to this non-trivial empirical problem. The system was evaluated on a large-sized corpus.

Keywords: Adverb, Coordination, POS disambiguation, Parsing, Dependency.

1 Introduction

Adverbs are a significant part of the lexicon of many languages and they occur very frequently in texts. Table 1 shows the frequency of *-mente* ending adverbs (henceforward, *Adv-mente*) in two large, publicly available, corpora of Portuguese, namely the CETEMPúblico[20][1], for European Portuguese, and NILC/São Carlos [18][2], for Brazilian Portuguese. Even though they represent little more than 10% of all (simple) adverb occurrences in the corpora, *-mente* ending adverbs constitute the majority of the simple-word lemmas from this category [3]. When coordinated, Portuguese *-mente* ending adverbs drop the suffix and appear in the feminine-singular (*fs*) form of the base

[1] http://www.linguateca.pt/cetempublico/ [last access: 2012-01-12].
[2] http://www.linguateca.pt/acesso/corpus.php?corpus=SAOCARLOS [last access: 2012-01-12].
[3] Excluding compound adverbs, naturally, which are at least as numerous as simple adverbs, and also occur quite frequently in texts [11][16].

H. Caseli et al. (Eds.): PROPOR 2012, LNAI 7243, pp. 24–34, 2012.

adjective: (1a) *O Pedro leu isso lenta e atentamente* 'Peter read that slow_fs and attentively'; (1a)=(1b) *O Pedro leu isso lenta[mente] e [o Pedro leu isso] atentamente* 'Peter read that slow(ly)_fs and (Peter read that) attentively'. If there is a feminine-singular noun before the reduced adverb, it is very likely that the adverb would be considered as an adjective instead, and treated as a modifier of that noun, e.g. *a revista lenta*, 'the magazine slow' in: (2) *O Pedro leu a revista lenta e atentamente* 'Peter read the magazine_fs slow_fs and attentively'. Finally, as coordination can be iterated, longer chains of reduced adverb forms can be found: (3) *O Pedro leu isso lenta, pausada e atentamente* 'Peter read that slow_fs, pausing_fs and attentively'. However, in both corpora, longer chains are rare. In the European Portuguese corpus mentioned above, only 24 multiple coordinated adverbs were found, against 438 simple coordination cases.

Table 1. Adverbs in two Portuguese corpora: (l) lemmas, (w) words

	NILC/São Carlos	CETEMPúblico
lemmas (l)	397 K	1,2 M
words (w)	32,3 M	191,6 M
Adv (l)	2,867	5,361
Adv (w)	1,5 M	9,1 M
Adv-mente (l)	1,936	4,654
Adv-mente (w)	103,6 K	1,0 M

Because the reduced form of the adverb and the feminine-singular form of its base adjective are homographs, the POS of the word has to be disambiguated. However, without semantic (distributional) information on noun-adjective combinations, adverb combinations, or even verb-adverb pairs, any solution to this non-trivial problem is just an approximation. On the other hand, it would be useless (and eventually hampering to a system) to consider that all feminine-singular adjectives could be adverbs in every context. So this particular type of strictly local ambiguity should be solved prior to general parsing rules or statistical models be applied to the text.

The performance of statistical POS taggers depends on the granularity of the tag set used by the learning algorithms, and since many systems only use a coarse tag set, i.e., considering only the major POS category, but discarding the inflection, it is very difficult to train models sensitive to this particular phenomenon.

Finally, the coordination of adverbs, while a relatively common phenomenon in Portuguese, occurs very infrequently in texts. For the system here used, the statistical POS tagger [19], based on the Viterbi algorithm, uses a manually annotated corpus of 250K words. In this corpus only 10 instances occur of the pattern corresponding to the coordination of *Adv-mente* but only 4 are in fact coordinated *Adv-mente*. The sparsity of the phenomenon makes it an interesting challenge to NLP systems, difficult to tackle by a purely machine-learning approach. An alternative solution should be devised.

To the best of our knowledge, no assessment has been made for Portuguese on the accuracy of any disambiguation method in dealing with this specific linguistic phenomenon. The study of [1] was a preliminary survey aimed at developing the parsing rules to be implemented in the system PALAVRAS [3]. In the manually corrected and revised Portuguese treebank Bosque (version 8)[4], only 6 instances were found among 9,368 sentences. In those sentences, both the reduced form, tagged as an adjective, and the adverb are coordinated modifiers of the the same word; apparently, the only two instances of the compound *pura e simplesmente* (purely and simply) are treated as coordinated simple words; in most cases the syntactic dependency ADVL (adverbial adjunct) applies to both items, the adjective is linked to the adverb and this to the verb it modifies. While the rules themselves could not be consulted, the processing of examples (1) to (3) by the PALAVRAS parser[5] correctly yields the syntactic dependency ADVL both for (1) and (2) reduced adverb forms (lema *lento*, with the tag *mente*) modifying the verb in both (1) and (2); as for (3), only the second reduced form is correctly analyzed, like in the latter examples, but *lenta* seems to be parsed as an ordinary adjective, and a PRED dependency on the root node is extracted. On the other hand, the LX-GRAM Dependency Parser[5][6], maybe due to an incorrect tagging of *lenta* as a (sg. masc.?) adjective, produces poorer results: it extracts a PRD dependency between the reduced form and the verb while the *Adv-mente* is linked by an M (modifier?) relation; naturally, the coordination between the reduced forms and the *Adv-mente* is not established.

This paper addresses the issues mentioned above in the context of the development of the STRING system [13], a Portuguese NLP chain developed at L2F/INESC ID Lisboa[7]. The system is composed of several modules, including a tokenizer, a morphological analyzer LEXMAN [7][9], a rule-based disambiguation module RuDriCo2 [8], a statistical POS tagger MARV [19], and a parser XIP (Xerox Incremental Parser) [2]. XIP is a cascade, finite-state, rule-based parser that analyzes sentences into chunks, extracts syntactic dependencies between chunks and it is also used for named entity recognition [12][15] and (partially) to co-reference resolution [14] and relation extraction [21]. The related problems of correct identification of the reduced adverbial form and of the parsing of coordinated *Adv-mente* is mainly a function of the morphological analyzer and of the chunking module of the parser, but it is set in the more general task of extracting the syntactic dependencies between the sentences' constituents.

The paper is structured as follows: Section 2 firstly sketches the integrated solution here proposed and then presents the methods used to implement it. These involve extending lexical coverage (2), building new linguistically motivated rules for POS disambiguation (2), and constructing specific chunking (2) and dependency extraction (2) rules. Finally, to evaluate the system's performance, a corpus has been built and manually annotated (2). Section 3 presents the evaluation of each one one of these main components of the system, while Section 4 discusses these results and projects future work.

2 Methods

The strategy for the disambiguation and parsing of the coordinated *-mente* ending adverbs consists in three steps: (i) at the lexical/morphological level, instead of considering all feminine adjectival forms as adverbs, extend the coverage of the existing lexicon

[4] http://www.linguateca.pt/Floresta/corpus.html#bosque [last access: 2011-11-04].
[5] http://beta.visl.sdu.dk/visl/pt/parsing/automatic/dependency.php [last access: 2012-01-12].
[6] http://lxcenter.di.fc.ul.pt/services/en/LXServicesParserDep.html [last access: 2012-01-12].
[7] http://string.l2f.inesc-id.pt

of adverbs, and associate them to their reduced forms; (ii) use this morphologic information with that of the environing words in order to build linguistically motivated rules and locally determine the patterns where a coordination of adverbs is likely to occur or, on the contrary, where a reduced adverbial form can reasonably be discarded; at the end of this rule-based disambiguation process, all the remaining ambiguous forms are tagged as adjectives; (iii) based on the results from previous steps, produce the adequate chunking and extract the syntactic-semantic dependencies between the sentence constituents. In the next subsections, these processing steps are described in detail. For the evaluation, a corpus with 1,132 sentences was collected from the CETEMPúblico, containing instances of coordination of a (surface) feminine-singular adjective or past participle with an *Adv-mente*. The corpus was parsed by the system and the output was manually corrected by two linguists. In the last subsection, corpus collection and annotation will be briefly presented.

Lexicon

The existing lexicon of the system has been systematically completed by adding all *Adv-mente* entries found in an orthographic vocabulary [4]. These correspond to 3,614 entries. Then, all valid *-mente* ending forms found in the European Portuguese corpus were manually perused and the adverbs selected. Duplicates from the first list were removed, thus yielding 3,636 new entries. For each entry, the feminine-singular form of the base adjective was automatically generated and the list was then manually revised for errors and for the insertion of orthographic variants, resulting from the new, unified Portuguese orthography. The final list consists of 7,250 *-mente* ending adverbs. For example, the entry for *abstratamente* ‘abstractly’ is associated with the orthographic variant *abstractamente*, and to the reduced forms *abstrata* and *abstracta* ‘abstract_*fs*’. This reduced form is then given the feature ‘`r`’ (for ‘reduced’). When analyzing a sentence where *abstracta* appears, at this morphologic stage, the system produces the following tags (format adapted for clarity): `abstracta: abstratamente Adv_r; abstrata Adj_fs`. In this way, only forms with attested *-mente* adverbial counterparts are validated.

It has been previously noted by [1] that compound adverbs (or colocational combinations), such as *única e exclusivamente* ‘uniquely and exclusively’ and *única e simplesmente* ‘uniquely and simply’ occurred quite often in the corpus. Other forms were added to the lexicon, v.g. *pura e simplesmente* ‘purely and simply’, *dire(c)ta ou indire(c)tamente* ‘directly or indirectly’, *explícita ou implicitamente* ‘implicitly or explicitly’ and *total ou parcialmente* ‘totally or partially’. These combinations occur 3,074 times in the CETEMPúblico. In our corpus, only *pura e simplesmente* occurs, 220 times.

Rule-Based Disambiguation

The next step in the system processing chain is a rule-based disambiguation module [7],[9]. The linguistically motivated disambiguation rules produced are at the core of the solution here presented. These rules are regular expressions that take the general form: |<left-context>| <pattern> |<right-context>| := <result> where <pattern> corresponds to the ambiguous target word and the different categories it may be associated with; <result> consists in selecting (+) or discarding

(−) a given category; the left and right contexts are facultative. For example, the general rule below selects the adverb reduced form when it appears coordinated with a *-mente* ending adverb:

```
0> [CAT='adv',SYN='red'][CAT='adj']
  |[surface='e'];[surface='ou'];[surface='mas'],
  [surfaceRegex='.+mente',CAT='adv']|
  := [CAT='adv']+.
```

This rule reads as follows: the left context is empty; the `<pattern>` consists of the ambiguous form adverb/adjective; the adverbial form must present the feature SYN with the value 'red' (for 'reduced'); then follows the right context, where the coordinative conjunctions and the *Adv-mente* are explicit; for the conjunctions, the surface form is sufficient; to define the adverb, a regular expression is used along with its POS.

Most rules have to be duplicated in order to deal with the feminine-singular form of past participles. This is the purpose of the rule below:

```
0> [CAT='adv',SYN='red'][MOD='par',GEN='f',NUM='s']
  |[surface='e'];[surface='ou'];[surface='mas'],
  [surfaceRegex='.+mente',CAT='adv']|
  := [CAT='adv']+.
```

Rule-order application is fixed, so more specific rules are stated before more general ones. For example, the pattern of coordinated adjectives, each modified by an adverb, is more constrained than the previous patterns and it is thus stated before the general rules above:

```
0> |[CAT='adv']|
  [CAT='adv',SYN='red'][CAT='adj',GEN='f',NUM='s']
  |[CAT='con',SCT='coo'],[surfaceRegex='.+mente',CAT='adv'],
  [CAT='adj',GEN='f',NUM='s'][MOD='par',GEN='f',NUM='s']|
  := [CAT='adv']-.
```

Some rules require lists of words to be spelled out, such as the next one, where a negation adverb in front of an ambiguous adjective is the context that allows to discard the reduced adverbial form; the negation adverb is provided by a list of words (at later stages, namely in the parser, this piece of information is encoded by way of feature-value pairs):

```
0> |[surface='não'];[surface='nem'];[surface='nunca'];
  [surface='jamais'];[surface='nada']|
  [CAT='adv',SYN='red'][CAT='adj']
  |[surface='e'];[surface='ou'];[surface='mas'],
  [surfaceRegex='.+mente',CAT='adv']|
  := [CAT='adv']-.
```

Finally, at the last stage of the process and for the remaining ambiguous forms, the tag corresponding to the reduced adverb form is discarded by a general "cleaning" rule:

```
0> [CAT='adv',SYN='red'][SYN= 'red']
  := [SYN='red']-.
```

So far, 16 rules have been devised, based on known cases of ambiguity. Our approach is conservative, in the sense that rules tend to be general in scope and as much precise as possible. During this process, some rules were devised but not yet implemented, for they are not linguistically well motivated even though the patterns appear often in text. This is the case of coordinated adjectives after a copula verb, where the second adjective is modified by an *Adv-mente*, as in: (4) *A crítica portuguesa foi agressiva e extremamente injusta* 'The Portuguese critic_fs (=the critics) was aggressive and extremely unfair'; or when both adjectives are modified, especially if a quantifying adverb is involved, as in: (5) *A corrida também é muito longa e fisicamente dura* 'The race also is very long and physically hard'. Strictly speaking, these patterns are grammatically ambiguous, but more often than not the adjective is found in this context.

Chunking

In the chunking stage, the XIP parser analyzes the sentence by splitting it into elementary constituents (or chunks). Ordinarily, a stand-alone adverb construes an adverbial phrase (`ADVP`). Chunks are formed according to chunking rules, such as the following, allowing up to three consecutive adverbs to form an `ADVP`:

```
ADVP @= (adv), (adv), adv.
```

At this stage, the system can make use of a rich set of lexicons, featuring syntactic and semantic information, as well as the information derived from the morphological analyzer. In the coordination of *Adv-mente*, an `ADVP` is construed. For example, for sentence (1) the following chunking is produced:

```
0>TOP{NP{O Pedro} VF{leu} NP{isso} ADVP{lenta e atentamente}.}
```

This `ADVP` results from the application of the following rule:

```
18> ADVP @=  |?[noun,fem,sg] |
    (adv[advquant];adv[advcomp];adv[neg])*,
    adv[reducedmorph],
    conj[lemma:e];conj[lemma:ou];conj[lemma:mas],
    (adv[advquant];adv[advcomp];adv[neg])*,
    adv[surface:"%c+mente"].
```

The chunking rule reads: an `ADVP` chunk is built with two coordinated adverbs, the first is a reduced form, indicated by the feature `[reducemorph]`, and the second an *Adv-mente*; only the conjunctions *e* (and), *ou* (or) and *mas* (but) are allowed; both adverbs can facultatively be further modified by a quantifying adverb, a comparative adverb or a negation adverb; these adverbs have been given in the lexicon the features `[advquant]`, `[advcomp]` and `[neg]`, respectively; this chunking is not made if there is a feminine-singular noun in the left context of the pattern. A similar rule is used for coordination of three (or more) *Adv-mente*.

Dependency Extraction

Finally, the parser extracts the syntactic relations between the chunks. Dependency extraction rules have the general format:

```
|<left-context>|<pattern> |<right-context>|
if <conditions>
   <dependencies>
```

Relevant for this paper are the *coordination* (`COORD`) and *modifier* (`MOD`) dependencies, which are now very briefly presented.

Coordination is a strictly local relation between a coordinative conjunction and two (or more) chunk heads. This dependency is extracted as early as possible in the parsing process, and before the modifier is calculated. In the case of the coordination of *Adv-mente*, the COORD dependency between the reduced form is given the feature c-mente. The basic rule for coordination extraction is provided below:

```
|ADVP{?*, adv#1,
conj#2[lemma:e];conj#2[lemma:ou];conj#2[lemma:mas],
(adv[advquant];adv[advcomp];adv[neg])*, adv#3[last]} |
if (~CLINK(#1,#3))
  CLINK(#1,#3),
  LCOORD[c-mente=+](#2,#1),
  RCOORD(#2,#3).
```

The rule reads: in an ADVP chunk with two coordinated adverbs, which are designated by variables #1 and #3, if no auxiliary dependency CLINK has yet been extracted between the two adverbs; then create that CLINK dependency between the adverbs, that is, the conjunction proper; and create two other dependencies, to produce an output: LCOORD (L=left) between the conjunction (variable #2) and the reduced form, which is then given the feature c-mente, and RCOORD (R=right) between the conjuncion and the *Adv-mente*. A facultative (quantifier, comparative or negation) adverb can occur between the conjunction and the *Adv-mente*; the modifier relation holding between these two adverbs is extracted by another rule.

For longer coordination chains, involving two or more reduced forms, the LCOORD dependency is propagated to the left by a similar rule:

```
|ADVP{?*, adv#1,
conj#2[lemma:e];conj#2[lemma:ou];conj#2[lemma:mas],
if (~CLINK(#1,#3) & CLINK(#3,?) & LCOORD(#2,#3))
  CLINK(#1,#3),
  LCOORD[c-mente=+](#2,#1).
```

The modifier dependency holds between two chunks. For *Adv-mente*, most of them modify a verb or an adjective. One of the basic rules for extracting the adverbial, right modifier of a verb is given below:

```
|#1[verb];sc#1, ?[verb:~ ,scfeat:~ ],
(AP;PP), (PUNCT[comma]), ADVP#2 |
if ( HEAD(#3,#1) & HEAD(#4,#2) & ~MOD(?,#4) & ~QUANTD(#3,#4))
  MOD[post=+](#3,#4)
```

Briefly, this rule reads: For a verb (or a subclause SC) #1 and an adverbial phrase #2, eventually admitting an adjectival or prepositional phrase, or a comma, in-between; if no modifier MOD has been extracted for the head of #2, nor a quantifier QUANTD dependency has been extracted between the heads of #1 and #2; then build the MOD dependency between the heads of the verb and the adverb phrases. Processing sentence (1) *O Pedro leu isso lenta e atentamente* 'Peter read this slowly and attentively', the system extracts the following dependencies:

```
COORD_C-MENTE(e,lenta),COORD(e,atentamente),
MOD_POST(leu,atentamente),MOD_C-MENTE_POST(leu,lenta).
```

The Evaluation Corpus

For the evaluation, a corpus with 1,132 sentences was retrieved from the CETEMPúblico. It consists of sentences presenting an adjective or past participle, one of the three main coordinating conjunctions – *e* (and), *ou* (or) or *mas* (but) –, and an *Adv-mente*. The sentences were obtained from the concordances retrieved using the AC/DC search system of Linguateca webpage. The corpus was then parsed by the system and the dependencies were manually corrected, each sentence being independently checked at least twice, by two linguists. The chunking was also corrected, when appropriate. For this paper, only the COORD and MOD dependencies involving *Adv-mente* or their reduced forms were kept from the system's output. Table 2 shows the breakdown of each dependency in the corpus. The difference between the number of COORD and COORD_C-MENTE dependencies is due to the cases of multiple coordination (i.e., more than two adverbs coordinated together). The large difference between MOD and MOD_C-MENTE consist of *Adv-mente* that, although occurring next to a conjunction and after a reduced form, are not coordinated with it, and modify some other constituent in the sentence.

Table 2. Dependencies in Reference Corpus

Dependency	#
COORD	438
COORD_C-MENTE	462
MOD	1,403
MOD_C-MENTE	462

3 Evaluation and Results

To assess the integrated solution implemented in the system, each of its three main steps was evaluated independently. A set of scripts were especially built to make the result-gathering process fully automatic.

Lexicon

First, the lexicon coverage is evaluated by computing the recall of the reduced forms. From the 462 reduced forms found in the reference corpus, only 10 (8 different) forms had not been previously encoded in the lexicon, thus yielding a recall of 0.978, *scilicet*: Two so-called point-of-view adverbs [17] (*bioquímica* 'biochemistry' and *iconográfica* 'iconographic'), four participle-based (*figurada*, *fundada*, *interpelada*, *zelada*), one numeral-based (dupla 'double'), and a spelling mistake (*massiça* = *maciça* 'massive'). The numeral-based form is clearly a lacuna, since not only the *Adv-mente* is already in the lexicon (*duplamente* 'doubly', but other reduced forms also have been encoded (*tripla*, *triplamente* 'triple, three times'). The remaining lacunae were also corrected. For the misspelled form and its variants, because this is a very frequent error in texts, a new entry, but with the correct lemma, was introduced in the lexicon[8].

[8] A finite-state morphological analyzer is currently under construction, to complement LexMan [7][9].

Disambiguation Rules
This step consists in assessing the impact of the disambiguation rules in selecting or discarding the POS tags corresponding to the adjective or the reduced adverbial form. Table 3 shows the results of the rule-based disambiguation module. From the 462 adverb reduced forms, the system fails to spot 21, while it incorrectly accords this tag to 316, therefore yielding a relatively low precision but high recall, contributing to the interesting F-measure result. This means that in spite of the conservative approach in devising the disambiguation rules and the final, "cleaning" rule that eliminates all remaining reduced forms not previously captured, the system still fails to recognize the cases where there is no coordination of adverbs.

Table 3. Results: Disambiguation Rules

Precision	Recall	F-Measure
0.583	0.955	0.724

Dependency Extraction
The next figures are a combined result of the chunking and of the dependency extraction modules. The purpose of parsing a text is to retrieve the syntactic-semantic relations between constituents, which (partially) express the text meanings. Table 4 shows the results for the dependency extraction module. In order to obtain a better perception of the system's performance, a set of experiments was carried out. The first line presents the overall performance of the system. In the next lines, each dependency is evaluated separately. Finally, the two coordination and modifier dependencies are evaluated in pairs. The overall performance of the system in the dependency extraction is promising. In general, the system is able to extract most of the modifier dependencies (92%), and only 39% of reduced adverbial forms are not adequately related to the element they modify. The system shows suboptimal performance in the extraction of coordination dependencies. There is a clear relation between the low precision in the `MOD_C-MENTE` and the low precision on `COORD` dependencies. When the system fails to extract the coordination, it also (partially) fails to extract the modifier dependency. The reason for this is to be found in the previous module of disambiguation rules, which often and inadequately selects the reduced adverb form instead of recognizing the coordination of adjectives.

Table 4. Results: Dependency Extraction

Experiment	Precision	Recall	F-Measure
All dependencies	0.754	*0.875*	0.810
MOD	*0.921*	0.852	*0.886*
MOD_C-MENTE	0.608	0.719	0.659
COORD	0.642	0.777	0.703
COORD_C-MENTE	0.646	0.805	0.717
2MOD	0.822	0.849	0.834
2COORD	0.644	0.858	0.736

4 Discussion and Future Work

These results confirm the difficulty of the task, sketched at the onset of this paper. The overall performance of the system may be considered satisfactory. The linguistic resources of the STRING natural language processing chain were systematically extended to be used in this paper and they proved to be comprehensive showing very good lexical coverage, and featuring a 0.98 recall.

Adequately capturing coordination is a difficult parsing task, mainly because of the different sentential levels at which it may operate, but also because of the many semantic constraints involved in the pairing of two constituents. In the case of reduced adverbs, ambiguity with another part-of-speech complicates matters even further, lowering results. Precision in the dependency extraction, while generally good (0.75), is directly related to the low precision of disambiguation rules (0.58), which needs to be improved.

The main cause for this low precision is the excessive tendency to analyze adjectives as adverbs in coordination. This could be avoided by using disambiguating rules which would be linguistically less motivated, but that would be more in accordance with the patterns frequently found in the corpus. For example, with coordinated adjectives, the presence of a copula verb often occurs (e.g. *a sua confusão é normal e provavelmente resolve-se com a experiência* 'his confusion is normal and probably can be solved with the experience'); the same happens in the presence of a quantifying adverb before the first adjective (e.g. *uma região bastante conservadora e notoriamente católica* 'a rather conservative and notoriously catholic region'), or the second, or both adjectives; there is also a tendency for the last adjective in a sequence of three coordinated adjectives to present an adverb modifier (e.g. *uma coisa horrível, ilegal e altamente reprovável* 'something horrible, illegal and highly reproachable'). Such solution, however, risks not to be easily adaptable to other domains or text genres. Another path to be tread would consist in using available semantic and syntactic information associated to *Adv-mente* [10], and collocational patterns they may show [22] to model the correct classification, using machine-learning techniques.

Acknowledgments. This work was partially supported by FCT (INESC-ID multiannual funding), through the PIDDAC Program, the FCT project REAP.PT (proj. ref. CMU-PT/HuMach/0053/2008) and the European Comission, Education and Training Erasmus Mundus Master Course Program (EMMC 2008-0083) in NLP&HLT.

References

1. Afonso, S.: Clara e sucintamente: um estudo em corpus sobre a coordenação de advérbios em -mente. In: XVIII Encontro Nacional da Associação Portuguesa de Linguística (APL 2002), Porto, Portugal, pp. 27–36 (2002)
2. Aït-Mokhtar, S., Chanod, J.P., Roux, C.: Robustness beyond shallowness: Incremental deep parsing. Natural Language Engineering 8, 121–144 (2002)
3. Bick, E.: The Parsing System PALAVRAS. Automatic Grammatical Analysis of Portuguese in a Constraint Grammar Framework. Arhus University Press (2000)

4. Casteleiro, J.M.: Vocabulário Ortográfico da Língua Portuguesa. Porto: Porto Editora (2009)
5. Costa, F., Branco, A.: LXGram: A Deep Linguistic Processing Grammar for Portuguese. In: Pardo, T.A.S., Branco, A., Klautau, A., Vieira, R., de Lima, V.L.S. (eds.) PROPOR 2010. LNCS, vol. 6001, pp. 86–89. Springer, Heidelberg (2010)
6. Costa, J.: O Advérbio em Português Europeu. Edições Colibri, Lisboa (2008)
7. Diniz, C.: RuDriCo2 - Um Conversor Baseado em Regras de Transformação Declarativas. M.Sc. Thesis, Lisboa: Instituto Superior Técnico/Universidade Técnica de Lisboa (2010)
8. Diniz, C., Mamede, N., Pereira, J.D.: RuDriCo2 - a faster disambiguator and segmentation modifier. In: II Simpósio de Informática (INForum), Universidade do Minho, pp. 573–584 (2010)
9. Diniz, C., Mamede, N.: LEXMAN - Lexical Morphological Analyser. Tech. rep., Lisboa L2F/INESC-ID Lisboa (2011)
10. Fernandes, G.: Automatic Disambiguation of -mente ending Adverbs in Brazilian Portuguese. M.A. Thesis, Universidade do Algarve/Universitat Autònoma de Barcelona, Faro/Barcelona (2011)
11. Gross, M.: Grammaire transformationnelle du français. 3 - Syntaxe de l'adverbe. ASSTRIL, Paris (1986)
12. Hagège, C., Baptista, J., Mamede, N.: Caracterização e processamento de expressões temporais em português. Linguamática 2(1), 63–76 (2010)
13. Mamede, N.: STRING - A Cadeia de Processamento de Língua Natural do L2F. Tech. Rep., Lisboa: L2F/INESC-ID Lisboa (2011)
14. Nobre, N.: Anaphora Resolution. M.Sc. Thesis, Lisboa: Instituto Superior Técnico/Universidade Técnica de Lisboa (2011)
15. Oliveira, D.: Extraction and Classification of Named Entities, M.Sc. Thesis, Lisboa: Instituto Superior Técnico/Universidade Técnica de Lisboa (2010)
16. Palma, C.: Expressões fixas adverbiais: descriçãco léxico-sintáctica e subsídios para um estudo contrastivo Português-Espanhol (M.A. Thesis). Faro, Univ. Algarve - FCHS, Faro, Universidade do Algarve/FCHS (2009)
17. Molinier, C., Levrier, F.: Grammaire des Adverbes. Description des formes en -ment. Librairie Droz, Genéve-Paris (2000)
18. Pinheiro, G.M., Aluísio, S.M.: Corpus NILC: Descrição e análise crítica com vistas ao projeto Lácio-Web. Tech. rep., NILC-TR-03-03, São Carlos, Brasil (Fevereiro 2003)
19. Ribeiro, R.: Anotação Morfossintáctica Desambiguada em Português, MSc Thesis, Lisboa: Instituto Superior Técnico/Universidade Técnica de Lisboa (2003)
20. Rocha, P., Santos, D.: CETEMPúblico: Um corpus de grandes dimensões de linguagem jornalística portuguesa. In: Actas do V Encontro para o processamento computacional da língua portuguesa escrita e falada, PROPOR 2000, Atibaia, São Paulo, Brasil, pp. 131–140 (November 2000)
21. Santos, D.: Extração de Relações entre Entidades Mencionadas. MSc Thesis, Lisboa: Instituto Superior Técnico/Universidade Técnica de Lisboa (2010)
22. Vieira, L.: Verb and -mente ending Adverb Collocations in Brazilian Portuguese: Extraction from Corpora and Automatic Translation into English. M.A. Thesis (in preparation)

Morphosyntactic Analysis of Language in Children with Autism Spectrum Disorder

Raquel Reis[1] and António Teixeira[2]

[1] Speech Therapist, Master in Speech and Hearing Sciences,
Universidade de Aveiro
rakel.reis@gmail.com
[2] Dep. Electrónica Telecomunicações e Informática/IEETA,
Universidade de Aveiro, 3810 193 AVEIRO, Portugal
ajst@ua.pt

Abstract. Autism is a neurodevelopmental disorder that involves severe and persistent deficits in social interaction, language and communication. This study aims to contribute to the characterization of the language of children with ASD (Autism Spectrum Disorders) in regard to morphology and syntax (morphosyntactic classes, most frequent words, use of personal pronouns, inflections of verbs: tense and person and syntactic structure). The four selected children, diagnosed with ASD, were recorded during activities that promoted speech. The records were transcribed and processed by PALAVRAS parser. The more frequent morphosyntactic classes detected were verbs and names. The most used determinants were the definite articles and, as far as adverbs are concerned, the most frequently used ones were adverbs of place. The syntactic structures were simple and short. The automated tools used for processing and analysis proved to be extremely functional and practical, with fast and effective results, opening the possibility of its usage for evaluation purposes in speech therapy practice.

1 Introduction

The Autism Spectrum Disorders (ASD) is a highly challenging area of intervention and research.

The relative scarcity of studies, particularly on expressive communication and language development [1, 2], happens due to the difficulties in studying aspects of language in children with ASD characteristics [3].

In fact, it may be considered that the diversity of language skills of these children is their most obvious feature of language [4] and, despite more than five decades of research, there is still much to discover about language in autism [5] .

The present work aims to contribute to characterization of the language of children with ASD in regard to morphology and syntax. It is intended to analyze the morphosyntactic classes, the most frequent words in classes with major representativeness, the use of personal pronouns, inflections of verbs: tense and person and the syntactic structure of utterances.

H. Caseli et al. (Eds.): PROPOR 2012, LNAI 7243, pp. 35–45, 2012.

2 Autism

The term "autism" comes from the Greek *autos*, meaning "self" and was preached for the first time in 1911 by psychiatrist Bleuler [6].

Currently, autism is considered a neurodevelopmental disorder that involves severe and persistent deficits, which result in signs and symptoms of varying intensity in different areas of functioning, including social interaction, language and communication, with restricted and stereotyped behavior patterns, interests and activities [1, 2, 7-12].

2.1 Communication in Autism

Communication is one of the three axes of the autism frame [13] and, in this area, it is present a serious qualitative impairment [7, 14], regardless of language level, since the problem lies on the use of existing skills, i.e., functionality [15].

This communication gap is present in both receptive and expressive aspects [1]. Children with ASD are less receptive to the communicative acts of others, which often leads them to don't respond to their own name [16]. Moreover, their communication initiatives are rare and occur more as a regulatory function than as a declarative one [11, 16].

In regard to the pre-requirements of communication, these children demonstrate absence of eye contact (diminutive interest in the human face), absence of social smile in their first years of life, difficulty in joint attention and in the use of gestures as a primary way of communication (e.g. pointing, greet with the hands, deny with the head, etc.) [11, 14, 16]. In other words, these children don't know how, and are not able to compensate for the absence of language with nonverbal communication [14, 17].

2.2 Language in Autism

There is no specific language symptom in autism [17], however, the existence of autism implies, necessarily, language impairment [4], thus the greatest difficulties these children experience are related to social interaction and communication, areas in which language plays a key role.

The most striking feature in the language of an autistic person is, perhaps, its diversity [16], as the language skills present in autism can range from mutism, with little or non-functional communication, to relatively well-developed syntactic capabilities and functional speech, although it exists a peculiar use of language (disturbed, repetitive and meaningless) [4, 5, 7, 18].

Language acquisition in these children is delayed [15] and it is estimated that only about one half of them will acquire some speech as a mode of communication [16] and even then, functional communication skills will never fully develop for the majority of those children who begin to speak [19], maintaining a linguistic level below standard [2].

When there is no language development, the most affected areas are characteristically the pragmatic, semantic [14] and suprasegmental phonology [17], however, other aspects of phonology, as well as morphology and syntax are also impaired [18, 20].

In terms of receptive language, which is often literal, an autistic person doesn't take into account the intention of the issuer, and has great difficulties in understanding indirect speech acts [21]. Thus, the act of understanding, in some cases, may be limited to names of familiar objects and contextualized orders [15].

In regard to expressive language, several pioneer studies revealed characteristics such as atypical intonation and vocal quality, idiosyncratic use of words and stereotyped phrases, echolalia and pronoun reversal [5, 11].

2.3 Morphology in Autism

Several studies on the acquisition of grammatical morphemes in children with autism found no differences when comparing to children with normal development. However, a Bartolucci's study, in 1980, showed that children with autism produced less grammatical morphemes, especially verb tense and articles [7]. Simultaneously, other authors suggest that these children tend to have peculiar difficulty in properly using morphemes and auxiliaries, such as tense markers [18].

In regard to verbs, the use of uninflected forms (in the infinitive) is typical of an autistic person, as well as the difficulties in other flections, mainly in the past tense[18] [5, 12]. On the other hand, a 2007 study has verified the existence of difficulties with comprehension of verbs [2].

In it is also common that children with ASD experience difficulties with the production of little linking words (or connectors), such as: "in" "on" "under" "before" and "because", which are often omitted or used in a wrong way[15].

During early language development, these children tend to use specific nouns more frequently than closed class words (like auxiliary verbs, conjunctions, determinants, prepositions, and pronouns). In fact, they have difficulties in using prepositions, avoiding them [18], and using a few pronouns, referring to people by their own name or another name that describe them [18] or, when they use it, frequently reverse them [22].

2.4 Syntax in Autism

Syntax deficits are not specific to ASD [4], but several recent publications suggest the presence of more substantive syntactic impairments in autism [7]. Although the development of structural aspects of the language may be appropriate in these children [9, 17], there are serious limitations in the social usage of structures [9].

The delay in this area is reflected in an atypical developmental pathway, slower and less complex, with difficulties with pronouns (reversal) and verb endings, change in the order of words in the sentence, fixation on certain sentence structures, use of short and simple syntactic structures, as well as the omission of grammatical elements (such as prepositions and conjunctions) and concordance of gender and number [4, 7, 22].

Children with ASD have difficulties in evolving from the combination of words into the construction of structured sentences [7] and tend to use a more restricted set of syntactic structures, mainly simple, that is, they tend to rigidly depend on a particular sentence structure, even though they are able to employ greater variety in speech [18].

3 Method

3.1 Sample

Among children with ASD accompanied on speech and language therapy by the first author of this work, three children were selected from a public school in (omitted for blind review) and one child from a clinic on the same district, being the selection criteria the diagnosis of ASD and the presence of oral communication.

The four selected children (3 males) are between the ages of 6 and 13 years old (3 of them with 6 and 1 with 13) and they have an autism degree between mild and severe. Two of the children present a global IQ (intelligence quotient) within the average and the other two have cognitive impairment. All of them have oral communication that varies in complexity, frequency and functionality.

3.2 Procedure

A session of about 45 minutes with each one of the children was recorded in audio and video. These sessions included different activities with the objective of promoting speech. The activities were: 1) presentation of thumbnails of animals; 2) presentation of color and black and white A4 images, for description; 3) presentation of a book with images of isolated actions; 4) presentation of material of symbolic game, namely a medical bag and a cosmetic bag.

The audio records were transcribed with the Transcriber software (v 1.5.1) [23] and, subsequently the annotations were reviewed and rectified based on the video display (Fig. 1). Intelligible productions were duly transcribed and the unintelligible ones were marked and excluded from the analysis. The same was done with the echolalia, the stereotyped productions and environmental noise.

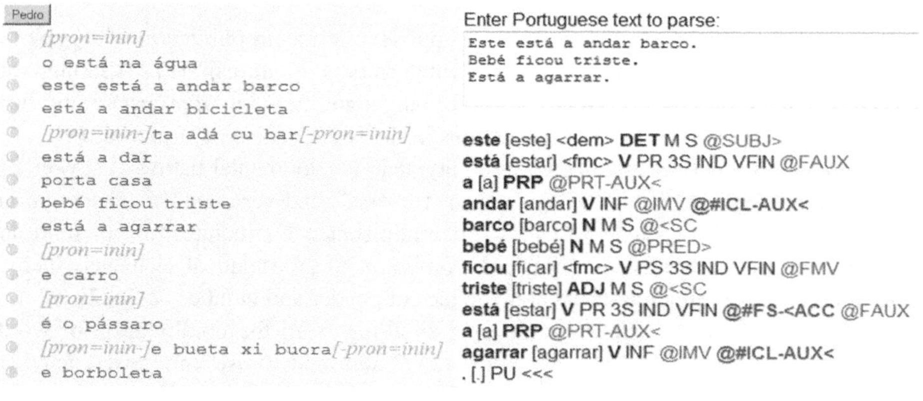

Fig. 1. Examples of transcription and result from PALAVRAS processing

The files resulting of the transcriptions were treated and submitted to processing, in a system of grammatical analysis available online (PALAVRAS), inserted in the VISL project (Visual Interactive Syntax Learning) of the Institute of Language and Communication of the University of Southern Denmark [24]. This analysis was carried out in terms of morphology and syntax, namely morphosyntactic classes, words most frequently present in classes with greater representativeness, use of personal pronouns, and inflections of verbs: tense and person, as well as the syntactic structure. An example is shown in Fig. 1. Besides the referred aspects, were also analyzed the length of utterances, the occurrence of echolalia and idiosyncratic or stereotyped speech, not included in this paper due to the limited space (see [25] for more details, available online).

4 Results

4.1 Morphosyntactic Classes

The morphosyntactic classes of the most frequent words (Fig. 2) are verbs and names, followed by determinants, prepositions, adverbs, and interjections with lesser extent.

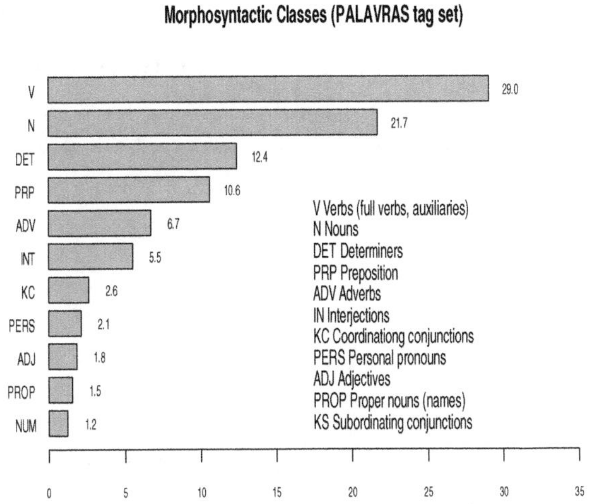

Fig. 2. Most frequent morphosyntactic classes

4.2 Detailed Analysis of the Most Frequent Morphosyntactic Classes

In this analysis, as well as in the following charts, we have excluded the occurrences below 1%.

The names with greater occurrence (Fig. 3a) are related to a subject, person or animal ("menino" (boy), "menina" (girl) and "cavalo" (horse)). The verbs used are all finite and the great majority (about 97%) arises in the indicative. The most frequent ones are "estar" (being) and "ser" (be) (Fig. 3b). In regard to the determinants, the most used are the definite articles, followed on a much smaller number of indefinite ones. The most commonly prepositions used are "a", followed on a much smaller number by "de" (from), "em" (in), "para" (to/for) and "com" (with). The adverbs most frequently used (Fig. 3c) are adverbs of place: "aqui" (here) and "ali" (there). Comparing the adverbs "sim" (yes) and "não" (no), "não" is the most used. The interjections more used (Fig. 3d) are "sim" (yes) and "no" (no), being the first one far superior (63.6%). The class of personal pronouns (Fig. 3e) has little incidence on the analyzed sample. As for the different pronouns used, the most frequent is "eu" (I), followed by "ele" (he) and "tu" (you).

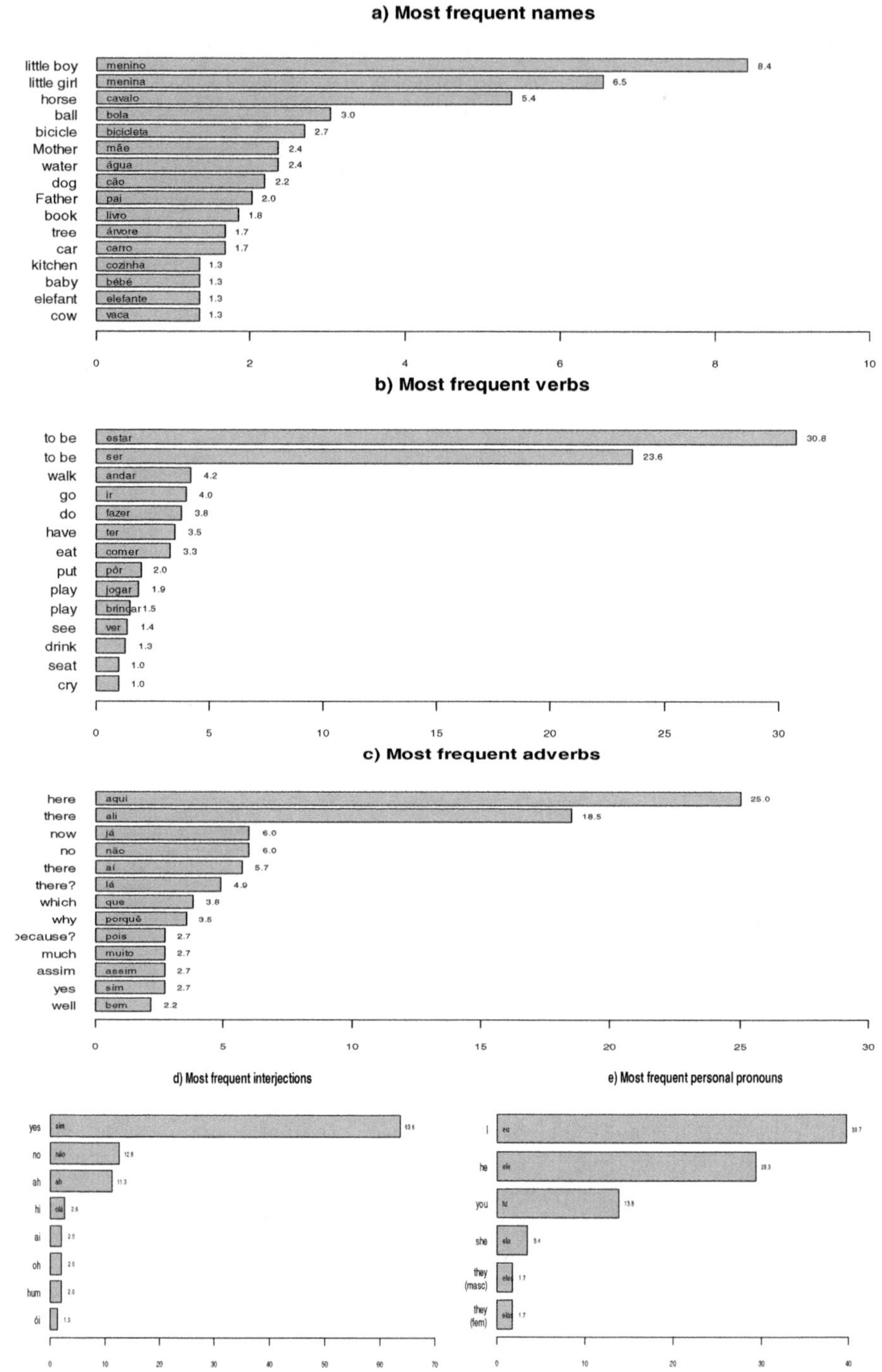

Fig. 3. Information on the most frequent words for the classes Noun, Verb, Adverb, Interjection and Personal Pronouns

4.3 Verb Tense and Person

The most frequently used tense (Fig. 4a) is the present, followed by the impersonal infinitive.

The person (Fig. 3b) more used is the third singular person, followed, with much lower values, by the third plural and the first singular persons. There is a high incidence of undefined person (0/1/3S), with 27.78%.

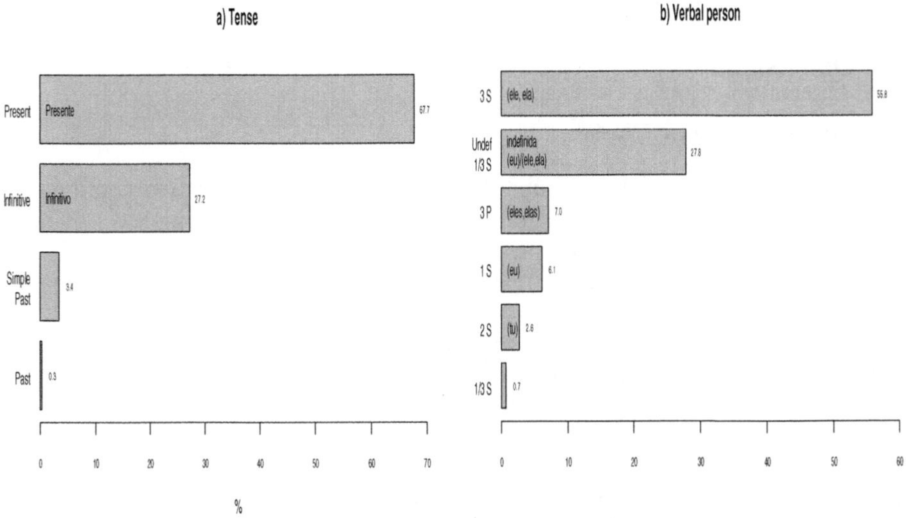

Fig. 4. Verbs tense and person usage (in percentage)

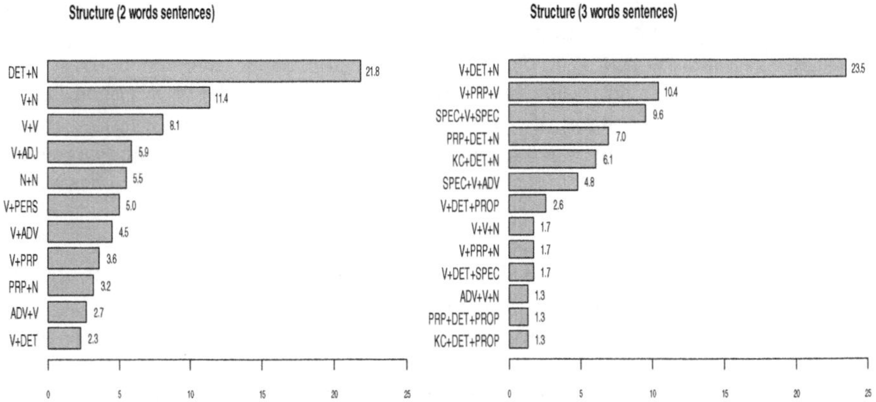

Fig. 5. Syntactic structures of utterances of 2 (at left) or 3 words (at right). N – Noun, V – Verb, DET – Determiners, PRP – Prepositions, ADV – Adverbs, KC – Coordinative Conjunctions, PROP – Proper Nouns, PERS – Personal Pronouns, ADJ – Adjective, SPEC – Specifiers.

In this analysis, we have excluded the occurrences below 1 or 2%. The most common syntactic structures (apart from the isolated words: interjections, names, verbs and adverbs) are: [verb + determinant + name] and [determinant + name].

By analyzing the most prominent structures in each size of utterance, it is possible to see that the ones with 2 words (Fig. 5a) present mostly simple structures [determinant + name], whereas the ones with 3 words (Fig. 5b) present, in its more frequent structure, an occurrence of verb and complement formed by a determinant and a name: [verb + determinant + name].

5 Discussion

5.1 Morphosyntactic Classes

Generally, the low occurrence of subordinate conjunctions (0.4%), and coordinating conjunctions (2.6%), can point out to a very low occurrence of complex sentences, consisting of more than one clause.

The most frequent names used, of people or animals, suggest that they are mainly used as the subject of the action.

The definite articles are the most used determinants, which represents a contrast to Bartolucci's study, in 1980, which showed that children with autism produced less grammatical morphemes, namely articles [7].

As far as the occurrence of prepositions is concerned, the frequency detected in the use of "a" as not expected, due to the difficulties described with their use [4, 7, 18, 22]. It is also described in the bibliography, the difficulty with small linking words [15] that include, apart from some prepositions, some adverbs relatively common in the analyzed sample.

In the class of adverbs, the adverb of negation "não" (no) comes up more often than "sim" (yes), however this relation reverses in a very sharp way when the "sim" and the "não" are considered to be interjections, with far superior percentage of "sim", result supported by the bibliography [18].

Face to this dual morphosyntactic classification, the data were manually reviewed, showing that the words "sim" and "não" are classified as adverbs or interjections when, respectively, are in the context of a sentence or isolated. Thus, in isolation, the word "yes" occurs more frequently than "no" and the opposite occurs in the sentence context.

5.2 Inflections of Verbs

The prominent use of verbal forms with undefined person (0/1/3S) refers to the use of the infinitive form of the verb, which is more recurring in this analysis.

This fact is supported by bibliography that suggest that the use of verbs without tense inflection is typical of ASD [5, 12, 18].

The same sources claim the existence of difficulties in verbal inflection, particularly in the past tense. The results of this study are in agreement with those claims, because the use of past tense is reduced.

Also in this context, it was noted that the vast majority of verbal productions express, predominantly, only two tenses: present and infinitive, which reveals little inflection of verbs. This is agreement with the Bartolucci's study (1980) which argues that children with autism produce less grammatical morphemes, in particular those relating to tense [7].

5.3 Personal Pronouns

Although the difficulty in using the pronoun "eu" (I) is defined as characteristic of ASD [10], there was a widespread use of this pronoun, followed by "ele" (he), which according to the same author is often used as compensation for the difficulty in using "eu" (I). This unexpected result can be related to another aspect referred in the bibliography, which focus on understanding personal pronouns that appear to be intact in children with ASD, without differences when compared with other children [4].

The pronoun "tu" (you) was long used in the sample, which can be related to the children's familiarity with their interlocutor, who is a therapist that has intervening with them for a few months, on a weekly basis.

5.4 Syntax

The tendency to use short and simple syntactic structures, referred to in the bibliography [22] [4] [7] [18], is confirmed by the high frequency of short utterances, predominately those with 1 or 2 words and decreasing the frequency of incidence as the number of words increases. In addition, in the utterances with more than one word, the most frequent syntactic structures are simple, i.e. with no more than one clause. In fact, it is visible the fixation on certain sentence structures [4, 7, 22], which arise with relatively emphasized occurrence.

The high incidence of single words can be justified by the difficulties, referred in the bibliography, in passing from combination of words to construction of structured sentences [7].

Among the structures of 2 words, the most repeated ([determinant + name]) suggest a frequent occurrence of enumeration, and only the utterances with more than 3 words have structures that include verbs.

5.5 Limitations

The sample size was limited, particularly in regard to the analysis that was possible, without, for example, comparing reliably within the sample variables (age, degree of autism and IQ). Due to this limitation, we also avoided statistical tests, concentrating on the use of descriptive statistics.

The automated tool used for analysis (PALAVRAS) proved to be extremely functional and practical, with fast and effective results making possible this study and allowing the extension of the language aspects discussed.

However, there are limitations inherent to the processing that, being automatic, has pre-defined analysis aspects. Therefore, some application settings were restrictive, for example, the analyzing of the incidence of the words "yes" and "no", that were classified in two categories (adverbs and interjections) depending on the context of occurrence, which hampered the analysis of the general occurrence, independently of the syntactic context (in sentence or isolated).

6 Conclusion

Aiming at providing Speech Therapists more detailed information on the morphosyntactic aspects of the language on children with ASD, this paper presents a method for obtaining such information and first results from a small group of subjects. Essential to the method is the use of automatic parsing and transcription of the productions.

Our study shows that the most frequent morphosyntactic classes are the names and verbs, with relatively frequent presence of determinants, prepositions and adverbs, the opposite of what would be expected.

The most widely used is the present tense, followed by the not inflected verb form (infinitive), with reduced use of verbs in the past tense.

The most regular verbal person is the third singular. The use of personal pronouns is reduced, but it wasn't verified the expected difficulty in the use of the pronoun "eu" (I), which is the most widely used. There was also not found prominent occurrence of pronoun reversal.

There is a majority use of simple and short syntactic structures, with a high occurrence of isolated words (utterances of a word). The few complex structures that occur are mostly coordinate clauses, instead of subordinate ones.

It would be interesting to expand the analysis to other aspects of morphosyntax that were not part of the study, in particular the grammar correction of productions and its deeper syntactic analysis.

Another suggestion of continuity is the possible expansion of the analysis made in this study to a wider sample of children with ASD, with broader age ranges, in order to obtain a descriptive analysis of language patterns present in this population. It is also very important to extend this analysis to a population of children with normal language development, in order to draw a complete profile comparison in normative terms.

Finally, authors consider that the method presented in this paper can be incorporated in the professional practice of Speech Therapists having ASD patients, to provide more quantitative evaluations.

References

1. Chiang, H.-M., Lin, Y.-H.: Expressive Communication of Children with Autism. J. Autism Dev. Disord 38, 538–545 (2008)
2. Foudon, N., Reboul, A., Manificat, S.: Language Acquisition in Autistic Children: A Longitudinal Study. CamLing, 72–79 (2007)
3. Tager-Flusberg, H.: The Challenge of Studying Language Development in Children With Autism. In: Menn, L., Ratner, N.B. (eds.) Methods for Studying Language Production, pp. 313–332. Lawrence Erlbaum Associates, Mahwah (2000)
4. Wilkinson, K.M.: Profiles of Language and Communication Skills in Autism. Mental Retardation and Developmental Disabilities Research Reviews 4, 73–79 (1998)
5. Tager-Flusberg, H.: Strategies for Conducting Research on Language in Autism. Journal of Autism and Developmental Disorders 34(1), 75–80 (2004)

6. Spengler, C.D., Fischer, J.: Distúrbios da Linguagem da Criança Autista. Revista de Divulgação Técnico-Científica do ICPG 3(12), 33–36 (2008)
7. Eigsti, I.-M., Bennetto, L., Dadlani, M.B.: Beyond Pragmatics: Morphosyntactic Development in Autism. J. Autism Dev. Disord 37, 1007–1023 (2007)
8. Oliveira, G., et al.: Epidemiology of autism spectrum disorder in Portugal: prevalence, clinical characterization, and medical conditions. Developmental Medicine & Child Neurology 49, 726–733 (2007)
9. Mulas, F., et al.: El lenguaje y los trastornos del neurodesarrollo. Revisión de las características clínicas. Revista de Neurología 42(supl. 2), pp. S103–S109 (2006)
10. Artigas, J.: El lenguaje en los trastornos autistas. Rev. Neurol., 28(supl. 2), pp. S118–S123 (1999)
11. Klin, A.: Autismo e síndrome de Asperger: uma visão geral. Rev. Bras. Psiquiatr 28(supl. I), pp. S3–S11 (2006)
12. Seung, H.K.: Linguistic characteristics of individuals with high functioning autism and Asperger syndrome. Clinical Linguistics & Phonetics 21(4), 247–259 (2007)
13. Monfort, I., Comunicación y lenguaje: bidireccionalidad en la intervención en niños con trastorno de espectro autista. Rev. Neurol. 48(supl. 2), S53–S56 (2009)
14. Rondal, J.A.: Teoria de la mente y lenguaje. Revista de Logopedia, Foniatria y Audiologia 27(2), 51–55 (2007)
15. Wing, L.: The Autistic Spectrum - new updated edition. Robinson, London (1996)
16. Calderón, R.S.: Comunicación y lenguaje en personas que se ubican dentro del espectro autista. Actualidades Investigativas en Educación 7(2), 1–16 (2007)
17. Martos, J., Ayuda, R.: Comunicación y lenguaje en el espectro autista: el autismo y la disfasia. Revista de Neurología 34(supl. 1), S58–S63 (2002)
18. Manookin, M.B.: A formal semantic analysis of autistic language: the quantification hypothesis, in Department of Linguistics, Brigham Young University (2004)
19. Lerman, D.C., et al.: A methodology for assessing the functions of emergins speech in children with developmental disabilities. Journal of Applied Behavior Analysis 38(3), 303–316 (2005)
20. Geurts, H.M., Embrechts, M.: Language Profiles in ASD, SLI, and ADHD. J. Autism Dev. Disord 38, 1931–1943 (2008)
21. Telmo, I.C., Ajudautismo, E.D.: Formautismo - Manual de formação em autismo para professores e famílias. APPDA-Lisboa (2006)
22. Mota, C., Bravo, P.: Autismo e Comunicação, Aveiro (2007)
23. Boudahmane, K., et al.: Transcriber. DGA (1998-2008)
24. Bick, E.: PALAVRAS. Institute of Language and Communication, University of Southern Denmark (1996-2009)
25. Reis, R.: A linguagem em crianças com perturbações do espectro do autismo: análise morfossintáctica. Universidade de Aveiro (2009)

Lince, an End User Tool for the Implementation of the Spelling Reform of Portuguese

José Pedro Ferreira[1], António Lourinho[2], and Margarita Correia[1]

[1] Instituto de Linguística Teórica e Computacional – ILTEC, Lisbon, Portugal
{jpf,mcf}@iltec.pt
[2] Knowledgeworks, Lisbon, Portugal
antonio.lourinho@knowledgeworks.pt

Abstract. This paper outlines the design choices of *Lince*, a multi-platform application that updates the textual contents of documents according to a recent spelling reform of Portuguese. We took advantage of an existing lexical relational database of which we mimicked the structure for homogeneity purposes and to streamline the updating process. Being a tool aimed both at the general public and professional users, a balance between simplistic interface design and wide-enough customization options proved essential. In this paper, we start by describing the background and rationale underlying the tool's development; the initial requirements are then presented and the design choices framed, with an emphasis on the conversion algorithm and customization options.

Keywords: NLP tools and resources, machine translation, spelling reform.

1 Introduction

A spelling reform is currently taking place in the Portuguese speaking countries. Affecting more than 200 million speakers, this reform changes only a relatively small number of spelling rules and word forms, but it raises a number of challenges and, as any similar initiative, it has a very public profile [1,2]. One of the problems mentioned most frequently is the perceived cost and slowness of updating existing books, documents and Internet contents. This paper outlines the architecture and criteria that underlined the creation of *Lince* [3], a tool that helps tackle this issue and which was adopted officially by the Portuguese State for that purpose.

Lince is a free multi-platform tool that allows for the mass conversion of any number of documents in the most common text storage and Internet editing formats. While it was conceived to be very simple to use by anyone, it tries to cater to the needs of more advanced users, such as professionals and enterprises, by allowing some customization where needed and allowing for GUI and CLI interaction.

Despite having been developed in Portugal and, so far, only attaining official status there, *Lince* accounts for all the national varieties of Portuguese, which is reflected by the fact that a large portion of its users come from countries other than Portugal. Released in June 2010, the tool has been downloaded over 300,000 times until January 2012. Besides this, it has been successfully used to convert large-scale databases by several institutions.

H. Caseli et al. (Eds.): PROPOR 2012, LNAI 7243, pp. 46–55, 2012.

In this paper, we start by quickly describing the goals of the tool and the background and rationale under which it was developed, so as to justify the design options and emphasize its very public profile. We then explain the underlying data structure and conversion algorithm, and outline the GUI interface design and customization options. Other similar-purpose conversion tools that have been simultaneously developed by other teams are mentioned, but their exhaustive comparison with *Lince* falls out of the scope of this paper.

2 Background

2.1 The Spelling Reform of Portuguese

Since the late 1800s, several attempts at imposing a reform over the spelling of Portuguese at state level took place. After the Portuguese republican revolution of 1910, a first reform was passed, putting into letter of law the way words should be written. This reform affected only Portugal, with Brazil, at the time the only other independent country that had Portuguese as its official language, not adopting it in the long run.

Over the past 100 years there have been different international proposals to unify the spelling rules [4]. After several failed attempts throughout the 20th century, an agreement was finally reached in 1990 [5] by the then seven countries of the Community of Portuguese Language Countries (CPLP). Although every CPLP member signed the agreement over the following years, it wasn't until the late 2000s, after Timor-Leste gained its independence and enlarged the CPLP to the current eight members, that it effectively started being implemented, with Cape Verde, Brazil, Guinea-Bissau, Portugal, and São Tomé and Príncipe all starting the migration process, and the remainder of the countries (Angola, Mozambique, Timor-Leste) expected to do the same soon. The reform changes about 1% of the dictionarized word forms in Brazil and around 1.5% in the other countries, although in many cases as little as 0.4% word forms are effectively changed in running text [6].

Before this reform, there were two different, national-level legal documents determining the spelling of Portuguese: one for Brazil and another one for Portugal and its former colonies. After a decades-long period of approximation and several failed attempts, the CPLP countries decided not to completely unify the spelling of all the words, but instead unify only the spelling rules, accounting for some degree of variation between the forms used in different countries. While this limits the number of affected words greatly, it also raises some additional challenges, as both the words that undergo changes and the result of those changes vary depending on where a text was written.

The changes implemented by the reform affect mainly four aspects of the Portuguese spelling: diacritics, hyphenation rules, usage of initial capital letters and the writing of consonants that are not phonetically produced. While some rules are context dependent, making the task to computationally model them easier, some depend on either semantic or phonetic knowledge, which relevantly lie outside of the orthographic realm. Another problematic issue is the fact that the latter of these sets of rules allows for some variation within a single country, hence resulting in several optional forms. These came to be the biggest challenges for the computational implementation of the reform.

2.2 Rationale

There is a century-long tradition for the spelling to be officially sanctioned and legally imposed, under the form of laws, in Portuguese speaking countries. Other than merely instituting the reforms, such laws are, at the same time, usually the documents that state the rules that should be followed to write correctly. Being legal documents, and often resulting, as in the case of this reform, from hard and lengthy international negotiating processes, such documents typically leave some room for interpretation and do not cover all possible contexts. On the other hand, while providing words that serve as examples for the application of the rules, these are usually limited in number and scope.

Given this, past spelling reforms were effectively applied mostly through the publication of an official vocabulary, large enough in dimension to cover all the possible spelling contexts and cases. While this might have been the best one could do at the time of the last similar reforms, which took place in the 1940s and 1970s, it is no longer so now that most of the written information is consulted, stored and exchanged in digital formats. These days, digital linguistic resources and applications such as easily accessible word lists, spell-checkers and conversion tools are actually of more use to the non-specialist end user than a vocabulary.

Several tools and resources are needed to implement such a reform: a large word-list that can serve as the base dictionary for the other resources that are to be created, with both lemmas and all of their inflected forms, an equivalence list with all the words that undergo changes and all their counterpart new forms, dictionaries for spell-checkers, and a conversion tool that allows for the quick migration of the existing textual resources and contents.

A text converting tool is especially important, as it helps overcome what are perceived as some of the bigger problems of a spelling reform: that it would be a costly and slow process; that it would imply redoing every existing resource; that it would technically be very hard to implement in the developing countries that have Portuguese as their national language.

2.3 Requirements

We established that all the resources and tools we were to develop for the spelling reform should follow the same principles: *Lince* should be free, based on an open-access web platform, and help not just professionals but also the general, unspecialized user in his everyday life. As such, it should run under all the most commonly used operating systems and support all the most common formats for text edition and textual data storage as well as for web edition.

Any lack of homogeneity in the interpretation of the spelling reform would be very damaging in such a process, and as such the converter tool should be fully integrated with the other developed resources, first and foremost the vocabulary. As such, a common and integrated lexical maintenance routine should be developed for the tool's base lexicon.

The tool should also be able to withstand any eventual changes of interpretation during the long six year transition period the Portuguese Government established, ending in May, 2015. In parallel, since the supporting lexical resources had an incremental development period, the data in the tool should be easy to update.

Since its main aim is to convert large amounts of text, the conversion process should be as quick as possible, and it should be possible to convert many different files simultaneously. At the same time, the hardware requirements should be as low as possible, and the tool should run in below-par machines and be independent from any other programs, such as text editors.

Being thought of for the general public, the tool should require very little previous technical knowledge and be able to fulfill its purpose, convert text-containing documents, in as few steps as possible, with little or no previous configuration and end-user interaction. At the same time, it should cater to the needs of more professional users, allow for some degree of customization and reusability for other purposes. In addition to this, in such cases where variation is created by the reform, the tool should allow its users to choose their preferred variants.

Last, and of foremost importance, the tool should allow for the easy adaptation to other national variants besides Portuguese, making the application of the spelling reform in other countries easy to implement. Technological constraints in those countries hence needed to be taken into account.

Being for the most part developed with public funding, all the resources for the implementation of the spelling reform should be made available freely, over a web interface, and, while resulting from an R&D project and aiming at being a resource for other researchers, an effort should be made to make all the resources as useful to the general public and non-scientist professionals as possible.

Other similar-purpose tools were developed simultaneously to *Lince*, most notably by the Natura group of the Minho University [7], Priberam [9,10], and Porto Editora [11,12]. These tools are distinct in nature to *Lince*, being more limited in scope and not fulfilling all or some of the requirements identified in this section: they either don't support all the file formats and operating systems *Lince* does, are only freely available over a limited Internet interface with a paid-for counterpart, can't be installed as a standalone application, can't simultaneously convert several files of any size, or don't present all the customization options *Lince* does. An evaluation of these tools (not including *Lince*) was done in [7]. In a small comparison ([13]) between *Lince* and the tool with the highest score in [13], based on European Union translation memories, *Lince* achieved better results, leading us to think this tool has the best accuracy and performance for this task, although ideally a common evaluation with the direct involvement of all the teams responsible for such tools should be conducted.

3 Tool Description

3.1 Design Options

Due to the social and technological profile of several Portuguese speaking countries - broadband Internet is not yet widely available in every region -, and to the need to convert large-scale databases, which would be impractical to send back and forth to a server, we opted for a stand-alone application,. The application, along with its description and documentation is made available freely over the Portal da Língua Portuguesa [8].

Java was chosen as the development language as it made the need for the application to be multi-OS easier to comply with. Furthermore, some available Java text parsing libraries for several popular text storage formats were perfect for our needs. As an additional reason, the possible development, in the future, of a server-side deployment to make the application available in our Portal via a web applet reinforced the arguments in favor of the choice. The selected parsing libraries allow us to support ODT, HTML, XML, DOC, DOCX and RTF, as well as any simple text or files with metadata enclosed by brackets. Due to format constraints, PDF has limited support, the output being an HTML file with little original formatting retained. Quick installers were developed for Apple, Linux, and Windows-family operating systems to make the installation process easier for general users. While being thought of mostly to convert documents of the most commonly used formats, *Lince* can convert any plain text input or any text with metadata enclosed by brackets. This makes it possible to, in a format-independent way, convert large bodies of texts, namely those contained in large-scale enterprise databases. To make it easier to integrate into scripts or routines, the tool can be invoked directly from a command line, including the options that can be defined in the GUI.

Lince converts the text in files through several steps and making use of different resources, namely a lexical database, as illustrated in greater detail over the next two sections, where the conversion algorithm and data structure are presented. Upon execution, the tool loads the entire lexical database, presented below, as a hash table into the memory, which makes initial load time somewhat high but conversion performance extremely good (usually taking just a few seconds to convert even large files). The modeling of the rules and the conversion algorithm try to limit the size of the lexicon as much as is possible without losing accuracy.

3.2 Conversion Algorithm

As the rules being changed present different challenges, the conversion algorithm includes several types of strategies. While some of the rules could apparently be modeled by mere contextual replacement rules, as they apply universally to given orthographic sequences (e.g., diacritics should no longer be used in the stressed diphthongs <ei> and <oi> in paroxytone words), several reasons discouraged us from going that way, the most important being the potential existence of foreign loan words that have concurring orthographic contexts. Other changes are completely impossible to predict from orthography alone (e.g., a <c> will no longer be written before a <c>, <ç> or <t> if it is not produced phonetically as a plosive, hence making *acção*, 'action', become *ação*) and in such cases affected forms really need to be listed, encouraging the explicit storage and organization of the data in a lexical database.

Conversely, while by their nature most spelling rules affect a potentially infinite set of forms, some are particularly general and affect many non-registered words, making the explicit storage of the affected forms unfeasible, such as the fact that hyphenation rules change for most of the productive prefixes. Fully automatic conversion based merely on pattern is not possible either, though, as most hyphenated morphological compounds are not affected by the reform, and others only change in given semantically dependent contexts (e.g., every previously hyphenated phrase loses its hyphen unless it corresponds to the name of a zoological or botanical species). To limit the size of the dictionary, we adopted a mixed strategy for these cases, storing only the affected prefixes explicitly in a discrete list and simultaneously applying rules to the words that contain them (e.g., for *mini-dicionário* there is an entry for *mini-* in the prefixes dictionary, along with a rule in the hyphenation rules stating hyphenated words with a registered prefix ending in a vowel should drop the hyphen if followed by any character other its last character – in this case, <i> – or <h>). The scope and effect of these rules is defined by a set of ordered regular expressions. To model these changes, we reached a set of 11 ordered rules of varying complexity.

Since there can be concurring rules affecting a given word, the algorithm needed to account for several changes, potentially of different kinds, applying to a given form (e.g. *co-protector* loses both the hyphen and a <c>, becoming *coprotetor*). Rule ordering was essential, as a rule being applied too early in the conversion process would render the temporary form inexistent (**coprotector*) in the explicit-form lexical database, making it go unnoticed by the final replacement steps. On the other hand, very productive morphological processes, such as compounding, generate hyphenated forms that cannot be predicted effectively without extremely large dictionaries. To sort out this issue, *Lince* takes hyphens to be word boundaries and converts only the elements of the compound that change, leaving the hyphenation unaffected until later stages in the conversion. For example, a hypothetical word *anti-activo* would first be converted to *anti-ativo*, given that *activo* is in the lexical conversion table, and later to *antiativo*, given that *anti* is in the prefixes list, and is affected by a rule that drops the hyphen whenever a registered prefix ends in a vowel different from the one that starts the element after the hyphen.

The conversion algorithm is illustrated below in language independent pseudo-code:

```
if (is_in_form_conversion_table(word)) {
word = replace_with_form_conversion_table(word)
end
}
var hyphen_index = number_of_hyphens(word)
while (hyphen_index > 0) {

    #for a word 'auto-info-didacta' and hyphen_index=2 subword will be 'info-
didacta'
var subWord = prefix_and_rest_of_word(word, hyphen_index)
hyphen_index = hyphen_index - 1

    if (is_in_form_conversion_table(hyphen_suffix(subWord))) {

#replace subWord suffix
replace_with_form_conversion_table(hyphen_suffix(subWord))

} else if (is_in_hyphen_prefix_list(hyphen_prefix(subWord)) {

#apply hyphen related regular expressions to subWord
apply_regex_hyphen_related_match_and_replace_rules(subWord)
}

    }

end
```

3.3 Data Structure

One of the most challenging aspects of the conversion system was the structure of the lexical database, the dictionary that contains the explicitly stored lexical data. Input words change depending on their national variety of origin (e.g., due to different pronunciation, *dactilografia* changes in Portugal, becoming *datilografia*, but not in Brazil) and there are some rules that only affect certain national varieties (e.g., the umlaut was only dropped in Brazil). This required for the explicit storage not just of the old forms and corresponding new forms, but also of the national variety that undergoes change.

One of the spelling rules that was changed allows for some degree of variation within the same country for around two hundred lemmas, and the user should be given the chance to choose his preferred word form in such cases. Given the highly inflectional nature of Portuguese, though, those two hundred lemmas represent too high a number of word forms in the dictionary to present to the end user. One can assume that if a speaker doesn't pronounce a given character in a word, he doesn't do so either in the word forms of the same inflectional paradigm. Furthermore, it is not uncommon for a speaker to follow the same pronunciation option for every word in the same word family, in the sense of [14], so there should be a way to define a preference for an entire word family (e.g., if the speaker's pronunciation corresponds to 'acupuntura' and not 'acupunctura', the most likely is that 'acupunturar' will also be the choice).

To cater for this, we associated word form dictionary entries to a lemma index, which constitutes a different dictionary relationally associated with the main lexical base through a lemma id that is common to the entries in both tables. This way, the user only defines his options for each paradigm, and the option is assumed by the tool for all the related word forms. In the preference list, the words are grouped by collapsible word families, allowing for the setting up of a preference for a group of related entries - and not overburdening the user with having to set his preference for several tens of cases, although he still can if needed be. This proved to be highly useful not just for structuring the data in the application, but also for its syncing with the main, central databases [15] which feed all the spelling reform resources, as it mimics the structure of those databases. As a result, whenever an entry is updated in the central database and ported to the tool's database, every word form associated with it is updated as well, using a purposefully built module in the lexical resources administration system [16].

The stand-alone character of the application presents some challenges concerning updating the data. We opted to deal with data and application updates separately. When started, the tool reads the data version it contains and, whenever an Internet connection is available, automatically checks the server for the data version currently made available there, downloading it automatically in a compressed format in case there is a data update. Software component updates are not installed automatically; the user is instead prompted with a message informing of the existence of a new version and of how to install it.

3.4 Interface, Customization Options and Documentation

The interface was designed to be as simple and intuitive as possible while still allowing for the needed configuration options and providing enough information about the conversion process. When being ran for the first time, a conversion can be done after just four clicks in three simple steps. After a quick splash with the credits, aimed at filling the loading time, a simple presentation screen outlining the tool and its objectives is shown (along with a checkbox to not show it anymore on future executions). As a second step, the user picks the files to convert. A screen prompting for some more general options ensues, after which the process is complete. Among these options is the desired destination folder for the output, the possibility to keep track of the changes made by the tool (via text revision comments in the converted document itself) and to automatically add an end note stating the document is written according to the new spelling rules.

The progress of the conversion is shown in a final screen from which the conversion files can be directly opened in the associated editor. *Lince* never replaces the file it creates, instead writing the converted text to a new file with a suffix added. All the original formatting is kept.

To be able to stick to these simple steps, a good deal of the options is defined by default, but can all be changed by the user manually. Once installed, the application looks for the locale of the system and tries to guess the most likely input variety, defining which words in the dictionary will be converted, although the user can manually select the input text variety. The preferred variants in cases of variation are also pre-defined (from studies based on user preference, frequency in corpora and reference sources that were conducted for the reform implementation process), but can easily be changed through a selection screen mentioned in the previous section.

Other customization options are aimed more at professional users. Lince includes a list which works as an exceptions dictionary, for cases such as anthroponyms and toponyms that contain words that stop being written with an initial capital letter (e.g. *janeiro*, 'January', is now written with an initial lower case letter, but the spelling of *Rio de Janeiro* is obviously left unchanged, so that particular sequence features in the exceptions dictionary). Expressions can be managed, added and removed by the end user to this exclusion list through a purposefully built customization screen included in the options section. Some cases are very hard to account for, given their coexistence as both proper nouns, which shouldn't be converted, and homonym common nouns, which should. The character of a word can at times not be guessed from context (e.g. *Baptista*, a common anthroponym, shouldn't be converted when it is a person's name, but very often it occurs in texts as an adjective), and capitalization alone can't be relied upon either, given words' usage in titles or contexts where a common noun is part of a multiword expression that acts as a name (e.g. names of publications or brands, which don't change unless the registry owners decide so).

It is possible for the user to define personalized markers to exclude entire sections from the conversion process, for cases such as literal quotes from old writings of which the spelling should be kept or text segments in other languages. For example, in some countries, *actual*, 'current', is changed to *atual*, so by converting a text containing sections in English one risks incorrectly converting occurrences of *actual* in that language as well, unless it is marked for exclusion. After selecting the exclusion markers, the user has to actually (manually) tag the text with them so that the tool doesn't convert the text in sequences that are to be excluded.

All the preferences are exportable to a configuration file that can be reimported back at a later time. This is particularly useful for options that are shared through an organization for homogeneity purposes. The tool includes a small help section, along with the text of the spelling reform and a simplified explanation of what is being changed by the reform.

4 Conclusion and Future Work

We have developed a tool for the mass conversion of text to a new spelling reform being implemented for Portuguese. Biggest achievements included having a multiplatform lightweight application that was still effective enough for demanding industrial mass conversion tasks, and catering to both more advanced users' need for customization and general users' need for simplicity and easiness of usage. The major hurdles we had to tackle pertain to the need to use a mixed data based and rule based system, both to, on the one hand, keep the dictionary size small enough for lightweight deployment, short load time in out of date systems and quick server-based updating, and on the other hand, to account for word forms that are not registered in dictionaries, namely with very productive prefixes for which hyphenation rules were changed.

We achieved a highly customizable tool that has been in successful usage both by a very large number of individual end-users (over 300,000) from several countries and for industrial mass conversion of large-scale databases, websites and other systems by enterprises and State agencies and companies, such as the national broadcast company (RTP), the statistics institute (INE), or the post office company (CTT), among others. The application data is regularly updated and can account for several countries' varieties, making it possible to adapt it and successfully deploy it in countries where the spelling reform application criteria are still being defined.

Future work includes a more detailed evaluation of the tool's performance, namely in comparison with other existing similar-purpose tools, some of them not freely available, a study that should be conducted independently, under common criteria and with the involvement of the teams responsible for the development of those tools.

References

1. Jacobs, D.: Alliance and Betrayal in the Dutch Orthography Debate. Language Problems & Language Planning 21(2), 103–118(16) (1997)
2. Johnson, S.A.: Spelling trouble? Language, ideology and the reform of German orthography. Multilingual Matters, LTD., Clevedon (2005)
3. Ferreira, J.P., Correia, M.: Lince – Conversor para a nova ortografia (2010), `http://www.portaldalinguaportuguesa.org/lince.php`
4. Resolution 26/91of the Portuguese Assembley of the Republic, of 23 of August, Annex II – Nota explicativa do Acordo Ortográfico da Língua Portuguesa (1991)
5. Resolution 26/91of the Portuguese Assembley of the Republic, of 23 of August, Annex I: Acordo Ortográfico da Língua Portuguesa (1991)
6. Ferreira, J.P., Correia, M.: O Vocabulário Ortográfico Português: critérios, ferramentas e resultados. In: Boletim da Academia Galega de Língua Portuguesa, 5 (TBP)
7. Almeida, J.J., Santos, A., Simões, A.: Bigorna – A Toolkit for Orthography Migration Challenges. In: Calzolari, N., et al. (eds.) Proceedings of the Seventh International Conference on Language Resources and Evaluation (LREC 2010), pp. 227–232 (2010)
8. Portal da Língua Portuguesa, `http://www.portaldalinguaportuguesa.org`
9. Conversor Priberam para o Acordo Ortográfico, `http://www.flip.pt/FLiP-Online/Conversor-para-o-Acordo-Ortografico.aspx`
10. FliP, `http://www.flip.pt/`

11. Porto Editora - Conversor Ortográfico de Texto, http://www.portoeditora.pt/acordoortografico/conversor-texto/
12. Porto Editora – Conversor Ortográfico de Ficheiros, http://www.portoeditora.pt/acordoortografico/conversor-ficheiros/
13. Direcção-Geral da Tradução da Comissão Europeia: Conversores ortográficos e vocabulário das memórias de tradução, A Folha 36, pp. 21–27 (2011)
14. Bauer, L., Nation, P.: Word Families. International Journal of Lexicography 6(4), 253–279 (1993)
15. Janssen, M.: Open Source Lexical Information Network. In: Proceedings of the Third International Workshop on Generative Approaches to the Lexicon, Geneva, Switzerland (2005)
16. Ferreira, J.P., Barbosa, S., Janssen, M.: MorDebe Admin: a lexicon management system. In: Proceedings of the XIII EURALEX International Congress (2008)

Searching a Mixed Corpus in the Light of the New Portuguese Orthographic Norm

Gracinda Carvalho[1,2,3], Isabel Falé[1], David Martins de Matos[2,4], and Vitor Rocio[1,3]

[1] Universidade Aberta, Rua da Escola Politécnica, 147 1269-001 Lisboa, Portugal
{gracindac,imsfale,vjr}@uab.pt
[2] L2F/INESC-ID Lisboa, Rua Alves Redol 9 1000-029 Lisboa, Portugal
david.matos@inesc-id.pt
[3] CITI - FCT/UNL
[4] Instituto Superior Técnico/UTL

Abstract. A mixed corpus of Portuguese is one in which texts of different origins produce different spelling variants for the same word. A new norm, which will bring together the written texts produced both in Portugal and Brazil, giving then a more uniform orthography, has been effective since 2009, but what happens in the perspective of search, to corpora created before the norm came into practice, or within the transition period? Is the information they contain outdated and worthless? Do they need to be converted to the new norm? In the present work we analyse these questions.

1 Introduction

Searching a corpus based on a user's information request requires the words used on the query to be matched against those of the corpus. In the present work we analyse search effectiveness in the case of a corpus containing different variants of Portuguese, which we call a mixed corpus. The two variants correspond to the variant used in Portugal[1] (pt_PT) and the variant used in Brazil (pt_BR). If a query is made in one variant, all answers occurring in other variants are not retrieved.

For a number of decades, countries with Portuguese as official language, have been involved in negotiations with the purpose of uniformising the orthography of both variants. An agreement was reached in 1990, which became known as "Acordo Ortográfico da Língua Portuguesa 1990", or Portuguese Language Orthographic Agreement 1990 (PLOA). The former orthographic norm was in use until 2009, when the adoption of the new norm was effective in countries covering the vast majority of the Portuguese native speakers. The new norm will bring together the written texts produced both in Portugal and Brazil, giving then a more uniform orthography, but what happens to the corpus that were created before the norm came into practice, or within the transition period? Are they

[1] As well as African Portuguese speaking countries, Macau and East Timor.

H. Caseli et al. (Eds.): PROPOR 2012, LNAI 7243, pp. 56–62, 2012.

worthless? Do they need to be converted to the new norm? In the present work we analyse these questions.

In the present text, we describe our motivation and existing related work in Sect. 2. Sect. 3 is dedicated to the orthography describing the PLOA 1990 and its effects. In Sect. 4 we describe our solutions for the treatment of multiple spellings in QA@CLEF corpus, and in Sect. 5 we analyse the impact of the PLOA and our solutions in this corpus. To finalise, in Sect. 6 we present our conclusions and future work.

2 Motivation and Related Work

For an application that has the main goal of facilitating the access to information, it is important to consider as many sources of information as possible. Therefore, for a given language, a restriction to a specific variant would represent that a lot of useful information would be left out. That was the idea behind the creation of the corpus for Portuguese of the Cross Language Evaluation Forum (CLEF), an international campaign to evaluate several information retrieval related tasks. This corpus was created in 2004, consisting of the newspaper articles from Portuguese newspaper "Público" and the Brazilian "Folha de São Paulo". This collection became known as "CHAVE" and was organized by Linguateca[2]. The edition of the Portuguese Wikipedia was added to the collection in the 2007 edition, with a frozen version of the HTML edition of the Portuguese Wikipedia from November 2006. The usefulness of the mixed corpus is referred in [1], that attributes the improvement of the performance by the addition of the pt_BR variant corpus, while in [2] the good performance of the lexical tools designed for the pt_PT variant in the scope of the mixed corpus is emphasised.

Studies have been made on mixed corpora of written Portuguese, particularly to determine the degree of convergence between them. That is for instance the idea behind the creation of the CONDIV (CONVergence DIVergence) corpus, though this work focused on semantic aspects[3].

The current orthographic norm has also been subject of a study in which the freely available lexical resources were combined and used for variant detection and conversion[4].

However we have no knowledge of a prior study dedicated to determining the best conditions in which to make documents belonging to both variants, before and after the 1990 norm, relevant and searchable. That is the purpose of the current work, together with the description of the particular techniques we find most appropriate to each particular case.

3 Portuguese Orthography: PLOA 1990 Rules

Officially sanctioned negotiations have taken place with the objective of creating a uniform intercontinental spelling of the Portuguese language, which culminated

[2] `http://www.linguateca.pt/chave/`

in the current orthographic norm for the Portuguese language. The official text can be consulted in [5].

The PLOA 1990 is organised in a set of 21 bases for the orthography of Portuguese, accompanied by an appendix with a set of explanatory notes. The first 3 bases define general aspects of the Portuguese language as its alphabet (Base I), the rules for the use of "h" in the beginning and end of words (Base II) and the homophonic graphemes (Base III). These bases are uniform across the two variants of Portuguese, contrary to Base IV, that is dedicated to consonant sequences. This base is divided in two groups of letters' pairs the first being more representative, and this group the first letter in the pair is kept when it is pronounced, eliminated otherwise, or a double spelling is allowed for words that maybe pronounced both ways. The second group defines groups of pairs in which the first consonant maybe dropped according to if it is pronounced or not. These rules take into consideration the pronunciation of a speaker with a standard cultural level, without getting into further details. Bases V to VII are again generic rules on the tonicity of vowels and diphthongs, and bases VIII to XIII establish accentuation rules. Base XIV treats the special case of diaeresis, that is to be used exclusively for words of foreign origin. Bases XV to XVII deal with the case of the hyphen, and the following bases are Base XVIII for the apostrophe, Base XIX for the use of capital letters, Base XX for syllabic division and Base XXI for signatures and commercial names.

3.1 Summary of Differences between the Two Variants of Portuguese

The differences of the pt_PT variant and the pt_BR variant are summarized in Table 1. This table identifies three main groups of differences in the spelling of the two variants: 1. Pairs of consonants that are simplified in the Brazilian variant to only the second consonant, because the first one is mute; Portugal should adapt its spelling to eliminate this mute consonants; Differences will remain because double spelling is allowed for words that are liable to be pronounced both ways. Examples can be opção [option] and aptidão [aptitude]; 2. Double pronunciation of "e" and "o" as tonic vowels at the end of a syllable, followed by a nasal consonant "m" or "n". In this case differences in timbre make the pt_PT spelling to use an acute accent, while the pt_BR spelling uses a circumflex accent. This situation will be kept. Examples are económica/econômica [economical] and ténis/tênis [tenis]; 3. Suppression of graphic accents, in cases of "éi", "ói", "êe", "ôo" and "ü" that are in use in Brazil, which has to adapt its spelling to remove them; this will no longer be a difference after the PLOA. Examples of such cases are Europeia/Européia [European], estreia/estréia [première], comboio/combóio [train],leem/leêm [(they) read], voo/vôo [flight], and consequência/conseqüência [consequence] and frequentador/freqüentador [regular (visitor)].

The number of words that will still present differences after the implementation of the PLOA is taken from the explanatory notes, and is based on the corpus of 100 000 words mentioned in the previous subsection.

Table 1. Differences between pt_PT and pt_BR variants of Portuguese

Rule	PLOA Base	Variant changes	Double spelling	Number of words (%)
Group of consonant pairs				
cc → c; cç → ç; ct → t; pc → c; pç → ç; pt → t				
bd → d; bt → t; gd → d; mn → n; tm → m				
2nd letter kept	IV,1.º,a	-	No	-
2nd letter dropped	IV,1.º,b	pt_PT	No	600(0.54)
Double spelling	IV,1.º,c and 2.º	-	Yes	575(0.5)
Double Accentuation				
pt_PT é, pt_BR ê; pt_PT ó, pt_BR ô; pt_PT o, pt_BR ô				
oxytone	VIII,1.º,a	-	Yes	little over 20
paroxytone	IX,2.º,a and b	-	Yes	little over 12
proparoxytone	IX,3.º	-	Yes	1400(1.27)
Accentuation Suppressed				
Acute accent suppressed from éi and ói				
éi and ói	IX,3.º	pt_BR	No	-
Circumflex accent suppressed from êe and ôo				
êe	IX,7.º	pt_BR	No	-
ôo	IX,8.º	pt_BR	No	-
Diaeresis accent suppressed from ü				
ü	XIV	pt_BR	No	-

4 Our Approach in the Treatment of a Mixed Corpus

The treatment of the corpus to be searched is done following a pre-processing stage according to the work of [6].

Our approach for searching a mixed corpus is done through the use of a root form for a word and use this root in the search. For words that are different in the two variants of Portuguese we use a rule based replacement of the root corresponding to the simplest form of both variants, for instance a group of two consonants is replaced by a single consonant.

We determine the root of each word via lemmatization, using a lexicon for Portuguese, pt_PT variant, with lemma information. This lexical knowledge base is named POLLUX (POrtuguese Lexical Largely Usable and eXtensible) [7].

Another technique is to use a synonym between equivalent words with the same spelling, and expand at query time to find both forms of the word. We rely upon a resource built upon the Wikipedia to obtain this type of information. However for alternative spelling variants of Portuguese words the information from Wikipedia is quite limited, so we will need to add information from other sources, for instance the official vocabulary of the PLOA[3], so that the information becomes more complete.

To summarise, the set of possible techniques is presented, with an analysis of advantages and disadvantages for each, with the option we find most appropriate: **Normalization** - Building equivalence classes choosing a representative to

[3] `http://www.portaldalinguaportuguesa.org/index.php?action=vop&page=info`

replace the occurrence of all elements in the class. The representative may not be a correctly spelt word, however it is used only for internal representation and the original text is not affected. It may have the disadvantage of join two unrelated classes, for instance in the case of pacto [pact] and *pato* [duck] that both have *pato* [duck] as root. The case of the representative of a class being another meaningful word is very small. This technique is adequate for replacing consonant pairs and normalize the suppressed accentuation in Brazil; **Equivalences** - This technique consists of explicitly relating the two version of the word, as the pt_BR variant sinônimo and pt_PT variant sinónimo of word synonym. This technique has the advantage of also preserving the original text, but equivalences must be expressed for all word pairs, and performance is slightly worse. Some equivalences already exist in the Wikipedia, and we automatically take them into consideration; **Conversion** - This technique corresponds to a selective process based on a given vocabulary and updates the orthography to the current norm, replacing the original version. As disadvantages of this method is the fact that some divergences persist, which is addressed by previous options, and not all converters produce exactly the same results as shown in[4]. The original version of the text must be known in advance, or wrong results may happen as words that should be converted being kept; **Do nothing** - This technique must be considered if the cases are very rare or the impact will be felt mainly on words of foreign origin. We choose to treat the "ü" this way.

5 Analysis of the Impact of the New Orthographic Rules in the Corpus

To assess the impact of the rules used, we computed, for each of the three sub-corpora, the number of distinct words where each rule was applicable, as well as the total number of occurrences of these words (tokens). Then we converted the corpus to the new orthography by using the freely available Lince converter[8], and analysed the changes that were made to the original text. Finally, we compared the result of the conversion with the set of words to which the rules apply.

This information was computed separately from the pt_PT variant of the corpus (news articles from "Público"), from the pt_BR variant of the corpus (news articles from "Folha de São Paulo") and for the Wikipedia, which is a mixed corpus. In the case of Wikipedia, we did two conversions with Lince, one according to the pt_PT variant and another according to pt_BR variant. Since there is no overlap on the set of converted words, we combined the results of both conversions for the Wikipedia sub-corpus. The results obtained are shown in Table 2.

We observed that there is a high number of words converted that fall into the category of other bases, concerning mostly with capitalization and hyphenation ($\approx$ 200 000 occurrences in the whole corpus). Such changes do not have implication in our system since, in order to perform a search, the text is all converted to lower case and the hyphen is removed, with the constituent words of a hyphenated compound kept as separate words.

Table 2. Impact of the Rules in CLEF Corpus

	Conversion						% total occurences converted		
	Distinct words			Total occurrences					
Rule	Público / Folha		Wikipedia	Público / Folha		Wikipedia	Público	Folha	Wikipedia
(1) cc → c	P	355	212	P	13 250	1 649	66.1	-	8.2
(2) cç → ç	P	287	192	P	80 416	14 149	87.6	-	72.5
(3) ct → t	P	2 656	1 592	P	332 875	67 051	77.0	-	42.7
(4) pc → c	P	60	63	P	2 302	958	10.0	-	18.7
(5) pç → ç	P	41	40	P	11 022	4 205	48.0	-	47.5
(6) pt → t	P	418	351	P	21 932	9 933	43.2	-	19.2
(7) ü → u	F	19	352	F	23	11 101	-	8.6	38.0
(8) éi → ei	F	162	272	F	26 498	17 892	-	75.7	67.3
(9) ói → oi	F	168	526	F	2 239	7 527	-	25.8	41.9
(10) ôo → oo	F	18	26	F	1 976	3 240	-	98.4	91.1
(11) êe → ee	F	27	25	F	1 387	828	-	99.3	95.5

For the "Público" (P) sub-corpus, most words in the scope of rules 1 to 3 are converted correctly, as we can see in the last column. There are of course exceptions, especially in rules 4 to 6. The rate of conversion for rule 4 is particularly low, so we didn't include it in the actual system. The application of the rules to those exceptions has a high probability of generating non-words, which doesn't affect the system. Troublesome cases occur only when the application of a rule leads to word sense ambiguity. For instance, by transforming "facto" [fact] into "fato" (fact in pt_BR, costume in pt_PT), we are enlarging the search to include meanings that were not originally intended.

As for results for the "Folha de São Paulo" (F), We can see that rules 7 and 9 have many exceptions. Almost all of rule 7 exceptions correspond to words of foreign origin mostly of German origin, as "Führer" or "Müller", however, for reasons we weren't able to ascertain, there are many word occurrences in the original corpus that don't follow the previous pt_BR orthography (e.g. freqüência [frequency] is written as frequência). This is in contrast with the Wikipedia sub-corpus, where many more words with "ü" exist.

Wikipedia, being a pt_PT and pt_BR mixed corpus, is converted to the current orthography using all 11 rules. For rules 2 to 6, results are mostly consistent with those obtained for the "Público" sub-corpus. Unexpectedly, rule 1 presents a much lower conversion rate, which can be explained by the higher occurrence in Wikipedia of proper names and foreign words, especially of Italian origin as "Puccini", "Rocco", etc. Similarly, results for rules 8 to 11 are consistent with those obtained for the "Folha de São Paulo" sub-corpus, the exception being rule 7, which we already addressed.

6 Conclusions and Future Work

We have analysed the convergence and divergence of the spelling of pt_PT and pt_BR variants of Portuguese before and after the application of the current norm. Although the divergence is not an extensive one in terms of vocabulary, some of the different words occur very frequently. The application of the current norm minimizes the divergences, but does not eliminate them totally, making it worthwhile to create ways to address this divergence.

Based on the analysis reported in this paper, we found that the application of the proposed rules and techniques allows a uniform search on a mixed corpus, where the pt_PT variant coexists with the pt_BR variant.

Since our work contributes to aggregate related words that would otherwise be considered separately, it is also applicable to other frequency based applications. We intend to apply it in the near future to the area of topic detection and clustering.

Acknowledgements. The present work was partly supported by FCT project CMU-PT/005/2007.

References

1. Costa, L.: 20th Century Esfinge (Sphinx) Solving the Riddles at CLEF 2005. In: Peters, C., Gey, F.C., Gonzalo, J., Müller, H., Jones, G.J.F., Kluck, M., Magnini, B., de Rijke, M., Giampiccolo, D. (eds.) CLEF 2005. LNCS, vol. 4022, pp. 467–476. Springer, Heidelberg (2006), DOI: `http://dx.doi.org/10.1007/11878773_52`
2. Amaral, C., Figueira, H., Martins, A., Mendes, A., Mendes, P., Pinto, C.: Priberam's Question Answering System for Portuguese. In: Peters, C., Gey, F.C., Gonzalo, J., Müller, H., Jones, G.J.F., Kluck, M., Magnini, B., de Rijke, M., Giampiccolo, D. (eds.) CLEF 2005. LNCS, vol. 4022, pp. 410–419. Springer, Heidelberg (2006), DOI: `http://dx.doi.org/10.1007/11878773_46`
3. Soares da Silva, A.: Integrando a variação social e métodos quantitativos na investigação sobre linguagem e cognição: para uma sociolinguística cognitiva do português europeu e brasileiro. Revista de Estudos da Linguagem 16, 49–81 (2008), `http://relin.letras.ufmg.br/revista/upload/02-Augusto_Soares.pdf`
4. João Almeida, J., Santos, A., Simões, A.: Bigorna – a toolkit for orthography migration challenges. In: Proceedings of the Seventh International Conference on LREC 2010, Valletta, Malta, ELRA, pp. 227–232 (May 2010), `http://www.lrec-conf.org/proceedings/lrec2010/pdf/898_Paper.pdf`
5. Diário da República - 1 Série-A: Decreto da Presidência da República 43/91 de 23 de Agosto de 1991 - Ratifica o Acordo Ortográfico da Língua Portuguesa de 1990. Imprensa Nacional, Lisboa (1991), `http://dre.pt/pdf1sdip/1991/08/193a00/43704388.PDF`
6. Carvalho, G., de Matos, D.M., Rocio, V.: Document Retrieval for Question Answering: A Quantitative Evaluation of Text Preprocessing. In: Proceedings of PIKM 2007, Lisboa, Portugal, November 5-10, pp. 125–130. ACM (2007) ISBN: 978-1-59593-832-9, DOI: `http://dx.doi.org/10.1145/1316874.1316894`
7. Alves, M.A.: Engenharia do Léxico Computacional: princípios, tecnologia e o caso das palavras compostas. Master's thesis, Faculdade de Ciências e Tecnologia da Universidade Nova de Lisboa, Lisboa, Portugal (2002)
8. Lince - Conversor para a nova ortografia: (ILTEC - Instituto de linguística teórica e computacional) (October 20, 2011), `http://www.portaldalinguaportuguesa.org/?action=lince&page=main`

Extraction of Bilingual Cognates from Wikipedia*

Pablo Gamallo and Marcos Garcia

Centro de Investigação em Tecnologias da Informação (CITIUS)
Universidade de Santiago de Compostela, Galiza, Spain
{pablo.gamallo,marcos.garcia.gonzalez}@usc.es

Abstract. In this article, we propose a method to extract translation equivalents with similar spelling from comparable corpora. The method was applied on Wikipedia to extract a large amount of Portuguese-Spanish bilingual terminological pairs that were not found in existing dictionaries. The resulting bilingual lexicons consists of more than 27,000 new pairs of lemmas and multiwords, with about 92% accuracy.

1 Introduction

A comparable corpus consists of documents in two or more languages, which are not translation of each other and deal with similar topics. The use of comparable corpora to acquire bilingual lexicons has been growing in the last years [4,5,14,3,17,16,9,20,19,15]. However, the number of these studies is not so large in comparison to those using a strategy based on aligned, parallel texts. A small but representative sample of extraction methods based on parallel texts is the following: [6,12,1,18,11].

The main advantage of comparable corpora is that they are easily available using the Web as a huge resource of multilingual texts. By contrast their main drawback is the low performance of the extraction systems based on them. According to [13], bilingual lexicon extraction from comparable corpora is a too difficult and ambitious objective, and much more complex than extraction from parallel, and aligned corpora. In fact, we can conceive a continuum of comparability between two poles: completely unrelated corpora (non comparable) and fully related parallel texts. The degree of comparability is directly related to the quality of the extracted lexicons. The more comparable is the corpus, the more precise the extracted lexicons are.

In this paper, we propose a method to learn bilingual lexicons from comparable corpora that try to overcome the low precision reached by most of the methods relying on comparable corpora. Two aspects will be taken into account to improve precision:

The Degree of Comparability of the Corpus: We will discover articles in the Wikipedia with very high degree of comparability (pseudo-parallel texts).

* This work has been supported by Ministerio de Ciencia e Innovación, within the project OntoPedia, ref: FFI2010-14986.

H. Caseli et al. (Eds.): PROPOR 2012, LNAI 7243, pp. 63–72, 2012.

The Extraction of Bilingual Cognates: The extraction will be focused on bilingual pairs of words with similar spelling (*cognates*), which are in fact true translation equivalents and not false friends nor false cognates. *Bilingual cognates* are considered here as those words in two languages with similar spelling and similar meaning.

We assume that it is possible to build high quality bilingual lexicons if the extraction is performed on very comparable corpora considering only bilingual cognates. To minimize the low coverage of the lexicons acquired by this method, it is convenient to use it on family related languages sharing many cognates. So, our experiments were performed on Portuguese and Spanish, two very close Latin languages.

Moreover, among the different web sources of comparable corpora, Wikipedia is likely the largest repository of similar texts in many languages. We only require the appropriate computational tools to make them comparable. The proposed method is based on the Wikipedia structure, even if it can be easily generalized to be adapted to other sources of comparable corpora. In this paper, we will use the method to enlarge an existing Portuguese-Spanish bilingual dictionary with new bilingual cognates, most of them representing domain-specific terminology found in Wikipedia.

This article is organized as follows. In the next two sections, 2 and 3, we describe the extraction method. Then, in section 4 some experiments are performed in order to learn a large Portuguese-Spanish bilingual lexicon. Finally, some conclusions are presented in 5.

2 Method Overview

Our extraction method relies on the multilingual structure of Wikipedia. It consists of the following four steps:

Corpus Alignment: First, we identify the Wikipedia articles in two languages whose titles are translations of each other.

Degree of Comparability: Then, to calculate a degree of comparability between two aligned articles, we apply a similarity measure and select the most comparable pairs of bilingual articles.

Candidates for Translation Equivalents: From each very comparable pair of articles, we calculate the Dice similarity between lemmas and select the most similar ones, which are considered as being candidates for translation equivalents.

Selecting Cognates: Finally, using the Edit distance, we check whether the candidates are *cognates* and select the most similar ones as true translation equivalents.

This whole method runs in time linear in the size of its input, and thus scales readily as the corpus grows. In the following section, we will describe in detail the four steps of our method.

3 Method Description

The method is based on the following assumption:

> The use of distributional similarity to extract bilingual cognates from very comparable corpora should generate high quality bilingual correlations.

Following this assumption, we develop a strategy adapted to the Wikipedia structure. The output is a bilingual dictionary containing many bilingual pairs of domain-specific terms.

3.1 Alignment of Bilingual Wikipedia Articles

The input of our strategy is CorpusPedia[1], a friendly and easy-to-use XML structure, generated from Wikipedia dump files. In CorpusPedia, all the internal links found in the text are put together in a vocabulary list identified by the tag *links*. In addition, the tag *translations* codifies a list of interlanguage links (i.e., links to the same articles in other languages) found in each article. Both internal and interlanguage links are very useful features to build comparable corpora.

For this purpose, the first task is to extract all pairs of bilingual articles related by interlanguage links. For instance, given the Portuguese article entitled *"Arqueologia"*, we search for its Spanish counterpart within the list of Spanish *translations* associated to the Portuguese article. If the Spanish translation, *"Arqueología"*, is in the list, we select the article entitled *"Arqueología"* within the Spanish Wikipedia, and build a small comparable corpus with the pair of articles. This algorithm is applied article by article and results in a large set of small, comparable, and aligned pairs of bilingual texts.

3.2 Wikipedia-Based Comparability Measure

The next step is to measure the degree of comparability for each pair of bilingual texts, such as it was described in [8]. The measure of comparability between two Wikipedia articles is defined as follows.

For a comparable corpus $\mathcal{C}$ built from the Wikipedia, and constituted for instance by a Portuguese article $\mathcal{C}_p$ and a Spanish article $\mathcal{C}_s$, a comparability coefficient can be defined on the basis of finding, for each Portuguese term t_p in the vocabulary $\mathcal{C}_p^v$ of $\mathcal{C}_p$, its interlanguage link (i.e., translation) in the vocabulary $\mathcal{C}_s^v$ of $\mathcal{C}_s$. The vocabulary of a Wikipedia article is the set of "internal links" found in that article. Those internal links are the key words and terms representing the content of the article. So, the two articles, $\mathcal{C}_p$ and $\mathcal{C}_s$, tend to have a high degree of comparability if we find many internal links in $\mathcal{C}_p^v$ that can be translated (by means of interlanguage links) into many internal links in $\mathcal{C}_s^v$. Let $Trans_{bin}(t_p, \mathcal{C}_s^v)$

[1] The software to build CorpusPedia, as well as CorpusPedia files for English, French, Spanish, Portuguese, and Galician, are freely available at `http://gramatica.usc.es/pln/`

be a binary function which returns 1 if the translation of the Portuguese term t_p is found in the Spanish vocabulary $\mathcal{C}_s^v$. The binary Dice coefficient, $Dice_{comp}$, between the two articles $\mathcal{C}$ is then defined as:

$$Dice_{comp}(\mathcal{C}_p, \mathcal{C}_s) = \frac{2\sum_{t_p \in \mathcal{C}_p^v} Trans_{bin}(t_p, \mathcal{C}_s^v)}{|\mathcal{C}_p^v| + |\mathcal{C}_s^v|} \quad (1)$$

It follows that two texts in two languages have a high degree of comparability if the main terms found in one text are translations of the main terms found in the other text. For instance, the pair of articles *"Arqueologia/Arqueología"* has a low comparability degree because the two texts only share two bilingual pairs of terms (e.g., *"sociedade/sociedad"*, *"antropologia/antropología"*) out of 95 (i.e., $Dice_{comp} = 0.04$). By contrast, the degree of comparability of the Portuguese and Spanish entries for the scientist *Brian Goodwin* reaches a $Dice_{comp}$ coefficient of 0.77, since the two articles share 13 out of 43 terms. Given that some authors of Wikipedia articles translate (part of) their contributions from the English version, the two articles on Brian Goodwin are likely to be translations from the English entry.

This comparability measure is applied to each pair of bilingual articles, and those pairs whose degree of comparability is higher than a specific threshold (empirically set to >= 0.3) are finally selected. It results in a set of very comparable pairs of bilingual texts. The whole set of very similar bilingual texts can be perceived as a pseudo-parallel corpus, since we observed that many pairs are constituted by paragraphs that are either (pseudo) translations of each other or translations sharing a common source.

3.3 Identifying Translation Equivalent Candidates

In this step, we identify all pairs of translation candidates within each pair of bilingual texts selected in the previous step. For this purpose, we apply a distributional-based strategy defined in [9]. The starting point of this strategy is as follows: lemma l_1 is a candidate translation of l_2 if the lemmas with which l_1 co-occurs are translations of the lemmas with which l_2 co-occurs. This strategy relies on a bilingual list of lemmas (called *seed words*) provided by an external bilingual dictionary. So, l_1 is a candidate translation of l_2 if they tend to co-occur with the same seed words. In our strategy, co-occurrences are defined by means of syntactic contexts, which are identified with a robust dependency parser, DepPattern [7]. Only three PoS categories of lemmas were taken into account: common nouns, adjectives, and verbs. In this experiment, proper names were not considered. They will be processed using a simpler strategy we will describe later.

Similarity between lemmas l_1 and l_2 is computed using the following version of the *Dice* coefficient:

$$Dice_{distrib}(l_1, l_2) = \frac{2\sum_i min(f(l_1, s_i), f(l_2, s_i))}{f(l_1) + f(l_2)} \tag{2}$$

where $f(l_1, s_i)$ represents the number of times the lemma l_1 co-occurs with seed s_i, and $f(l_1)$ the total frequency of l_1 in the corpus. As a result, each lemma of the source language is assigned a list of candidate translations. Potential candidates are restricted to be of the same category: nouns are compared with nouns, verbs with verbs, and adjectives with adjectives.

It is worth mentioning that this strategy takes into account multiwords. Before computing similarity, the most representative multiword terms are identified with GaleXtra[2], a multilingual term extractor that uses both patterns of PoS tags and association measures to select term candidates. So, Dice similarity is computed on pairs of bilingual lemmas containing single lemmas and/or multiword terms.

Since our objective is to enlarge the existing bilingual dictionary, only bilingual pairs of lemmas that are not in that source dictionary are considered as candidate translation equivalents.

3.4 Identifying Bilingual Cognates

The final step is to identify *cognates* out of the set of translation equivalent candidates. For this purpose, we use a spelling similarity measure based on Edit distance. The spelling based similarity, noted $Dice_{eds}$, between two lemmas, l_1 and l_2, is defined by means of the following equation:

$$Dice_{eds}(l_1, l_2) = 1 - \frac{2\ eds(l_1, l_2)}{length(l_1) + length(l_2)} \tag{3}$$

where $eds(l_1, l_2)$ is the Edit distance of lemmas l_1 and l_2, and $length(l_i)$ represents the number of characters of l_i. It means that the $Dice_{eds}$ similarity between the spelling of two lemmas is a function of their Edit distance and their string lengths. This similarity measure allows us to select those pairs of translation candidates that share similar spelling, i.e., that can be perceived as being bilingual cognates. We assume that the final selected bilingual cognates are very probably correct translations.

4 Experiments

We performed an experiment aimed at learning a large set of new bilingual correlations from the Portuguese and Spanish versions of Wikipedia.

[2] `htpp://http://gramatica.usc.es/~gamallo/gale-extra/index2.1.htm`

4.1 Existing Dictionaries

Our method requires a list of seed words taken from existing bilingual resources. We used two different dictionaries:

OpenTrad. The general purpose bilingual dictionary Portuguese-Spanish integrated in the open source machine translation system, OpenTrad-Apertium [2]. The dictionary is freely available[3]. For our experiment, we selected all bilingual pairs containing nouns, verbs, and adjectives.

Wikipedia. We created a new Portuguese-Spanish dictionary using the inter-language links of Wikipedia. Since Wikipedia is an encyclopedia dealing with named entities and terms, this new dictionary only contains proper names and domain-specific terminology.

Table 1 shows the size of the two existing dictionaries and the total union of them. We count the different number of bilingual correspondences (not the number of entries). A bilingual correspondence is, for instance, the pair *"sociedade/sociedad"*. The total size obtained by the union of both resources is 263, 362 different bilingual correspondences.

Table 1. Existing lexical resources

	nouns	**adject.**	**verbs**	**total**
OpenTrad	4, 210	1, 428	4, 226	9, 854
Wiki	253, 367	-	-	253, 367
Union	257, 708	1, 428	4, 226	**263,362**

Note that the two dictionaries are complementary: we only found 131 entries in common.

4.2 Extraction

After applying our method on the whole Wikipedia in Portuguese and Spanish, we extracted 27, 843 new bilingual correspondences. None of them were in the two input dictionaries. Preliminary tests led us to set the thresholds of $Dice_{comp}$,

Table 2. Results of the extraction

	nouns	**adject.**	**verbs**	**total**
single lemmas	9, 374	5, 725	2, 215	17, 314
multiword terms	9, 585	-	944	10, 529
all lemmas	18, 959	5, 725	3, 159	**27,843**

[3] `http://sourceforge.net/projects/apertium/files/`

$Dice_{distrib}$, and $Dice_{eds}$ to 0.3, 0.6, and 0.6, respectively. Table 2 depicts the final results. In the first row, we show the extractions of single lemmas, while the second row is just focused on multiwords. The total extractions considering both multiword terms and single lemmas are shown in the third row.

Notice that the size of the new bilingual dictionary, 27843 lemmas, is much larger than that of the general purpose dictionary of OpenTrad, which only contain 9,854 bilingual correspondences.

4.3 Evaluation

To evaluate the quality of the extracted dictionary, a test set of 450 bilingual pairs were randomly selected. The test set was created with the aim of obtaining these three balanced subsets:

- 150 bilingual pairs of nouns
- 150 bilingual pairs of verbs
- 150 bilingual pairs of adjectives

Results are depicted in Table 3. Accuracy is the number of correct pairs divided by the number of evaluated pairs. The best performance was achieved by the extraction of adjectives, since it reaches 95% accuracy. By contrast, verb extraction only achieves 89%. The total accuracy is still high: about 92%. This performance is much better than state-of-the-art work on extraction from comparable corpora, whose best scores were about 70% accuracy [14]. The good quality of the generated translation equivalents allows us to reduce the time spent in manual correction. It follows that our method permits to minimize the effort to build a new bilingual dictionary of two related languages.

Table 3. Accuracy of the extracted bilingual pairs

	accuracy
nouns	91%
verbs	89%
adjectives	95%
total	**92%**

4.4 Error Analysis

We found 39 errors out of 450 evaluated extractions. Most of them (58%) were due to foreign words, namely English words appearing in the input text as part of titles or citations. For instance, the translation pair "*about/about*" was incorrectly learnt from two Portuguese and Spanish texts containing such a word within a non translated English expression. It would not be difficult to avoid this kind of problem if we use a language identifier to find parts of the input text written in other languages.

The second type of most common errors (8%) were caused by prefixes appearing in one of the two correlated words, for instance:

americanismo / **anti**-americanismo
anti-fascista / fascista
hispanorabe / **neo**-hispano-árabe

Note that it would be possible to filter out those cases by making use of a list of productive prefixes.

In Table 4, we show some types of errors found in the evaluation. As the two most common errors (foreign words and prefixes), which represent 66% of the total number of errors, can be easily filtered out, the total accuracy of our system could achieve 97%.

Table 4. Types of errors

	frequency
foreign words	58%
prefixes	8%
typos	8%
multiwords	5%
PoS-tagging	3%

4.5 Further Experiments: The Case of Proper Names

As proper names are less ambiguous than common nouns, verbs, and adjectives, we consider it is not necessary to grasp high quality bilingual correspondences by using distributional-contextual similarity. So, we defined a simpler strategy to extract translation equivalents of proper names:

Given two bilingual pairs of articles in Wikipedia, for instance the Portuguese and Spanish articles entitled *"Arqueologia/Arqueología"*, all proper names found in those articles are identified, then a list of bilingual pairs is created, and the $Dice_{eds}$ similarity is computed between them. If two pairs of bilingual proper names with similar spelling has a $Dice_{eds}$ similarity higher than 0.9, they are selected as being a translation equivalent candidate. Notice that we do not compute the degree of comparability between the two Wikipedia articles nor the distributional $Dice_{distrib}$ similarity between the two proper names. In this case, we assume that two proper names with very similar spelling in two languages, and appearing in two texts with similar or identical titles, are strong candidates to be true translation equivalents.

This strategy led us to extract 818, 797 new bilingual pairs of proper names that were not found in the two existing dictionaries. This kind of bilingual pairs can be integrated into rule-based machine translation systems, as OpenTrad. As this system is not provided with a Named Entity Recognition module, it proposes bilingual translations of proper names on the basis of large lists of bilingual correspondences of proper names.

5 Conclusions

We have proposed a method to extract new bilingual terminology from Wikipedia-based comparable corpora, achieving more than 90% accuracy. The proposed strategy was adapted to the internal structure of Wikipedia, but it could be applied to other types of comparable text corpora with minor changes.

One of the main drawbacks of our strategy is due to the inherent limitations of Edit distance to measure the spelling based similarity between lemmas that have undergone systematic changes in the past. For instance, action nouns in Portuguese may contain the suffix *-ção* while the equivalent suffix in Spanish is *-ción*. If we consider that these two suffix represent the same abstract concept, the distance between two words such as *organização* and *organización*, whose dissimilarity is just due to the suffix, must be close to 0, i.e., they should be taken as identical cognates. However, the Edit distance between these two words does not consider the strong relationship between the two suffixes. To address these cases, [10] proposed a new spelling similarity based on the generalization of substitution patterns. A different strategy, based on linguistic knowledge, could be the use of a list of equivalent bilingual pairs of prefixes and suffixes. In future work, we will make use of both strategies to enlarge the coverage of the extracted dictionaries.

References

1. Ahrenberg, L., Andersson, M., Merkel, M.: A Simple Hybrid Aligner for Generating Lexical Correspondences in Parallel Texts. In: 36th Annual Meeting of the Association for Computational Linguistics and 17th International Conference on Computational Linguistics, Montreal, pp. 29–35 (1998)
2. Armentano-Oller, C., Carrasco, R.C., Corbí-Bellot, A.M., Forcada, M.L., Ginestí-Rosell, M., Ortiz-Rojas, S., Pérez-Ortiz, J.A., Ramírez-Sánchez, G., Sánchez-Martínez, F., Scalco, M.A.: Open-Source Portuguese–Spanish Machine Translation. In: Vieira, R., Quaresma, P., Nunes, M.d.G.V., Mamede, N.J., Oliveira, C., Dias, M.C. (eds.) PROPOR 2006. LNCS (LNAI), vol. 3960, pp. 50–59. Springer, Heidelberg (2006)
3. Chiao, Y.-C., Zweigenbaum, P.: Looking for candidate translational equivalents in specialized, comparable corpora. In: 19th COLING (2002)
4. Fung, P., McKeown, K.: Finding terminology translation from non-parallel corpora. In: 5th Annual Workshop on Very Large Corpora, Hong Kong, pp. 192–202 (1997)
5. Fung, P., Yee, L.Y.: An IR Approach for Translating New Words from Nonparallel, Comparable Texts. In: Coling 1998, Montreal, Canada, pp. 414–420 (1998)
6. Gale, W., Church, K.: Identifying Word Correspondences in Parallel Texts. In: Workshop DARPA SNL (1991)
7. Gamallo, P., González, I.: A grammatical formalism based on patterns of part-of-speech tags. International Journal of Corpus Linguistics 16(1), 45–71 (2011)
8. Gamallo, P., González, I.: Measuring comparability of multilingual corpora extracted from wikipedia. In: Workshop on Iberian Cross-Language NLP tasks (ICL 2011), Huelva, Spain (2011)

9. Gamallo Otero, P., Pichel Campos, J.R.: Learning Spanish-Galician Translation Equivalents Using a Comparable Corpus and a Bilingual Dictionary. In: Gelbukh, A. (ed.) CICLing 2008. LNCS, vol. 4919, pp. 423–433. Springer, Heidelberg (2008)
10. Gomes, L., Lopes, G.P.: Measuring Spelling Similarity for Cognate Identification. In: Antunes, L., Pinto, H.S. (eds.) EPIA 2011. LNCS (LNAI), vol. 7026, pp. 624–633. Springer, Heidelberg (2011)
11. Kwong, O.Y., Tsou, B.K., Lai, T.B.: Alignment and extraction of bilingual legal terminology from context profiles. Terminology 10(1), 81–99 (2004)
12. Melamed, D.: A Portable Algorithm for Mapping Bitext Correspondences. In: 35th Conference of the Association of Computational Linguistics (ACL 1997), Madrid, Spain, pp. 305–312 (1997)
13. Nakagawa, H.: Disambiguation of single noun translations extracted from bilingual comparable corpora. Terminology 7(1), 63–83 (2001)
14. Rapp, R.: Automatic Identification of Word Translations from Unrelated English and German Corpora. In: ACL 1999, pp. 519–526 (1999)
15. Rubino, R., Linarès, G.: A Multi-view Approach for Term Translation Spotting. In: Gelbukh, A. (ed.) CICLing 2011, Part II. LNCS, vol. 6609, pp. 29–40. Springer, Heidelberg (2011)
16. Saralegui, X., San Vicente, I., Gurrutxaga, A.: Automatic generation of bilingual lexicons from comparable corpora in a popular science domain. In: LREC 2008 Workshop on Building and Using Comparable Corpora (2008)
17. Shao, L., Ng, H.T.: Mining New Word Translations from Comparable Corpora. In: 20th International Conference on Computational Linguistics (COLING 2004), Geneva, Switzerland, pp. 618–624 (2004)
18. Tiedemann, J.: Extraction of Translation Equivalents from Parallel Corpora. In: 11th Nordic Conference of Computational Linguistics, Copenhagen, Denmark (1998)
19. Yu, K., Tsujii, J.: Bilingual dictionary extraction from wikipedia. In: Machine Translation Summit XII, Ottawa, Canada (2009)
20. Yu, K., Tsujii, J.: Extracting bilingual dictionary from comparable corpora with dependency heterogeneity. In: NAACL HLT 2009, Boulder, Colorado, pp. 121–124 (2009)

Corpus-Based Acquisition of Support Verb Constructions for Portuguese

Britta D. Zeller and Sebastian Padó

Department of Computational Linguistics,
Heidelberg University, Germany
{zeller,pado}@cl.uni-heidelberg.de
http://www.cl.uni-heidelberg.de

Abstract. We present a resource-poor approach to automatically acquire Support Verb Constructions (SVCs) for European Portuguese with a two-stage procedure. First, we apply a cross-lingual approach with a bilingual parallel corpus: starting with a Portuguese full verb, we use the translations into another language and the corresponding backtranslations to identify Portuguese verb-noun pairs with the same meaning. Since not all of these are SVCs, the candidates are ranked and filtered in a second, monolingual step based on association statistics. We discuss two parametrisations of our procedure for a high-precision and a high-recall setting. In our experiments, these parametrisations achieve a maximum precision of 91% and a maximum recall of 86%, respectively.

Keywords: lexical acquisition, support verbs, multi-word expressions, parallel bilingual data, word alignment, association measures.

1 Introduction

Support Verb Constructions (SVCs), like *dar um passeio* 'to take a walk', are verb-noun complexes which occur in many languages. They form a syntactic and semantic unit and act as a multi-word predicate. Their meaning is mainly reflected by the nominal predicate, while the support verb (SV) is often a semantically impoverished verb, e.g., a light verb [3]. The distinction of SVCs from other complex predicates (CPs) or arbitrary verb-noun combinations is not a simple task. On the syntactic level, the difficulty is that SVCs occur in different forms – e.g. with direct object (*dar esperança* 'to give hope') or prepositional object (*estar na dúvida* 'to be in doubt') – and there are exceptions for most syntactic criteria [1]. Semantically, it is challenging to capture the difference between SVCs and a fully compositional construction in a corpus-driven fashion.

SVCs play a role in many natural language processing (NLP) tasks, such as anaphora resolution. Consider the following mini-discourse from Storrer [25], where the nominal of the SVC acts as antecedent of a pronoun: *One should only* ***provide [assistance]***$_1$ *to the children when they need [it]*$_1$. *[It]*$_1$ *can take the form of questions (...)* This construction would not be possible when the full verb *to assist* is used. Similarly, semantic role labelling works differently for full verbs

H. Caseli et al. (Eds.): PROPOR 2012, LNAI 7243, pp. 73–84, 2012.

(where the verb introduces the event and its dependents are arguments) and for SVCs (where the noun introduces the event and arguments are distributed) [22].

In this paper, we present a two-stage approach for the acquisition of SVC lists for Portuguese, a relatively resource poor language. We presuppose only a part-of-speech (POS) tagger and a parallel corpus. We concentrate on SVCs formed with a direct object, a very productive SVC pattern for Portuguese whose SVCs can often be paraphrased with a full verb [8].

2 Related Work

There are many studies about SVCs and other CPs, ranging from manual linguistic and lexicographic work to automatic NLP-oriented studies. On the manual side, Hanks et al. discuss dictionary representations of SVCs [11]. Hendrickx et al. develop a specific annotation layer for Portuguese SVCs on the CINTIL corpus[1], and carry out studies on the manually annotated data regarding syntactic and semantic aspects [12,7][2]. Cinková et al. take a step towards automatisation by developing a component to extract Swedish SVCs semi-automatically [5].

On the automatic side, Duran et al. use POS patterns to identify CPs in Brazilian Portuguese and extract productive patterns for SVCs [8]. Grefenstette and Teufel extract argument structures for SVs [10] by searching for nominalisations of full verbs, e.g. *to appeal* → *appeal*, and then locating the corresponding SV, e.g. *make + appeal*. Krenn and Evert [14,9] and Wermter and Hahn [27] compare association measures regarding their ability to establish rankings for collocations, including SVCs. Generally, the studies find that the choice of the association measure is crucial, but their performance varies across collocations.

Other studies have used bilingual parallel corpora. Villada Moirón and Tiedemann distinguish literal from idiomatic multi-word expressions (MWEs) [26]. Mukerjee et al. detect Hindi CPs as multi-word units aligned to English verbs in an English-Hindi parallel corpus [17]. Sinha first determines Hindi light verbs employing parallel data and subsequently uses them to retrieve CPs [24]. Bannard and Callison-Burch acquire within-language paraphrases from parallel corpora by observing which expressions share the same translation, which they call *pivot* [2]. Zarrieß and Kuhn apply this idea to the acquisition of MWEs but require dependency parses in both languages [28]. In sum, parallel data can provide strong clues to the identification of MWEs, but comes with problems inherited from the reliance on word alignments (e.g., bad performance for infrequent words).

3 A Two-Stage Strategy for the Acquisition of SVCs

Our goal in this paper is to generate lists of *non-prepositional* SVCs which semantically correspond to a given full verb. Our assumption is that there are full verbs which approximately correspond to the meaning of one or several SVCs, as in

[1] `http://catalog.elra.info/product_info.php?products_id=1102`

[2] These annotations could be used in the future to evaluate SVC extraction methods.

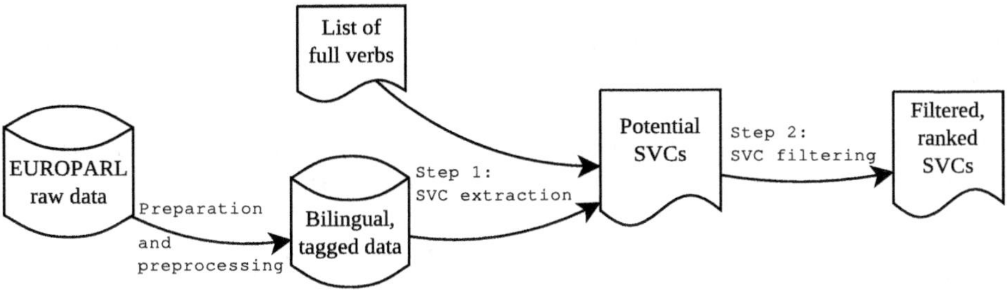

Fig. 1. Overall structure of the SVC acquisition procedure

Responda-me! 'Answer me!' and *Dá-me uma resposta!* 'Give me an answer!' [1]. The resulting lists can be used, for example, to combine statistics collected for different surface forms of the same underlying predicate, or conversely, to generate alternative surface forms for a predicate. To do so, we combine the two main approaches introduced in Section 2: the monolingual and the cross-lingual one:

- From cross-lingual data, we obtain information about *semantic equivalence*, i.e. whether two expressions have (approximately) the same meaning;
- From monolingual data with part-of-speech tags, we obtain information about the *strength of correlation* and *syntactic status* of a given expression.

Combined, these complementary types of information allow us to identify SVCs reliably even in the absence of deeper linguistic analysis, which makes it suitable for languages with few resources like Portuguese.

Figure 1 shows the overall structure of our extraction procedure. The first step is a cross-lingual one, inspired by Bannard and Callison-Burch's proposal to use translations in parallel corpora as pivots for paraphrase extraction [2], adopting their setup specifically to SVCs (cf. Section 3.2 for details). The resulting list contains many SVCs, but also other types of paraphrases that are not SVCs (i.e., that are false positives). The second, monolingual step applies association measures that encode our assumptions about the nature of SVCs to filter out the 'true' SVCs. Our results show that a combination of bi- and monolingual approaches leads to sizable improvements over just the cross-lingual method.

3.1 Data Preparation and Alignment Analysis

For the bilingual step, we use the Portuguese and German (PT–DE) portion of EUROPARL[3] [13]. We expect that this language pair shows sufficient typological differences so that direct 1-to-1 translation (which would lead to low variation) is unlikely but still close enough so that word alignment is still reliable.

We first align the PT–DE EUROPARL data using sentence alignment scripts provided on EUROPARL's web site[4] and the word alignment toolkit GIZA++ [18].

[3] Version 3 (September 2007)

[4] `http://www.statmt.org`

The word alignment is subsequently symmetrised into a single alignment with the 'grow-diag heuristic' [19]. Then we conduct POS tagging and lemmatisation. For Portuguese, we use FreeLing, version 2.2 [4,20] and TreeTagger [23] for German. We take special care of retokenisation issues occurring with FreeLing, i.e. decomposition of contractions and composition of MWEs. These procedures leave us with a parallel corpus of 982,039 sentence pairs.

Qualitative Evaluation of the Alignment. Our bilingual step retrieves Portuguese SVCs through word alignments to German. For pairs of full verbs and SVCs, this involves 1-to-n alignments, which are notoriously unreliable (Zarrieß and Kuhn [28] fall back to syntactic information for this reason). To assess the quality of these alignments, we perform a manual analysis of typical alignments between Portuguese full verbs and their German translations (1-to-1 and 1-to-n). We concentrate on the Portuguese verbs *apoiar* 'to support', *perguntar* 'to ask' and *ler* 'to read' since they are expected to lead to synonymous SVCs. We extract 17,943 sentences, each containing at least one of these full verbs.

We first consider the effect of alignment symmetrisation. It establishes many links which previously do not exist in at least one of the unidirectional alignments, e.g. *apoiar* $\rightarrow \emptyset$ becomes *apoiar* $\rightarrow$ *Beihilfe* 'aid'. Although it also leads to unnecessarily or incorrectly aligned tokens, filling these alignment gaps is strongly desirable. We count 22.9% differences between the symmetrised and the unidirectional alignment for the three full verbs mentioned above. In 10.6%, an alignment is created for an unaligned token. Since after symmetrisation, over 97% of the Portuguese full verbs are aligned with one to four German words and most remaining instances are wrong, we disregard all 1-to-5 (or more) alignments.

We then analyse the translation and word alignment patterns that we find between full verbs and SVCs. Full verbs are often translated as full verbs, mirrored in an 1-to-1 alignment. In the case of a translation as an SVC, the full verb is mostly aligned with the SVC's noun, e.g. *fragen* 'to ask' $\rightarrow$ *pergunta* 'question'. The SVC's verb frequently remains unaligned, which means that one cannot easily effect a large-scale SVC extraction solely from word alignments. The situation is similar for SVC-SVC translations. In most cases, the noun of one SVC is aligned, either to the noun of the corresponding SVC or with the whole SVC. In contrast, the semantically impoverished SV often remains unaligned or is aligned to an SV in the other language. For example, the noun *pergunta* in *fazer uma pergunta* 'to ask a question' is *always* aligned (in 77.5% of cases to a noun), whereas the support verb *fazer* is unaligned at 21.1% and aligned to a verb at 63.9%.

In terms of relative frequencies, about 30% of 1-to-n alignments align a Portuguese verb with a noun-verb combination, i.e., SVC candidates. Most of the remaining 1-to-n alignments are either rejected (since $n > 4$) or due to the fact that Portuguese verbs can incorporate more information than German (as well as English) verbs. E.g. they incorporate person information which must be added in German by a personal pronoun, leading to an 1:2 alignment.

In sum, this analysis suggests that if there is a proper SVC equivalent for a full verb, there are enough and reliable alignments to reveal this equivalence. Unfortunately, we cannot straightforwardly use them to acquire *complete* SVCs. However, the frequent alignments between full verbs and the SVC nouns can serve as a starting point: a heuristic extension of these alignments can be hoped to improve the retrieval of SVCs. Thus, the acquisition of SVCs, starting from a full verb, is reasonable and promising, even though some effort additionally to the automatic alignment is necessary. We return to this point in Section 3.2.

3.2 Step One: Bilingual SVC Extraction

Our cross-lingual SVC extraction method is an adaptation of Bannard and Callison-Burch's pivot approach for paraphrase extraction [2]. We start with a quick review of their method, using s to denote source language phrases and t for target language phrases. Their algorithm takes as input an initial phrase s_1 (to be paraphrased). It then locates all target language phrases t aligned with s_1 (*first pivot step*). Next, it gathers all instances of the t phrases and collects their backtranslations into the source language, resulting in a list of source phrases s_2 (*second pivot step*). An example for the language pair English–German: the initial phrase s_1 = *under control* is aligned with t = *unter Kontrolle*, which is backtranslated into s_2 = *in check*. Assuming that a translation is (largely) meaning-preserving, the source language phrases s_2 are considered as candidate paraphrases for s_1 and ranked using probabilities based on relative frequency. An extended version of the model that included word sense disambiguation achieved 70.4% accuracy in an evaluation for correct meaning for English–German.

We apply this model to full verbs as the inputs s_1. For our purpose, we however believe that it makes sense to concentrate on two different parameters of the model than those investigated in detail by Bannard and Callison-Burch.

Occurrence Thresholds. First, instead of using probabilities, we apply some simple occurrence thresholds which indicate how many times an alignment pair must occur to be considered. They are sufficient to counteract the effect of misalignments and overly context-specific translations, both of which are rather infrequent. We use four different thresholds: two each for the first and the second pivot step, respectively. Since there are 1-to-1 as well as 1-to-n translations, both pivot steps contain unigrams (single words) and n-grams (multiple words). We require n-grams to occur at least 6 times in the first and 9 times in the second pivot step. Unigrams are naturally more frequent than n-grams, so that we define a higher threshold for them, i.e. 300 in the first pivot step, and exclude them completely in the second pivot step, for SVCs always consisting of two or more words. Unlike Zarrieß and Kuhn, we do not encounter the problem of losing many n-grams by virtue of these thresholds [28]. Instead, this restriction reliably rejects many arbitrary verb-noun combinations, while not overly lowering recall.

Word Alignment Extension. Our analysis in Section 3.1 has shown that symmetrised alignments provide translations for almost all full verbs and the nominal parts of SVCs but are incomplete with regard to the SVs themselves. Since it is reasonable that the cross-lingual step should focus on recall – precision can be increased in the subsequent filtering step, if desired – we will focus exclusively on the symmetrised word alignment rather than the unidirectional ones. Furthermore, we strive to further extend the word alignment to support verbs using linguistically motivated rules.

To be able to phrase these rules concisely, we focus on word alignments between parts of speech that are supposed to participate in SVCs, i.e. nouns and verbs (recall that we ignore prepositional SVCs), discarding all others.[5] This leaves us with word alignments of the three following basic structures:

(1) X→ NOUN + VERB (2) X→ VERB (3) X→ NOUN

Alignments of type (1) are already complete. Correct alignments of type (2) occur almost exclusively in the first pivot step and connect Portuguese and German full verbs. Thus, they did not yet lead to extracted SVCs but there is a chance to find an SVC in the second pivot step. Alignments of type (3) are expanded in both directions (i.e., for both pivot steps). The expansion procedure is as follows: if a token X is 1-to-1-aligned with a single noun N, check the tokens in the neighbourhood of N. This neighbourhood is defined as 3 following tokens in German and 6 preceding tokens in Portuguese, respectively, reflecting the different syntactic structures in the two languages: while Portuguese has a rather strict word order and a broader neighbourhood can be considered, the German word order is more flexible; to avoid spurious extensions of N, we consult only a narrow word window. If a verb V occurs within this window, add V to the alignment. We assume that prepositional phrases cannot be inserted into an SVC[6] and that SVCs cannot split across sentences. Hence, the search is stopped after the closest verb is found or as soon as a preposition or a sentence boundary is reached. Finally, we added one lexical restriction: for Portuguese, we exclude occurrences of the verb *ser* ('to be'); according to the literature, *ser* does not form SVCs with direct objects, but it frequently occurs in the corpus.

An exemplary analysis shows that this heuristic increases the recall as intended. We even encounter unexpected SVCs, e.g. *dar assistência* 'to assist' for *apoiar* 'to support'. However, many false positives remain, since the pivoting extracts not only synonymous SVCs but also their antonyms, e.g. *exigir apoio* 'to demand support' for *apoiar*. The second step, filtering, attempts to eliminate these errors.

3.3 Step Two: SVC Filtering with Association Measures

As stated above, the purpose of the monolingual filtering is to increase the precision of the SVC candidate list created by the cross-lingual extraction step.

[5] If more than one verb or noun are co-aligned, only the first hit is kept.

[6] This is an oversimplification, but serves successfully to identify clear true positives.

There are at least two possible approaches to this task: either with linguistic heuristics or statistically. In line with our strategy in Step 1, we first adopted a linguistically informed strategy that checked whether extracted candidates were likely paraphrases for the initial full verbs. The goal of our strategy was to first detect the candidates' arguments through POS patterns which typically surround SVCs, and then to compare the candidates' argument heads with the argument heads for the full verbs. Very similar arguments indicate similar meaning [15], which we would expect for SVCs but not for compositional noun-verb combinations. Unfortunately, we found that the actual corpus occurrences of the SVC candidates showed too much variance, and we were unable to make reliable decisions based on the shallow linguistic information available to us.

We therefore adopted a statistical approach, more specifically one based on association measures (AMs). AMs model the common information of two words, that is, how predictable one word is given the other. We expect that SVCs will be recognisable by a predictability between verb and noun that is higher than for compositional verb-noun combinations.

The rest of this section discusses the two main design decisions for this step. The first one is the choice of association measures. A number of AMs have been investigated by Krenn and Evert [14], among which *(relative) frequency*, *pointwise mutual information (PMI)* and *student's t-test*. We decided to experiment with these three measures, the latter two of which are defined as:

$$\text{PMI} = \frac{p(v,n)}{p(v)p(n)} \qquad \text{t-test} = \frac{p(v,n) - p(v)p(n)}{\sqrt{s^2/N}}$$

where $p(v,n)$ is the *observed* co-occurrence probability of verb and noun and $p(v)p(n)$ can be interpreted as the *expected* co-occurrence probability. s^2 is the sample variance and N is the sample (corpus) size. Not surprisingly, all AMs involve co-occurrence frequencies. We only count directly adjacent noun-verb co-occurrences, since we found that intervening words degrade results.

The second design decision is the optimisation of the filtering step for either precision or recall. We indicated in Section 3.2 that the filtering step can be used to improve precision, which corresponds to aggressive filtering. However, for some settings (e.g. for manual post-processing), it might be better to filtering only leniently in order to keep recall high. We define two settings: a high-recall setting (*hiRec*) and a high-precision setting (*hiPrec*). The parameters of the filtering procedure that are varied between the two settings are as follows:

- Since many AMs are known to be oversensitive to low-probability (i.e., unreliable) events, we introduce a minimum verb-noun co-occurrence threshold and discard unfrequent pairs. Specifically, we set it to 2.5 co-occurrences per million words for *hiPrec* and to 1 for *hiRec*.
- Other studies show that there are two categories of SVCs [8,25]: The first one consists of SVCs where the SVs are light verbs, which have a very high context diversity, e.g. *dar apoio* 'to give support', *dar resposta* 'to give an answer', *dar um passo* 'to take a step'. The second category contains SVCs of nearly idiomatic meaning where the SV has a very low context diversity.

An example of this type is *correr um risco* 'to run a risk', whose SV *correr* 'to run' occurs in no other verb-noun pair with a co-occurrence frequency >2.5 per million words. In contrast, verbs which cooccur with an average number of nouns are not likely to be part of an SVC. To capture this fact, we compute the 'diversity' for each verb as the number of different noun lemmas it occurs with in the complete corpus. For the *hiPrec* setting, we retain only those SVC candidates whose diversity is either 1, or higher than the median diversity. For *hiRec*, no filtering takes place.

4 Evaluation

4.1 Creating the Gold Standard

To evaluate our approach, we need a gold standard of SVCs. Since it is impossible to determine how many 'gold SVCs' exist for a given full verb, we took the output of the cross-lingual step to be the basis for manual annotation. Against this gold standard, we can compute precision and (relative) recall, i.e., recall relative to the extraction procedure as defined in Pantel et al. [21].

For the annotation, we concentrated on six Portuguese full verbs: *ameaçar* 'to threaten', *apoiar* 'to support', *faltar* 'to lack', *perguntar* 'to ask', *prometer* 'to promise' and *responder* 'to answer'. Each of them has approximately the same meaning as at least one SVC. The retrieved candidate expressions were annotated by two native speakers with professional linguistic knowledge, judging for each expression *i)* whether it was an SVC and *ii)* whether it semantically corresponded to the initial full verb. A total of 84 candidate SVCs have been annotated, ranging from 1 to 64 expressions per verb. The main criterion provided to the annotators was whether the verb can be interpreted as a semantically impoverished SV in the given expression.

We computed inter-annotator agreement (IAA) with Cohen's κ [6] and obtained a value for *i)* of 0.60 and for *ii)* of 0.74. The first κ value is lower than the second one because the decision if an expression is an SVC or not is more general and thus more difficult. These are fairly good IAA rates, regarding the fact that SVC determination is a difficult task because of the fuzziness of SVs [10]. Since other SVC acquisition studies either do not provide IAA rates or have a fairly different setting, we cannot compare our IAA rates. However, Landis and Koch consider these rates as moderate and substantial, respectively [16]. The final gold standard was formed from the intersection of the two annotations, and for all cases in which the evaluators did not agree, we classified the expression by ourselves. This procedure leads to 22 SVCs judged as true positives.

4.2 Results after the Extraction Step

Table 1 shows the results of the cross-lingual extraction step.[7] As noted in section 4.1, the list of candidate SVCs resulting from the pivoting serves as basis

[7] All results presented in Section 4 refer to our automatically processed corpus. Unfortunately, no numbers are available on the quality of the preprocessing components.

Table 1. Results for the extraction step

	ameaçar	apoiar	faltar	perguntar	prometer	responder	all
Precision	1.00	0.16	1.00	0.71	0.33	0.43	**0.26**
Recall	1.00	1.00	1.00	1.00	1.00	1.00	**1.00**
F_1	1.00	0.27	1.00	0.83	0.50	0.60	**0.42**

Table 2. Overall results of the two-step procedure

	PMI		*Frequency*		*t-test*	
	hiPrec	hiRec	hiPrec	hiRec	hiPrec	hiRec
Precision	**0.91**	0.61	**0.91**	0.61	0.90	0.60
Recall	0.45	**0.86**	0.45	**0.86**	0.41	0.81
F_1	0.61	**0.72**	0.61	**0.72**	0.56	0.69

for the gold standard. Hence, recall is always 100%. However, precision varies considerably between verbs: the SVC lists for *ameaçar* and *faltar* are already perfect, which speaks to the efficacy of our alignment extension (Section 3.2), but the results for other verbs are far from perfect. *apoiar* is especially bad: while the other verbs lead to a maximum of 7 candidate SVCs, *apoiar* results in 64 candidates with many false positives. This outlier seems corpus-specific: *apoiar* is strikingly more frequent in EUROPARL than the other full verbs, and two commonly aligned nouns, *apoio* 'support' and *ajuda* 'help', are very frequent as well. The verb-noun pairs in which they occur are often arbitrary, e.g. *encontrar apoio* 'to find support', albeit frequent enough to overcome the thresholds defined in section 3.2. Thus, many false positives slip into the results of Step 1. This also explains the rather low overall precision and f-scores. In sum, the quality of the results of the cross-lingual step depends on the properties of the initial verb.

4.3 Final Results Including the Filtering Step

Recall from Section 3.3 that the filtering step had two parameters: the choice of the AM measure (*PMI*, *frequency* and *t-test*) and the choice between a high-recall and a high-precision setting. Table 2 shows the results for these combinations. We discuss both parameters in turn.

High Precision vs. High Recall. The figures in Table 2 indicate that the filtering step indeed improves substantially over the results of the cross-lingual extraction step: from an f-score of 0.42, we reach an f-score of 0.72 in the optimal case, corresponding to an error reduction of 50%. The Table also demonstrates that the filtering step can be tuned to the requirements of a particular setting. If high precision is required, the filtering mechanisms we introduced can produce a precision of above 90%, at the cost of a recall of slightly below half. At the same time, the high recall setting can still substantially improve precision (from 26% to 61%) within a rather small loss in recall (from 100% to 86%).

Consider the the verb *perguntar* 'to ask'. The *hiPrec* setting correctly retrieves the SVC *fazer pergunta*. For the *hiRec* setting, the following expressions are found: *fazer pergunta*, *levantar questão*, *colocar pergunta*, *colocar questão*, *apresentar pergunta*, and *formular pergunta*. According to our gold standard, only the last expression is a false positive. All SVCs contained in the gold standard are found.

Association Measures. Krenn and Evert [14] did not find any single measure to consistenly outperform the others across all tested collocations. For SVCs, *t-test* and *frequency* worked best, while *PMI* performed poorly, and the authors even suggested to use a modified version of PMI.

In contrast, on our data *PMI* performs very well and does not show the idiosyncrasies observed by Krenn and Evert. It shows essentially identical results to *frequency* in a precision/recall evaluation, while *t-test* performs consistently worse. We also evaluated the lists with average precision, i.e., took the ranking within the lists into account (not shown in the tables). In that case, *PMI* substantially outperforms *frequency* with an AP of 0.33 compared to 0.11 for *frequency*. This indicates that *PMI* does a better job at ranking.

We attribute this difference to the fact that Krenn and Evert re-rank a list of all verb-noun combinations from a corpus, while we only consider the candidates extracted by the cross-lingual step, which are typically located within a fairly narrow range for all AMs. We see this as a further validation of our two-step approach, dividing the work between the cross-lingual alignment-based and the monolingual association-based approach. In sum, the joint application of mono- and cross-lingual methods leads to a very satisfactory overall result.

5 Conclusions and Outlook

This paper has presented a resource-poor two-stage approach to acquire Support Verb Constructions, applied to the Portuguese language. We explored whether cross-lingual techniques are suitable for the extraction of syntactically correct SVCs which semantically correspond to a given full verb, and whether monolingual methods can further improve the cross-linguistically obtained results.

Within the limits of our evaluation, our results indicate that this is indeed the case: word alignment-based extraction is perfectly applicable to the SVC acquisition task without the need for complex preprocessing, while the computation of association measures is capable of ranking and refining the expressions found in the first step. Our approach provides adjustment possibilities for both solid precision and recall values, depending on which focus the user intends.

The main caveat of our approach is that it depends crucially on acquiring reliable translations for the initial full verb. Full verbs which occur in heterogeneous contexts and are translated in many different ways will give rise to noisy candidate lists which cannot be re-ranked successfully. In future work, we plan a corpus-based evaluation (using CINTIL [12]) on a larger number of full verbs, assessing also the distribution of the SVs involved in SVCs.

Another direction for future research is the generalisation of our method to *prepositional* SVCs or a large-scale acquisition of different CPs. This will presumably require better extraction and filtering methods.

References

1. Athayde, M.F.: Construções com Verbo-suporte (Funktionsverbgefüge) do Português e do Alemão. Cadernos Do Cieg 1, 5–68 (2001)
2. Bannard, C., Callison-Burch, C.: Paraphrasing with Bilingual Parallel Corpora. In: Proceedings of the 43rd Annual Meeting on Association for Computational Linguistics, Ann Arbor, MI, pp. 597–604 (2005)
3. Butt, M.: The Light Verb Jungle. Harvard Working Papers in Linguistics 9, 1–49 (2003)
4. Carreras, X., Chao, I., Padró, L., Padró, M.: FreeLing: an Open-Source Suite of Language Analyzers. In: Proceedings of the 4th International Conference on Language Resources and Evaluation, Lisbon, Portugal (2004)
5. Cinková, S., Pecina, P., Podveský, P., Schlesinger, P.: Semi-automatic Building of Swedish Collocation Lexicon. In: Proceedings of the 5th Conference on International Language Resources and Evaluation, Genoa, Italy (2006)
6. Cohen, J.: A Coefficient of Agreement for Nominal Scales. Educational and Psychological Measurement 20(1), 37–46 (1960)
7. Duarte, I., Gonçalves, A., Miguel, M., Mendes, A., Hendrickx, I., Oliveira, F., Cunha, L.F., Silva, F., Silvano, P.: Light Verbs Features in European Portuguese. In: Proceedings of the 2nd Interdisciplinary Workshop on the Identification and Representation of Verb Features and Verb Classes, Pisa, Italy (2010)
8. Duran Sanches, M., Ramisch, C., Aluísio, S.M., Villavicencio, A.: Identifying and Analyzing Brazilian Portuguese Complex Predicates. In: Proceedings of the Workshop on Multiword Expressions: from Parsing and Generation to the Real World, Portland, USA, pp. 74–82 (2011)
9. Evert, S., Krenn, B.: Methods for the Qualitative Evaluation of Lexical Association Measures. In: Proceedings of the 39th Annual Meeting of the Association for Computational Linguistics, Toulouse, France, pp. 188–195 (2001)
10. Grefenstette, G., Teufel, S.: Corpus-Based Method for Automatic Identification of Support Verbs for Nominalizations. In: Proceedings of European Chapter of the Associaton of Computational Linguistics, Dublin, Ireland, pp. 98–103 (1995)
11. Hanks, P., Urbschat, A., Gehweiler, E.: German Light Verb Constructions in Corpora and Dictionaries. International Journal of Lexicography 19(4), 439–457 (2006)
12. Hendrickx, I., Mendes, A., Pereira, S., Gonçalves, A., Duarte, I.: Complex Predicates Annotation in a Corpus of Portuguese. In: Proceedings of the 4th ACL Linguistic Annotation Workshop, Uppsala, Sweden, pp. 100–108 (2010)
13. Koehn, P.: Europarl: a Parallel Corpus for Statistical Machine Translation. In: Proceedings of the 10th Machine Translation Summit, Chiang Mai, Thailand, pp. 79–86 (2005)
14. Krenn, B., Evert, S.: Can We Do Better than Frequency? A Case Study on Extracting PP-Verb Collocations. In: Proceedings of the ACL Workshop on Collocations, Toulouse, France (2001)

15. Lin, D., Pantel, P.: Discovery of Inference Rules for Question Answering. Journal of Natural Language Engineering 7(4), 343–360 (2001)
16. Landis, J.R., Koch, G.G.: The Measurement of Observer Agreement for Categorical Data. Biometrics 33(1), 159–174 (1977)
17. Mukerjee, A., Soni, A., Raina, A.M.: Detecting Complex Predicates in Hindi Using POS Projection across Parallel Corpora. In: Proceedings of the Workshop on Multiword Expressions: Identifying and Exploiting Underlying Properties, Sydney, Australia, pp. 28–35 (2006)
18. Och, F.J., Ney, H.: A Systematic Comparison of Various Statistical Alignment Models. Computational Linguistics 29(1), 19–51 (2003)
19. Och, F.J., Tillmann, C., Ney, H.: Improved Alignment Models for Statistical Machine Translation. In: Proceedings of the Joint SIGDAT Conference on Empirical Methods in Natural Language Processing and Very Large Corpora, College Park, MD, pp. 20–28 (1999)
20. Padró, L., Collado, M., Reese, S., Lloberes, M., Castellón, I.: FreeLing 2.1: Five Years of Open-Source Language Processing Tools. In: Proceedings of the 7th Conference on International Language Resources and Evaluation, Valleta, Malta (2010)
21. Pantel, P., Ravichandran, D., Hovy, E.: Towards Terascale Knowledge Acquisition. In: Proceedings of the 20th International Conference on Computational Linguistics, Geneva, Switzerland, pp. 771–777 (2004)
22. Ruppenhofer, J., Ellsworth, M., Petruck, M.R.L., Johnson, C.R., Scheffczyk, J.: FrameNet II: Extended Theory and Practice (2010), `https://framenet2.icsi.berkeley.edu/docs/r1.5/book.pdf`
23. Schmid, H.: Probabilistic Part-of-Speech Tagging Using Decision Trees. In: Proceedings of the International Conference on New Methods in Language Processing, Manchester, UK (1994)
24. Sinha, R.M.K.: Mining Complex Predicates in Hindi Using a Parallel Hindi-English Corpus. In: Proceedings of the Workshop on Multiword Expressions: Identification, Interpretation, Disambiguation and Applications, Singapore, pp. 40–46 (2009)
25. Storrer, A.: Corpus-based Investigations on German Support Verb Constructions. In: Fellbaum, C. (ed.) Collocations and Idioms: Linguistic, Lexicographic, and Computational Aspects, London, pp. 164–188.
26. Villada Moirón, B., Tiedemann, J.: Identifying Idiomatic Expressions Using Automatic Word-Alignment. In: Proceedings of the EACL Workshop on Multiword Expressions in a Multilingual Context, Trento, Italy (2006)
27. Wermter, J., Hahn, U.: Collocation Extraction Based on Modifiability Statistics. In: Proceedings of the 20th International Conference on Computational Linguistics, Geneva, Switzerland (2004)
28. Zarrieß, S., Kuhn, J.: Exploiting Translational Correspondences for Pattern-Independent MWE Identification. In: Proceedings of the Workshop on Multiword Expressions: Identification, Interpretation, Disambiguation and Applications, Singapore, pp. 23–30 (2009)

Improving Portuguese Term Extraction

Lucelene Lopes and Renata Vieira

Faculdade de Informática, FACIN, PUCRS
Porto Alegre, RS, Brazil
{lucelene.lopes,renata.vieira}@pucrs.br

Abstract. This paper presents the evaluation of a set of heuristics to improve the quality of extracted terms from an annotated domain corpus written in Portuguese. The proposed heuristics start from part-of-speech and grammatical functional annotation of texts, identifying nouns and noun phrases that are the best candidates to be considered terms of the domain. These nouns and noun phrases are submitted to a set of approximative rules (heuristics) that may either discard some, accept others (removing words or not), or even discover implicit terms that can be inferred. The effectiveness of these heuristics is verified through a corpus experiment, on the basis of a reference list for which usual metrics are computed.

1 Introduction

The importance of correctly extract terms from corpora to build language resources such as an ontology is a known fact [11,3,6]. The automatic term extraction from corpus efforts are usually based on statistical or on linguistic approaches.

The statistical approaches, *e.g.*, NSP [1], Ogma [12], are language independent, but they offer solutions that are usually less precise than linguistic-based term extractors [8]. Linguistic approaches, *e.g.*, OntoLP [13], ExATOlp [7], need more sophisticated language tools, including a high quality POS-tagging and solid linguistic information to identify morpho-syntactic constructions of interest.

Ogma applied to a Portuguese corpus [12] delivered precision around 35% for a list of reasonable size (over 1,000 terms). NSP, also applied to a Portuguese corpus [8], delivered precision a little above 30% to a list of approx. 2,000 terms.

OntoLP applied to a Portuguese corpus [10] delivered precision around 6% for large lists (nearly 9,000 terms). However, following a similar approach, but using linguistic heuristics to refine extracted terms, ExATOlp delivered precision values around 50% for lists of approx. 1,300 terms [8].

Motivated by these observations, in this paper we propose a set of linguistic-based heuristic rules to improve the quality of automatic term extraction from Portuguese corpus. We assume the availability of an annotated corpus with POS-tagging and grammatical function of words. To such annotated corpus, we perform an extraction based on the detection of noun phrases (NPs) followed by the application of 11 heuristics. The effectiveness of the proposed heuristics is verified through the comparison with a hand made reference list.

Section 2 describes the domain corpus used in the experiments and its reference list. Sections 3 and 4 present the proposed heuristics and their evaluation. The conclusion summarizes the contribution of this paper and suggests future work.

H. Caseli et al. (Eds.): PROPOR 2012, LNAI 7243, pp. 85–92, 2012.

2 Domain Corpus and Reference List

The domain corpus used in this paper experiments is composed of 281 Portuguese articles extracted from the bilingual Pediatrics Journal (*Jornal de Pediatria* - `http://www.jped.com.br`), with a total of 835,412 words distributed in 27,724 sentences. This corpus was assembled by Coulthard [5], and the main reason to choose it to our experiments was the availability of a reference list of terms to be used as gold standard.

The reference list was build up by TEXTECC project (`http://www.ufrgs.br/textecc`). The primary goal of this list was to create glossaries for translation support. To identify items for these glossaries, terms were extracted from plain texts from the corpus. In this process, terms with less than 4 occurrences in the corpus were discarded, as well as terms composed with more than 3 words and less than 2. A list of 4,194 terms was obtained, with 1,534 bigrams and 2,660 trigrams.

The Pediatrics corpus was linguistically annotated by the parser PALAVRAS [2] and submitted to an extraction procedure based on identifying noun phrases (NPs)[1]. As result, 189,146 NPs were extracted, being 126,491 different ones.

Classifying the extracted NPs according to the number of words we obtain:

n-gram	unigrams	bigrams	trigrams	4-grams	5-grams	≥6-grams	total
NPs	2,210	17,407	15,577	13,572	14,312	63,413	126,491
total	5,583	58,500	25,480	17,154	16,993	65,436	189,146

3 Proposed Heuristics

The proposed heuristics are divided in Adjustment, Discard, and Inclusion groups.

3.1 Adjustment Heuristics

The adjustment heuristics are meant to remove words that do not carry significance to the term represented under the NP. The proposed rules are the removal of: articles at the beginning (and anywhere inside) of a NP, pronouns at the beginning (and anywhere inside) of a NP. Statistical approaches offer some similar, but less precise, means to adjust extracted terms with "stop word" lists.

It s important to notice that the adjustment heuristics changes the number of words of NPs, *i.e.*, an adjusted trigram may become a bigram, or even a unigram.

A1 - Adjustment Rule 1 – Removal of Articles at the Beginning of NPs. Despite the important role of articles as determinant, the removal of the first article serves the purpose of term extraction for semantic resources. While the NP "o leite materno" is ("the mother's milk" in English) distinct from the NP "um leite materno" ("a mother's milk" in English), they include the same domain concept candidate, "leite materno". Applying heuristic A1 to the 189,146 NPs of the Pediatrics corpus, change nearly 43% of them (81,031 NPs).

[1] This parser was chosen due to its robustness and high performance, but any other parser tool could be used to annotate the corpus, since the proposed heuristics deal on the final annotation output only.

A2 - Adjustment Rule 2 – Removal of All Articles of NPs. The second heuristic is the removal of all articles inside a NP, and not only the article who is the first word. In fact, this rule is a generalization of the first one, and all considerations about the change of meaning made previously also apply to this second heuristic. However, this second heuristic is harder to implement in a statistical approach, since it does not consider only contiguous words to form an extracted term.

Applying the second heuristic to the NP "o leite da mãe" ("the milk of the mother" in English) result in the term "leite de mãe" ("milk of mother"). This heuristic changes nearly half (92,754 NPs) of the Pediatrics corpus NPs.

A3 - Adjustment Rule 3 – Removal of Pronouns at the Beginning of NPs. Similar to the first heuristic, this third heuristic intends to keep the generalized form of the extracted NP. This heuristic is only applied when the pronoun to be removed is not the head of the NP[2]. For example, the NP "nossa margem oceânica" ("our oceanic margin" in English) becomes "oceanic margin". The application of this heuristic changes 12,793 NPs of the total 189,146 NPs of Pediatrics corpus.

A4 - Adjustment Rule 4 – Removal of All Pronouns of NPs. The same extension made for articles in A2 is proposed here for pronouns and the same considerations made for rule A3 are also valid here. The NP "o caso de seu paciente" ("the case of his/her patient" in English) becomes "o caso de paciente" ("the patient case"). The application of this heuristic to the NPs extracted from the Pediatrics corpus changes 18,230 NPs from the 189,146 total extracted NPs.

3.2 Discard Heuristics

The discard heuristics are rules to refuse annotated noun phrases that are not likely to be representative terms of a domain. These heuristics are the refusal of NPs with: numerals, symbols, a pronoun as head, and an adverb at the beginning.

The discard heuristics, unlike the adjustment ones, do not change the number of words that a NP has, but it may significantly reduce the number of NPs. Considering the application of all discard heuristics over the NPs extracted from the Pediatrics corpus, the original total number of 189,146 NPs (126,491 different ones) drops to 133,250 NPs (69,907 different ones).

D1 - Discard Rule 1 – Refusal of NPs with Numerals. The first discard heuristic rule out the extracted NPs that contains numerals, either in their written form or using numeric characters. Despite being a restrictive rule because it ignores terms like "the seven wonders", this heuristics is often valid to discard NPs that express quantities that are common in scientific texts. Successful examples of this rule application in the Pediatrics corpus are "três meses" ("three months" in English) and "ano 2000" ("year 2000" in English). Applying heuristic D1 over the 189,146 NPs of the Pediatrics corpus resulted in the refusal of 30,969 NPs.

---

[2] In Portuguese, the head of a NP is usually a noun, but it can also be an adjective, a participle past verb, or a pronoun (an anaphora).

D2 - Discard Rule 2 – Refusal of NPs with Symbols. This rule discards NPs that contains symbols, *i.e.*, it considers only NPs composed by the letters (with and without diacritics) and digits. It is also accepted the dash symbol, since in Portuguese it is common to have composed words written with dash characters between the original basic words, *e.g.*, "recém-nascido" ("new born" in English).

Many of the refused NPs with symbols also have numerals, as percentile values, *e.g.*, 46%, but it is also common to find other symbols, like electronic address, *e.g.*, "dilma@presidencia.gov.br" and ordinal numbers abbreviations, *e.g.*, "2°" (2nd in English). The application of rule D2 to the 189,146 NPs extracted from the Pediatrics corpus resulted in the refusal of 40,989 NPs.

D3 - Discard Rule 3 – Refusal of NPs with Pronoun Head. Usually a NP head is a common or proper noun. However, the head may also be an adjective, a participle past verb or a pronoun. This third discard rule is meant to accept only NPs where the head has a self-contained meaning, *i.e.*, the head is either a common or proper noun, an adjective or a participle past verb. Therefore, when the NP head is a pronoun (usually an anaphora) the NP is discarded.

For example, in the sentence "Aqueles que sabiam, pergutaram." ("Those who knew, have asked" in English), the subject of the predicate "perguntaram" ("have asked") is the NP "Aqueles que sabiam". The NP head is the pronoun "Aqueles" ("those"), then this NP does not carry a meaningful information to be considered a term of the corpus. The application of the rule D3 over the 189,146 NPs extracted from the Pediatrics corpus resulted in the refusal of 6,109 NPs.

D4 - Discard Rule 4 – Refusal of NPs Starting with an Adverb. This rule deals with NPs that make comparisons with previously mentioned terms. In such cases, usually the NP starts with an adverb and its head is not a noun. For example, in the Pediatrics corpus the NP "mais frequente" ("more frequent" in English) was found 11 times, but in these occurrences it was employed 5 times to refer the frequent use of a particular medicine drug, and 6 times to refer to the frequent adoption of a particular patient habit. However, it is meaningless to consider the NP "mais frequente" as a term, since only reading the context in which the term was employed it is possible to identify if term is referring to a medicine drug or a patient habit.

The application of heuristic D4 over the 189,146 NPs of the Pediatrics corpus resulted in the refusal of 650 NPs.

3.3 Inclusion Heuristics

The inclusion heuristics are the more linguistic sophisticated rules, since these rules aim to consider NPs that are not written as such, but may be inferred from the annotation. These heuristics are the detection of implicit NPs by: removal of adjectives, multiple predicates, and conjunction of adjectives.

The practical effect of the inclusion heuristics is to increase the number of NPs. The 133,250 NPs (69,907 different ones) extracted from the Pediatrics corpus after applying all discard rules increases to 179,867 NPs (82,975 different ones).

I1 - Inclusion Rule 1 – Detection of Implicit NPs by Adjectives Removal. The first inclusion heuristic is based on the detection of smaller NPs by the successive removal of adjectives. For example, the NP "parede abdominal anterior" ("frontal abdominal wall" in English) has two other implicit NPs: "parede abdominal" ("abdominal wall") and "parede" ("wall") that are not annotated as such by the parser. While extracting the NP "parede abdominal anterior", rule I1 will produce the NP "parede abdominal" by removing the adjective "anterior", them it will produce the NP "parede" by removing the adjective "abdominal".

In the Pediatrics corpus, 40,156 NPs ends with at least one adjective. Applying heuristic I1 to these NPs included 44,020 implicit NPs.

I2 - Inclusion Rule 2 – Detection of Replicated NPs by Multiple Predicate. The second inclusion heuristic replicates the occurrences of an NP when it appears as subject or object of a multiple verb predicate.

For example, the sentence "Crianças saudáveis sofrem e curam-se de doenças sazonais." ("Healthy children suffer and heal themselves from seasonal diseases." in English) has one explicit occurrence of the NP "crianças saudáveis" ("healthy children"), the subject of the double predicate "sofrem e curam-se" ("suffer and heal themselves"). For this example, the application of I2 rule considers 2 occurrences of the NP "crianças saudáveis", *i.e.*, one occurrence as subject of the verb "sofrem" and another as subject of the verb "curam-se". Still in this example, the object "doenças sazonais" ("seasonal diseases") is also duplicated by I2, since this NP is the object of both verbs composing the predicate. Applying I2 to NPs extracted from the Pediatrics corpus adds 3,472 implicit occurrences.

I3 - Inclusion Rule 3 – Detection of Implicit NPs by Adjective Conjunction. The last heuristic is also based of multiple structures. Rule I3 detects implicit NPs when a same noun is qualified by two or more adjectives. For example, the sentence "Crianças normais e prematuras foram observadas." ("Normal and premature children were observed." in English) has two implicit NPs: "crianças normais" ("normal children") and "crianças prematuras" ("premature children").

This heuristic deals with NPs ending by two or more adjectives connected by a conjunction and it considers each adjective with the NP head. The application of I3 over the NPs extracted from the Pediatrics corpus added 861 NPs.

4 Experiments and Quantitative Results

In order to quantify the benefits brought by each heuristic, we start analyzing each group of heuristics at time comparing the extraction result with the reference list.

To each heuristic, we compare the reference list of bigrams and trigrams with a list of the more frequent terms. Since the original number of extracted bigrams and trigrams is respectively 17,407 and 15,577, we will consider the more frequent terms the top 10% terms organized according to the absolute term frequency [9]. Therefore, we consider the top 1,741 bigrams and the top 1,558 trigrams after applying each heuristic, and compare it with the 1,534 bigrams and 2,660 trigrams of the reference lists, computing precision (P), recall (R) and f-measure (F) [15].

4.1 Adjustment Heuristics

Table 1 illustrates the benefits brought by the adjustment heuristics to the extracted NPs of the Pediatrics corpus. Besides P, R and F values, the last column indicates how many terms in the top 10% of extracted list was found in the reference list. In the first row (*none*) the results achieved without the heuristics are indicated. The following four rows indicate the results applying each adjustment heuristics alone. The last row (A) indicates the results for all these heuristics.

Table 1. Benefits brought by the adjustment heuristics

Bigrams					Trigrams				
adjustment heuristics	P	R	F	matches in top 1,741	adjustment heuristics	P	R	F	matches in top 1,558
none	12%	13%	13%	206	*none*	13%	7%	9%	202
A1	38%	43%	40%	653	A1	55%	32%	40%	852
A2	38%	43%	40%	654	A2	59%	35%	44%	914
A3	14%	16%	15%	252	A4	15%	9%	11%	229
A4	15%	17%	16%	257	A3	16%	9%	12%	242
A	48%	55%	51%	839	A	60%	35%	44%	934

The results without heuristics (*none*) are low for all values. These values are similar to those found by other extraction procedures made over PALAVRAS annotation [13]. However, applying A1 and A2, there is a large improvement (between 25% and 46%) in both precision and recall. Pronouns removal heuristics (A3 and A4) were less effective, nonetheless, they improve the values of at least 2%.

4.2 Discard Heuristics

To analyze the discard heuristics we start considering the previous results. Table 2 indicates in its first row (A) the baseline result with all adjustment and none of the discard rules. The following rows show the benefits of each discard heuristics alone. The last row (AD) summarize all adjustment and discard heuristics.

Table 2. Benefits brought by the discard heuristics

Bigrams					Trigrams				
discard heuristics	P	R	F	matches in top 1,741	discard heuristics	P	R	F	matches in top 1,558
A	48%	55%	51%	839	A	60%	35%	44%	934
D1	52%	60%	56%	914	D1	61%	36%	45%	947
D2	57%	65%	61%	993	D2	64%	38%	47%	995
D3	48%	55%	51%	842	D3	61%	36%	45%	953
D4	48%	55%	51%	840	D4	60%	35%	45%	936
AD	57%	65%	61%	1,001	AD	65%	38%	48%	1,006

Observing Table 2, it is possible to observe that most of the benefits brought by the discard heuristics is due to the refusal of NPs with symbols (D2). The refusal of NPs with numerals (D1) was also relevant for bigrams, improving the precision (4%), but less effective for trigrams (1%). The heuristics D3 and D4, even though affecting a large number of NPs, do not deliver a relevant increase in precision or recall. Nevertheless, for both bigrams and trigrams, the heuristics D3 and D4 still contribute with some matches. The combined use of all discard heuristics delivered an improvement between 4% and 10% for f-measure.

4.3 Inclusion Heuristics

Analogously to the discard heuristics, the quantitative evaluation of the inclusion heuristics is made considering the use of all heuristics of the previous two groups. The first row (AD) in Table 3 presents the result of extraction using all adjustment and discard, and none of the inclusion heuristics. The next three rows shows the benefits obtained by each inclusion heuristics applied alone. The last row (*all*) indicates the results of the extraction using all 11 heuristics.

Table 3. Benefits brought by the inclusion heuristics

Bigrams					Trigrams				
inclusion heuristics	P	R	F	matches in top 1,741	inclusion heuristics	P	R	F	matches in top 1,558
AD	57%	65%	61%	1,001	AD	65%	38%	48%	1,006
I1	59%	67%	63%	1,027	I1	67%	39%	50%	1,044
I2	58%	65%	61%	1,004	I2	65%	38%	48%	1,011
I3	58%	66%	62%	1,010	I3	65%	38%	48%	1,009
all	60%	68%	64%	1,041	*all*	68%	40%	50%	1,052

Observing Table 3 it is possible to notice that all three inclusion heuristic seem numerically less impressive, since even all of them together increase f-measure in 2% or 3%. However, it is important to keep in mind that after applying all other heuristics, the values are reasonably high, when compared to other similar results [10], when the same Pediatrics corpus, also annotated by PALAVRAS, delivered nearly 10% precision. Therefore, even the improvement of 1% in precision is non-negligible when we increase the precision from 57% to 58%.

5 Conclusion

This paper shows the evaluation of a set of linguistic-based heuristics that allows significative improvements in quality for automatic extracted terms. These improvements were analyzed through a comparison of extracted bigrams and trigrams from a Pediatrics corpus with a previously available reference list of terms. The proposed heuristics raised precision from 12% (approx.) to more than 60%, and recall was also improved from approx. 10% to at least 40%.

Despite the good results obtained, a future work is to extend the heuristics analysis using different annotation tools, *e.g.*, the parser LX-Center [14]. It is also possible to analyze the benefits of the heuristics to other corpora, and even to propose and test new heuristics in order to further improve the extraction quality.

However, it is important to observe that the experiments were made with lists of terms sorted according to their absolute frequency. Therefore, a natural future work is to observe the effectiveness of the proposed heuristics considering more sophisticated criteria [4] to sort the list of extracted terms.

Finally, we want to stress that the quality improvement brought by the heuristics made possible to foresee more ambitious natural language applications. The high precision values indicate that the extracted terms are good domain concept candidates, so it is possible to use them to build good quality ontological structures, as concept hierarchies and even to extract complex semantic relations.

References

1. Banerjee, S., Pedersen, T.: The design, implementation and use of the ngram statistics package. In: 4th ITPCL, pp. 370–381 (2003)
2. Bick, E.: The parsing system PALAVRAS: automatic grammatical analysis of portuguese in constraint grammar framework. PhD thesis, Arhus University (2000)
3. Buitelaar, P., Cimiano, P., Magnini, B.: Ontology learning from text: An overview. In: Buitelaar, P., Cimiano, P., Magnini, B. (eds.) Ontology Learning from Text. Front. in Art. Intel. and Apllic., vol. 123. IOS Press (2005)
4. Chung, T.M.: A corpus comparison approach for terminology extraction. Terminology 9, 221–246 (2003)
5. Coulthard, R.J.: The application of Corpus Methodology to Translation: the JPED parallel corpus and the Pediatrics comparable corpus. Master's thesis, UFSC (2005)
6. Fortuna, B., Lavrač, N., Velardi, P.: Advancing Topic Ontology Learning through Term Extraction. In: Ho, T.-B., Zhou, Z.-H. (eds.) PRICAI 2008. LNCS (LNAI), vol. 5351, pp. 626–635. Springer, Heidelberg (2008)
7. Lopes, L., Fernandes, P., Vieira, R., Fedrizzi, G.: ExATO lp – An Automatic Tool for Term Extraction from Portuguese Language Corpora. In: Proc. of the 4th Language & Tech. Conf., LTC 2009, pp. 427–431. Adam Mickiewicz Univ. (2009)
8. Lopes, L., Oliveira, L.H., Vieira, R.: Portuguese term extraction methods: Comparing linguistic and statistical approaches. In: PROPOR 2010 (2010)
9. Lopes, L., Vieira, R., Finatto, M.J., Martins, D.: Extracting compound terms from domain corpora. Journal of the Brazilian Computer Society 16, 247–259 (2010)
10. Lopes, L., Vieira, R., Finatto, M.J., Zanette, A., Martins, D., Ribeiro Jr., L.C.: Automatic extraction of composite terms for construction of ontologies: an experiment in the health care area. RECIIS 3(1), 72–84 (2009)
11. Maedche, A., Staab, S.: Learning ontologies for the semantic web. In: SemWeb (2001)
12. Maia, L.C., Souza, R.R.: Uso de sintagmas nominais na classificação automática de documentos eletrônicos. Perspec. em Ciência da Inform. 15, 154–172 (2010)
13. Ribeiro, L.C.: OntoLP: Construção semi-automática de ontologias a partir de textos da língua portuguesa. Master's thesis, UNISINOS (2008)
14. Silva, J., Branco, A., Castro, S., Reis, R.: Out-of-the-Box Robust Parsing of Portuguese. In: Pardo, T.A.S., Branco, A., Klautau, A., Vieira, R., de Lima, V.L.S. (eds.) PROPOR 2010. LNCS, vol. 6001, pp. 75–85. Springer, Heidelberg (2010)
15. van Rijsbergen, C.J.: Information Retrieval. Butterworths, London (1975)

A Method for Automatically Extracting Domain Semantic Networks from Wikipedia

Clarissa Castellã Xavier and Vera Lúcia Strube de Lima

PPGCC – PUCRS, Av. Ipiranga, 6681 – Prédio 32, Porto Alegre, Brazil
{clarissa.xavier,vera.strube}@pucrs.br

Abstract. This paper describes a method for automatically extracting domain semantic networks of concepts connected by non-specific relations from Wikipedia. We propose an approach based on category and link structure analysis. The method consists of two main tasks: concepts extraction and relations acquisition. For each task we developed two different implementation strategies. Aiming to identify what strategies have the best performances we conducted different extractions for two domains and we analyze their results. From this evaluation we discuss the best approach to implement the extraction method.

Keywords: Wikipedia, semantic networks, knowledge acquisition.

1 Introduction

Wikipedia is a free, collaborative encyclopedia with an enormous amount of information in approximately 250 languages. Relevant research has been developed by the scientific community, regarding the extraction of semantic information from Wikipedia English version. However we note that the versions in different languages have their own contents, directly related to the culture of each of these communities[1]. For this reason it is important to exploit the material of each specific version when the goal is to obtain semantic knowledge in those languages.

In this context, our goal is to develop a method to extract semantic information from Wikipedia that works in languages other than English. Our particular focus is Portuguese. In this paper we present a method for automatically extracting from Wikipedia domain semantic networks of concepts connected by non-specific relations.

The semantic network consists of a graph where the vertices represent concepts and the edges represent the relations among then. We are not classifying the relations between the concepts. As in [1], we are labeling them with a "related-to" tag.

The proposed solution is based on the category structure and link analysis and consists of two main tasks: concepts extraction and relations acquisition. Additionally we present different strategies to implement the method and we identify what strategies achieve the best results for each task.

---

[1] For instance, see `http://en.wikipedia.org/wiki/Banda_Oriental`, `http://es.wikipedia.org/wiki/Banda_Oriental` and `http://pt.wikipedia.org/wiki/Banda_Oriental`

H. Caseli et al. (Eds.): PROPOR 2012, LNAI 7243, pp. 93–98, 2012.

This text is organized as follows. Section 2 discusses related work. Section 3 describes our extraction method. Section 4 presents strategies to implement the method, their prototyping and results. Section 5 concludes and points to future work.

2 Related Work

Wikipedia has proved to be a very interesting source for semantic information extraction. Several studies have been conducted with this purpose, quite varying the encyclopedia's components used as source for extraction, the extraction techniques applied and the types of ontologies generated. Yago[2] [2], DBpedia[3] [3] and Wikinet [4] are examples of projects that use Wikipedia as source for the production of data repositories in the form of ontological structures.

Wikipedia categories are widely explored as a semantic source [5]. Wikipedia's category structure provides a schema where the subjects are previously associated, although these associations do not have a specific nature. For these characteristics the categories have been used in studies like [2, 5, 4, 6, 7, 8, 9] as a source for data extraction.

Wikipedia articles can be interconnected through internal links. The link structure creates a network between pages which facilitates the navigation and the understanding of concepts [5], being used as a source for extraction of relations between concepts [4, 5, 7].

Infoboxes are templates that provide standardized information across related articles[4]. They are extensively used for the extraction of simple facts [1, 6, 2, 8].

The use of WordNet[5] in conjunction with Wikipedia as a data source appears in several works [1, 2, 6, 7, 8] though fitting the method to languages with robust versions of this resource.

The majority of these studies [1, 2, 5, 6, 7] extract data from Wikipedia English version. However, the multilingual nature of Wikipedia is already being explored in studies like [4, 8, 9].

A method that generates structures from Wikipedia, independent of other resources, seems to be an initiative that will foster the creation of machine-readable semantic information in languages with large demand for this type of data, as Portuguese.

3 Extraction Method

Here we detail our method to automatically extract a semantic network of related concepts from Wikipedia. Our approach does not make use of external resources enabling the extraction of semantic structures in languages as Portuguese, which do not present currently available resources like a widely populated WordNet.

---

[2] `http://www.mpi-inf.mpg.de/yago-naga/`
[3] `www.dbpedia.org`
[4] `http://en.wikipedia.org/wiki/CAT:INFOBOX`
[5] `http://wordnet.princeton.edu/`

The extraction method shown in Figure 1 is based on the category structure and link analysis and consists of two main tasks: Concepts Acquisition (CA) and Relations Acquisition (RA).

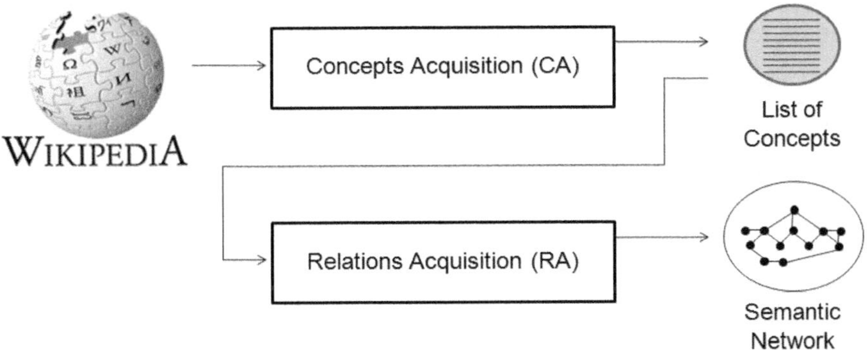

Fig. 1. Extraction method consists of two tasks: concept extraction and relations extraction

The CA task aims to acquire the concepts that populate the semantic network describing a specific domain. We assume that articles and categories names represent concepts. We respectively call source article and source category, the article and category describing the domain, for example Tourism article and category for the Tourism domain. This task is performed in two steps. The first step explores the category structure. The concepts are acquired by selecting all source category subcategories. The second step explores the article internal links.

The RA task aims to indicating how the previously extracted concepts relate to each other in the context of this selected domain. Relations between the concepts extracted in the first task are acquired through a link analysis process.

4 Implementation and Results

We propose two different implementation strategies for each task of the method. The first one is based on weak links[6] and the second one is based on strong links[7].

The CA 1st and 2nd strategies implement the 1st step selecting all the subcategories of the source category. The 2nd step in the 1st strategy selects all weak links to articles in the source article. For the 2nd strategy this step selects all strong links to articles in the source article.

The RA 1st Strategy is based on the following rule: one concept is related to another if its corresponding article or category has a link to the article or category that corresponds to the other concept. The 2nd Strategy is based on the rule: one concept is related to another if its corresponding article or category has a strong link to the article or category that corresponds to the other concept.

---

[6] A weak link exists if in page P_o there exists a link to page P_d.

[7] A page P_o has a strong link to page P_d if in P_o there exists a link to P_d and in P_d there is a link back to P_o [5].

In order to choose the best strategies to implement the method we have developed a prototype. The input is Wikipedia's database[8], particularly the tables containing data related to categories and links structure. The output is a semantic network with concepts connected by "related_to" relationships.

Table 1 presents the size of the semantic network obtained for each of the extractions for 2 different domains, in number of concepts and in number of relations.

Table 1. Number of concepts and relationships obtained for Tourism and Philosophy domains using the three mentioned implementation strategies

	Tourism		Philosophy	
	Concepts	Relations	Concepts	Relations
Extraction 1	449	6322	392	2895
Extraction 2	449	2716	392	821
Extraction 3	269	1102	285	573

4.1 Concepts Acquisition (CA)

To perform the evaluation, we have used two samples, one for each domain, namely Tourism and Philosophy. The samples are composed of 50 randomly selected concepts that are present in the network generated by the first extractions and not present in the network generated by the 3rd extraction. As 1st and 2nd extractions use the same strategy to acquire concepts, they generate the same concepts. So, we may compare the concepts in the network generated by 1st and 2nd extraction with the concepts generated by the 3rd extraction.

Each concept in the samples was manually classified as belonging or not to the domain. Table 2 presents the results of this evaluation: 78% of the removed concepts in the Tourism network and 94% of the concepts in the Philosophy network do belong to the domains. We understand that such removal compromises the quality of the generated structure because it excludes from the network concepts that belong to the domain.

We notice that the use of strong links did not increase the quality of the results. This let us conclude that the first implementation strategy for the concept acquisition task achieves better results.

Table 2. Evaluation in case the concepts and relations in the sample belong or not to the domain

	CA		RA	
	Tourism	Philosophy	Tourism	Philosophy
Yes	78%	94%	59.5%	95.5%
No	22%	6%	40.5%	4.5%

[8] Wikipedia Portuguese edition dump comes from
`http://dumps.wikimedia.org/ptwiki/20110410/`

4.2 Relations Acquisition (RA)

We have investigated the differences between 1st and 2nd extractions, regarding to RA. To perform this evaluation, we have used two samples, one for each domain. The samples are composed by relations associated with 10 randomly selected concepts. Table 2 presents the results for this evaluation.

The Philosophy sample contains 67 relationships for 10 concepts. The evaluation result was 64 correct relations and 3 incorrect relations. The Tourism sample contains 153 relationships for 10 concepts. The evaluation result was 91 correct relations and 62 incorrect relations.

In both networks more than a half of the relationships in the networks generated by the 1st extraction and not present in the networks generated by the 2nd extraction were correct. This percentage is even more representative in the Philosophy domain.

Given this results, we point out that, as well as in CA, the strategy that obtains more elements presents a better performance. Thus, the combination of the 1st strategy to CA and the 1st strategy to RA could be chosen to implement the extraction method with better results.

5 Conclusion

We have presented an automatic approach for extracting domain semantic networks of concepts connected by non-specific relations from Wikipedia. This method consists of two main tasks: concepts extraction and relations acquisition.

For each task we have proposed two different implementation strategies. To determine what strategies have the best performance we conducted three extractions, each with one combination of strategies. We have extracted networks for two domains: Tourism and Philosophy.

Considering this first study, we may conclude that the strategies that extract the highest number of concepts and relations obtain the best results, i.e., the largest amount of relevant information about the domain. Apparently, the use of strong links did not increase the quality of the result.

The prototype was used to perform the evaluation of the extracted semantic networks from the Portuguese version of Wikipedia. Despite working with Portuguese, our method is language independent and can be used in any version of the encyclopedia.

As future work, we plan to perform more accurate evaluations of the method, including assessment in different languages. We also intend to develop strategies to label the semantic relations, i.e. discover the nature of the concepts relationship in the network.

References

1. Szumlanski, S.R., Gomez, F.: Automatically acquiring a semantic network of related concepts. In: Huang, J., Koudas, N., Jones, G.J.F., Wu, X., Collins-Thompson, K., An, A. (eds.) CIKM, pp. 19–28. ACM (2010)
2. Suchanek, F., Kasneci, G., Weikum, G.: YAGO: A Large Ontology from Wikipedia and WordNet. Web Semantics Science Services and Agents on the World Wide Web 6(3), 203–217 (2008)

3. Auer, S., Bizer, C., Kobilarov, G., Lehmann, J., Cyganiak, R., Ives, Z.: DBpedia: A Nucleus for a Web of Open Data. In: Aberer, K., Choi, K.-S., Noy, N., Allemang, D., Lee, K.-I., Nixon, L.J.B., Golbeck, J., Mika, P., Maynard, D., Mizoguchi, R., Schreiber, G., Cudré-Mauroux, P. (eds.) ASWC 2007 and ISWC 2007. LNCS, vol. 4825, pp. 722–735. Springer, Heidelberg (2007)
4. Nastase, V., Strube, M., Boerschinger, B., Zirn, C., Elghafari, A.: Wikinet: A very large scale multilingual concept network. In: Proceedings of the Seventh Conference on International Language Resources and Evaluation (LREC 2010), Valletta, Malta (2010)
5. Fogarolli, A.: Wikipedia as a Source of Ontological Knowledge: State of the Art and Application. In: Caballé, S., Xhafa, F., Abraham, A. (eds.) Intelligent Networking, Collaborative Systems and Applications. Studies in Computational Intelligence, vol. 329, pp. 1–26. Springer, Heidelberg (2010)
6. Syed, Z., Finin, T.: Unsupervised techniques for discovering ontology elements from Wikipedia article links. In: Proceedings of the NAACL HLT 2010 First International Workshop on Formalisms and Methodology for Learning by Reading (FAM-LbR 2010), pp. 78–86. Association for Computational Linguistics, Stroudsburg (2010)
7. Navigli, R., Ponzetto, S.P.: BabelNet: building a very large multilingual semantic network. In: Proceedings of the 48th Annual Meeting of the Association for Computational Linguistics (ACL 2010), pp. 216–225. Association for Computational Linguistics, Stroudsburg (2010)
8. de Melo, G., Weikum, G.: MENTA: inducing multilingual taxonomies from wikipedia. In: Huang, J., Koudas, N., Jones, G.J.F., Wu, X., Collins-Thompson, K., An, A. (eds.) CIKM, pp. 1099–1108. ACM (2010)
9. Xavier, C.C., de Lima, V.L.S.: A Semi-automatic Method for Domain Ontology Extraction from Portuguese Language Wikipedia's Categories. In: da Rocha Costa, A.C., Vicari, R.M., Tonidandel, F. (eds.) SBIA 2010. LNCS, vol. 6404, pp. 11–20. Springer, Heidelberg (2010)

Extracting Temporal Information from Portuguese Texts

Francisco Costa and António Branco

University of Lisbon
{fcosta,Antonio.Branco}@di.fc.ul.pt

Abstract. This paper reports on experimenting with the extraction of temporal information from Portuguese texts and presents LX-TimeAnalyzer, a tool that annotates a text with the temporal information conveyed by it. This tool is the first of its kind being reported for Portuguese, and its performance is similar to the state-of-the-art for other languages.

1 Introduction and Related Work

Extracting the temporal information present in a text is relevant to many Natural Language Processing applications, including question-answering, information extraction, and even document summarization, as summaries may be more readable if the information is presented in chronological order.

The two recent TempEval challenges [9,10] focused on extracting the temporal information conveyed in written text and provided data that can be used to develop and evaluate systems that can automatically annotate a natural language text with the temporal information conveyed in it. Figure 1 shows an example of similarly annotated data.

```
<s>Em Washington, <TIMEX3 tid="t53" type="DATE"
value="1998-01-14">hoje</TIMEX3>, a Federal Aviation Administration <EVENT
eid="e1" class="OCCURRENCE" stem="publicar" aspect="NONE" tense="PPI"
polarity="POS" pos="VERB">publicou</EVENT> gravações do controlo de tráfego
aéreo da <TIMEX3 tid="t54" type="TIME" value="1998-XX-XXTNI">noite</TIMEX3>
em que o voo TWA800 <EVENT eid="e2" class="OCCURRENCE" stem="cair"
aspect="NONE" tense="PPI" polarity="POS" pos="VERB">caiu</EVENT>.</s>
<TLINK lid="l1" relType="BEFORE" eventID="e2" relatedToTime="t53"/>
<TLINK lid="l2" relType="OVERLAP" eventID="e2" relatedToTime="t54"/>
```

Fig. 1. Sample of Portuguese data with temporal annotations, corresponding to the fragment: *Em Washington, hoje, a Federal Aviation Administration publicou gravações do controlo de tráfego aéreo da noite em que o voo TWA800 caiu.*
The English equivalent is: *In Washington today, the Federal Aviation Administration released air traffic control tapes from the night the TWA Flight eight hundred went down.*

H. Caseli et al. (Eds.): PROPOR 2012, LNAI 7243, pp. 99–105, 2012.

Terms denoting events, such as the event of releasing the tapes that is described in that text, are annotated using `EVENT` tags, and temporal expressions, such as *today*, are enclosed in `TIMEX3` tags. The attribute `value` of time expressions holds a normalized representation of the date or time they refer to (e.g. the word *today* denotes the date `1998-01-14` in this example). The `TLINK` elements at the end describe temporal relations between events and temporal expressions. For instance, the event of the plane going down is annotated as temporally preceding the date denoted by the temporal expression *today*.

The first TempEval challenge focused solely on the temporal relations. TempEval-2 additionally included tasks related to the identification and normalization of event terms and temporal expressions. Identification is concerned with classifying words in a text as to whether they are event terms or part of temporal expressions or none of these. Normalization is about determining the value of the various attributes of `EVENT` and `TIMEX3` elements, specially the `value` attribute of `TIMEX3` elements. By combining the outcome of all these tasks, it is possible to fully annotate raw text with temporal information (event terms, temporal expressions and temporal relations) in a way similar to what is shown in the example above. Table 1 shows the scores obtained by the best participant for each of these problems. The evaluation measures used were the f-measure for the problems of identifying the extents of event and time expressions and accuracy for the tasks dealing with the attributes. Full details can be found in [10].

Table 1. Best system results for the various tasks of TempEval-2, according to [10]

Temporal expressions			Events		
Task	English	Spanish	Task	English	Spanish
Extents	0.86	0.91	Extents	0.83	0.88
`type`	0.98	0.99	`class`	0.79	0.66
`value`	0.85	0.83	`tense`	0.92	0.96
			`aspect`	0.98	0.89
			`polarity`	0.99	0.92

2 Approach and Evaluation

The data that was used for the first TempEval has recently been adapted to Portuguese, as reported in [3]. The documents that make up this corpus were translated to Portuguese, and the annotations adapted to the language. The fragment presented above in Figure 1 is taken from this corpus. The training subset contains 68,351 words, 6,790 events, 1,244 temporal expressions and 5,781 temporal relations.

These data allow for the training and evaluation of temporal processing systems for Portuguese. In Table 2 we include information about the performance of our system LX-TimeAnalyzer, evaluating each subtask that was evaluated in TempEval-2 (with the exception of temporal relation classification, which is reported in [2,4]). We use the same evaluation measures as in TempEval-2 (f-measure for extent identification and accuracy for the tasks dealing with the attributes). It must be noted that: (i) the Portuguese data are an adaptation of the English data used in the first TempEval, (ii) the results in Table 1 refer to TempEval-2, (iii) the English data of TempEval and TempEval-2 are not identical, although there is a large overlap between them. For the data of the first TempEval there are unfortunately no published results that we know of concerning the identification and normalization of temporal expressions and event terms, as TempEval-1 focused only on temporal relations. It is thus important to note that our results are fully not comparable to the results for English (and they are even less comparable to the results for Spanish, as those are based on completely different data).

Table 2. Evaluation of LX-TimeAnalyzer on the test data

Temporal expressions		Events	
Task	Score	Task	Score
Extents	0.85	Extents	0.72
`type`	0.91	`class`	0.74
`value`	0.81	`tense`	0.95
		`aspect`	0.96
		`polarity`	0.99

The document to be processed is initially tagged with a morphological analyzer [1]. This tool annotates each word with its part-of-speech category (noun, verb, etc.), its lemma (i.e. its dictionary form), and a tag describing its inflection features.

For the tasks we addressed via machine learning techniques, we employed Weka's [11] implementation of the C4.5 algorithm, using the training data for training and the held-out test data for evaluation.

2.1 Event Identification and Normalization

A simple solution to identifying event terms in text is to classify each word as to whether it denotes an event or not. This strategy is not very efficient, since (i) some very frequent words cannot possibly denote events (e.g. determiners, conjunctions etc.), and (ii) most event terms are verbs or nouns (92% according to the training data). Nevertheless, for the sake of reproducibility, we followed this straightforward approach.

The classifier features employed are:

- **Features about the Last Characters of the Lemma**
  A Boolean attribute represents whether the lemma ends in one of several word endings from a hand-crafted list. This list includes suffixes such as

-mento. The motivation is that this information may be useful especially to separate eventive nouns from non-eventive nouns. There are additional attributes that include information about the last two characters of the lemma and the last three characters of the lemma; they are intended to capture suffixes not covered by the list of suffixes.
- **The Part-of-Speech and the Inflection Tag Assigned by the Tagger** As argued above, information about part-of-speech can rule out many words in a document. The inflection tag may also be relevant. For instance, even though singular forms are more common than plural forms for both eventive and non-eventive nouns, this difference is sharper in the case of eventive nouns (since these denote multiple or repeated events).
- **The Part-of-Speech and the Inflection Tag of the Preceding Word Token, the Following Word Token, the Preceding Word Token Bigram, the Following Word Token Bigram**
  These attributes are used in order to capture some contextual information.
- **Whether the Preceding Token was Classified as an Event**
  The intuition is that adjacent event terms are infrequent.

Our result for this task (0.72 f-measure) is worse than the best systems of TempEval-2 for both English (0.83) and Spanish (0.88).

We believe that the major cause of this differences is that these systems used features based on WordNet, which we were unable to experiment with as there is no available WordNet for Portuguese verbs.

The task of event normalization is concerned with the annotation of the several attributes appropriate for `<EVENT>` elements. The values of many of the attributes of `<EVENT>` elements are already provided by the morphological analyzer: `stem` (the term's dictionary form), `tense` (tense) and `pos` (part-of-speech). Three attributes are not, however: `aspect`, `polarity` and `class`.

For the `polarity` attribute, we simply check whether one the three preceding words is a negative word—*não* "not", *nunca* "never", *ninguém* "nobody", *nada* "nothing", *nenhum/nenhuma/nenhuns/nenhumas* "no, none", *nenhures* "nowhere"— and there is no other event intervening between this n-word and the event that is being annotated. The accuracy for this heuristic is 0.99, considering all annotated events in both the training and the test data. On the test data, the accuracy of this simple heuristic is also 0.99, which is identical to the best score in TempEval-2 for English (0.99) and better than the one for Spanish (0.92).

In the Portuguese data, the attribute `aspect` only encodes whether the verb form is part of a progressive construction. This attribute is also computed symbolically, and the implementation simply checks for gerund forms (e.g. *fazendo*) or constructions involving an infinite verb form immediately preceded by the preposition *a* (*a fazer*). Once again considering all the data (both training and testing data), this approach has an accuracy of 0.99. On the evaluation data, its accuracy is 0.96, in between the TempEval-2 best scores for English (0.98) and Spanish (0.89).

The most interesting and hardest problem of event normalization is determining the value of the `class` attribute of `<EVENT>` elements. This attribute includes some information about the semantic class of event terms, distinguishing `REPORTING`, `PERCEPTION` and `ASPECTUAL` terms from the others, and also includes some aspectual distinctions in the spirit of [8,5], distinguishing `STATE` situations from non-stative events, marked as `OCCURRENCE`s. It is thus sensitive to both lexical and contextual (i.e. syntactic) information. For this attribute, a specific classifier was trained, with a very limited set of features:

- **The Lemma of the Event Term Being Classified**
  This type of information is highly lexicalized, so it is expected that the lemma of the word token can be quite informative.
- **Contextual Features**
  These attributes encode the part-of-speech of the previous word and that of the next word, and the following bigram of inflection tags.

We experimented with more features, similar to the ones used for event detection, but they did not improve the results. We obtained a result of 0.74.

2.2 Temporal Expression Identification and Normalization

In order to identify temporal expressions, we trained a classifier that, to each word in the text, assigns one of three labels: `B` (begin), `I` (inside), `O` (outside). The features employed were:

- **Features about the Current Token**
  These include the token's part-of-speech and its inflection tag. Additionally, there is an attribute that checks whether the current token's lemma is part of a list of temporal adverbs. This is specially useful for the `B` class, which is the one with the highest error rate.
- **Features about the Previous Token and the Following One**
  These features are taken from the morphological analyzer and encode part-of-speech and inflection tag.
- **The Classification for the Previous Token**
  Tokens classified as `I` cannot directly follow tokens classified as `O`.
- **Whether There Is White Space Before the Current Token and the Previous One**
  The reason behind this attribute is to treat punctuation and special symbols in a special manner (they are tokenized separately; e.g. a time expression of the form `XXXX-XX-XX` is tokenized into five word tokens).
- **Whether (i) the Current Token's Lemma was Seen in the Training Data at the Beginning of a Temporal Expression, or (ii) It was Seen inside a Temporal Expression, or (iii) the Bigram of Lemmas Formed by the Current Token's Lemma and the Next One's was Seen inside a Temporal Expression**
  Instead of using an attribute encoding the lemma directly, we used a series of Boolean attributes capturing distinctions that are expected to help classification.

As shown in Table 2, this component shows an f-measure of 0.85 for the `B` and `I` classes.

The task of temporal expression normalization consists in identifying the value of the `TIMEX3` attributes `type` and `value`. LX-TimeAnalyzer solves it symbolically. The normalization rules take as input the following parameters:

- The word tokens composing the temporal expression, and their morphological annotation
- The document's creation time
- An anchor. This is another temporal expression that is often required for normalization. An expression like *the following day* can only be normalized if its anchor is known. We use the previous temporal expression that occurs in the same text and that is not a duration, a simple heuristic similar to previous approaches found in the literature.
- The broad tense (*present*, *past*, or *future*) of the closest verb in the sentence where it occurs, with the distance being measured in number of word tokens from either boundary of the time expression. For example, all past tenses are treated as *past*. This is used to decide whether an expression like *February* refers to the previous or the following month of February (relative to the document's creation time).

These rules are implemented by a Java method. It takes approximately 1600 lines of code and is recursive: e.g. when normalizing an expression like *terça de manhã* "Tuesday morning", the expression *terça* "Tuesday" is normalized first, and then its normalized `value` is changed by appending `TMO` (with `T` being the time separator and `MO` the way to represent the vague expression "morning"); its `type` is also changed from `DATE` to `TIME`. The same method fills in both the `value` and the `type` attributes of `TIMEX3` elements. This implementation was conducted by looking at the examples in the training data, and additionally at a small set (c. 5000 words) of news reports taken from on-line newspapers.

The accuracy of LX-TimeAnalyzer at predicting the value of the `value` attribute of `TIMEX3` elements is 0.81 on the test data. For the `type` attribute this is 0.91.

3 Concluding Remarks

Full temporal information processing is fairly recent. Only in the TempEval-2 challenge, last year in 2010, were there systems capable of fully annotating raw text with temporal information (e.g. [7,6]).

LX-TimeAnalyzer is the first fully-fledged temporal analyzer for Portuguese. It performs in line with the state-of-the-art for other languages, although (i) the data used for evaluation are not fully comparable, and (ii) event detection is somewhat worse, but can possibly be improved by incorporating information similar to that in WordNet.

References

1. Branco, A., Silva, J.: A suite of shallow processing tools for portuguese: LX-Suite. In: Proceedings of the 11th Conference of the European Chapter of the Association for Computational Linguistics (EACL 2006), Trento, Italy (2006)
2. Costa, F.: Processing Temporal Information in Unstructured Documents. Ph.D. thesis, Universidade de Lisboa, Lisbon (to appear)
3. Costa, F., Branco, A.: Temporal information processing of a new language: Fast porting with minimal resources. In: Proceedings of the 48th Annual Meeting of the Association for Computational Linguistics, ACL 2010 (2010)
4. Costa, F., Branco, A.: LX-TimeAnalyzer: A temporal information processing system for Portuguese. Tech. rep., Universidade de Lisboa, Faculdade de Ciências, Departamento de Informática (to appear)
5. Dowty, D.R.: Word Meaning and Montague Grammar: the Semantics of Verbs and Times in Generative Semantics and Montague's PTQ. Reidel, Dordrecht (1979)
6. Llorens, H., Saquete, E., Navarro, B.: TIPSem (English and Spanish): Evaluating CRFs and semantic roles in TempEval-2. In: Erk, K., Strapparava, C. (eds.) Proceedings of the 5th International Workshop on Semantic Evaluation, SemEval 2010, pp. 284–291. Uppsala University, Uppsala (2010)
7. UzZaman, N., Allen, J.F.: TRIPS and TRIOS System for TempEval-2: Extracting temporal information from text. In: Erk, K., Strapparava, C. (eds.) Proceedings of the 5th International Workshop on Semantic Evaluation, SemEval 2010, pp. 276–283. Uppsala University, Uppsala (2010)
8. Vendler, Z.: Verbs and times. In: Linguistics in Philosophy, pp. 97–121 (1967)
9. Verhagen, M., Gaizauskas, R., Schilder, F., Hepple, M., Pustejovsky, J.: SemEval-2007 Task 15: TempEval temporal relation identification. In: Proceedings of SemEval 2007 (2007)
10. Verhagen, M., Saurí, R., Caselli, T., Pustejovsky, J.: SemEval-2010 task 13: TempEval-2. In: Strapparava, C., Erk, K. (eds.) Proceedings of the Workshop 5th International Workshop on Semantic Evaluation, SemEval 2010, pp. 51–62. Uppsala University, Uppsala (2010)
11. Witten, I.H., Frank, E.: Data Mining: Practical Machine Learning Tools and Techniques with Java Implementations, 2nd edn. Morgan Kaufmann, San Francisco (2005)

It Is the Time for Portuguese Texts!

Olga Craveiro[1,2], Joaquim Macedo[3], and Henrique Madeira[2]

[1] School of Technology and Management, Polytechnic Institute of Leiria, Portugal
[2] CISUC, Department of Informatics Engineering, University of Coimbra, Portugal
{marine,henrique}@dei.uc.pt
[3] Department of Informatics, University of Minho, Portugal
macedo@di.uminho.pt

Abstract. In this work, we introduce a software testbed for temporal processing of Portuguese texts, composed by several building blocks: identification, classification and resolution of temporal expressions and temporal text segmentation. Starting from a simple document, we can reach a set of temporally annotated segments, which enables the establishment of relationships between words and time. This temporally enriched information is then placed into an Information Retrieval system. This work represents a step forward for Portuguese language processing, with notorious lack of tools. Its main novelty is temporal segmentation of texts. Even with target application in temporal aware Information Retrieval, the described software tools can be used in other application scenarios.

1 Introduction

Time is an important dimension in understanding the text information. However, there is still much to do to achieve its full integration in the most popular retrieval models [1]. Our focus is the design, implementation and evaluation of a temporal aware Information Retrieval (IR) models.

As we work also with Portuguese text collections, our first concern was to find the available tools that allow us to reach as soon as possible the focused subject related tasks. However, with almost an inexistence of temporal information extraction tools from Portuguese texts, we decided to develop a system from scratch.

We had to start by identifying the temporal expressions in the text. Later it was necessary classify them, in order to facilitate their resolution into standard dates, where possible. Thus, we have a series of *chronons* associated with each text. A *chronon* is a normalized date which is anchored in a calendar/clock system. Finally, and as our aim is to associate the time with the words that describe entities and events or facts, we decided to do the temporal segmentation of the text. The result is a set of text segments, each with one or more dates, used to create time enriched indexes.

In the following, we present related work, with special emphasis on Portuguese language processing. Subsequently, we present the testbed software architecture, with a concise description of the components and their relationships. At the end, we have a more detailed evaluation and the conclusions and further work.

H. Caseli et al. (Eds.): PROPOR 2012, LNAI 7243, pp. 106–112, 2012.

2 Related Work

There are several tools for extraction of temporal expressions from English texts. Most of them process term by term, using linguistic features for their identification [2]. An annotation scheme, using finite state automata, represents dates and times based on a diverse set of manual defined and automatically discovered rules [3]. In [4], it is described a different approach based on Context-Scanning Strategy. The TARSQUI is also a very popular toolkit [5]. There are several temporal expressions resolution strategies [3,6]. Unfortunately, these strategies or tools are not directly applicable to the Portuguese language. For Portuguese, the temporal information extraction area is not yet focus of significant amount of work. We found only the XTM tool, recently developed by Hagège et al. [7]. It is rule-based and processes word by word, using a deep analysis approach to extract temporal information from texts. XTM is not available in the public domain.

In terms of topic segmentation of texts, the literature is quite extensive [8-9]. But, about the temporal one it is very limited. To the best of our knowledge, there is no research work focused on Portuguese text temporal segmentation. Bramsen [10] also worked on the temporal segmentation of texts. However, despite being mainly interested in the chronological order of the segments, the application domain is the clinical narratives.

In temporal extraction, the novelty of our proposal is to be supported by the existence of lexical patterns generated automatically from Portuguese texts. It also uses an inductive approach that starts from the data to the knowledge, unlike above referenced work. The distinction of our segmentation approach is the division of the text into temporally coherent units. It is not concerned with segment chronological order, but with association between segments and accurate dates, for later use of this information. The use of a rule based algorithm, instead of machine learning one, is justified by the lack of suitable Portuguese training collections.

3 Testbed System Architecture

The objective of our work is the temporal enrichment of the IR system. So, we want to establish a relationship between words and time. We are building up a testbed system which extracts temporal information from Portuguese texts. Despite we are used the four modules together, each one can be used individually. It can be used for instance for temporal characterization of a text collection. In testbed system design, we are searching for a tradeoff between simplicity and efficiency, while maintaining a suitable effectiveness level.

In the Fig. 1 is shown a tool used to extract relevant temporal information from Portuguese documents which is composed by three modules: Co-Occurrence Processor (henceforth COP), Annotator and Resolver. The recognition task is carried out by the COP and the Annotator modules. The interpretation and normalization of the temporal expressions into *chronons*, normalized dates which are anchored in a calendar/clock system, are performed by the Resolver module (see section 4).

Fig. 2 shows the module for the temporal segmentation of the Portuguese texts. Based on content and metadata temporal information of documents, this single module partitions texts into temporally coherent segments, like in topical segmentation. These segments are also tagged with timestamps which are the *chronons* collected from the text segment (see section 5).

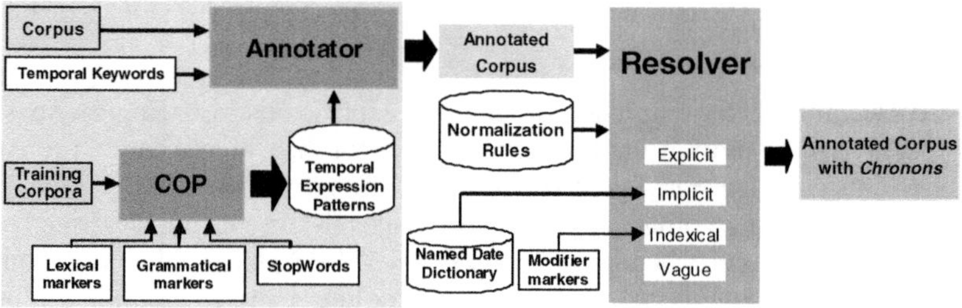

Fig. 1. System Architecture for Temporal Information Extraction

4 Temporal Expressions Extraction

Due to space limitations, we could not include the details of our extraction tool which are reported in [11,12]. A brief description of their modules is presented below.

The temporal information extraction task is divided into recognition and resolution of temporal expressions. The identification and semantic classification of temporal expressions are in the recognition part. Temporal expressions are classified as date, hour, interval, duration, and frequency, following the guidelines defined in [13].

Our recognition method is based on a two-stage approach, each stage being carried out by a different module (see Fig. 1). Firstly, the COP module produces semantically classified temporal patterns, based on regular expressions. The patterns are created using word co-occurrences, determined from training corpora and a pre-defined set of seed keywords derived from the used language temporal references. These reference words are divided in lexical [1] and grammatical [2] markers. An example of the COP output (pattern, classification) is: (*In the (last | following)* ***year***, DATE).

Then, these patterns are used by the Annotator module to annotate Portuguese temporal expressions. The Annotator processes each sentence to determine whether it matches any of the temporal patterns, and, if so, the sentence is annotated with semantic classification corresponding to the matching pattern in the original text. An example of the Annotator:

```
I run <EM ID="2" CATEG="TIME" TYPE="FREQUENCY">every day</EM>.
```

The resolution comprises interpretation and normalization of the temporal expressions. The interpretation of temporal expressions consists in the inference of a new date, using information from document. The normalization is the transformation of the dates into a standard format. Our definition of temporal expressions as explicit, implicit, relative and vague is supported by Schilder et al. [6]. Relative references are expressions that need one point in time to be completely resolved and anchored in calendar/clock system and can be classified as *deictic timexes* or *anaphoric timexes* [14].

The Resolver module maps temporal expressions found in documents' content into a discrete representation of time, denoted *chronons* by Alonso [15]. A *chronon* is a normalized date which is anchored in a calendar/clock system by timelines defined by points.

---

[1] Some examples: months, seasons, weekdays, Christmas/*Natal*, day/*dia*, today/*hoje*, ...
[2] Some examples: prepositions {in/*em*, since/*desde*, ...}, ordinal adjectives {next/*próximo*, ...}.

Each timeline has a different level of granules, such as year, month, week, day and hour. This module relies on a set of rules, used to interpret temporal expressions previously annotated by the Annotator module. It starts with the document timestamp normalization, a time related metadata, such as the creation or publication data of the document. This date is used in *deictic timexes* resolution. In the *anaphoric timexes* resolution is used a date evoked in the text. Since indexical references can mention a past, present or future event, it is also required the modifiers of these references. The correct rule to be applied is chosen by these modifiers, such as *next*, *after*. For example, *next year* is resolved with the rule *"document timestamp + 1 time_Unit(y)"*.

Our system has also a named date dictionary used to resolve the implicit references. For example, *Christmas Day 2011* is normalized as *2011-12-25*.

5 Temporal Segmentation of Text

Our segmentation algorithm makes use of temporal information extracted from the text to divide text into temporally coherent segments. These segments are tagged with a timestamp to obtain an association between time and document terms. A temporal segment is defined as a set of adjacent sentences that shares the same temporal focus. The segment length ranges from a single sentence to a multi-paragraph text. Thus, adjacent sentences with the same *chronons* must belong to the same segment. Each segment is tagged with all the different *chronons* found in their sentences. The document timestamp is also associated to each segment.

Some examples are presented below. The segment of the first example is composed by two sentences. The second sentence will also belong to this segment, since the topic is the same. The other example shows a segment tagged with two *chronons*, because these two normalized date are in the same sentence.

(1) <SEGMENT DN="2011-10-31">**Sunday**'s storm caused some problems in electricity networks. The company of electricity received about 31,000 calls.</SEGMENT>

(2) <SEGMENT DN="2011-11-10 2011-11-11">It rained on **Friday** and **Saturday**.</SEGMENT>

Fig. 2 shows the temporal segmentation module. The segmentation process begins at the sentence level. Each sentence is a candidate to start a new segment and it is compared with the *current segment*, composed by the previous adjacent sentences of the text. There are two approaches defined by the temporal information analysis of the sentence: sentence with *chronons* and sentence without *chronons*.

If the *chronons* of the sentence are equal to the *chronons* of the *current segment*, the sentence must belong to this segment; otherwise, this sentence starts a new segment.

If the sentence has no *chronon,* a temporal mark in the three first words of the sentence determine if the sentence starts or not a segment. These marks are used to express temporal relation between successive actions or events and can signal the topic continuity or discontinuity [16]. Thus, if the sentence has a continuity marker, e. g. *and*, it remains within the *current segment*. If the marker expresses discontinuity, e.g. *next*, this sentence starts a new segment. If there is not any marker in the sentence, the sentence must be processed with the similarity calculation. Before this calculation, the sentence and the *current segment* are pre-processed for removing punctuation marks and stopwords. Based on the vector space model, the topic cohesion of the sentence and the *current segment* is computed using the cosine measure, according to the approach used in some topic segmentation methods [17]. A threshold value is also defined to decide if the sentence starts or not a new segment.

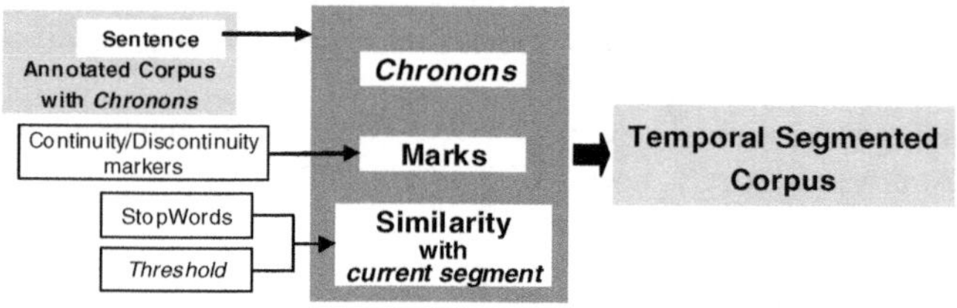

Fig. 2. System Architecture for Temporal Segmentation

6 Evaluation and Results

In this evaluation, we intended to verify the effectiveness of the system. Indeed, such evaluation has become an hard task, namely for the Temporal Segmentation module as we explain below. Each module was evaluated using a Portuguese text collection. Unfortunately, due to space limitations, we could not include the details of temporal information extraction modules, namely COP, Annotator and Resolver, which are reported in [11-12].

The collections used in our experiments are subsets of the Second HAREM Collection[3] (henceforth HC), properly detailed in [13]. Since there is no Portuguese collection for temporal segmentation we created two HC subsets: Temporal Segmentation Training Set, composed by 4 documents with 82 sentences and 1,195 words, and, Temporal Segmentation Evaluation Set which has 28 documents with 401 sentences and 6,678 words.

The evaluation of the Temporal Segmentation module was a complex task. The selection of the reference segmentation is difficult since the detection of the boundaries of topics involves an inherent subjectivity. In order to solve this difficulty we compared our algorithm against a manual segmentation corpus based on human judgments. Since human judges do not always agree, we measured the agreement between judges removing the probability of chance agreement using a commonly used measure, *Kappa coefficient.*

The documents of the corpus were manually annotated and segmented by two human judges. We observed an agreement of 0.91 and the *Kappa* value was equal to 0.82. So, the corpus is appropriated for the evaluation [18].

For this evaluation, our algorithm was implemented in Perl Language and named as *Time4Word* (henceforth *T4W*). The *T4W* input are an annotated and normalized corpus, a list of continuity[4] and discontinuity[5] marks, a stopwords list composed by prepositions, conjunctions, articles and pronouns of the Portuguese language, and, a threshold value for the similarity measure, as is shown in Fig. 2.

---

[3] Available at Linguateca site: http://www.linguateca.pt/HAREM.

[4] Some examples: e/*and*, também/*also*, (n)este/*(in) this*, (n)esse/*(in) that*, eles/*they*, *(…)*.

[5] Some examples: após/*after*, antes/*before*, depois/*next*, mais tarde/*later*, então/*then*, *(…)*.

Besides the traditional measure of accuracy, we decided to use also the WindowDiff (*WD*) measure. This measure, a variation of the *Pk* metric, was proposed in [19] as the suitable measure for the evaluation of the text segmentation tasks. *T4W* was evaluated considering not only the boundaries of the segment but also the timestamp of the segment. As the segment length is variable, the width of window (*k*) used in the calculation of *WD* was set to the average of the segment length in the reference segmentation, using the sentence as the unit. The average of the segment length in the evaluation set is 1.18, *k* was set to 1. We made several evaluations varying the threshold of the cosine similarity between 0.01 and 0.35. Table 1 shows the minimum and the maximum values of *WD*. The best result (0.15) was obtained when the threshold was set to 0.04. As the threshold increases, the number of false positives increases as well.

Since a segment timestamp can have more than one *chronon*, we calculated the agreement, overlapping and disagreement, analyzing the matching between the timestamps of the segments obtained by *T4W* and the reference segmentation. Table 1 shows such values considering the same variation of the cosine similarity threshold. Indeed, the difference between the minimum and maximum values is not very significant.

The results obtained show a good effectiveness, even with some limitations. Although the accuracy was about 78%, the *WD* was not so penalized (0.*15*). This means that some incorrect boundaries of the segment are within the *k*-sentence window used by *WD*. The incorrect boundaries of the segments have been determined in sentences without dates and where it was applied the similarity calculation. The use of synonyms and a stemmer will certainly improve these results.

Table 1. Agreement, overlapping and disagreement of the segment timestamp

WD, *k=1*	Agreement	Overlapping	Disagreement
0.15 – 0.193	76% - 79%	1.75% - 2%	19.5% - 22.5%

7 Conclusion and Future Work

We introduced a software testbed for temporal processing of Portuguese texts. Even with a set of limitations and simplifications, our system has shown promising results in the effectiveness evaluation of each module (a precision in the range of 78%-84% and a recall in the range of 64%-75%).

The main contribution of this paper is the original text segmentation method which uses temporal information of documents (metadata and contents), for divide the text into temporal coherent parts, allowing a relationship between words and time which will be used to enrich a temporal aware IR index. Supported by a simple rule-based algorithm, despite the auspicious result of 0.15 for *WD*, the method can be improved using phrases as minimal segment size and, for instance, thesaurus and stemming for topic change detection.

References

1. Alonso, O., Strötgen, J., Baeza-Yates, R., Gertz, M.: Temporal Information Retrieval: Challenges and Opportunities. In: 1st International Temporal Web Analytics Workshop (TWA-WWW 2011), pp. 1–8 (2011)

2. Mani, I.: Recent developments in temporal information extraction. In: RANLP, Borovets, Bulgaria, pp. 45–60 (2003)
3. Mani, I., Wilson, G.: Robust temporal processing of news. In: 38th Annual Meeting on Association for Computational Linguistics, Morristown, NJ, USA, pp. 69–76 (2000)
4. Vazov, N.: A system for extraction of temporal expressions from French texts based on syntactic and semantic constraints. In: ACL 2001 Workshop on Temporal and Spatial Information Processing, Toulouse, France (2001)
5. Verhagen, M., Pustejovsky, J.: Temporal processing with the TARSQI toolkit. In: COLING, ACL, Morristown, USA, pp. 189–192 (2008)
6. Schilder, F., Habel, C.: From temporal expressions to temporal information: Semantic tagging of news messages. In: ACL 2001 Workshop on Temporal and Spatial Information Processing, Toulouse, France, pp. 65–72 (2001)
7. Hagège, C., Baptista, J., Mamede, N.J.: Caracterização e processamento de expressões temporais em português. Linguamática 2(1), 63–76 (2010)
8. Misra, H., Yvon, F., Jose, J.M., Cappe, O.: Text segmentation via topic modeling: an analytical study. In: CIKM 2009, pp. 1553–1556. ACM, New York (2009)
9. Misra, H., Yvon, F., Cappé, O., Jose, J.: Text segmentation: a topic modeling perspective. Information Processing and Management 47(4), 528–544 (2011)
10. Bramsen, P., Deshpande, P., Lee, Y.K., Barzilay, R.: Finding temporal order in discharge summaries. In: AMIA 2006, Washington DC, USA, pp. 81–85 (2006)
11. Craveiro, O., Macedo, J., Madeira, H.: Use of Co-occurrences for Temporal Expressions Annotation. In: Karlgren, J., Tarhio, J., Hyyrö, H. (eds.) SPIRE 2009. LNCS, vol. 5721, pp. 156–164. Springer, Heidelberg (2009)
12. Craveiro, O., Macedo, J., Madeira, H.: Leveraging temporal expressions for segmented-based information retrieval. In: ISDA, pp. 754–759. IEEE (2010)
13. Mota, C., Santos, D. (eds.): Desafios na avaliação conjunta do reconhecimento de entidades mencionadas: O Segundo HAREM. Linguateca (2008)
14. Ahn, D., Adafre, S.F., de Rijke, M.: Extracting temporal information from open domain text: A comparative exploration. In: DIR 2005, pp. 3–10 (2005)
15. Alonso, O., Gertz, M., Baeza-Yates, R.: Clustering and exploring search results using timeline constructions. In: CIKM 2009, pp. 97–106. ACM, New York (2009)
16. Bestgen, Y., Vonk, W.: The role of temporal segmentation markers in discourse processing. Discourse Processes 19, 385–406 (1995)
17. Hearst, M.A.: Multi-paragraph segmentation of expository text. In: 32nd Annual Meeting on Association for Computational Linguistics, pp. 9–16. ACL (1994)
18. Carletta, J.: Assessing agreement on classification tasks: the kappa statistic. Computational Linguistics 22, 249–254 (1996)
19. Pevzner, L., Hearst, M.A.: A critique and improvement of an evaluation metric for text segmentation. Computational Linguistics 28, 19–36 (2002)

A Large Portuguese Corpus On-Line: Cleaning and Preprocessing

Michel Généreux, Iris Hendrickx, and Amália Mendes

Centro de Linguística da Universidade de Lisboa,
Av. Prof. Gama Pinto, 2,
1649-003 Lisboa, Portugal
{genereux,iris,amalia.mendes}@clul.ul.pt
http://www.clul.ul.pt

Abstract. We present a newly available on-line resource for Portuguese, a corpus of 310 million words, a new version of the Reference Corpus of Contemporary Portuguese, now searchable via a user-friendly web interface. Here we report on work carried out on the corpus previous to its publication on-line. We focus on the processes and tools involved for the cleaning, preparation and annotation to make the corpus suitable for linguistic inquiries.

Keywords: Corpus, Cleaning, Linguistic Preprocessing.

1 Introduction

The aim of this paper is to present our work in preparing a large Portuguese corpus, the Reference Corpus of Contemporary Portuguese (CRPC[1]), into a suitable format for on-line publication and the enrichment of the corpus with automatically assigned pos-tags and lemmas. We hope that sharing our experience in preparing a Portuguese corpus for on-line querying can be of genuine general interest, given the predominance of platforms designed and developed mostly for English. The approaches, practices and techniques described are not novel, although we present them in such a way as to underline key points and potential pitfalls. Language technologists engaged in preparing and publishing corpora on-line may find here some useful insights. The CRPC [3] has been developed at the *Centro de Linguística da Universidade de Lisboa* (CLUL[2]) for more than two decades. This is an electronically based linguistic corpus of written and spoken materials, with a total of 311 million tokens. The written part of this corpus covers 309,812,943 tokens, 1,146,189 types, compiled from 356,208 documents and it is now available online. The corpus covers essentially the chronological period between 1970 to 2008, although texts from 1850 forward are also included (mainly fiction books and parliamentary debates).

1 A full description can be found here: http://www.clul.ul.pt/en/research-teams/408-crpc-description
2 http://www.clul.ul.pt/

H. Caseli et al. (Eds.): PROPOR 2012, LNAI 7243, pp. 113–120, 2012.

Our main focus is European Portuguese (see table 1), but other varieties of Portuguese are represented. These sub-parts are not comparable in size since they depend on the availability of data (we try to assure that only texts from native speakers of these varieties with no external linguistic influences are included). The corpus materials are taken by sampling from several types of written texts, chosen to assure as much text diversity as possible, but also according to the availability of the materials. Texts were obtained from different sources and this will reflect strongly on the cleaning procedure presented in the following section. Most recent newspapers, books and magazines were downloaded from the internet, others were obtained directly in digital format from their owners, like the parliamentary debates. But to assure diversity in terms of time period, text type, technical and didactic texts, and Portuguese varieties, we needed to use original texts in paper format. These were prepared in a time-consuming three-step process: digitalization with OCR, manual correction and final revision by a different team member. Our objective when compiling the corpus was closer to the notion of a monitor corpus and, for this online version, although we excluded some of the data, we decided to make as much of this material available as possible. Text diversity and corpus balance are yet aspects to improve in future versions.

Table 1. Text and Token distribution of the CRPC

Country	Texts	Tokens	**Type**	Texts	Tokens
Portugal	93.3%	289,840,619	Newspaper	50.8%	110,503,376
Angola	5.5%	10,744,627	Politics	45.9%	163,267,089
Cape Verde	0.3%	1,449,269	Magazine	1.4%	7,581,850
Macau	0.3%	2,086,763	Various	1.2%	4,806,176
Mozambique	0.2%	1,126,299	Law	0.3%	2,927,953
Sao Tome and Principe	0.2%	537,600	Book	0.3%	20,557,296
Brasil	0.2%	3,539,770	Correspondence	0.03%	88,370
Guinea Bissau	0.04%	364,421	Brochure	0.01%	80,833
Timor	0.0008%	123,575	–	–	–
Total	100%	309,812,943	Total	100%	309,812,943

The corpus and its access through its web-interface[3] provide an important resource for linguistic studies and NLP research on Portuguese especially because it is the largest and diversified corpus of European Portuguese to be made available on-line. The platform provides extensive search options for concordances of word forms, sequences of words and POS categories. It allows for restricted query per variety and text type (and other meta-data if one uses the CQP query syntax), and provides collocations using different statistical measures. The full set of options is described in the CRPC manual on the platform. This new platform is already proving extremely useful for ongoing projects.

[3] `http://alfclul.clul.ul.pt/CQPweb/`

2 Related Work

We refer to [15] for a full overview of the history of corpus development for Portuguese. Here we only discuss corpora which are available online and which share similar features and purpose with CRPC.

The Lácio-Web project[4] [1] was a 2.5 year project aimed at developing a set of corpora for contemporary written Brazilian Portuguese, namely a reference corpus of size 8,291,818 tokens, a manually verified portion of the reference corpus tagged with morpho-syntactic information, a portion of the reference corpus automatically tagged with lemmas, syntactic and POS-tags [2], two parallel and comparable corpora of English-Portuguese and a corpus of non-revised texts. In total, the Lácio-Web corpora together comprise around 10 million words. These corpora can be accessed online and are a follow-up of the NILC Corpus, a corpus of 32M tokens, developed at NILC and available at the Linguateca site, in the scope of the AC/DC project.

The Portuguese Corpus[5] contains 45 million words from Brazilian and European Portuguese taken from the 14th to the 20th century. It includes texts from other corpora, such as the Tycho Brahe corpus[6] and the above mentioned Lácio-Web reference corpus. The corpus is available online via a web interface that allows users to search for word lemmas, pos-tags, frequencies, collocations and restrict their queries for registers, countries or time periods.

The AC/DC[7] project (Acesso a Corpos/Disponibilização de Corpos) aims at having one website where many different corpora are available under a practical user interface. The web interface is based on the same architecture underlying the CRPC, the IMS Open Corpus Workbench (CWB). CETEMPúblico [16] is the largest of the available corpora and contains around 190 million words from the Portuguese newspaper *Público*.

The Bank of Portuguese[8] [17] is a result of joining several corpora together to form one large corpus of nearly 230 million words. A small part of the corpus, 1.1 milion words, is available for online search of concordances.

3 Cleaning

The CRPC is composed of documents from various sources, including internet (88.75% of the documents), which makes it challenging to clean automatically. It seemed therefore appropriate for cleaning the corpus to focus our efforts on a two-step approach, the first designed to get rid of metatags, and the second addressing directly lexical content. This two-step approach allows specialized algorithms to work more efficiently, as it proves much more difficult to process data coming from diverse sources in one single pass.

[4] `http://www.nilc.icmc.usp.br/lacioweb/`
[5] `http://www.corpusdoportugues.org/`
[6] `http://www.tycho.iel.unicamp.br/`
[7] `http://www.linguateca.pt/ACDC/`
[8] `http://www2.lael.pucsp.br/corpora/bp/`

The removal of meta-tags does not require extensive processing, as these labels usually follow a specific structure easily modelled by simple rules. In contrast, the cleaning of the remaining lexical content requires a more sophisticated approach, including methods based on learning lexical models from annotated content according to whether it is relevant or not (such as advertising or spam). In this context, the tool NCleaner [11] appears well suited for cleaning the corpus. This tool has proven very successful on a task aimed at cleaning web page content (*CLEANVAL* 2007). In addition, NCleaner automatically segments the text into short textual units, mainly paragraphs. To our knowledge, NCleaner has not been evaluated for a language other than English, so we provide a comparative evaluation of its application to Portuguese. For details of the approaches used in NCleaner, the reader is referred to [11].

NCleaner requires the creation of an annotated corpus to learn to distinguish "relevant" from "not relevant" segments. In [11], 158 documents (about 300,000 words and 2 million characters) were used to create a model of English vocabulary. For our Portuguese model, we have annotated 200 documents (about 200,000 words and 1.7 million characters) randomly selected among all the 359k documents included in the corpus. These 200 documents were first stripped of meta-tags and segmented by NCleaner. These documents were then handed over to an annotator. The task of our annotator, who was already familiar with the corpus and work in corpus linguistics in general, was to identify typical irrelevant segments that should be removed from the final corpus. This work has produced 1,474 irrelevant segments among the 6,460 segments included in the 200 documents. The most frequent classes of irrelevant segments we found were *titles*, *web navigation controls*, *copyrights* and *dates*. Some examples of irrelevant segments:

- *OUTROS TÍTULOS EM SOCIEDADE* [Title]
- *Regresso à página anterior* [Web navigation control]
- *Copyright 1998 Sojornal. Todos os direitos reservados.* [Copyright]
- *TERA-FEIRA, 30 DE JULHO 1996* [Date]

Regardless of the category to which they belong, these segments share a common characteristic: they do not represent a typical use of language within a collection of texts of a specific genre and on a defined subject, and distort the analysis of language that human experts, but especially NLP tools, could produce. However, we recognize that this definition of *noise* in the corpus is rather schematic and may be advantageously complemented by a more comprehensive list of general categories.

We also wanted to compare the lexical cleaning phase of NCleaner with two other approaches. The first approach of [8] originally designed to identify the language of a text is based on a comparison of the statistical distribution of words and groups of letters (N-grams). The second approach is that of SVM (*Support Vector Machine*) [13] and deemed successful for text classification tasks[9]. The results of this comparison with NCleaner are presented in Table 2.

[9] See also BeautifulSoup: `http://www.crummy.com/software/BeautifulSoup/`

Table 2. Comparative evaluation (at the level of the segment) of three approaches for cleaning the corpus

Approach	Parameters setting	F-score
N-GRAMS	Sequences of 5 letters or less	82%
SVM	500 Most frequent words	89%
NCLEANER	We keep accented letters	90%

All of the 6,460 annotated segments were used for the evaluation, 75% (4,845) dedicated to learning and 25% (1,615) for testing. We see that NCleaner performs best with an F-score comparable to the results obtained for English during *CLEANVAL* 2007 (91.6% at the word level). Applied to the entire corpus corpus, NCleaner reduced the number of tokens from 433 to 310 millions, a reduction of about 28%. The number of documents decreased from 359k to 356K[10].

4 Linguistic Preprocessing

We conducted the following types of automatic linguistic preprocessing: tokenization, POS-tagging and lemmatization. For tokenization we applied the LX-tokenizer [6] which splits punctuation marks from words and detects sentence boundaries. This tokenizer is developed specially for Portuguese and can handle typical Portuguese phenomena such as contracted word forms and clitics (including middle clitics).

For POS-tagging we compared two POS-taggers against each other and used the most performant to tag the full corpus. We compared MBT [10], a memory-based tagger to the LX-tagger [7]. The LX-tagger is a state-of-the-art tagger and has been applied to Portuguese with a reported accuracy of 96.87%.

We used the written CINTIL[11] corpus for training and evaluating the taggers, this corpus consists of a mixture of newspaper and fictional texts and is annotated with POS and lemma information and manually verified [4]. The tagset used in the CINTIL corpus is the result of extensive testing of different annotation options in previous projects and seems to best fit requirements for linguistic analysis. For the evaluation experiment we split the data in a 90% training set and a 10% test set. As MBT has features and parameters to be set, we ran ten-fold cross-validation experiments on the training set for finding a suitable setting. The LX-tagger was used without any modification. We measured the performance on the test set of 86,078 tokens in accuracy and F-score as shown in Table 3. We can observe that MBT outperforms the LX-tagger and therefore we used MBT to tag the full corpus. MBT splits the data in two categories: known and unknown words. Known words are the words present in the training data. For the known words only a limited set of POS labels is available, and the tagger can pick a label from this set. However, for unknown words, all labels

[10] Some documents having been completely emptied of their contents.

[11] `http://cintil.ul.pt/cintilfeatures.html`

need to be considered. Here the context of the unknown word as well as prefix and suffix information are useful features to predict the POS-label. For the known words, MBT achieves an accuracy of 96% on the test set while for the small subset of unknown words (5,664 tokens), it has an accuracy of 88.2%. The most frequent errors that are made by MBT match the type of decisions that are also difficult for humans to make such as distinguishing if the word "que" is a relative or a conjunction, and if a word like "italiano" is a common noun or an adjective. Also, proper names are often a source of errors as they are the same as normal dictionary words.

For the labelling of the full corpus, we trained the POS-tagger on a slightly adapted version of the CINTIL data. In CINTIL, contracted word forms are all split, while for our purpose we want to keep the contraction in the corpus. For example the contraction *das* (*de* "from" and *as* "the") receives a double tag 'PREP+DA' indicating that it is both a preposition and a definite article. After studying the CINTIL annotation we noticed that the multi-word units (MWU) as labelled in CINTIL were problematic for the tagger as they have a low frequency and are easily confused with other POS tags, let alone the fact that they are not always consistently annotated within the CINTIL corpus. CINTIL contains 900 different MWU types of which 425 occur once and many are genuine idiomatic expressions. We decided to decompose the MWU. The POS-tagger was trained on this adapted version of the CINTIL corpus containing 643,697 tokens and 30,344 sentences. We used the main POS-tag labels in CINTIL[12], which can be considered as a "simple" version of the tags that leaves out the more detailed information about genre, number, time, etc.

Table 3. Results on the test set of the two POS-taggers MBT and LX-tagger

Evaluation	MBT	LX-tagger
Accurary	95.48	94.06
F_micro	95.42	93.92
F_macro	70.10	66.40

As freely available Portuguese lemmatizers are scarce, we decided to convert an existing lemmatizer to the Portuguese language. We chose MBLEM [5], a lemmatizer developed by the ILK research group[13] with a very good performance for Dutch and English. MBLEM combines a dictionary lookup with a machine learning algorithm to produce lemmas. The classifier is a memory-based learning algorithm [9] which is very well suited for lemmatization as all previous seen cases are stored in memory. In practice, this means that the full dictionary is stored and all exceptional cases (e.g. irregular verb forms) are kept. The learning algorithm learns to associate transformation rules that map word forms to their lemmas. As basis for the dictionary we used a list of wordform - POS-tag combinations mapped to lemmas. This list was produced in-house. The dictionary used in

[12] We refer to the CINTIL annotation manual for details on the tag set.
[13] `http://ilk.uvt.nl`

MBLEM contains 102,196 word forms combined with 27,860 lemmas, leading to 120,768 wordform-lemma combinations.

To evaluate the performance of MBLEM we used a sample of 50,000 words from the written CINTIL as test set. As CINTIL has been tagged with a different set of POS-tags (80 different main tags) than the set of tags listed in the dictionary (31 tags), we asked a Portuguese linguist to create a mapping between the two POS-tag sets. The mapping was quite straight-forward, almost all CINTIL tags could be mapped against a suitable coarse-grained dictionary tag, although in some cases (e.g. numeral adjectives) categories were treated differently in the two tag sets. Note that MBLEM predicts a lemma for each token in the file, but in CINTIL not all tokens have a lemma, function words such as prepositions and adverbs do not. In our test set, 17,117 word forms have a gold-standard annotated lemma. MBLEM achieves an accuracy of 96.7% on this test set.

5 Conclusion and Perspectives

We have presented the on-line publication of the written sub-part of the CRPC, a large and diverse Portuguese corpus. We have discussed its internal constitution, available resources for cleaning and preprocessing such a corpus, and the new platform used for online queries. The current version of the corpus can be used for lexical studies as well as a resource for NLP applications. The corpus has already been used in many projects and studies, the most recent being a study of comparable subcorpora of Portuguese varieties [14] and a computational study that compares lexicons from different time period [12]. Future work includes a second phase of cleaning that will focus on improving segmentation and consolidating our lexical model, as well as adding more searchable meta-data tags and introducing a language spotter for the few remaining pockets of foreign languages present in the corpus. We are planning the development of a constituent chunker for Portuguese so that the corpus can be enriched with syntactic annotations. We also plan to enlarge the corpus annotation to cover information on nominal and verbal inflection (genre, number, person, tense, etc.) based on the CINTIL annotation schema and to address the issue of MWU. Since its publication online, the platform has been visited and used extensively by users from all over the world.

Acknowledgement. We thank Luísa Alice Santos Pereira and Paulo Henriques for their help. This work is financed by FCT for the project PEst-OE/LIN/UI0214/2012, by Fundação Calouste Gulbenkian and by the FCT Doctoral program Ciência 2007/2008.

References

1. Aluísio, S., Pinheiro, G.M., Manfrin, A.M.P., de Oliveira, L.H. M., Genoves Jr., L.C., Tagnin, S.E.O.: The lacio-web: Corpora and tools to advance brazilian portuguese language investigations and computational linguistic tools. In: Proceedings of 4th Conference on International LREC, pp. 1779–1782 (2004)

2. Aluísio, S.M., Pelizzoni, J.M., Marchi, A.R., de Oliveira, L., Manenti, R., Marquiafável, V.: An Account of the Challenge of Tagging a Reference Corpus for Brazilian Portuguese. In: Mamede, N.J., Baptista, J., Trancoso, I., Nunes, M.d.G.V. (eds.) PROPOR 2003. LNCS, vol. 2721, pp. 110–117. Springer, Heidelberg (2003)
3. Bacelar do Nascimento, M.F., Pereira, L., Saramago, J.: Portuguese Corpora at CLUL. In: Second International Conference on Language Resources and Evaluation (LREC 2000), Athens, vol. II, pp. 1603–1607 (2000)
4. Barreto, F., Branco, A., Ferreira, E., Mendes, A., Bacelar do Nascimento, M.F.P., Nunes, F., Silva, J.: Open resources and tools for the shallow processing of portuguese. In: Proceedings of the 5th International Conference on Language Resources and Evaluation (LREC 2006), Genoa, Italy (2006)
5. van den Bosch, A., Daelemans, W.: Memory-based morphological analysis. In: Proceedings of the 37th Annual Meeting of the Association for Computational Linguistics, ACL 1999. pp. 285–292 (1999)
6. Branco, A., Silva, J.: Contractions: Breaking the Tokenization-Tagging Circularity. In: Mamede, N.J., Baptista, J., Trancoso, I., Nunes, M.d.G.V. (eds.) PROPOR 2003. LNCS, vol. 2721, pp. 167–170. Springer, Heidelberg (2003)
7. Branco, A., Silva, J.: Evaluating solutions for the rapid development of state-of-the-art pos taggers for portuguese. In: Proc. of LREC 2004, pp. 507–510 (2004)
8. Cavnar, B., Trenkle, J.M.: N-gram based text categorization. In: Proceedings of the Third Annual Symposium on Document Analysis and Information Retrieval, pp. 161–175 (1994); UNLV Publications/Reprographics
9. Daelemans, W., Van den Bosch, A.: Memory-Based Language Processing. Cambridge University Press, Cambridge (2005)
10. Daelemans, W., Zavrel, J., Berck, P., Gillis, S.: Mbt: A memory-based part of speech tagger generator. In: Proceedings of the 4th ACL/SIGDAT Workshop on Very Large Corpora, pp. 14–27 (1996)
11. Evert, S.: A lightweight and efficient tool for cleaning web pages. In: 6th International Conference on Language Resources and Evaluation (LREC 2008), Marrakech, Morocco (2008)
12. Générreux, M., Mendes, A., Pereira, L.A.S., do Nascimento, M.F.B.: Lexical analysis of pre and post revolution discourse in portugal. In: Proceedings of the 3rd Workshop on Building and Using Comparable Corpora, LREC 2010 (2010)
13. Joachims, T.: Learning to Classify Text Using Support Vector Machines. Ph.D. thesis, Cornell University, USA. Kluwer Academic Publishers / Springer (2002)
14. do Nascimento, M.F.B., Estrela, A., Mendes, A., Pereira, L.: On the use of comparable corpora of african varieties of portuguese for linguistic description and teaching/learning applications. In: Proceedings of the Workshop on Building and Using Comparable Corpora, LREC 2008 (2008)
15. Santos, D.: Linguateca's infrastructure for portuguese and how it allows the detailed study of language varieties. Oslo Studies in Language 3(2) (2011)
16. Santos, D., Rocha, P.: Evaluating CETEMPublico, a Free Resource for Portuguese. In: Proceedings of 39th Annual Meeting of the Association for Computational Linguistics (ACL), pp. 450–457. Association for Computational Linguistics, Toulouse (2001)
17. Sardinha, T.B.: History and compilation of a large register-diversied corpus of Portuguese at CEPRIL. The Especialist 28(2), 211–226 (2007)

Dicionário-Aberto: A Source of Resources for the Portuguese Language Processing

Alberto Simões[1], Álvaro Iriarte Sanromán[1], and José João Almeida[2]

[1] Center for Humanistic Studies, Minho's University
{ambs,alvaro}@ilch.uminho.pt
[2] Computer Science and Technology Center, Minho's University
jj@di.uminho.pt

Abstract. In this paper we describe how Dicionário-Aberto, an online dictionary for the Portuguese language, is being used as the base to construct diverse resources that are relevant in the processing of the Portuguese language.

We will briefly present its history, explaining how we got here. Then, we will describe the resources already available to download and use, followed by the discussion on the resources that are being currently developed.

Keywords: dictionary, lexicography, open resource, thesauri.

1 A Brief History

Dicionário-Aberto[1] started in June 2005, when a few people felt that the Portuguese language was missing an open dictionary for use (any use). The process of creating a dictionary from scratch is difficult and expensive. When most of the interested persons are engineers and computer scientists, this task gets more difficult. As one member of this group was responsible for the transcription of Portuguese books for the Project Gutenberg [1], the idea of transcribing a full dictionary appeared. An old dictionary with expired copyright was searched and chosen[2], digitalized and the transcription process started using the Project Gutenberg Distributed Proofreaders web interface. A detailed description of this process is described in [8].

The transcription was performed by volunteers, using a simple textual syntax, very similar to a subset of common wiki syntaxes. In March, 2010, the full transcription was concluded (different validation rounds were performed for each page). This textual document was converted to a more formal syntax, based on XML.

[1] Available at http://www.dicionario-aberto.net/
[2] The chosen dictionary was "Novo Diccionário da Língua Portuguesa, Cândido de Figueiredo, 1913." It was not chosen by its lexicographic quality, but only because of a set of circumstantial facts.

H. Caseli et al. (Eds.): PROPOR 2012, LNAI 7243, pp. 121–127, 2012.

A subset of the TEI [3] (Text Encoding Initiative) format for dictionaries was chosen. Simões and Almeida [9] describe this conversion process.

The chosen dictionary used an old orthographic form (prior to 1943/45 agreements). To be part of Project Gutenberg the books must be transcribed in original form, so the transcribed documents needed orthographic modernization to be useful. This task was automated and at the present moment, a set of volunteers are approving the modernized entries[3] Dicionário-Aberto has now 128 521 entries, and about 8% of the entries were verified for modernization errors.

The new orthographic agreement (1990) will require a new modernization process. Fortunately, this will be easier to automate, as there are a couple of good conversion tools available [2].

In the future, Dicionário-Aberto will be open as a dictionary Wiki, where the community can edit corrections or add new words. To guarantee quality (and a somewhat controlled language) a two tier process will be implemented: a change or addition (or even deletion) will be available right away, but in a "non official" status, until a moderator approves the change.

2 Currently Available Resources

With the current status of Dicionário-Aberto as described in the previous section, there are some resources that can be downloaded and used to learn about the Portuguese language, its history and to process automatically using natural language processing techniques.

Original TXT and TEI Transcriptions

The more basic resources are the plain text files with the original transcriptions, both in wiki or TEI formats. These resources are available in 28 separate files, one for each letter, plus a geographic and an onomastic appendixes. These documents (specially the TEI version as it is annotated and is easier to process automatically) are mostly useful to study the Portuguese language before the 1943/45 orthographic agreement.

Current Database Snapshot

A view of the Dicionário-Aberto web-site database is available to download and use. It is an SQL document that can be imported in any MySQL database server (and probably in other tools with minor changes). Figure 1 shows the structure of this view (further tables might be available in the future, accordingly with new resources being developed).

This is the better way to use the current database in offline mode, as it enables the user to access the current versions for all dictionary entries, as well as the previous versions (before orthographic modernization). Therefore this format can be used not just to perform text-mining in the dictionary but also

[3] Unfortunately the modernization process had some false positive substitutions.

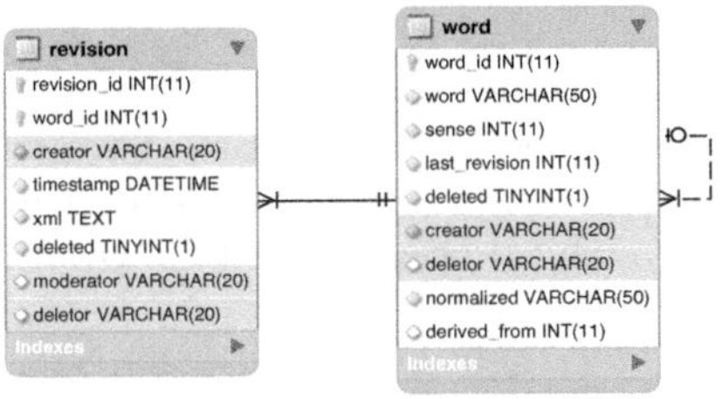

Fig. 1. Dicionário-Aberto database view

for contrastive studies. The database is regularly checked for quality control, checking the database for consistency and validating all XML snippets. Another test, that validates the dictionary completeness (that all cross-references have a valid target) is currently disabled given the modernization process.

Modernization Rules

With the modernization process we obtained two versions of the dictionary in two different versions of Portuguese. Although they are similar in great extend, they can be considered two different languages, and therefore all typical approaches to align and extract bilingual dictionaries automatically can be applied.

Tables 1 present two types of orthographic modernization rules that can be extracted (these lists were constructed using `lexdiff` [2]). The first one maps full words and the second maps letter sequences. The latter is easier to apply in documents with words that are not in the dictionary (but that share a subsequence of letters with some other word), but the first is better for accuracy. The third column is the rule confidence about its transformation.

Table 1. Word and pattern rules for orthographic modernization

Full-word Rules		
gênero	género	100%
aquelle	aquele	100%
effeito	efeito	100%
fórma	forma	100%
pessôa	pessoa	100%
camillo	camilo	100%
póde	pode	99%
sôbre	sobre	98%
ás	às	90%

Pattern Rules		
sôa	soa	99%
ffe	fe	95%
ph	f	95%
lle	le	90%
llo	lo	88%
aes	ais	87%
gên	gén	87%
bôa	boa	80%
cci	ci	40%

Note, however, that these dictionaries were calculated with the current version of Dicionário-Aberto, where some hundred words were not modernized correctly by the automatic process[4].

[4] Being a dictionary, it includes very odd words that are not easily modernized by general rules.

Morphologic Dictionary

All the dictionary entries include partial morphological information (at least its main category, and in some situations the genre or type of verb), making it possible to extract a list of words with associated morphologic information. Table 2 resumes the size of this dictionary and distribution by main morphologic categories.

Table 2. Morphologic distribution of Dicionário-Aberto entries

Nouns		Adjectives	Adverbs	Verbs		Locutions
masc.	*fem.*			*transitive*	*intransitive*	
45 657	38 488	30 469	2 962	10 147	4 016	579

REST API

To enable the use of the dictionary in the cloud a simple web service API based on REST principles is available. It supports queries both in XML and JSON, and lets the user query for definitions, given a specific word, or search for near misses, prefixes and suffixes. This enables the development of mobile applications. An example of such application is the iPhone interface to Dicionário-Aberto[5]

3 Resources under Development

Further work is being done to make the Dicionário-Aberto experience more interesting, enabling new services in the web site, but also to develop new resources that can be used by natural language processing researchers.

Reverse-Order Dictionary

Reverse-Order Dictionaries are not very common (do not confuse with Reverse Dictionaries, discussed below). They let the user browse the dictionary searching by the end of the word, instead of its beginning (looking up for suffixes instead of prefixes).

One of their applications is the construction of a rhyme dictionary (notice however that this will be a partial rhyme dictionary, as some words with different orthography have the same sound, like *doce* and *fosse*).

Another use of these kind of dictionaries is the study of the morphology of a language [6], like the study of suffix productivity (productivity for some scientific terminology suffixes — *-ato, -eto, -ito* — the productivity of effect/result words — *-data, -ção, -são, -ança, -ância,* etc.).

[5] Developed by log.oscon, available from the Apple AppStore. Further details can be found at `http://log.pt/dicionarioaberto/`

Reverse Search or Reverse Dictionary

In a standard (paper) dictionary, the query can only be performed browsing the list of alphabetically sorted lemmas. In machine-readable dictionaries, the search should not be limited to the lemmas. The user should be able to search the full entries, including its definition, examples, etymology or any other section.

This reverse search capabilities transforms electronic dictionaries in ideological or conceptual databases, also known as analogical or onomasiological dictionaries. This feature is available for some languages like Spanish[6] and English[7].

There is a long tradition of onomasiological dictionaries for the European languages. Some ideological dictionaries were prepared during XIX and XX centuries[8], that allowed the reader to search an idea or concept in a descriptors structure similar to a thesaurus [11], or a structured list of concepts sorted by subjects (summary tables) together with a list of hypernyms or broader terms (categories, general ideas) that lead the reader to the searched word.

The reverse search functionality can help surpassing some of the limitations present in paper versions of these dictionaries. This functionality can not be only seen as a simple query tool. Imagine the potentiality of reverse search if the dictionary authors use a controlled language to write their definitions, just like a descriptors thesaurus.

Its main usage is to search a word that we know adequate for a specific situation, but that we can not remember at that moment, or to search for a more specific word, or even to check if there is some word to express some concept [4].

Dicionário-Aberto will not only offer this functionality for end-users through the web interface, and for cloud applications through the REST API, but also make available the reverse index, for offline processing.

Ontology View

Hugo Oliveira [7] has been working in the creation of Onto.PT, a lexical ontology for the Portuguese Language. Oliveira performed some experiments with different resources, and Dicionário-Aberto was also covered. A similar approach was also performed by Simões et al [10]. With these experiments in mind, and the interesting results obtained, a new view for Dicionário-Aberto is being developed: a (lexical) ontology view.

This view will enable the user to query the dictionary, using the standard search or one of the two new methods described earlier, and together with the definition, the examples, and the etymology, consult a thesaurus-like structure. This structure includes a set of relations (like synonymy, antonymy, hyperonymy,

---

[6] Reverse dictionary search in the Dictionary of the Real Spanish Academy, by Gabriel Alberich: `http://dirae.es/`

[7] OneLook Reverse Dictionary, by Doug Beeferman, that searches more than one thousand indexed dictionaries: `http://www.onelook.com/reverse-dictionary.shtml`

[8] Some examples are listed by Martínez de Sousa [5].

instances/species/genres, actor/action, etc) to other dictionary entries (in case of multi-word expressions each component word will have its own link).

The extraction uses a set of patterns, just like the methods described by the mentioned authors. We decided not to reuse the extracted data because in the future, as stated in the introduction, Dicionário-Aberto will be a Wiki, making it crucial to have an automatic method to recalculate the ontology. The extraction method will run everyday in an unsupervised way.

The ontology completeness will be guaranteed by a set of completion rules. For example, the synonymy relation is symmetric, the hyperonymy relation is inverse of the hyponymy relation, the hyperonymy relation can be seen as a special case of a transitive relation, antisymmetric relations, anti-reflexive, etc. These properties can be described in a mathematical notation and used to ensure that entries that do not have a reference to related words can still get the relation information.

This kind of feature will make the dictionary much more interesting for the end user but also for the natural language processing researcher. For the first, it will make entries browsable by concept. For the second, will be a complement of a Portuguese word-net, as concepts will also include definitions.

4 Conclusions

In this document we described the current status of Dicionário-Aberto, an open project to the development of a knowledge and feature rich dictionary for the Portuguese language, both to be used as a standard dictionary but also as a resource for natural language processing tasks. We defend that resources construction should be automatic, especially in a case like Dicionário-Aberto that will be open for the community to cooperate. This guarantees that the extracted resources can grow in size and quality at the same time as its main resource.

References

1. Project Gutenberg. Project Gutenberg Literary Archive Foundation (November 2011), `http://www.gutenberg.org/`
2. Almeida, J.J., Santos, A., Simões, A.: Bigorna – a toolkit for orthography migration challenges. In: Calzolari, N., et al. (eds.) Seventh International Conference on Language Resources and Evaluation (LREC 2010), Valletta, Malta, pp. 227–232. European Language Resources Association (May 2010)
3. TEI Consortium, editor. TEI P5: Guidelines for Electronic Text Encoding and Interchange, chapter 9. Dictionaries. TEI Consortium (January 2012), `http://www.tei-c.org/release/doc/tei-p5-doc/en/html/DI.html`, Version 2.0.1 edition, December 22, 2011
4. Porto da Pena, J.A.: Dicionário de uso del Español (2003), `http://cvc.cervantes.es/actcult/mmoliner/diccionario/` (retrieved November 3, 2011)
5. Martínez de Sousa, J.: Diccionario de lexicografía práctica. Biblograf, Barcelona (1995)

6. Millán, J.A.: Zigzag, gong, ping-pong, iceberg. donde se descubre que hay diccionarios inversos, y su utilidad manifiesta para el progreso de la humanidad (1999), `http://jamillan.com/inverso.htm` (retrieved November 3, 2011)
7. Oliveira, H.G., Gomes, P.: Onto.PT: automatic construction of a lexical ontology for Portuguese. In: 5th European Starting AI Researcher Symposium (STAIRS 2010) (August 2010)
8. Simões, A., Farinha, R.: Dicionário Aberto: Um novo recurso para PLN. Vice-versa (16), 159–171 (2011)
9. Simões, A., Almeida, J.J.: Processing XML: a rewriting system approach. In: Simões, A., da Cruz, D., Ramalho, J.C. (eds.) XATA 2010 — 8ª Conferência Nacional em XML, Aplicações e Tecnologias Aplicadas, Vila do Conde, Maio, pp. 27–38 (2010)
10. Simões, A., Almeida, J.J., Farinha, R.: Processing and extracting data from Dicionãrio Aberto. In: Calzolari, N., et al. (eds.) Seventh International Conference on Language Resources and Evaluation (LREC 2010), Valletta, Malta, pp. 2600–2605. European Language Resources Association (May 2010)
11. van Slype, G.: Les langages indexation: conception, construction et utilisation dans les systmes documentaires. Les Editions d'Organisation, Paris (1987)

Towards a Common Sense Base in Portuguese for the Linked Open Data Cloud

Vládia Pinheiro, Vasco Furtado, Tarcisio Pequeno, and Caio Ferreira

Programa de Pós-Graduação em Informática Aplicada,
Universidade de Fortaleza (UNIFOR)
Av. Washington Soares, 1321, Fortaleza, Ceará, Brasil
{vladiacelia,vasco,tarcisio}@unifor.br,
caioferreirax@gmail.com

Abstract. The Linked Open Data (LOD) cloud is a promising reality since the major content producers are offering their data on an open and linked network, through RDF (Resource Description Framework), with the aim of providing Semantic Web applications with a single global database for retrieval of related content and to perform inferences over the network. However, bases with Portuguese-language content are still incipient. In this paper we present the process of inclusion of the InferenceNet – the first resource with common sense and inferentialist knowledge in Portuguese language – on the LOD. Our main goal is to leverage the use and development of Semantic Web applications by content producers in Portuguese language. We develop and evaluated a platform, called *SemWidgets*, for the creation and execution of widgets able to access and reason over InferenceNet and the open linked data, like DBPedia, Yago, and Article Search API of the New York Times.

Keywords: Linguistic Resource, Portuguese Language, Web Semantic, Linked Open Data.

1 Introduction

In the context of Semantic Web (SW) technologies, the publication of interconnected content in RDF on the Linked Open Data (LOD) network is a promising reality. Major content producers are offering their data on an open and linked network, through RDF, with the aim of providing SW applications with a base for retrieval of related content and to perform inferences over the network, among other services. As an example of large knowledge bases of the LOD network, we can cite DBpedia [1], Yago [2], and WordNet [3]. For example, based on LOD, content producers can link their texts (news, blog posts, etc.) to these large bases, retrieve related content (wikipedia pages, photos, movies, news, ontologies, etc.), and present them to their users or make inferences. The aim of LOD is precisely to provide a base where content can be linked.

However, the representativeness of bases with Portuguese-language content on the LOD is still incipient and there are no linguistic and semantic resources. Imagine a

H. Caseli et al. (Eds.): PROPOR 2012, LNAI 7243, pp. 128–138, 2012.

scenario in which a web-based newspaper (in Portuguese) wishes to make its news items available so that other newspapers can retrieve them in contextualized manner and present them as related links. In another sense, imagine this producer wishes to use a crawler on the Internet to automatically retrieve related content and present it to its readers. Approaches of syntactic searches on the Web or approaches that search content on the LOD network using Portuguese-English translation can be used, but will be ineffective in capturing semantically related content. For example, they would find it difficult to associate a news report that speaks of the death of Gaddafi to another one speaking of the assassination of Gaddafi. This would only be minimally possible if the crawler had access to the knowledge that semantically related the concepts "death" and "assassination."

Our work was developed precisely to fill this gap and is aimed at leveraging the use of the Semantic Web for Portuguese-language content producers. We present the process of inclusion of the first semantic-linguistic resource for the Portuguese language – InferenceNet.Br [4] – on the LOD, the heuristics used, and the major challenges. Besides being the first resource in Portuguese on the LOD, InferenceNet has the distinguishing feature of expressing common sense and inferentialist knowledge, which enables Natural Language Processing (NLP) systems to make richer inferences [5]. For example: based on the knowledge that "politicians are able to propose laws," news reports that deal with laws and politicians might be interesting for users who want to know about politicians.

As a scenario for the application of InferenceNet in LOD, we developed a platform, called *SemWidgets*, which uses InferenceNet for the creation and execution of piece of programs able to access and reason over open linked data. This application provided a scenario of assessing how a resource with semantic content on the LOD can be used as a "bridge" in the process of content retrieval and inference through the Semantic Web.

2 InferenceNet.BR

The motivation and the process of construction of InferenceNet.Br[1] (www.inferencenet.org) are described in [4]. In the context of this work, InferenceNet's Conceptual Base is the most important since it contains the inferential and common-sense content of concepts of the Portuguese and English languages, defined and agreed upon in a community or area of knowledge. According to the inferentialist view, the content of a concept must be expressed, becoming explicit, through the use of it (the concept) in inferences, as premises or conclusions of reasoning. Moreover, what determines the use of a concept in inferences or potential inferences in which this concept may participate are: (i) its pre-conditions or premises of use: what gives someone the right to use the concept and what could exclude such a right, serving as premises for utterances and reasoning; and (ii) its post-conditions or conclusions of use: what follows or what are the consequences of using the

[1] Research project funded by FUNCAP (*Fundação Cearense de Apoio ao Desenvolvimento Científico e Tecnológico Ceará* [Foundation for Support to Scientific and Technological Development], Proc. 9145/08 – Edital Nº 06/2008 – Information Technology.

concept, which let one know what someone is committed to by using a particular concept, serving as conclusions from the utterance *per se* and as premises for future utterances and reasoning. Formally, this base is represented in a directed graph $G_c(C,Rc)$. Each inferential relationship $rc_j \in Rc$ (set of inferential relations of the concept *c*) is represented by a tuple (*relationName*, c_i, c_k, *type*), where *relationName* is the name of an InferenceNet semantic relation (*Capableof*, *PropertyOf*, *EffectOf* etc.), c_i and c_k are concepts of a natural language, and *type* = "Pre" or "Pos" (pre-condition or post-condition for using the concept c_i). Figure 1 presents part of the conceptual base for the concept **crime**.

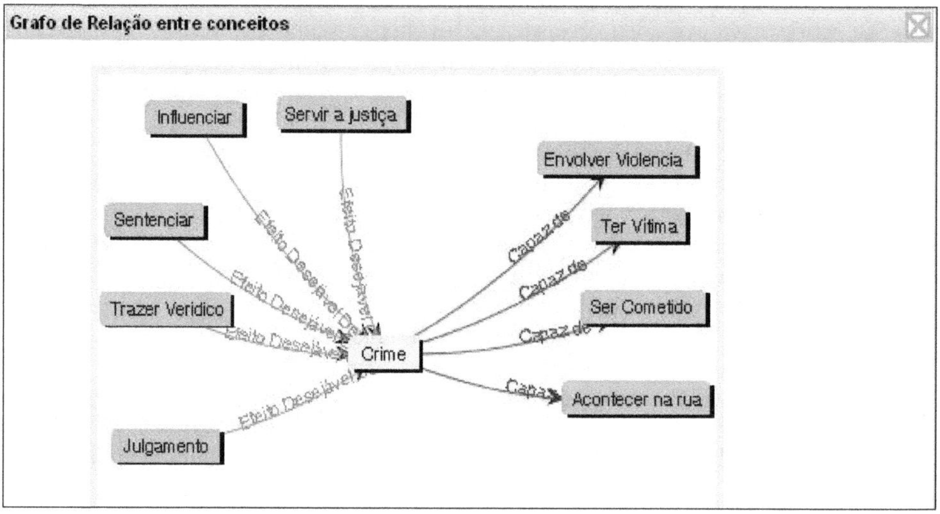

Fig. 1. Part of the conceptual base for the concept **crime** with some pre-conditions (incoming arrows) and post-conditions (outgoing arrows)

3 Linked Open Data (LOD)

Linked Open Data[2] (LOD) relates to a series of practices and standards adopted for the publication of information on the Web so that machines are able to process them efficiently. LOD is, principally, about using the Web like a single global database. Moreover, LOD is about using the Web to connect related data that wasn't previously linked, or using the Web to mitigate the difficulties to linking data currently linked via a strong interference of computer experts. Key technologies that support Linked Data are URIs (a generic means to identify entities or concepts in the world), HTTP (a simple yet universal mechanism for retrieving resources, or descriptions of resources), and RDF (a generic graph-based data model with which to structure and link data that describes things in the world) [6].

[2] See `http://linkeddata.org/`

Tim Berners-Lee defined four guidelines [7] that are characterized as best practices of publishing LOD: (i) Use of URIs to identify things; (ii) attribution of an http address to an URI in order to be easily recovered on the web; (iii) Use of RDF as a standard to express information; and (iv) use of data linked via URIs enabling the search for new data. The concept description and their relationship are made via ontologies. The concepts are described from classes and properties. LOD is strongly based on the reuse of properties of ontologies.

Many initiatives to describe free content on the web via ontologies in RDF have emerged, such as DBpedia and Freebase (http://www.freebase.com). However, the recent initiatives of large media conglomerates such as The New York Times and the BBC to do the same massively boosted the area.

4 Towards InferenceNet for LOD

In order for content producers in Portuguese take advantage of the LOD, we have generated a version of InferenceNet in RDF that allowed us to connect it to Yago, DBPedia and WordNet, and other open data networks. The InferenceNet concepts and its inferential content were translated to RDF and linked to Yago by the name of the concept in English (property "hasEnglishName") that in its turn is linked to DBpedia. In Figure 2, we show a piece of the RDF describing the link between the InferenceNet concept of politician ("**http://inferencenet.org/rdf/conceito41738**") with Yago's concept "**http://www.mpii.de/yago/resource/wordnet_politician_109772277**".

The content expressed in InferenceNet and its connection with the LOD allows rich inferences to be made, since the base expresses common-sense knowledge and the inferential import of the concepts in the reasoning.

```
<rdf:RDF
   xmlns:rdf="http://www.w3.org/1999/02/22-rdf-syntax-ns#"
   xmlns:in="http://www.inferencenet.org/rdf/inference#">
   <rdf:Description rdf:about="http://inferencenet.org/rdf/conceito41738">
     <rdf:type
rdf:resource="http://www.mpii.de/yago/resource/wordnet_politician_109772277"/>
     <in:hasEnglishDescription>politician</in: hasEnglishDescription>
     <in:hasPortugueseDescription>político</in: hasPortugueseDescription>
     <in:hasEnglishName>politician</in: hasEnglishName>
     <in:hasPortugueseName >político</in: hasPortugueseName>
     <in:CapableOfReceivingAction
rdf:resource="http://inferencenet.org/rdf/concept_5147"/>
           <in:CapableOf rdf:resource="http://inferencenet.org/rdf/concept_45058"/>
    </rdf:Description>
</RDF>
```

Fig. 2. Piece of the RDF of the concept "politician" linked to Yago's concept "wordnet_politician_109772277"

The process of inclusion of InferenceNet's Conceptual Base in LOD included the following steps:

1. **Definition of the Core of the Concept.** For each concept of InferenceNet, its main word or core of the corresponding phrase is defined, in English and Portuguese. For example, for the concept "light pollution," the main word is "pollution." In this process we used the parser PALAVRAS [8] to recover the root of the parse tree of the phrase. The main word is used as a search alternative on the LOD, in step 2.
2. **Search for InferenceNet Concepts Based on the LOD.** We used WordNet as a hub for linking InferenceNet to the LOD. Initially, the file "wordnet-synset.n3" was indexed by SIREn [9]. Then, URIs corresponding to the following items are searched in the indexed file: (1) the description of the InferenceNet concept in English; (2) the main word (as defined in step 1); and (3) synonyms of the concept in question, in this order of priority. For example, as no URI corresponding to the description of the concept "light pollution" was found, the URI that matches the main word of this concept ("pollution") is searched. In the example, the URI found was "http://www.w3.org/2006/03/wn/wn20/instances/synset-pollution-noun-1." Finally, based on the URIs of WordNet, the respective matching URIs are retrieved from the Yago base (in the example, "http://www.mpii.de/yago/resource/wordnet_pollution_113690156").
3. **Generation of RDF.** This step comprises the generation of the conceptual graph of each concept of InferenceNet in RDF triples, which relate each resource to the value of their properties. Each concept is published as a Web resource and is identified by a URI (e.g., the URI of the concept "politician" is "http://inferencenet.org/rdf/conceito41738"). The properties of the InferenceNet concepts in RDF are described in Table 1. Figure 2 shows an example of a description in RDF of the resource "politician."

Figure 3 shows a graph with the percentage and number of concepts linked to the LOD, in each n-gram category, and concepts that were not linked. Currently, the InferenceNet base contains 182,172 concepts, of which 69% are linked to the LOD.

The importance of a semantic-linguistic resource of the Portuguese language in the LOD results from the connectivity with billions of RDF triples, which allow an exponential growth of the resource's expressive and inferential power. InferenceNet expresses common-sense and inferentialist knowledge, and in the LOD it can take advantage of the knowledge expressed in other ontologies and taxonomies, as well as knowledge bases such as DBPedia, Foaf [10], Freebase, WordNet, and Yago.

Table 1. Piece of the RDF of the concept "politician" linked to Yago's concept "wordnet_politician_109772277"

RDF Property	Content	Example
rdf:Description rdf:about	URI of the InferenceNet concept	rdf:about="http://inferencenet.org/rdf/conceito41738"
rdf:type	URI of the Yago concept	rdf:resource="http://www.mpii.de/yago/resource/wordnet_politician_109772277"
hasEnglishDescription	Description of the concept in English	politician
hasPortugueseDescription	Description of the concept in Portuguese	político
hasEnglishName	Name of the concept in English	politician
hasPortugueseName	Name of the concept in Portuguese	político
<Semantic_relation *n*>	URI of the InferenceNet concept related by <semantic relation *n*>	CapableOf rdf:resource="http://inferencenet.org/rdf/concept_45058"

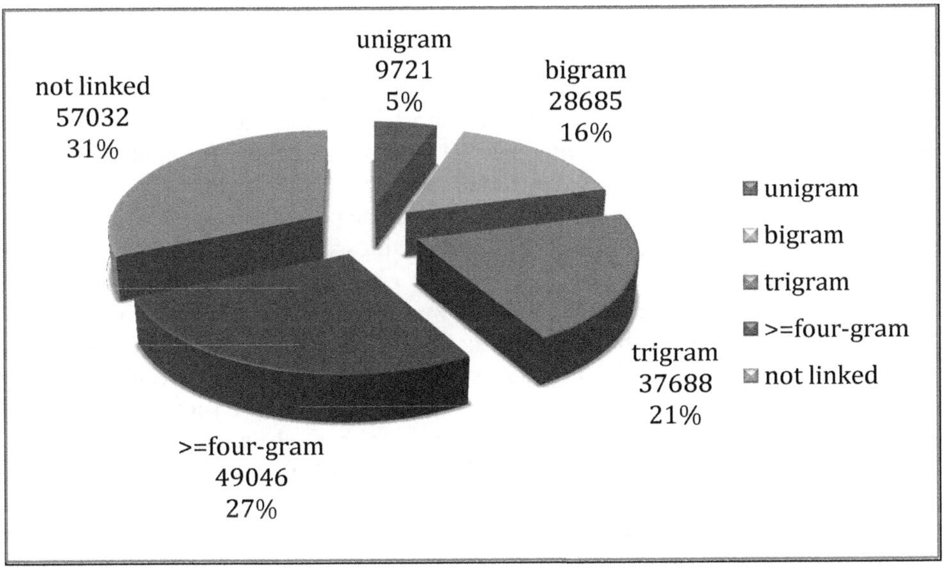

Fig. 3. Graph with the percentage and number of InferenceNet concepts linked and not linked to the LOD

5 Using InferenceNet and LOD in a Semantic Web Application

5.1 Overview

In this project, we developed a platform, called *SemWidgets*, for the creation and execution of piece of programs able to access and reason over open linked data. The main innovation of *SemWidgets* is that it provides to the content producer of a website a way to attain semantic characterization of what s/he intends to have in a body of the widget. Examples of these are politicians, football teams, places where garbage is piling up, or specific persons (e.g. Obama), institutions (e.g. Stanford, CNN), places (e.g. Rio de Janeiro), etc. In short, in the "semantic characterization process," the content producer – by informing a linguistic expression that designates a concept or an instance of a concept – is guided in a process that aims to associate concepts from the InferenceNet base to this linguistic expression. Thus, *SemWidgets* starts to have knowledge over common-sense inferential relations that define the semantic value of the concept and is able to make inferences about the network of linked data and retrieve related information. At the end of this interactive process of semantic characterization, a linguistic expression that designates a concept or an instance of a concept is linked to InferenceNet, which, in turn, is connected with the LOD. Doing so, we enable *SemWidgets* to retrieve related content arranged in the LOD (News, Articles, Videos, Photos, Social Network Profiles, etc.), while the site is being used.

When the linguistic expression refers to an instance of a concept (i.e. Pelé – instance of soccer player, or Barack Obama – instance of politician), *SemWidgets* searches at DBPedia to discover the Yago classes that best represents the instance. The most specific concept that is and associated to InferenceNet is retrieved from the navigation into the complete graph of concepts for a specific instance. For example, as "Senator," "African American US Senators," "leader," and "Current National Leaders" are not represented in InferenceNet, *SemWidgtes* chooses "Politician" as the concept that best represents the instance "Barack Obama."

Figure 4 shows examples of inferences that can be performed based on the concept "politician," which have been associated to the expression "Barack Obama," informed by the user. In this example, the user (content producer) is asked about the subject s/he wants to retrieve content while using a website: (i) Barack Obama; (ii) laws with which Barack Obama has been involved; (iii) cases of corruption in which Barack Obama has been involved; (iv) Politicians; (v) laws with which politicians have been involved; (vi) cases of corruption in which politicians have been involved. Then the content producer is asked about the media type that should be retrieved: (i) News; (ii) Articles; (iii) Videos; (iv) Photos; (v) Profiles in Social Networks (Facebook, Twitter, Google, etc).

The subjects proposed by *SemWidgets* to the user are defined from InferenceNet's concept associated to the linguistic expressions. In the example shown in Figure 4, the possible subjects were authorized by the inferential content of the concept "politician," expressed in InferenceNet: "a politician is able to propose laws" and "a politician has the property of being corrupt." This common-sense knowledge is expressed in InferenceNet through the relationships (*CapableOf*, "politician", "propose law", "Pre") and (*PropertyOf*, "politician," "corrupt," "Pre"), respectively. Furthermore, the concepts "politician," "law," and "corruption" are associated – via LOD – to Yago and DBpedia resources.

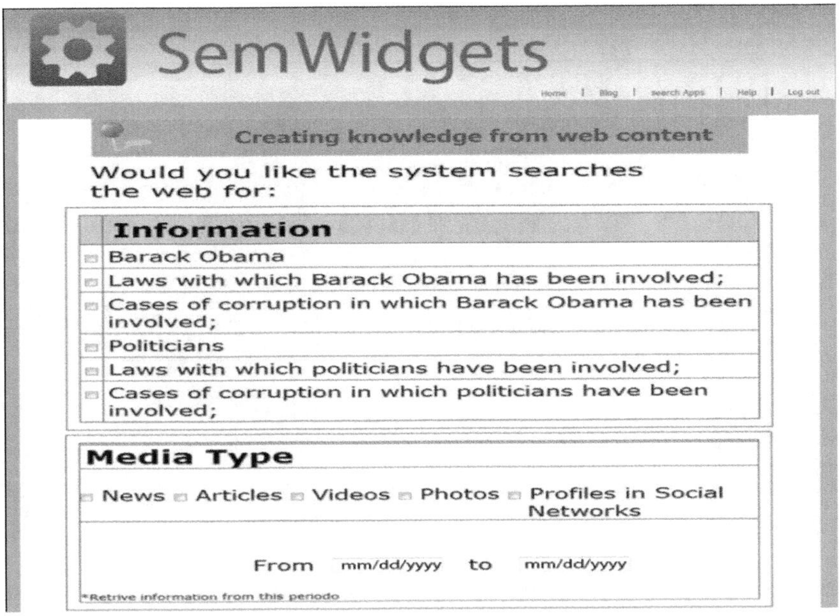

Fig. 4. Screenshot of the *SemWidgets* interaction with the user about the subject s/he wants to retrieve content while using a website

5.2 Putting Semantic to Work

To exemplify how the knowledge represented on a Widget generated from *SemWidgets* works, let's assume the existence of a blog called "Politics." On this blog, the content producer, using *SemWidgets*, creates a widget from the linguistic expression "Politician," which was associated with InferenceNet's "politician" concept that in turn is linked to instances of Yago and DBpedia, as described in the previous section. The blog's owner has chosen the following subject during the process of interaction with *SemWidgets*:

- laws with which politicians have been involved;
- cases of corruption in which politicians have been involved.

Also, s/he has selected to receive news and photos as media content.

Figure 5 presents the results of a query made by the Widget, whereby a news item on a law in which Obama as a politician was involved was retrieved from the New York Times service. The News text highlighted is about a law that addresses re-election and in which Obama was involved. The widget also presents photos related to Obama and to the Law that are retrieved from DBpedia. In this example, the request was made to The Article Search API of the New York Times[3], sending: (i) the URI of the concept "politician" in DBpedia, retrieved from the link between InferenceNet and Yago, and (ii) the URI of the concept "law" (related to "politician") which was also recovered from InferenceNet and Yago.

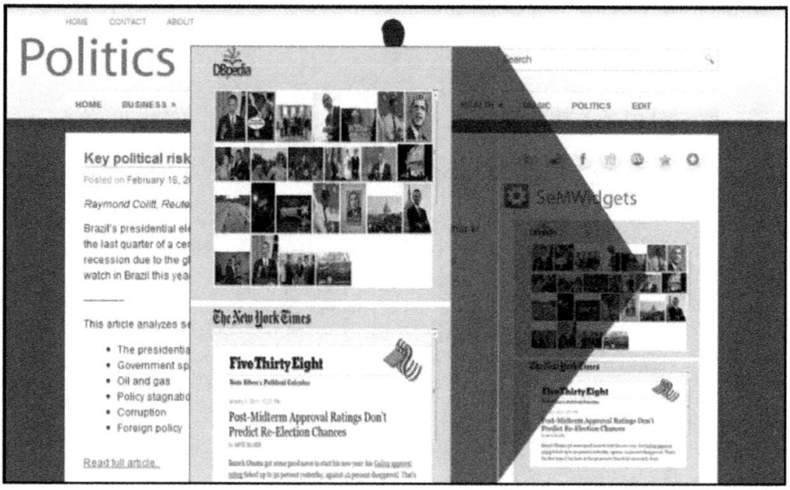

Fig. 5. Example of search and content retrieval in LOD by *SemWidgets*, powered by InferenceNet knowledge

5.3 Empirical Evaluation

Our basic hypothesis to be verified is that as the widget understands the concept that is being manipulated, it can retrieve and organize linked open data in a more informative, targeted and easy way. For this, we made a qualitative comparative analysis between the results of using a website with a widget generated from *SemWidgets* (a knowledge widget) and the same website embedding a Google search in the content of Wikipedia and New York Times sites.

The sample consisted of 10 users accustomed to the use of information systems and social media (blogs and social networks). These users participated in two sessions using two collaborative maps whose theme was mapping politicians. In one map, here called semantic, the widget shows information to the user about politicians coming from Wikipedia and the New York Times using LOD and InferenceNet. The other contains a Google widget, here called "syntactic," that accesses Wikipedia and New York Times with the parameters "politicians Illinois." In the first session of use, five users accessed the "semantic" map and the other five accessed the "syntactic" map. In the second session, the positions were inverted, those who had used the semantic map accessed the syntactic map and vice versa.

---

[3] See `http://developer.nytimes.com/docs/article_search_api/`

Based on the observation of users, data collected in the maps used and the questionnaire responses, we could note the following. As expected, the information searched and presented to users through the map with semantic widgets proved to be better contextualized, facilitating the consumption of information. First of all, we understand that this is due to the fact that the semantic widget brought information about persons rather than general information about politics, as the syntactic one did. The syntactic map shows information about political institutes, parties, etc. This fact was perceived by the users and was considered to be the main difference between the maps. Another observation found in the user's answers to the questionnaire was that the information on the semantic map was well structured, and the information shed light on relevant aspects they needed to consider in the task.

Our interpretation of the answers to the open questions was reinforced by a quantitative analysis of the log of user interaction. When interacting with the semantic map, users performed much more interactions than when using the syntactic map. On the semantic map, the 10 users marked 24 politicians, and out of 120 fields (24 markers x 5 fields per marker) likely to be filled in, 73 of them were filled in by the users (rate of 0.60). For the syntactic scenario, 21 markers were created by the users and, out of 105 fields to be filled in, only 20 ones received information (rate of 0.20).

But the difference is not just the quantity of information added. The field *additional facts*, when completed on the syntactic map, always had information of a general nature such as the name of the political party, the age of the politician, etc., which was often redundant with what was already described in other fields. On the map customized with the semantic widget, it was most common to find information about political scandals, corruption or positive initiatives that exemplify the ethical character of politicians being marked. Information of this type seems to be more informative for the purpose of helping voters to choose a candidate.

6 Conclusion

In this paper we present the process of towards InferenceNet – the first resource with common sense and inferentialist knowledge in Portuguese language – for the Linked Open Data cloud. Our main goal is to leverage the use and development of Semantic Web applications by content producers in Portuguese language. In order to show how InferenceNet in LOD can be used, we develop a platform, called *SemWidgets*, for the creation and execution of piece of programs able to access and reason over Inferencenet and the open linked data, like DBPedia, Yago, and Article Search API of the New York Times, and others. We evaluated empirically the use of a semantic widget in one collaborative map and, as expected, the information searched and presented to users through the map with semantic widgets proved to be better contextualized, facilitating the consumption of information.

As future work, we are conducting an intrinsic assessment of the links between InferenceNet and Yago and DBPedia and conducting studies on how to take advantage of recent developments in knowledge base ConceptNet 5, which already has concepts related to Wikipedia.

References

1. Auer, S., Bizer, C., Kobilarov, G., Lehmann, J., Cyganiak, R., Ives, Z.G.: DBpedia: A Nucleus for a Web of Open Data. In: Aberer, K., Choi, K.-S., Noy, N., Allemang, D., Lee, K.-I., Nixon, L.J.B., Golbeck, J., Mika, P., Maynard, D., Mizoguchi, R., Schreiber, G., Cudré-Mauroux, P. (eds.) ISWC/ASWC 2007. LNCS, vol. 4825, pp. 722–735. Springer, Heidelberg (2007)
2. Suchanek, F.M., Kasneci, G., Weikum, G.: Yago: a core of semantic knowledge. In: Proceedings of the 16th International Conference on World Wide Web, Banff, Alberta, Canada, May 08-12 (2007)
3. Fellbaum, C. (ed.): WordNet: An electronic lexical database. MIT Press (1998)
4. Pinheiro, V., Pequeno, T., Furtado, V., Franco, W.: InferenceNet.Br: Expression of Inferentialist Semantic Content of the Portuguese Language. In: Pardo, T.A.S., Branco, A., Klautau, A., Vieira, R., de Lima, V.L.S. (eds.) PROPOR 2010. LNCS (LNAI), vol. 6001, pp. 90–99. Springer, Heidelberg (2010)
5. Pinheiro, V., Pequeno, T., Furtado, V., Nogueira, D.: Information Extraction from Text Based on Semantic Inferentialism. In: Andreasen, T., Yager, R.R., Bulskov, H., Christiansen, H., Larsen, H.L. (eds.) FQAS 2009. LNCS (LNAI), vol. 5822, pp. 333–344. Springer, Heidelberg (2009)
6. Bizer, C., Heath, T., Berners-Lee, T.: Linked Data - The Story So Far. In: International Journal on Semantic Web and Information Systems, Special Issue on Linked Data (2011) (in press)
7. Berners-Lee, T.: Linked Data - Design Issues (2006), `http://www.w3.org/DesignIssues/LinkedData.html`
8. Bick, E.: The Parsing System "Palavras". Automatic Grammatical Analysis of Portuguese in a Constraint Grammar Framework. Aarhus University Press (2000)
9. Delbru, R., Campinas, S., Tummarello, G.: Searching Web Data: an Entity Retrieval and High-Performance Indexing Model. Journal of Web Semantics (2011), `http://siren.sindice.com/`
10. Brickley, D., Miller, L.: FOAF Vocabulary Specification. Technical report, RDFWeb FOAF Project (2003)

Weak Object Pronouns in Brazilian Portuguese: An LFG Analysis

Ana R. Luís*

University of Coimbra/CELGA
aluis@fl.uc.pt

Abstract. This paper argues that weak object pronouns in Brazilian Portuguese (BP) are best viewed as verbal morphology. However evidence also indicates that they can take wide scope over coordinated verbs, a property which entails that syntax can have access to the internal structure of words. To account for the mismatch between the morphological and syntactic properties of weak object pronouns in BP, we propose a mapping algorithm, formulated within LFG, that allows morphological sequences to be mapped onto more than one phrase-structure node.

Keywords: LFG, Brazilian Portuguese, object pronouns, morphology, coordination.

1 Introduction

An important claim about the grammar of the Romance languages is that weak object pronouns, such as the French third singular object *le* (as in *Martine le lit* 'Martine reads it'), are integrated into the morphology of the verb, rather than treated as independent phrase-structure nodes. This observation has been formalized, within non-derivational formalisms like HPSG and LFG, by Andrews (1990), Miller & Sag (1997), Monachesi (1999), Crysmann (2002), Miller & Monachesi (2003), Luís (2004), among other.

Within the parallel architecture of LFG, the morphological status of weak object pronouns is captured by mapping the OBJ feature onto a verbal node, as in (1). For nominal complements, the OBJ features are mapped onto an independent phrase-structure node, as in (2). Thus, the nature of the mapping between functional information and configurational/phrase-structural information is determined by the grammatical status of the objects.

(1)

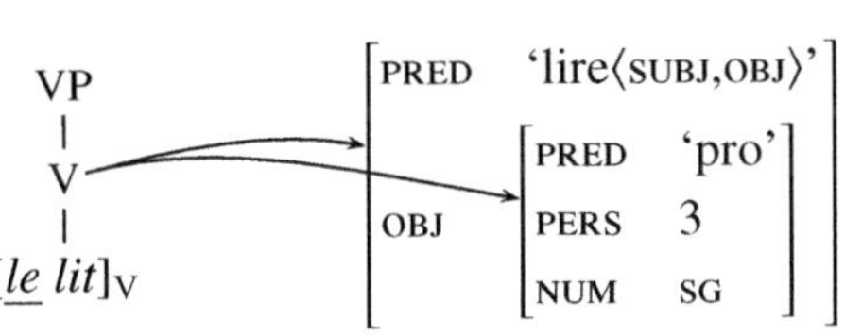

* I would like to thank Ryo Otoguro for his much appreciated support and one anonynmous reviewer for helpful comments.

H. Caseli et al. (Eds.): PROPOR 2012, LNAI 7243, pp. 139–145, 2012.

(2)

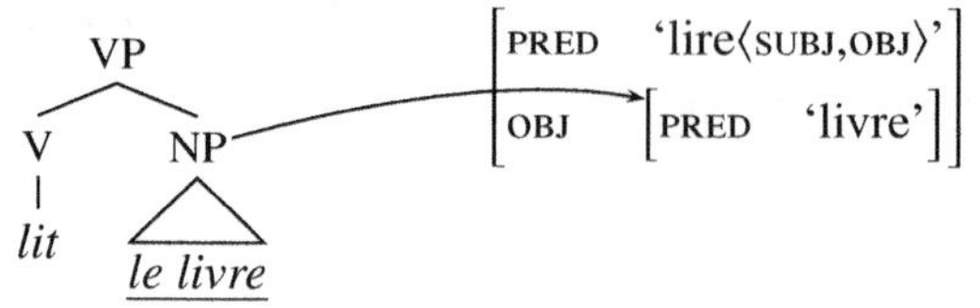

Whereas the morphological status of weak object pronouns has found support in a number of Romance languages, non-derivational theories of grammar have paid hardly any attention to Brazilian Portuguese (BP). Therefore, the goal of this paper will be to investigate the grammatical status of BP weak pronouns such as *me* (e.g., *Uma amiga me deu uma carona* 'A friend gave me a lift) and to argue that they do not exhibit the syntactic autonomy of function words but behave like morphologically dependent units.

We also show that weak pronouns in BP take wide scope over VPs, a property which entails that syntax can have direct access to the internal components of words (see also Crysmann 2002, for European Portuguese). Scopal transparency poses a serious challenge to the Principle of Lexical Integrity given that syntactic processes are not allowed to refer to parts of a word (Lapointe 1980, Bresnan 2001). To account for the mismatch between the morphological and the syntactic behaviour of weak object pronouns in BP, we propose a mapping algorithm within LFG that allows morphologically well-formed sequences to be mapped onto more than one syntactic atom, in line with insights by Wescoat (2002), Luís (2004), Otoguro (2006) and Luís & Otoguro (2011). From our analysis, it follows that weak object pronouns in BP, despite being integrated into the morphology of the verb, remain nonetheless transparent to syntactic operations and accessible to coordination.

The structure of our paper is as follows. Section 2 lays out the basic empirical data supporting the morphological status of weak object pronouns in BP. Section 3 provides evidence which shows that they can have wide scope over coordinated verbs. Section 4 formulates an LFG account of the morphology-syntax of weak object pronouns and section 5 provides a short conclusion.

2 Weak Object Pronouns in Brazilian Portuguese: Empirical Survey

We start this section by looking at distributional criteria which show that weak object pronouns in BP do not behave like function words but are integrated into the morphology of the verb. We then provide morphological evidence.

In analogy to other Romance languages, weak object pronouns in BP cannot be coordinated (cf.(3a)), cannot bear constrastive stress (cf.(3b)) and cannot form an utterance on their own (cf.(3c)). This behaviour clearly shows that weak pronouns do not have the syntactic autonomy of function words.

(3) a. *O menino me e te convidou.
'He has invited me and you.'

b. *Ela está ME ou TE convidando?
'Is he inviting me or you?'

c. Quem você está convidando? * Me?
'Who are you inviting? Me?'

On the contrary, function words, such as the personal pronouns *ele* '3SG.MASC' and *ela* '3SG.FEM', can be coordinated (cf.(4a)), can bear contrastive stress (cf.(4b)) and can occur in isolation (cf.(4c)):

(4) a. Certo dia, encontrei ele e ela voltando do trabalho.
'One day I found him and her returning from work.'

b. E a conta quem paga? ELE ou ELA?
'And who pays the bill? He or she?'

c. Quem está falando? Ele?
'Who is speaking? He?'

Further reinforcing our claim that weak object pronouns are not syntactically independent words is their selective nature: they can only be followed by verbs. As shown in (5a), they occur in strict adjacency with the verb and nothing can intervene between the weak pronoun and the verb (5b). Function words, on the contrary, do not show such selectivity: e.g., the preposition *com* 'with' can precede nouns (e.g., *com crianças* 'with children'), determiners (e.g., *com essas* 'with those.FEM'), quantifiers (*com muitas* 'with many'), among other word classes.

(5) a. Ontem ele me perguntou.
'Yesterday he asked me.'

b. *Ele me ontem perguntou.

We will now provide morphological evidence in support of the morphological status of weak object pronouns in BP. An important piece of evidence is the fact that they occur invariably in preverbal position (Galves 2001, Duarte et al 2005), as shown in (5a). In the other Romance languages, on the contrary, the position of weak pronouns may be either preverbal or postverbal, depending on the properties of the verb (as in Spanish, Italian or French) or on the properties of the clause (as in European Portuguese).

Another piece of evidence is that weak object pronouns show idiosyncratic gaps in their paradigm. While the Romance languages use their complete inventory of forms (including all number and gender combinations), BP excludes several forms, including the 3rd singular and plural forms (i.e., *o* 'OBJ.3SG.MASC', *a* 'OBJ.3SG.FEM', *os* 'OBJ.3PL.MASC', *as* 'OBJ.3PL.FEM', *lhe* 'OBJ2.3SG', *lhes* 'OBJ2.3PL') and the 1st and 2nd plural forms (i.e., *nos* '1PL' and *vos* '2PL') (Araújo Ramos 1992, Galves et al 2005). The allowed inventory of weak object pronouns in BP is therefore restricted to the pronouns *se* REFL, *me* 1SG, *te* 2SG and, also, the weak pronoun *lhe/s* (but with a 2SG/PL reading) (Nunes 1995, Cyrino et al. 2000, among other).

It is difficult to see how such exclusion might be accounted for on general syntactic grounds, not least because there appears to be no syntactic principle that would license one set of pronominal features over another. However, under the assumption that weak pronouns are integrated into the morphology of the verb, the exclusion of weak pronouns can be viewed simply as the result of idiosyncratic gaps in the morphological paradigms of BP verbs. This is what defines defective paradigms, namely the fact that not all morphosyntactic combinations have a morphological form. Hence, under our view, the BP data illustrates morphological idiosyncrasies that are not at all uncommon in verbal paradigms.

One final (and crucial) property concerns the behaviour of weak object pronouns in complex tenses and predicates. As shown in (6a), weak pronouns systematically precede the thematic verb (i.e., the verb whose argument-structure they realize) rather than the auxiliary verb (i.e., the verb carrying the finite features). This property, which has been previously observed by a number of scholars (Kato 1993, Galves et al. 2005) is absent from the Romance languages. In Italian, for example, weak pronouns will 'climb' to the vicinity of the finite verb, as shown in (6b). In BP, however, they are inserted into phrase-structure as part of the morphology of their thematic verb.

(6) a. Você vai [me esquecer]$_V$.
a′. *Você me [vai esquecer].
'You will forget me.'
b. Giovanni l'[ha letto].
'Giovanni has read it.'

Summing up: The data provided in this section has shown that the behaviour of weak pronouns in BP is dependent on the word class of the adjacent word and on the morphosyntactic features of the pronouns. A number of idiosyncrasies and local restrictions determine their behaviour, thus indicating that weak object pronouns in BP should be assigned morphological status.

3 Weak Object Pronouns in BP and Coordination

The ability to show wide scope over coordination is a property that is generally expected of autonomous lexical items, such as function words and content words (Villavicencio et al. 2005, Wälchli 2005, Kuhn & Sadler 2007). For example, a quantifier such as *muito* 'many' may be shared by two coordinated nouns (e.g., *carro e avião* 'car and plane') and, thus, take wide scope over the whole NP, as in *montei muito carro e avião da 2ª guerra* 'I assembled many cars and planes from World War II'.

Surprisingly, weak object pronouns in BP also exhibit wide scope. As the data in (7–9) illustrates, the weak pronoun *me* is shared by two coordinated verbs in (7), and by both *levava à escola* 'took to school' and *buscava* 'picked up' in (8); in (9), *se* takes wide scope over the verb *olharam* 'looked' and the verb phrase *cumprimentaram com um sinal de mãos* 'greeted with a hand-wave'.

(7) A menina me viu e abraçou e não queria me soltar.
'The girl saw and hugged me and didn't want to let go.'

(8) ...que durante alguns meses [ele] me levava à escola e buscava.
'...that for several months he took me to school and picked me up.'

(9) Eles se olharam e cumprimentaram com um sinal de mãos.
'They looked at each other and greeted each other with a hand-wave.'

It is interesting to observe that, while weak object pronouns can be shared by coordinated verbs, they still preserve the strict adjacency to the verb. The first member of the conjunct must be a verb and nothing can intervene between the weak pronoun and the first verb. Function words, on the contrary, do not necessarily show any specific adjacency requirements when they are shared under coordination.

The fact that weak object pronouns in BP can be shared by two coordinated verbs clearly indicates that weak object pronouns in BP exhibit mixed properties: while distributional and morphological evidence (see section 2) shows that they do not have the autonomy of typical function words, scopal evidence seriously challenges the word-internal status of weak pronouns in BP, not least because it violates one fundamental tenet of LFG, namely the Principle of Lexical Integrity. Under this principle, syntax cannot have access to the internal structure of words (Lapointe 1980, Bresnan and Mchombo 1995). Formally, then, an analysis is needed which accounts for both the syntactic transparency and the morphological dependency of these weak pronouns.

In this paper, we will argue that both the morphological dependency and the syntactic transparency of weak object pronouns in BP follows from the fact that they are phrasal affixes. This effectively means that, while they constitute morphological units, they do not attach to roots (like word-level affixes) but to phrasal hosts (V′ or VP). The phenomenon of phrasal affixation has been previously examined, for a wide range of language, by scholars such as Klavans (1985), Anderson (1992, 2005), Miller (1992), Spencer (2000), Crysmann (2002), among other, who have shown that it is not uncommon for languages to have linguistic units with an intermediate grammatical status.

4 Representing Weak Object Pronouns in Phrase-Structure

From our discussion so far, it is clear that BP weak objects cannot receive the c-structure representation given to the weak pronoun *le* and the verb *lit*, in (1) above, which shows that both the French pronoun and the verb are dominated by the same c-structure node. This indicates that French weak pronouns are opaque to syntax and, hence, cannot have wide scope, unlike preverbal weak pronouns in BP, which are visible to phrase-structure rules and, hence, should appear on an independent node.

To capture the fact that pronominal affixes in BP are both morphologically dependent and syntactically transparent, we explore a new approach to wordhood, within LFG, following insights by Luís (2004), Otoguro (2006) and Luís & Otoguro (2011). In particular, we assume that morphological strings cannot be inserted directly into the phrase structure. Instead, at the interface between morphology and c-structure, we propose that morphological tokens must be put in correspondence with syntactic atoms. We define morphological tokens and syntactic atoms, as follows:

(10) a. Morphological token: Each morphological token corresponds to a well-formed stem-affix string which is defined by morphology-internal principles.

b. Syntactic atom: Syntactic atoms are leaves on c-structure trees; each leaf corresponds to one and only one terminal node; the insertion of syntactic atoms into c-structure is subject to standard phrase structure constraints.

At the morphological level, we propose that inflectional strings are defined by principles of inflectional morphology. We adopt Stump's (2001) theory of Paradigm Function Morphology (PFM), more specifically a revised version of it, called Generalised Paradigm Function Morphology (GPFM) (Spencer ms). The morphological sequence generated by the morphology, however, cannot be directly inserted into phrase-structure. We therefore propose that the stem-affix string *me deu* constitutes one morphological token, which is represented as [*me*, *deu*] in (11b).

At the c-structure level, syntactic atoms are the leaves of c-structure trees. In most cases, one morphological token corresponds to one syntactic atom (cf. (1) and (11a), for French). However, the correspondence between morphological tokens and syntactic atoms may also be non-isomorphic, as in BP, where the sequence *me deu* corresponds to 'two' syntactic atoms and the weak object pronoun carries the category label PRO_{CL}, as shown in (11b):

(11) a. [*le*, *lit*] = $\textit{le-lit}_{\text{V}}$

b. [*me*, *deu*] = $\textit{me}_{\text{PRO}_{\text{CL}}}$ $\textit{deu}_{\text{V}}$

In c-structure, the morphologically well-formed sequence *me deu* is mapped onto two c-structure terminals, in harmony with standard phrase structure principles (such as immediate domination, linearisation and instantiation) and BP phrase structure, as given in (12). Under this view, morphological strings cannot be inserted directly into the phrase structure.

(12)

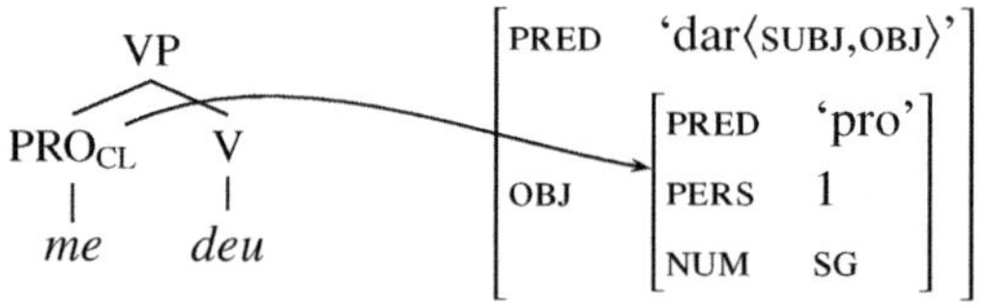

5 Conclusion

The puzzle posed by BP weak object pronouns can be attributed to their partly affixal and partly phrasal properties. By separating inflectional strings from their phrase structural properties, our mapping algorithm captures these mixed properties. The key goal of our analysis is to allow single morphological tokens (i.e., stem-affix combinations) to be mapped onto one or more syntactic atoms, at the interface between morphology and c-structure, without incurring any violation of lexical integrity. Our proposal is based on the view that 'integrity' can be defined as a condition on morphological tokens (i.e. complete and well-formed stem-affix sequences). This constitutes a challenge to the traditional view of 'integrity' which is defined in terms of phrase structure terminals.

References

Anderson, S.R.: A-Morphous Morphology. Cambridge University Press, Cambridge (1992)
Anderson, S.R.: Aspects of the Theory of Clitics. Oxford University Press, Oxford (2005)
Andrews, A.D.: Unification and morphological blocking. Natural Language and Linguistic Theory 8, 507–557 (1990)

Bresnan, J.: Lexical-Functional Syntax. Blackwell, Oxford (2001)
Bresnan, J., Mchombo, S.A.: The lexical integrity principle: Evidence from Bantu. Natural Language and Linguistic Theory 13, 181–254 (1995)
Crysmann, B.: Constraint-based Coanalysis. Ph.D. thesis, Universität des Saarlandes and DFKI Gmbh (2002)
Cyrino, S., Duarte, M., Kato, M.: Visible subjects and invisible clitics in Brazilian Portuguese. In: Kato, M., Negrão, E. (eds.) Brazilian Portuguese and the Null Subject Parameter, pp. 55–73. Vervuert, Frankfurt (2000)
Duarte, I., Matos, G., Gonçalves, A.: Pronominal clitics in European and Brazilian Portuguese. Journal of Portuguese Linguistics 4(2), 113–141 (2005)
Galves, C.: Ensaios Sobre as Gramáticas do Português. Editora da Unicamp, Campinas, São Paulo (2001)
Galves, C., Ribeiro, I., Morais, T., Aparecida, M.: Syntax and morphology in the placement of clitics in European and Brazilian Portuguese. Journal of Portuguese Linguistics 4(2), 143–177 (2005)
Kato, M.: The distribution of pronouns and null elements in object position in Brazilian Portuguese. In: William Ashby, P., Raposo, E. (eds.) Linguistic Perspectives on the Romance Languages, pp. 225–235. John Benjamins, Amsterdam (1993)
Klavans, J.: The independence of syntax and phonology in cliticization. Language 61, 95–120 (1985)
Kuhn, J., Sadler, L.: Single conjunct agreement and the formal treatment of coordination in LFG. In: Butt, M., King, T.H. (eds.) Proceedings of the LFG 2007 Conference, pp. 302–322. CSLI Publications, Stanford (2007)
Lapointe, S.G.: A Theory of Grammatical Agreement. Ph.D. thesis, University of Massachusetts, Amherst (1980)
Luís, A.: Clitics as Morphology. Ph.D. thesis, University of Essex (2004)
Luís, A., Otoguro, R.: Inflectional morphology and syntax in correspondence: Evidence from European Portuguese. In: Galani, A., Tsoulas, G., Hicks, G. (eds.) Morphology and Its Interfaces, pp. 187–225. John Benjamins, Amsterdam (2011)
Miller, P.H.: Clitics and Constituents in Phrase Structure Grammar. Garland, New York (1992)
Miller, P.H., Monachesi, P.: Les pronoms clitiques dans les langues romanes. In: Godard, D. (ed.) Langues Romanes, Problemes de la Phrase Simple, pp. 67–123. CNRS Editions, Paris (2003)
Miller, P.H., Sag, I.: French clitic movement without clitics and movement. Natural Language and Linguistic Theory 15, 573–639 (1997)
Monachesi, P.: A Lexical Approach to Italian Cliticization. CLSI Publications, Stanford (1999)
Nunes, J.: Aninda o famigerado SE. D.E.L.T.A. 2(11), 201–240 (1995)
Otoguro, R.: Morphosyntax of Case: A Theoretical Investigation of the Concept. Ph.D. thesis, University of Essex (2006)
Spencer, A.: Verbal clitics in Bulgarian: A paradigm function approach. In: Grijzenhout, J., Gerlach, B. (eds.) Clitics in Phonology, Morphology and Syntax, pp. 355–386. John Benjamins, Amsterdam (2000)
Spencer, A.: Generalized Paradigm Function Morphology (ms), University of Essex
Stump, G.T.: Inflectional Morphology: A Theory of Paradigm Structure. Cambridge University Press, Cambridge (2001)
Villavicencio, A., Sadler, L., Arnold, D.: An HPSG account of closest conjunct agreement in NP coordination in Portuguese. In: Müller, S. (ed.) Proceedings of the HPSG 2005 Conference, pp. 427–447. CSLI Publications, Stanford (2005)
Wälchli, B.: Co-Compounds and Natural Coordination. Oxford University Press, Oxford (2005)
Wescoat, M.T.: On Lexical Sharing. Ph.D. thesis, Stanford University (2002)

Entropy-Guided Feature Generation for Structured Learning of Portuguese Dependency Parsing

Eraldo R. Fernandes[1,2] and Ruy L. Milidiú[1]

[1] PUC-Rio, Rio de Janeiro, Brazil
milidiu@inf.puc-rio.br
http://www.puc-rio.br
[2] IFG, Jataí, Brazil
eraldoluis@ifg.edu.br
http://www.ifg.edu.br

Abstract. Feature generation is a difficult, yet highly necessary, subtask of machine learning modeling. Usually, it is partially solved by a domain expert that generates complex and discriminative feature templates by conjoining the available basic features. This is a limited and expensive way to obtain feature templates and is recognized as a modeling bottleneck. In this work, we propose an automatic method to generate feature templates for structured learning algorithms. The method receives as input the training dataset with basic features and produces a set of feature templates by conjoining basic features that are highly discriminative together. We denote this method *entropy guided* since it is based on the conditional entropy of local decision variables given the feature values. We illustrate our approach on the Portuguese dependency parsing task and report on experiments with the Bosque corpus. We show that the entropy-guided templates outperform the manually built templates used by MSTParser, which was the best performing system on the Bosque corpus up to now. Furthermore, our approach allows an effortless inclusion of two new basic features that automatically generate additional templates. As a result, our system achieves a per-token accuracy of 92.66%, what represents a reduction by more than 15% on the previous smallest error rate for Portuguese dependency parsing.

Keywords: dependency parsing, machine learning, structured learning, entropy-guided feature generation.

1 Introduction

Dependency parsing (DP) is known to be useful in many applications, such as question answering, machine translation, information extraction, and natural language generation. Particularly, semantic role labeling greatly benefits from DP [13]. Accordingly, DP was part of the Conference on Natural Language Learning (CoNLL) shared task from 2006 to 2009.

H. Caseli et al. (Eds.): PROPOR 2012, LNAI 7243, pp. 146–156, 2012.

Dependency parsing is to derive a syntactic structure for an input sentence. The words and punctuation marks in a sentence are called tokens. The dependency structure is a rooted tree, called *dependency tree*, whose nodes are the sentence tokens. Each dependency edge from a *head* token to a *dependent* token defines the syntactic dependency of the dependent token on its head. Following CoNLL'2006 shared task definition [3], the dependency tree can be derived from the sentence constituent structure by recursively defining a head token for each constituent. Additionally, each dependency edge is labeled according to its type. For instance, the dependency of an adjective on a noun is labeled as *modifier*. However, in this work, we do not consider the dependency type and focus exclusively on the tree structure of a sentence. Moreover, most dependency parsers label the dependency edges only after finding the dependency tree structure.

Let $\boldsymbol{x} = (x_0, x_1, \ldots, x_T)$ be the input sentence, where x_i is the i-th token in $\boldsymbol{x}$ and x_0 is a dummy token which is always the root of the dependency tree. For Portuguese and many other languages, the dependency structure of a sentence is always a rooted tree with each token as a node. We define $\mathcal{Y}(\boldsymbol{x})$ as the set of all rooted trees whose nodes are tokens in the sentence $\boldsymbol{x}$ and the root node is x_0. For any dependency tree $\boldsymbol{y} \in \mathcal{Y}(\boldsymbol{x})$, we say that $(i, j) \in \boldsymbol{y}$ whenever token x_i is the head of token x_j in the tree $\boldsymbol{y}$. Figure 1 shows the sentence "*João viu Maria*", the corresponding token heads and the dependency tree $\boldsymbol{y} = \{(0, 2), (2, 1), (2, 3)\}$.

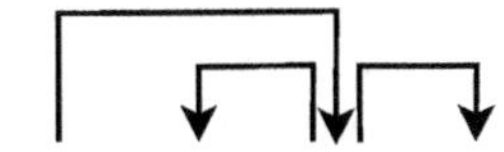

$\boldsymbol{y}$				
$\boldsymbol{i}$	0	1	2	3
$\boldsymbol{x}$	*root*	João	viu	Maria
head	–	2	0	2

Fig. 1. Dependency tree example

Several dependency parsing systems have been proposed and, in the last years, machine learning methods have achieved state-of-the-art performance on most of the available datasets [3]. For Portuguese, for instance, the main dependency treebank available is based on the Bosque corpus [12] and was derived for the CoNLL'2006 shared task. MSTParser [15–17] currently holds the best performance record on this corpus [3]. This system is based on a structured learning method and casts DP as a maximum branching problem on the graph whose nodes are the sentence tokens.

MSTParser makes use of basic features available in the Bosque corpus, like words and part-of-speech (POS) tags. However, these features are not used independently. They are conjoined to generate more complex and discriminative features. For instance, the system combines the dependent token POS tag and the head token POS tag to create a feature that aggregates information from both ends of a dependency relation. Such combinations are key to improve a system performance, and we call them feature templates or, simply, *templates*. MSTParser uses 21 manually generated templates.

Feature generation or, more specifically, template generation, is a difficult yet highly necessary subtask of machine learning modeling that generally requires a domain expert to formulate them. This is a limited and expensive way to obtain feature templates and has been recognized as a modeling bottleneck. In this work, we propose an automatic method to generate feature templates for structured learning algorithms. The method receives as input the training dataset with basic features and produces a set of feature templates by conjoining basic features that are highly discriminative together. We denote this method *entropy guided feature generation* (EFG) since it builds templates that minimize the conditional entropy of local decision variables given basic feature values. We illustrate our approach by applying it to the Portuguese dependency parsing task. We report on experiments with the Bosque corpus showing that EFG outperforms the manually built templates used by MSTParser.

We further demonstrate the value of the proposed feature generation method by automatically including two new basic features in our system, namely phrase chunks and clauses. The system unlabeled attachment score (UAS) is improved by more than one percent when including these features through EFG. Moreover, this system reduces the previous smallest error by more than 15%, achieving a UAS of 92.66% against the MSTParser score of 91.36%. Observe that to manually include new basic features in MSTParser, for instance, it is necessary that a domain expert builds discriminative feature templates. Usually, it is necessary to perform a time consuming procedure that iteratively adjusts the new templates in order to optimize the overall system performance, just as has been done with the Bosque original basic features. Additionally, as our experimental results indicate, the final performance achieved with manually built templates is not better than with EFG. These results demonstrate the great value of the proposed feature generation method.

The following section presents the structured learning modeling of dependency parsing and the machine learning algorithm used to train our DP system. Section 3 describes the entropy-guided template generation method. Experimental results on the Portuguese language corpus Bosque are discussed in Section 4 and, finally, we present our concluding remarks in Section 5.

2 Structured Learning

Structured learning (SL) [1, 7] comprises some machine learning methods that build a mapping from inputs to complex outputs by processing a given training set of input-output pairs. These approaches are very general and, compared to classic classification techniques, allow a more natural yet powerful modeling of tasks that involve learning complex and interdependent outputs like sequences, trees and even more general graphs. Many NLP tasks have been successfully modeled as SL problems. To name a few, sequence labeling [2, 7], word sense disambiguation [5], parsing [26], and machine translation [14].

2.1 Dependency Tree Learning

The current best performing system for Portuguese dependency parsing is MST-Parser, a SL method. MSTParser recasts the parsing problem as a maximum branching problem [24]. They apply an online algorithm to learn a scoring function $s(i,j)$ over all candidate head-dependent edges (i,j) such that the *prediction problem* $F(\boldsymbol{x})$, that is to predict a dependency tree for an input sentence $\boldsymbol{x}$, is to find the maximum-score tree among the valid rooted trees:

$$F(\boldsymbol{x}) = \arg \max_{\boldsymbol{y} \in \mathcal{Y}(\boldsymbol{x})} s(\boldsymbol{y}),$$

where $s(\boldsymbol{y}) = \sum_{(i,j)\in \boldsymbol{y}} s(i,j)$ is the score of a candidate tree $\boldsymbol{y} \in \mathcal{Y}(\boldsymbol{x})$, i.e., the sum of its edges scores. This is the well studied maximum branching problem that can be efficiently solved by Chu-Liu-Edmonds' algorithm [4, 10]. There is also an improved implementation of this algorithm by Tarjan [24].

By processing a given sample of correct sentence-tree pairs, MSTParser learns a scoring function that generalizes for unseen sentences. The scoring function takes the form

$$s(i,j) = \sum_{m=1}^{M} w_m \cdot \phi_m(\boldsymbol{x}, i, j),$$

where $\boldsymbol{w} = (w_1, \ldots, w_M)$ is the weight vector that defines a dependency parsing model and $\phi_m(\cdot,\cdot,\cdot) \in \mathbb{R}$ is the m-th feature function. The feature functions depend on a unique edge $(i,j) \in \boldsymbol{y}$, but can use any information directly derived from the whole input sentence $\boldsymbol{x}$.

By observing that

$$s(\boldsymbol{y}) = \sum_{(i,j)\in \boldsymbol{y}} \sum_{m=1}^{M} w_m \cdot \phi_m(\boldsymbol{x}, i, j) = \sum_{m=1}^{M} w_m \cdot \sum_{(i,j)\in \boldsymbol{y}} \phi_m(\boldsymbol{x}, i, j),$$

and defining $\phi_m(\boldsymbol{x}, \boldsymbol{y}) = \sum_{(i,j)\in \boldsymbol{y}} \phi_m(\boldsymbol{x}, i, j)$ as the m-th *global* feature function, that is just the sum of the m-th feature function over all edges in the tree, we obtain that

$$s(\boldsymbol{y}) = \langle \boldsymbol{w}, \boldsymbol{\Phi}(\boldsymbol{x}, \boldsymbol{y}) \rangle$$

where $\boldsymbol{\Phi}(\boldsymbol{x}, \boldsymbol{y}) = (\phi_1(\boldsymbol{x}, \boldsymbol{y}), \ldots, \phi_M(\boldsymbol{x}, \boldsymbol{y}))$ is the joint feature mapping function and $\langle \cdot, \cdot \rangle$ is the scalar product operator.

Thus, the learning problem consists in determining the feature weights, such that the resulting predictor $F(\boldsymbol{x})$ is accurate on the training data and, moreover, shows good generalization performance on unseen data.

2.2 Structured Perceptron

The structured perceptron algorithm [7] is analogous to its univariate counterpart [21]. Given a training sample $\mathcal{D}$ of correct sentence-tree pairs, the algorithm generates a sequence $\boldsymbol{w}^0 = \boldsymbol{0}, \boldsymbol{w}^1, \ldots, \boldsymbol{w}^m$ of models.

At each iteration t, the structured perceptron draws a training instance $(\boldsymbol{x}, \boldsymbol{y}) \in \mathcal{D}$ and performs two major steps:

(1) a prediction $\hat{\boldsymbol{y}} = F(\boldsymbol{x})$ is made using the current model $\boldsymbol{w}^t$;
(2) $\boldsymbol{w}^{t+1} \leftarrow \boldsymbol{w}^t + \boldsymbol{\Phi}(\boldsymbol{x}, \boldsymbol{y}) - \boldsymbol{\Phi}(\boldsymbol{x}, \hat{\boldsymbol{y}})$.

Note that, when the current model makes a correct prediction $\hat{\boldsymbol{y}} = \boldsymbol{y}$, the model does not change, that is $\boldsymbol{w}_{t+1} \leftarrow \boldsymbol{w}_t$. When the prediction is wrong, the update rule favors the correct output $\boldsymbol{y}$ over the predicted one $\hat{\boldsymbol{y}}$. Regarding binary feature functions, for instance, the update rule increases the weights of features that are present in $\boldsymbol{y}$ but missing in $\hat{\boldsymbol{y}}$ and decreases the weights of features that are present in $\hat{\boldsymbol{y}}$ but not in $\boldsymbol{y}$. The weights of features that are present in both $\boldsymbol{y}$ and $\hat{\boldsymbol{y}}$ are not changed. A simple extension of Novikoff's theorem [18] shows that the structured perceptron is guaranteed to converge to a zero loss solution, if one exists, in a finite number of steps [2, 6].

2.3 Large Margin Classifiers

The structured perceptron algorithm finds a classifier with no concern about its margin. However, it is well known that large margin classifiers provide better generalization performance on unseen data. MIRA is an extension of the structured perceptron algorithm that generates a large margin classifier. MSTParser's training algorithm is based on a generalization of the MIRA algorithm [9].

In this work, we use a large-margin generalization of the structured perceptron that is based on the well known margin rescaling technique for structural support vector machines [25]. For a training instance $(\boldsymbol{x}, \boldsymbol{y}) \in \mathcal{D}$, instead of $F(\boldsymbol{x})$, we use the following *loss-augmented* prediction problem in step 1 of the structured perceptron learning algorithm:

$$F_\ell(\boldsymbol{x}) = \arg \max_{\bar{\boldsymbol{y}} \in \mathcal{Y}(\boldsymbol{x})} s(\bar{\boldsymbol{y}}) + \ell(\boldsymbol{y}, \bar{\boldsymbol{y}}),$$

where $\ell(\cdot, \cdot) \geq 0$ is a given loss function that measures the difference between a candidate tree and the correct one. We use the most common loss function for dependency trees, which just counts, for all tokens within the input sentence, how many head tokens have been *incorrectly* assigned in the predicted tree.

By using the loss-augmented prediction, a training instance $(\boldsymbol{x}, \boldsymbol{y})$ implies a model update when the current model does not respect the following margin constraint:

$$s(\boldsymbol{y}) - s(\bar{\boldsymbol{y}}) \geq \ell(\boldsymbol{y}, \bar{\boldsymbol{y}}), \; \forall \bar{\boldsymbol{y}} \in \mathcal{Y}(\boldsymbol{x}).$$

If a model respects this constraint then the original prediction function $F(\boldsymbol{x})$, which is always used during test time, separates every candidate tree $\bar{\boldsymbol{y}} \in \mathcal{Y}(\boldsymbol{x})$ from the correct tree $\boldsymbol{y}$ by a margin as large as the loss between them $\ell(\boldsymbol{y}, \bar{\boldsymbol{y}})$.

3 Entropy-Guided Feature Generation

Feature generation is an important subtask of structured learning modeling. Usually, a task dataset includes basic features that are either naturally included in the very phenomenon of interest, like words; simply derived from other basic features, like capitalization information; or automatically generated by external systems, like POS tags. However, using basic features independently is not enough to achieve state-of-the-art results. Thus, it is necessary to conjoin basic features to improve performance. Frequently, a domain expert manually generates feature templates by conjoining the given basic features. In the structured learning modeling of dependency parsing, features are defined for a dependency edge that comprises a head token and a dependent token. Additionally, features can make use of any information from the input sentence, since this is fixed. The Bosque corpus includes the following basic features: head and dependent words, head and dependent POS tags, head and dependent morphological features, POS tags of tokens surrounding head and dependent tokens, and POS tags of tokens between head and dependent tokens. MSTParser uses 21 manually created feature templates that conjoin these basic features.

In this section, we describe the proposed entropy-guided feature generation method for structured learning that automatically creates feature templates by conjoining basic features that are highly discriminative together. EFG uses decision tree induction to incorporate the conditional entropy of local decision variables given basic features. Its key idea is to obtain new features by conjoining the selected feature combinations. The same strategy has been used for automatic template generation in Entropy Guided Transformation Learning [22].

3.1 Decision Tree Learning

Decision tree (DT) learning is one of the most widely used machine learning algorithms. This learning algorithm performs a partitioning of the training set using principles of information theory. It executes a general to specific search of the feature space, and, at each step of the search, the most informative feature, w.r.t the decision variable, is added to a tree structure. Information gain, which is based on the data entropy, is normally used as the informativeness measure. The objective is to construct a tree that uses a minimal set of features and efficiently partitions the training set into decision variable values. Usually, after the tree is grown, a pruning step is carried out in order to avoid overfitting.

Information gain is based on entropy and is a key strategy for feature selection. The most popular decision tree learning algorithms [20, 23] use this measure. Hence, they provide a quick way to provide entropy guided feature selection. One of the most used algorithms for DT induction is Quinlan's C4.5 [20]. We use C4.5 to obtain the required entropy guided selected features.

3.2 Template Generation

The first step of our automatic template generation method is to train a decision tree on a dataset, where each example comprises the basic features related to one *local decision variable* of the prediction problem. Local decision variables correspond to edges, since the prediction problem is to choose some edges from all candidate edges between tokens. Thus, given a sentence in the input dataset, we generate a DT example for each candidate edge. The *binary* decision variable indicates whether an edge is included in the correct dependency tree or not. Hence, the DT algorithm learns a decision tree by predicting whether an edge is correct or not.

From the learned DT, our method uses a very simple tree decomposition scheme to extract templates. The decomposition process is based on a depth-first traversal of the DT and thus can be recursively described as follows. For each visited node, a new template is created by conjoining the node feature with its parent template. To limit the maximum template length, we use pruned trees. Figure 2 illustrates the EFG method. The tree in the left side of the figure uses

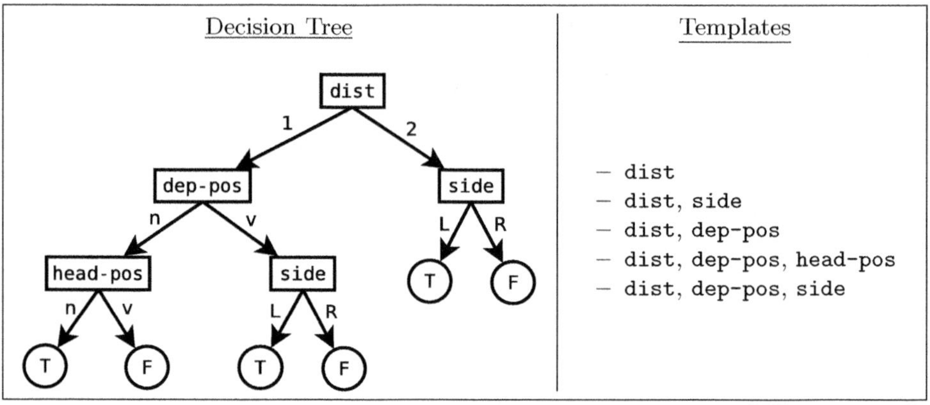

Fig. 2. Feature generation from a decision tree

four basic features (rectangular nodes): `dist` is the absolute distance between the head and the dependent token, `side` is the side of the head token relative to the dependent token, `dep-pos` and `head-pos` are the part-of-speech tag of the dependent and the head token, respectively. The round nodes are the decision variable values (`T` for true and `F` for false). The generated templates are listed in the right side of the figure. In other words, we create a template for each path from the root to every other DT node, ignoring the feature values at the DT edges and, thus, using only the node features.

4 Experimental Results

We perform several experiments on the Portuguese dependency treebank provided on the CoNLL'2006 shared task [3], which is derived from the Bosque corpus. This corpus is already divided into two parts: train and test. We additionally split the train part into two subsets – dev-train and dev-test – in order to perform model selection and choose the number of iterations of the perceptron algorithm. Since structured perceptron is an online algorithm and the order in which examples are processed influences the learned model, the perceptron performances are averages over 10 runs of the training algorithm with shuffled examples. We also report the standard errors of these values.

Our experiments aim at comparing the performance of EFG with the manual templates used in MSTParser. Thus, we use different template sets to train large-margin structured perceptron models and compare their performances on the Bosque test set. The first and second rows in Table 1 respectively present the performances using the manual templates from MSTParser and the entropy-guided templates. Both template sets are based on exactly the same set of basic features. In this table, UAS stands for unlabeled attachment score, that is the accuracy to assign heads to tokens; the parenthesis enclosed values indicate the observed standard errors over 10 runs; and the last column reports the percentage of examples where all tokens are assigned to their correct heads. EFG reduces the error rate by approximately 2.2%, when compared to the performance achieved by MSTParser manual templates. The observed standard errors indicate that these improvements are all statistically significant. Given that EFG avoids the need of an expensive resource, the domain expert, this result highlights EFG value.

Table 1. Test set performances for MSTParser and structured perceptron models (S-Perceptron) with manual and entropy-guided templates

Algorithm	Basic Features	Templates	UAS	Acc./Ex.
	1st order	Manual [15]	90.06 (0.007)	37.33 (0.06)
S-Perceptron	1st order	EFG	90.28 (0.006)	35.38 (0.06)
	1st order+ck+clause	EFG	**92.66** (0.010)	**45.38** (0.10)
MSTParser	1st+2nd order	Manual [16]	91.36	

The original Bosque corpus is a treebank of Portuguese sentences and includes several layers of linguistic annotations. In [11], the authors have used this treebank to extract phrase chunking and clause identification features. Phrase chunks and clauses are highly relevant to the syntactic structure of sentences. In order to further assess our method, we include phrase chunks, clause start tokens and clause end tokens as additional basic features to EFG. The third row in Table 1 presents the performance of the structured perceptron trained with these automatically generated templates. The resulting system achieves an impressive 15% reduction on error rate when compared to the performance of the previous best system, that is MSTParser.

It is important to notice that MSTParser uses some basic features that are not used by our systems. We use basic features that are based only on individual dependency edges, the so called *first order* features. MSTParser additionally makes use of *second order* features that depend on pairs of sibling edges. These more complex features greatly improve its performance, as can be observed when comparing the first and last rows in Table 1. MSTParser training algorithm is very similar to large margin structured perceptron and, thus, it is likely that our system would also benefit from second order features as well.

5 Concluding Remarks

We present a new method to automatically generate highly discriminative features for structured learning by conjoining the given basic features. The proposed method, called EFG, is based on the conditional entropy of local decision variables given basic features. Through experimental evaluation on the Bosque dependency treebank, we show that EFG outperforms manually generated templates when using exactly the same basic features. Moreover, EFG avoids the need of a domain expert which is an expensive resource. Additionally, we use EFG to automatically include two new basic features in our system. The achieved UAS score of 92.66% represents a reduction by more than 15% on the previous smallest error on Portuguese DP. Since dependency parsing plays an important role in several NLP-based systems, this significant improvement has a positive impact on the related applications.

The basic features used in this work are based only on individual dependency edges, the so called first order features. MSTParser additionally makes use of second order features that depend on pairs of edges. According to the results presented in [17], second order features drive an increase of more than one percentage point in the UAS score. We plan to include this type of features in our system in order to lift the performance for Portuguese DP even higher.

We also plan to apply our system to other languages. The proposed feature generation procedure does not make use of any language dependent information and, thus, can be directly applied on other DP corpora. In fact, our method is even more general. Since it does not make use of any *task* dependent information, it can be integrated on structured learning systems for other tasks. We plan to apply this method for other tasks, like POS tagging, phrase chunking, clause identification, named entity recognition and semantic role labeling, in Portuguese and any other language where we could find an annotated corpus.

Principal component analysis [19] and independent component analysis [8] are two classic feature combination measures that can also be explored in the context of feature generation. We plan to use these methods to generate feature combinations and compare them to EFG.

Acknowledgments. The first author is supported by a doctoral grant from Conselho Nacional de Desenvolvimento Científico e Tecnológico. Part of the large margin structured perceptron algorithm has been implemented during an internship at Yahoo! Research Barcelona under supervision of Dr. Ulf Brefeld. During this period, the first author received a doctoral internship grant from Coordenação de Aperfeiçoamento de Pessoal de Nível Superior. The authors also thank Carlos E. M. Crestana and Guilherme De Napoli for providing Bosque-related scripts.

References

1. Altun, Y., Hofmann, T., Tsochantaridis, I.: SVM learning for interdependent and structured output spaces. In: Machine Learning with Structured Outputs (2007)
2. Altun, Y., Tsochantaridis, I., Hofmann, T.: Hidden Markov support vector machines. In: Proceedings of the International Conference on Machine Learning (2003)

3. Buchholz, S., Marsi, E.: CoNLL-X shared task on multilingual dependency parsing. In: Proceedings of the Tenth Conference on Natural Language Learning. pp. 149–164 (2006)
4. Chu, Y.J., Liu, T.H.: On the shortest arborescence of a directed graph. Science Sinica 14, 1396–1400 (1965)
5. Ciaramita, M., Altun, Y.: Broad-coverage sense disambiguation and information extraction with a supersense sequence tagger. In: Proceedings of the 2006 Conference on Empirical Methods in Natural Language Processing, EMNLP 2006, pp. 594–602 (2006)
6. Collins, M.: Ranking algorithms for named-entity extraction: Boosting and the voted perceptron. In: Proceedings of the Annual Meeting of the Association for Computational Linguistics (2002)
7. Collins, M.: Discriminative training methods for hidden Markov models: theory and experiments with perceptron algorithms. In: Proceedings of the ACL 2002 Conference on Empirical Methods in Natural Language Processing. pp. 1–8 (2002)
8. Comon, P.: Independent component analysis, a new concept? Signal Processing 36(3), 287–314 (1994)
9. Crammer, K., Singer, Y.: Ultraconservative online algorithms for multiclass problems. Journal of Machine Learning Research 3, 2003 (2001)
10. Edmonds, J.: Optimum branchings. Journal of Research of the National Bureau of Standards 71B, 233–240 (1967)
11. Fernandes, E.R., dos Santos, C.N., Milidiú, R.L.: A Machine Learning Approach to Portuguese Clause Identification. In: Pardo, T.A.S., Branco, A., Klautau, A., Vieira, R., de Lima, V.L.S. (eds.) PROPOR 2010. LNCS, vol. 6001, pp. 55–64. Springer, Heidelberg (2010)
12. Freitas, C., Rocha, P., Bick, E.: Floresta Sintá(c)tica: Bigger, Thicker and Easier. In: Teixeira, A., de Lima, V.L.S., de Oliveira, L.C., Quaresma, P. (eds.) PROPOR 2008. LNCS (LNAI), vol. 5190, pp. 216–219. Springer, Heidelberg (2008)
13. Hacioglu, K.: Semantic role labeling using dependency trees. In: Proceedings of the 20th International Conference on Computational Linguistics (2004)
14. Liang, P., Bouchard-côté, A., Klein, D., Taskar, B.: An end-to-end discriminative approach to machine translation. In: Proceedings of the Joint International Conference on Computational Linguistics and Association of Computational Linguistics, pp. 761–768 (2006)
15. McDonald, R., Crammer, K., Pereira, F.: Online large-margin training of dependency parsers. In: Proceedings of the 43rd Annual Meeting on Association for Computational Linguistics, ACL 2005, pp. 91–98 (2005)
16. Mcdonald, R., Lerman, K., Pereira, F.: Multilingual dependency analysis with a two-stage discriminative parser. In. In: Proceedings of the Conference on Computational Natural Language Learning, CoNLL, pp. 216–220 (2006)
17. Mcdonald, R., Pereira, F.: Online learning of approximate dependency parsing algorithms. In: Proc. of EACL, pp. 81–88 (2006)
18. Novikoff, A.B.: On convergence proofs on perceptrons. In: Proceedings of the Symposium on the Mathematical Theory of Automata (1962)
19. Pearson, K.: On lines and planes of closest fit to systems of points in space. Philosophical Magazine 2(6), 559–572 (1901)
20. Quinlan, J.R.: C4.5: Programs for Machine Learning (Morgan Kaufmann Series in Machine Learning), 1st edn. Morgan Kaufmann (1992)
21. Rosenblatt, F.: The perceptron: A probabilistic model for information storage and organization in the brain. Psych. Rev. 65, 386–407 (1958), Reprinted in Neurocomputing. MIT Press (1988)

22. dos Santos, C.N., Milidiú, R.L.: Entropy Guided Transformation Learning. In: Hassanien, A.-E., Abraham, A., Vasilakos, A.V., Pedrycz, W. (eds.) Foundations of Computational, Intelligence Volume 1. SCI, vol. 201, pp. 159–184. Springer, Heidelberg (2009)
23. Su, J., Zhang, H.: A fast decision tree learning algorithm. In: Proceedings of the 21st National Conference on Artificial Intelligence, pp. 500–505 (2006)
24. Tarjan, R.E.: Finding optimum branchings. Networks 7, 25–25 (1977)
25. Taskar, B., Guestrin, C., Koller, D.: Max–margin Markov networks. In: Advances in Neural Information Processing Systems (2004)
26. Taskar, B., Klein, D., Collins, M., Koller, D., Manning, C.: Max–margin parsing. In: Proceedings of the Conference on Empirical Methods in Natural Language Processing (2004)

Bayesian Induction of Syntactic Language Models for Brazilian Portuguese

Daniel Emilio Beck and Helena de Medeiros Caseli

Department of Computer Science – LaLiC/NILC
Federal University of São Carlos (UFSCar) – São Carlos/SP – Brazil
{daniel_beck,helenacaseli}@dc.ufscar.br

Abstract. Recent approaches for building syntactic language models include the combination of Probabilistic Tree Substitution Grammars (PTSGs) and Bayesian learning methods. While PTSGs have appealing features for syntax modeling, Bayesian methods provide a framework for inducing compact grammars that do not overfit the training corpus. In this paper, we apply these approaches to learn syntactic language models from a Brazilian Portuguese treebank.

Keywords: language models, tree substitution grammars, Bayesian learning, grammar induction.

1 Introduction

Language models are key resources in Natural Language Processing tasks, being a very important component in speech recognition and machine translation systems, for example. Considered the state-of-the-art, statistical, ngram-based models are widely used to induce these models due to its relatively high performance, efficiency and simplicity.

The ngram-based models assign probabilities to sentences according to ngram frequencies usually trained from an input corpus. Although they have proved to be useful for the task, they have some drawbacks when dealing with non-local context between words. This inherited feature can cause the model to incorrectly give low probabilities to completely grammatical sentences and vice-versa.

An example of this inability to model non-local context can be found in the sentence *Os meninos da escola municipal foram ao jogo* (The boys at the municipal school went to the game). In this example, a trigram-based model would have a low value for the probability term $P(foram|escola, municipal)$ due to a local agreement error between the noun *escola* (singular) and verb *foram* (plural). So, a low probability is induced even though the source sentence is actually grammatical: the verb agrees with the noun *meninos* (plural) instead, which is a correct agreement.

Syntactic language models have arisen to tackle those issues. By using grammars and syntactic trees to model sentences, they have the power to find those long-distance dependencies and accurately treat them. The first approaches were based on classical parsers that extended Probabilistic Context-Free Grammars

H. Caseli et al. (Eds.): PROPOR 2012, LNAI 7243, pp. 157–167, 2012.

(PCFGs) with parent and head word annotation [10,4]. The results obtained from these first initiatives were mixed: they were successful in improving speech recognition tasks [7,20,5] but not machine translation systems [6,16]. Since then, new approaches have emerged taking into account the syntactic models [18,19] and the statistical frameworks behind them [11].

In this work, we build a syntactic language model for Brazilian Portuguese using the Bosque treebank[1]. Our model is based on an extension of PCFGs called Probabilistic Tree Substitution Grammars (PTSGs) which has been used widely in Data-Oriented Parsing (DOP) [3]. We also follow recent works in Bayesian methods for grammar inducing [19,9], which have been applied successfully in a number of other applications like word segmentation [13] and phrase-based statistical machine translation [12].

The remainder of this paper is organized as follows: Sect. 2 introduces PTSGs and its features. Our statistical framework for grammar inducing is shown in Sect. 3. Sect. 4 presents the experiments and results obtained so far and Sect. 5 gives the final remarks and ideas for future work.

2 Probabilistic Tree Substitution Grammars

Tree Substitution Grammars (TSGs) [14] are a generalization of standard context-free grammars that allows rewriting of non-terminals as tree fragments instead of sequences of symbols. Because of that, TSGs have larger rules which capture more context in a sentence and also generate fewer independence assumptions when parsing (since larger rules result in less rule applications). Since they expand non-terminals into tree fragments, they have also the natural ability to model lexicalization and parent dependencies, obviating the need to explicitly annotate them in the rules [17].

Formally, TSGs are defined as a quadruple $G = (T, S, N, R)$, where T is a set of *terminal symbols*, N is a set of *non-terminal symbols*, $S \in N$ is the *initial symbol* and R is a set of rules or *elementary trees*: tree fragments where each internal node is labelled with a non-terminal symbol and each leaf node is labelled with either a non-terminal or a terminal symbol. These elementary trees can also be viewed as rewriting rules for its root non-terminal (see Fig. 1).

Probabilistic TSGs (PTSGs), in turn, assign weights to each elementary tree following the restriction that the weights of all elementary trees with the same root must sum 1. Like probabilistic CFGs, the PTSGs provide a framework for ranking between sentence parses by assigning probabilities to each parse. However, unlike CFGs, a parse tree in a TSG can have multiple derivations (Figure 2 shows an example of two possible derivations for the same tree). So, the total probability of a parse must take into account the probabilities of each rule (*rule*) of each possible derivation (*deriv*), resulting in the following equation:

$$P(parse) = \sum_{deriv \in parse} \prod_{rule \in deriv} P(rule) \tag{1}$$

[1] Available at `www.linguateca.pt/floresta/corpus.html`

Elementary trees

S(NP|VP(V(dormiu)|ADV))	S → NP|VP(V(dormiu)|ADV)
NP(N(Maria))	NP → N(Maria)
NP(N(Tiago))	NP → N(Tiago)
ADV(tarde)	ADV → tarde
ADV(cedo)	ADV → cedo

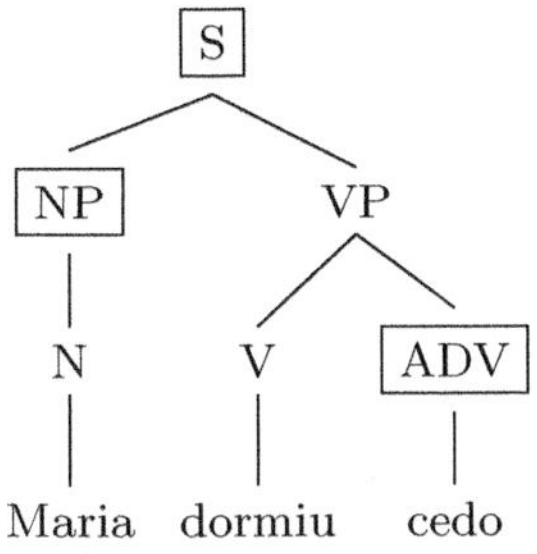

Fig. 1. Example of a TSG with a sample tree. The elementary trees are shown above in parenthesized format, on the left, and in its alternative representation as a rewriting rule, on the right. The boxed nodes in the sample tree represent elementary tree roots.

Determining the full set of derivations given a parse tree and a TSG is a NP-hard problem [21]. So, the probability is usually approximated by summing over the n-best derivations.

2.1 PTSGs as Language Models

To use a PTSG for language modeling we need to actually parse the input sentences according to the grammar. This can be done by transforming it into a PCFG: first, we add artificial non-terminal symbols that encode the internal tree representation, then we flat the elementary trees into 1-depth trees. For example, the first rule in Figure 1 would result in two PCFG rules:

S(NP|VP(V(dormiu)|ADV)) ⇒
S → "NP|VP(V(dormiu)|ADV)" ⇒ (artificial non-terminal)
"NP|VP(V(dormiu)|ADV)" → NP dormiu ADV

In a parsing task, it is possible to retrieve the original parse tree representation by "decomposing" the artificial non-terminals. However, when using a parser as a language model we are not actually interested in retrieving the syntactic structure of a sentence, only its probability [19]. In the light of this fact,

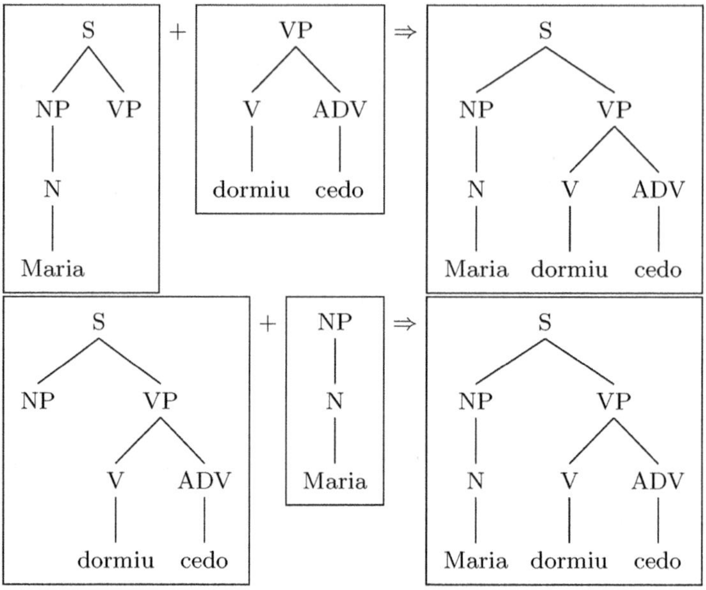

Fig. 2. Two possible TSG derivations for the same parse tree

when transforming the PTSG we do not create the artificial non-terminals and associated rules, instead we just flatten the elementary trees into 1-depth trees.[2]

3 Grammar Induction

In the grammar induction process, the weights of the rules are usually defined by training procedures which aim to maximize the likelihood of an input treebank (known as *Maximum Likelihood Estimate* – MLE). For PCFGs, the MLE is achieved by counting relative frequencies of rules but this strategy is not feasible in PTSGs due to the multiple derivations problem mentioned in Sect. 2. So, to be able to count the frequencies, it is necessary firstly to infer which is the correct derivation for each tree in the treebank.

In our experiments, we follow closely previous work on Bayesian grammar induction [18,9] defining our approach as a generative process (in which we suppose the treebank is *generated* by a sequence of elementary trees). Using the Bayes' Rule, our task becomes the maximization of the posterior distribution $P(E|T)$ (where T is the treebank and E is the sequence of elementary trees that generated the treebank):

$$P(E|T) \propto P(T|E) \times P(E) \tag{2}$$

[2] In the flattening step, if two or more elementary trees generate the same 1-depth tree, its probabilities are summed. This is done to keep the sum of all rules associated with each non-terminal equal to 1.

In this model, the likelihood term $P(T|E)$ is equal to 1 when the E generates T and 0 otherwise. This is because each sequence E of elementary trees define one and only one treebank T. For example, in Figure 1, the sequence composed of the 1st, 2nd and 5th elementary trees (in that order) would generate a treebank composed of a single tree (the one shown in the figure). Because of that, all the work in determining the best grammar is done by the prior $P(E)$.

The main problem in this task is to infer grammar rules that capture enough context but are still small enough to do not overfit the treebank. This can be achieved using a Dirichlet Process (DP) for each non-terminal in the grammar.

3.1 Dirichlet Process

The Dirichlet Process (DP) defines a distribution over the infinite space of possible elementary trees[3]. Formally, the distribution of an elementary tree e according to its non-terminal root r is defined as:

$$\begin{aligned} e|r &\sim G_r \\ G_r|\alpha_r, P_0 &\sim DP(\alpha_r, P_0(\cdot|r)) \end{aligned} \tag{3}$$

where $P_0(\cdot|r)$ (the *base distribution*) is a distribution over the infinite set of elementary trees with the non-terminal r as the root and α_r is the *concentration parameter* for r. Intuitively, the base distribution ($P_0(\cdot|r)$) defines which elementary trees we want to create in our generative treebank-building process while the concentration parameter (α_r) controls the tendency of creating new elementary trees or reusing existing ones.

Following the procedure described in [19,9], instead of representing the G_r distribution, we integrate over all its possible values, resulting in the following formula for determining the probability of an elementary tree e_i (given its root non-terminal r_i):

$$P(e_i|\mathbf{e}_{<i}, r_i, \alpha_{r_i}, P_0) = \frac{count(e_i) + \alpha_{r_i} P_0(e_i|r_i)}{count(r_i) + \alpha_{r_i}} \tag{4}$$

where $\mathbf{e}_{<i}$ is the set of all other elementary trees currently found in the treebank, $count(e_i)$ is the number of times e_i appears in $\mathbf{e}_{<i}$ and $count(r_i)$ is the number of times an elementary tree with root r_i appears in $\mathbf{e}_{<i}$. If α_{r_i} is set to 0, the formula above becomes the relative frequency of the elementary tree. So this formula can be seen intuitively as calculating the relative frequency but also taking into account the base distribution (with weight α_{r_i}).

In our generative process, the DP can be understood as a *cache* model in which every time a elementary tree is generated the model chooses between picking it from the cache of already existing trees or creating a new one based on the base distribution. As the treebank is generated, the cache becomes larger and, therefore, elementary trees tends to be picked from it. This is actually something that happens in natural language in general: although it is always

[3] For a more detailed explanation of the Dirichlet Process, see Appendix A in [13].

possible to create new tree fragments, we tend to use existing ones more and more frequently.

The base distribution governs the trees that should appear in our resulting grammar. As explained before, we prefer to have small elementary trees since large ones can overfit the treebank. To accomplish this, the base distribution P_0 is defined as another generative process on its own where we consider that each tree is created by alternating between expanding a non-terminal or not and, if positive, which PCFG rule is chosen to expand it. The expansion probability is defined as a Binomial distribution with parameter β_r while the PCFG is induced from the same treebank using MLE. With this base distribution, larger rules will have smaller probabilities (because the number of factors in the probability calculation will increase).

3.2 Gibbs Sampling

If we could enumerate all the treebank possible derivations it would be possible to simply apply (4) for all elementary trees, obtaining the best derivation for all parse trees and then extracting the grammar from the treebank. Unfortunately, as explained in Sect. 2, it is not feasible to find all derivations from a single tree. Because of that, we sample over a subspace of those possible derivations using a procedure called Gibbs sampling, a Markov Chain Monte Carlo (MCMC) method in which variables are sampled according to the values of all the other variables in the model [13].

In our model, the variables are all the non-terminal non-root nodes in each parse tree. Each of these nodes can be the root of an elementary tree or an internal node of some other elementary tree, which is indicated by a *substitution flag*. Within this schema, each node with its corresponding flag defines a possible TSG derivation for the tree. For example, in the parse tree of Fig. 1, the boxed nodes have its flags set to *True* while the other, non-boxed nodes, have its flags set to *False*. The Gibbs sampling then works by visiting each non-terminal, non-root node in each tree in the treebank and deciding to set it to *True* or *False*.

The two possible outputs generate two possible derivations for the entire treebank and therefore two possible posteriors. The decision made by the sampler is made proportionally to these posterior values. For example, if the posterior value for $Flag = True$ is 0.03 and for $Flag = False$ is 0.02, the sampler will set the flag to $True$ with 60% chance and to $False$ with 40% chance.

To calculate the posterior values, it would be necessary to take all the elementary trees in the treebank into account. But to decide the value of a flag, the Gibbs sampler do not need the posterior values, only their *proportion*. So, to calculate this proportion, the sampler considers only the elementary trees directly associated with the current node. As shown in Fig. 3, the current node defines three elementary trees: (i) the *merged* tree, where the current node's flag is not set (meaning that it is an internal node), and the (ii) *inside* and the (iii) *outside* trees, in which the flag is set (meaning that this node is a substitution point). The probabilities of those trees are calculated using (4). This is possible to do because the DP is *exchangeable*, meaning we can consider each elementary tree as the last generated in our treebank generative process. The resulting probabilities for setting or not the substitution flag are then calculated as follows:

$$P(flag = False) = P(merged)$$
$$P(flag = True) = P(inside) \times P(outside) \tag{5}$$

The Gibbs sampling starts by setting every substitution flag to some value (it may be random or predefined) and then repeats this process for each non-root non-terminal node in the treebank. If it decides to change the flag value, the counts in $\boldsymbol{e}_{<i}$ are updated accordingly. A full iteration of Gibbs sampling corresponds to visit each of these nodes once. After a reasonable number of iterations, the resulting treebank derivation tends to be closer to the true posterior distribution $P(G|T)$, from which we can extract the resulting grammar.[4]

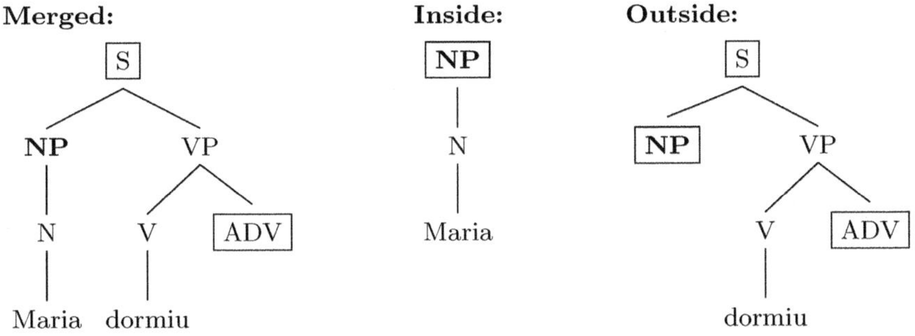

Fig. 3. Elementary trees defined by the current node (in this case, the NP node in bold) in a sampler iteration. Unboxed nodes have flags set to *False* and boxed nodes have flags set to *True*.

4 Experiments

4.1 Data

For training, we used the Brazilian Portuguese portion of the Bosque treebank, version 8.0. This set is composed of 4213 sentences from the *Folha de São Paulo* newspaper, analyzed by the parser PALAVRAS[1] and manually corrected. For testing, we used a sample of 100 sentences from the Mac-Morpho corpus with 20 or less tokens. This sample is composed of tokenized sentences from the same newspaper, POS-tagged by PALAVRAS and also manually corrected. We used the Mac-Morpho version available in the NLTK toolkit[5] [2].

[4] Theoretically, the Gibbs sampling procedure tends to achieve the true posterior distribution as the number of iterations tends to infinite.

[5] Available at `www.nltk.org`

In Bosque treebank, each node is composed by a pair $X : Y$ where X is the syntactic function and Y is a set of informations about the node. For the purpose of our experiments, we considered only the syntactic constituent, POS tag and word information in each node, as shown on Fig. 4.

```
SOURCE: CETENFolha n=17 cad="Ilustrada" sec="nd" sem="94b"
CF17-5 O panorama sofre prejuízos demais em favor da tese.
A1
STA:fcl
=SUBJ:np
==>N:art('o' <artd> M S) O
==H:n('panorama' <np-def> M S) panorama
=P:vp
==MV:v-fin('sofrer' PR 3S IND) sofre
=ACC:np
==H:n('prejuízo' <np-idf> M P) prejuízos
==N<:np
===H:pron-det('demais' <quant> M P) demais
=ADVL:pp
==H:prp('em_favor_de' <sam->) em_favor_de
==P<:np
===>N:art('o' <artd> <-sam> F S) a
===H:n('tese' F S) tese
=.
```

Fig. 4. A sentence from Bosque treebank with its syntactic tree. The number of '=' symbols indicates different levels within the tree. The underlined tags and words represent the actual information used when inducing the grammars.

We also made additional preprocessing steps:

- To deal with unknown words, we extracted a lexicon composed of all words with frequency ≥ 2 in Bosque. Words that weren't in this lexicon in both corpora were mapped to a set of unknown tags.
- Since the POS tag set in Bosque and Mac-Morpho are different, we mapped them to a common set.
- Finally, both corpora were downcased.

4.2 Methodology

The grammar induction scripts were written in Python. The parameters used in the grammar induction were the same used by [17, p. 70]: the concentration parameters α were fixed at 100 for all non-terminals, the β parameters for the base distribution were set to 0.7 and the sampler ran for 1000 iterations. Each substitution flag in the treebank was started with value *True*. We induced two grammars: one considering the full treebank and another considering a version with POS tags as leaves instead of words. For each treebank version, we also extracted a standard MLE PCFG and generated a trigram language model using the SRILM toolkit[6] [22] to use as baselines. To process new sentences with the induced grammars, we used a modified version of the NLTK chart parser.

Following common practice in language modeling evaluation, we calculated perplexity [15] scores for all six models on the test corpus. Perplexity measures the level of surprise when generating words in a sentence and lower levels obtained for a correct corpus indicate good models. To be conformant with the values returned by SRILM, we averaged perplexity by the total count of words and sentences, using the following formula:

$$\begin{aligned} H_{corpus} &= \frac{\sum_{sent \in corpus} - log(P(sent))}{\#words + \#sents} \\ Perplexity_{corpus} &= 10^{H_{corpus}} \end{aligned} \tag{6}$$

4.3 Results

Table 1 shows the perplexity values and other results for the test corpus.

Table 1. Perplexity results, parse errors and average parsing times (in seconds). The experiments were performed in a quad-core (Phenom II) machine with 4GB RAM.

	Words			POS tags		
	Perplexity	Errors	Avg. time	Perplexity	Errors	Avg. time
standard PCFG	97.929	1	1440s	7.1476	2	324s
flattened PTSG	112.303	1	94s	8.8519	4	64s
trigram	115.851	-	-	6.7623	-	-

For the POS tag case, the PTSG model is outperformed by the standard PCFG by 23%, while in the word case this difference drops to 14%. This suggests that the PTSG benefits more from lexical information than the standard PCFG, which corroborates to the theoretical assumption in Sect. 2 that it can capture this lexical information in a natural way. Further experiments will be carried out to investigate why PTSGs had worse performance than PCFGs, mainly we believe that better grammar inducing procedures can result in more accurate PTSGs.

Unlike previous work [19], both grammars outperformed the trigram baseline in the word case. A possible reason for this is due to language differences: while previous work used an English corpus, our experiments are focused on Brazilian Portuguese. Since we are dealing with a language with richer morphology, it has the potential to generate more infrequent and unknown words (in our experiments, 20% of the tokens in the test corpus were tagged as unknown). While our grammars deal with those words by simply replacing them by an unknown tag, the standard trigram model uses backoff models to cope with them and this must be the source of the problem. More advanced trigram models can give

[6] Available at `www.speech.sri.com/projects/srilm`

lower perplexities but nevertheless these results shows that syntactic language models have potential to improve the state-of-the-art word language models.

Table 1 also shows average parsing times. It is interesting to note that PTSGs performed much faster than PCFGs (15x faster in the word case). The reason for this performance improvement comes from the fact that we used a chart parser: PTSG elementary trees, after being flattened, result in rules with large right-hand sides, which generate less edges in the chart. We plan to investigate further if other parsing algorithms also result in faster times for PTSGs.

5 Conclusions and Future Work

In this work, we presented a Bayesian method to induce a PTSG-based language model from a treebank and the first experiments applying it to Brazilian Portuguese. Our first results suggest that syntactic language models can give better results than trigram-based models for this language. Although the induced grammars were unable to outperform standard PCFGs, the combination of PTSGs and Bayesian inducing methods are a rather new approach to language modelling, with large room for improvements. Specifically, our future work include refining those models in the following directions:

Statistical Models and Training Procedures: in recent work, [8] used a generalization of the DP called Pitman-Yor Process while using more advanced sampling procedures. Applying these models and procedures to the Bosque treebank is a natural extension of our work.

Dependency Grammars: our model can be easily applied to dependency trees instead of constituent trees. These can be obtained from Bosque but [8] also describes a method for gathering them from raw corpora.

Morphological Information: agreement is a key concept when dealing with rich morphological languages like Portuguese. Finding ways to efficiently encode those information in a TSG could improve agreement checking when building syntactic language models.

Evaluation: [19] shows perplexity results from experiments with ungrammatical test corpora. We plan to do similar experiments with Brazilian Portuguese, to investigate how these models perform in the presence of ungrammatical input. We also plan to incorporate these models in a Machine Translation (MT) system and evaluate them in terms of MT metrics.

Acknowledgements. We thank the financial support of FAPESP (projects 2010/03807-4 and 2010/07517-0) and CAPES.

References

1. Bick, E.: The Parsing System "Palavras": Automatic Grammatical Analysis of Portuguese in a Constraint Grammar Framework. Ph.D. thesis, Aarhus University (2000)
2. Bird, S., Klein, E., Loper, E.: Natural Language Processing with Python. O'Reilly Media (2009), `http://nltk.org/book`
3. Bod, R.: Do all fragments count? In: Natural Language Engineering, pp. 1–20 (2003)

4. Charniak, E.: A maximum-entropy-inspired parser. In: Proceedings of the 1st North American Chapter of the Association for Computational Linguistics Conference, pp. 132–139. Morgan Kaufmann Publishers Inc. (2000)
5. Charniak, E.: Immediate-head parsing for language models. In: Proceedings of the 39th Annual Meeting on Association for Computational Linguistics, pp. 124–131. Association for Computational Linguistics, Morristown (2001)
6. Charniak, E., Knight, K., Yamada, K.: Syntax-based language models for statistical machine translation. In: MT Summit IX. pp. 40–46 (2003)
7. Chelba, C., Jelinek, F.: Exploiting Syntactic Structure for Language Modeling. In: COLING-ACL (1998)
8. Cohn, T., Blunsom, P., Goldwater, S.: Inducing tree-substitution grammars. The Journal of Machine Learning 11, 3053–3096 (2010)
9. Cohn, T., Goldwater, S., Blunsom, P.: Inducing compact but accurate tree-substitution grammars. In: Proceedings of Human Language Technologies: The 2009 Annual Conference of the North American Chapter of the Association for Computational Linguistics, pp. 548–556. Association for Computational Linguistics, Morristown (2009)
10. Collins, M.: Head-driven statistical models for natural language parsing. Ph.D. thesis, University of Pennsylvania (1999)
11. Collins, M., Roark, B., Saraclar, M.: Discriminative syntactic language modeling for speech recognition. In: Proceedings of the 43rd Annual Meeting on Association for Computational Linguistics, pp. 507–514. Association for Computational Linguistics (2005)
12. DeNero, J., Bouchard-Côté, A., Klein, D.: Sampling alignment structure under a Bayesian translation model. In: Proceedings of the Conference on Empirical Methods in Natural Language Processing, pp. 314–323. Association for Computational Linguistics (2008)
13. Goldwater, S., Griffiths, T.L., Johnson, M.: A Bayesian framework for word segmentation: Exploring the effects of context. Cognition 112(1), 21–54 (2009)
14. Joshi, A., Schabes, Y.: Tree-adjoining grammars. Handbook of Formal Languages, Beyond Words 3, 69–123 (1997)
15. Jurafsky, D., Martin, J.H.: Speech and Language Processing. Prentice-Hall (2000)
16. Och, F., Gildea, D., Khudanpur, S., Sarkar, A., Yamada, K., Fraser, A., Kumar, S., Shen, L., Smith, D., Eng, K., et al.: A smorgasbord of features for statistical machine translation. In: Proceedings of HLT-NAACL, pp. 161–168 (2004)
17. Post, M.: Syntax-based Language Models for Statistical Machine Translation. Ph.D. thesis, University of Rochester (2010)
18. Post, M., Gildea, D.: Bayesian learning of a tree substitution grammar. In: Proceedings of the ACL-IJCNLP 2009 Conference Short Papers, pp. 45–48. Association for Computational Linguistics, Morristown (2009)
19. Post, M., Gildea, D.: Language modeling with tree substitution grammars. In: NIPS Workshop on Grammar Induction, Representation of Language, and Language Learning, pp. 1–8 (2009)
20. Roark, B.: Probabilistic top-down parsing and language modeling. Computational Linguistics 27(2), 249–276 (2001)
21. Sima'an, K.: Computational complexity of probabilistic disambiguation by means of tree-grammars. In: Proceedings of the 16th Conference on Computational Linguistics, pp. 1175–1180. Association for Computational Linguistics, Morristown (1996)
22. Stolcke, A.: SRILM - An Extensible Language Modeling Toolkit. In: Proceedings of the International Conference on Spoken Language Processing, Citeseer, pp. 901–904 (2002)

Automatic Generation of *Cloze* Question Stems

Rui Correia[1,3], Jorge Baptista[2], Maxine Eskenazi[3], and Nuno Mamede[1]

[1] INESC-ID Lisboa / IST, Portugal
[2] Universidade do Algarve, Portugal
[3] Language Technologies Institute, Carnegie Mellon University, USA
{Rui.Correia,Nuno.Mamede}@inesc-id.pt, jbaptis@ualg.pt, max@cs.cmu.edu

Abstract. *Fill-in-the-blank* questions are one of the main assessment devices in REAP.PT tutoring system. The problem of automatically generating the stems, i.e. the sentences that serve as basis to this type of question, has been studied mostly for English, and it remains a challenge for a language as morphologically rich as European Portuguese (EP), for which additional data scarcity problems arise. To address this problem, a supervised classification technique is used to model a classifier that decides whether a given sentence is suitable to be used as a stem in a *cloze* question. The major focus is put in the feature engineering task, describing both the development of new criteria, and the adaptation to EP of features already explored in the literature. The resulting classifier filters out inadequate stems, allowing experts to build and personalize their instruction focusing on a set of potentially good sentences.

Keywords: Question Generation, *Cloze* Questions, CALL.

1 Introduction

REAP.PT [9] (REAder-specific Practice for PorTuguese) is the Portuguese version of REAP [7], developed at Carnegie Mellon University. This Computer Assisted Language Learning (CALL) tutoring system aims at teaching vocabulary to L2 learners of European Portuguese (EP) having reading activities as a starting point. Students are presented with real texts, collected from the Web, in which a set of words from the Portuguese Academic Word List [2] (P-AWL) are highlighted. After each reading, there is an assessment phase, composed of questions targeting the vocabulary that was highlighted in the text.

Also known as *fill-in-the-blank*, *cloze* questions are one of the question types used in that assessment phase, testing the highlighted words in context by requiring the student to find the word that better fits a sentence. *Cloze* questions are composed of three elements: **stem** (sentence from which a word was deleted), **target word** (correct answer), and **distractors** (set of wrong answers).

To successfully create an adequate *cloze* question, the stem and the distractors must be in accordance, so that no ambiguity allows for more than one acceptable answer. Correia et al. [3] focused on generating coherent distractors for a set of 4,000 stems manually selected by linguists. This set of stems was produced according to a predefined set of criteria such as considering only full sentences and

H. Caseli et al. (Eds.): PROPOR 2012, LNAI 7243, pp. 168–178, 2012.

not fragmentary text, excluding paratextual elements or lexicographic context, excluding sentences where the target word is at the beginning or end, considering only sentences with length between 100 and 200 characters, and choosing sentences that constitute a non ambiguous environment for the correct identification of the target word. For each word in P-AWL, two or three stems were selected. Figure 1, taken from Pino et al. [13], exemplifies an inadequate stem, showing that not considering some of these criteria can lead to an undesired question formulation, where all the options fit in the blank slot.

"He used that word _____."

❑ quietly ❑ deliberately ❑ wildly ❑ carefully

Fig. 1. Example of an inadequate *cloze* question formulation

Relying on experts to generate this resource is expensive, time-consuming, and dependent on the level of expertise and common sense of each individual. Additionally, each time the target word list suffers changes or if REAP.PT is adapted to a more specific context (e.g. teaching medicine vocabulary), the process of generating stems manually has to be repeated. These issues motivated the development of automatic techniques to generate stems for *cloze* questions in the context of REAP.PT.

The remainder of this document is structured as follows: Section 2 presents some previous work, Section 3 describes the proposed solution, Section 4 presents the evaluation of the solution and the results achieved, and finally, Section 5 concludes and proposes future work. For better comprehension, all the Portuguese examples were translated to English, maintaining the aspects that they are intended to represent.

2 Related Work

One of the first attempts to automatic generation of *cloze* question stems came from Hoshino and Nakagawa [8], who trained a model to decide where to insert blank spaces, given a text. In their work, they used machine learning tools (Naive Bayes and K-Nearest Neighbors), a gold standard of questions for training, and a set of features (such as part-of-speech of the previous/next words, position of the target word in the sentence, and sentence length). The authors were able to successfully insert a blank 60% of the time, concluding that more features needed to be considered in order to achieve better results.

In the context of vocabulary tutoring systems, Pino et al. [13] generated stems using two different methods: a baseline technique, that directly extracts the example sentences from the WordNet [5] and a technique that uses linguistic features to decide on the suitability of a sentence from raw text corpora. These features include the length of the sentence, the number of clauses, co-occurrence scores and the grammaticality score given by the parser. The authors then manually tuned the weights of each feature on training data. The second method outperformed the baseline, producing high quality stems 66% of the time (an improvement of 26%). The authors stressed the importance of measuring the number of clauses, claiming that stems should be comprised of at least two clauses: one with the target word, and another to specify the context.

Finally, Skory and Eskenazi [16] used crowd-sourcing to prove the significance of the previously mentioned co-occurrence feature, finding a high correlation between that criterion and the number of correct answers given by native speakers.

3 Architecture of the Solution

The solution here proposed merges the main ideas mentioned in the previous section, using machine learning techniques to automatically classify a set of sentences extracted from real text corpora, and considering a set of features that mimic the experts' decisions.

Figure 2 presents the general architecture of the stem generation system. A corpus of real texts of European Portuguese is split into individual sentences and indexed. These sentences are then manually classified as positives or negative examples, forming a gold standard that will be used to train the model. A feature engineering task takes place in parallel, extracting information from the sentences that can act as predictors of sentence quality. The gold standard and the features are then used in a Support Vector Machine (SVM), producing the final stem classifier. The remainder of this section will focus on each one of these modules, giving particular emphasis to the feature computation task.

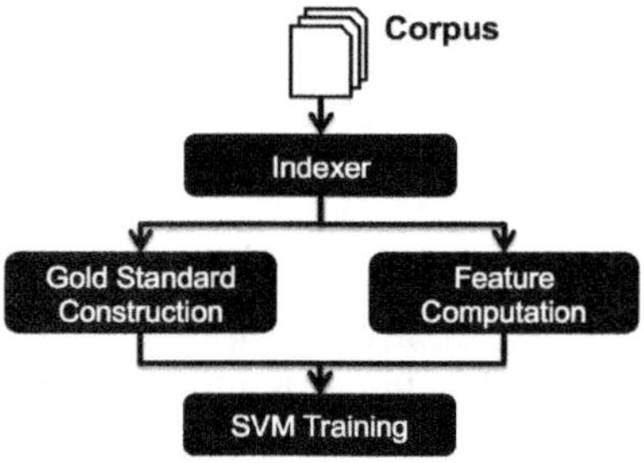

Fig. 2. General architecture of the solution

3.1 Corpus

CETEMPúblico[1] was used as a source of candidate stems. This corpus provides sentences with high quality since they have been extracted from newspaper articles. CETEMPúblico [15] is composed of about 7 million sentences collected from a Portuguese daily newspaper, between 1991 and 1998.

[1] `http://www.linguateca.pt/cetempublico/` (visited in Jan. 2012).

3.2 Indexer

In order to index the sentences from the corpus, the Apache Lucene^{TM}[6] text search engine was used. The resulting index allows searching for P-AWL's target words in CETEMPúblico's sentences, for which stems have to be generated, and it also provides an efficient way to compute some of the features (see Section 3.4). In Information Retrieval notation, each sentence of the corpus is a *document*, and each word is a *term*.

3.3 Gold Standard Construction

The motivation behind building a gold standard, and do not use the set of stems manually selected by linguists, has to do with the fact that this set only contains positive examples, i.e, it does not contain the type of sentences that are supposed to be filtered out by the final classifier. In order to build a model that is able to accomplish the task at hand, it is necessary to have both positive and negative instances, not biasing the classifier's decisions (for instance, by using only sentences already selected based on the length criterion).

Thus, all the sentences with the word *computer* (chosen because it had a substantial representation on the corpus, with over 2K examples) were annotated as being good or bad stems, by someone with deep knowledge of the task. It is important to notice that having as a training set sentences targeting only one word is not limiting, since the goal is to build a word-independent classifier.

As one would expect, the number of negative examples (2,057) is higher than the number of positive (180). For the held-out set, we reserved 30 negative and 30 positive examples, randomly chosen. The remaining 150 positive examples and a set of 150 randomly chosen negatives formed the train/test set.

3.4 Feature Computation

The success of a classification task highly depends on the features that are used and the information each one of them is able to represent. In this scope, feature engineering demands for Natural Language Processing (NLP) techniques capable of representing information that the experts use while selecting stems.

Before focusing in each feature, it is important to introduce the NLP tool that supports the computation of most of those criteria. The STRING NLP chain (Figure 3) endows each sentence with information resultant from each of its 4 modules:

- **LexMan** [10, 4] – assigns to each word all the possible POS tags;
- **RuDriCo** [12] – disambiguates the results from LexMan by applying transformation rules based on pattern matching;
- **MARv** [14] – statistic disambiguation of the results from RuDriCo;
- **XIP (Xerox Incremental Parser)** [1] – appends information of elementary syntatic constituents (*chunks*), syntactic dependencies, named entities and anaphoric relations.

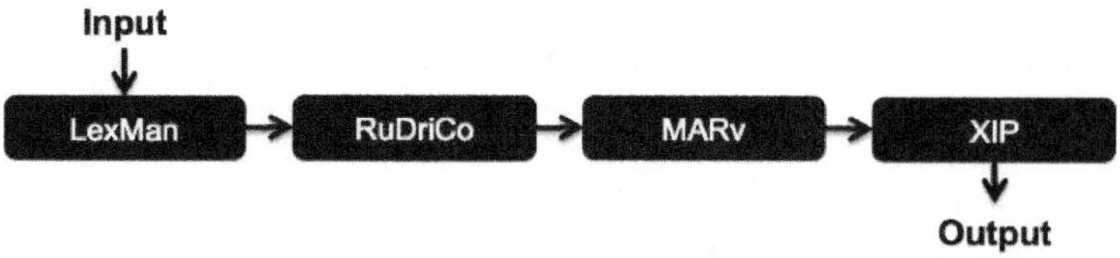

Fig. 3. Simplified STRING chain schema

The remainder of this section describes each one of the features that were used.

Sentence Length – Pino et al. [13] stresses that short sentences do not usually provide sufficient context to be used as stems in a *cloze* question, while long sentences may distract the student by adding unnecessary complexity.

Target Word Position – Additionally, Pino et. al [13] states that the position of the target word, and therefore the blank space, in the beginning or end of the example sentence, is typically an indicator of an inadequate stem (see Figure 1).

Secondary Idea and Secondary Idea Length – These two innovative features take into consideration that sometimes the target word is used in a complementary idea to the main sentence, ultimately representing only side information as in *"Through a services control center (**computer**), you can combine several call centers"*.

However, it may happen that the text between the secondary idea markers (in this case, the brackets) is able to define a context by itself as in *"He is dedicated to computational physics (physics research through **computer** simulation)"*. So, considering the length of the secondary idea is also important.

In Enumeration – Another criterion first explored in this work is the presence of the target word in a list, as an enumeration of objects, actions, etc. Typically, this particular sentence structure does not provide the necessary context. The *COORD* dependency provided by the STRING chain signals this structure.

Proper Names, Foreign Words, Acronyms and Numerical Expressions – Another innovation of the present work is the use of these phenomena. This set of criteria aims at penalizing sentences that require specific domain knowledge to be understood, such as in *"The Hollywood requires a personal **computer** IBM PS/2 or compatible with 80286 processor, operating with the DOS 3.3 operating system or higher and with Microsoft Windows 3.0"*. When these phenomena appear several times in one sentence the student may become distracted with the interpretation of these elements, instead of focusing in solving the exercise. It is worth noting that this feature was not part of the original set of criteria used by the linguists to build the aforementioned set of 4K sentences. It resulted instead from the need to discard the significant amount of sentences in CETEMPúblico with overwhelming content similar to the example above.

Co-occurrences – The literature that focus on stem's generation uses context windows around the target word as the main feature. The approach we present here formulates this feature in a different manner, focusing on the sentence as a whole, and not limiting the co-occurrences window.

The graph-based ranking model TextRank [11] has been used in Keyword Extraction to identify in a text the terms that best represent it. This concept could be used in our scope, finding sentences in which the extracted keywords were the words that should be tested. However, the computational cost of the algorithm and the insufficient context that a single sentence provides, excludes this as a solution.

To mimic TextRank's effect, the Skip Bigram Co-occurrence Counts (SBCC) was computed between each content word of the sentence and the target word. As the name states, this formulation considers occurrences of two words, that are not necessarily contiguous, and is computed according to the following:

$$SBCC(s, k, V) = \sum_{i=0}^{length(s)} f(s_i, k, V) \tag{1}$$

$$f(s_i, k, V) = \begin{cases} \frac{counts(s_i \cap k, V)}{counts(w, V)} & \text{if } s_i \text{ is a content word;} \\ 0 & \text{otherwise.} \end{cases} \tag{2}$$

The SBCC score for a given sentence, s, a target word k, and a set of sentences V, is the sum of the quotient between the number of times a content word, s_i, and the target word k occur in the same sentence in the corpus, and the number of times s_i occurs by itself.

If s_i always occurs in the same sentences that k, the value of f achieves its maximum, i.e., 1. Additionally, since s_i occurs at least once with k, f is never 0. To account for misspells, we assumed that if a word only occurs once in the corpus, it will not be considered in the computation.

In practice, to compute this score, we used the POS tags resulting from the STRING chain to determine the content words, and the index described in Section 3.2 to find the required counts.

Verb Domain – The *VDOMAIN* dependency (representing verb phrases in STRING chain) provided a measure of the sentence's complexity.

Level, Known Words and Unknown Words – In order to take into account the global difficulty of the stem in comparison with the target word that it aims to test, a bag-of-words approach was used to compute grade levels. Based on a corpus of 47 text books, exercise books and national exams, divided by grade levels, the algorithm considers unigrams to estimate the probability of a word being in a certain level. The probability of a word, w_i, being in the level l_j is:

$$P(w_i = l_j) = \frac{counts(w_i, l_j)}{\sum_{k=1}^{N_j} counts(w_k, l_j)} \tag{3}$$

with N_j being the number of different words in the level j.

This method computes a probability for each level (in our case, from 5 to 12) and, looking at the first level in which the word appears and the following 2 levels, assigns the level in which the probability is higher, amongst these three levels. However, this solution may not be entirely adequate in some cases. For instance, if a word occurs only once in the 5th level and then only occurs in the 10th, 11th and 12th levels it will probably be incorrectly classified in the 5th level, even if it appears more often in the higher levels.

This feature filters out candidate sentences where the level of the words in the stem is higher than the level of the target word itself, which is intuitively a bad formulation for a *cloze* question. On the other hand, this criterion boosts sentences with "easier" words to test target words of higher level of difficulty.

Contrary to previous solutions, the number of known and unknown words was also used as a feature. An unknown word was considered to be a word that was not found in any of the textbooks, exercise books or national exams.

3.5 SVM Training

The WEKA data mining tool[2] was used with the *LibSVM* classifier (Support Vector Machines), with a 20-fold cross-validation, using a radial kernel and a cost value of 1,000, increasing the cost of misclassifying points (parameters adjusted using the held-out set). WEKA provides the result of the classification of an instance along with the correspondent probability estimate, allowing for the ranking of the results.

4 Results

The results of the present work are divided in two subsections. Section 4.1 will present the main results of the classifier, while Section 4.2 presents an evaluation of the stems selected by linguists using the resulting classifier.

4.1 General Results

The first experiment used two features that already have been proved in literature to be relevant for the task: *length* and *co-occurrences*. This constitutes a baseline that will allow to compare the contribution of the remaining features.

Figure 4 presents the distribution of the *length* and the *co-occurrence* features, along the manual classification as good or bad stems.

Regarding the *length* criterion, one can see a weak distinction. However, it is interesting to notice that negative examples tend to spread more along the *length* axis, whereas good examples tend to have a lower standard deviation from the mean length. For the *co-occurrences* criterion, it is clear that higher values of this feature tend to constitute a good stem. The majority of the negative examples have a value between 0 and $\frac{1}{4}$ in the Y axis.

[2] `http://www.cs.waikato.ac.nz/ml/weka/` (visited in Jan. 2012).

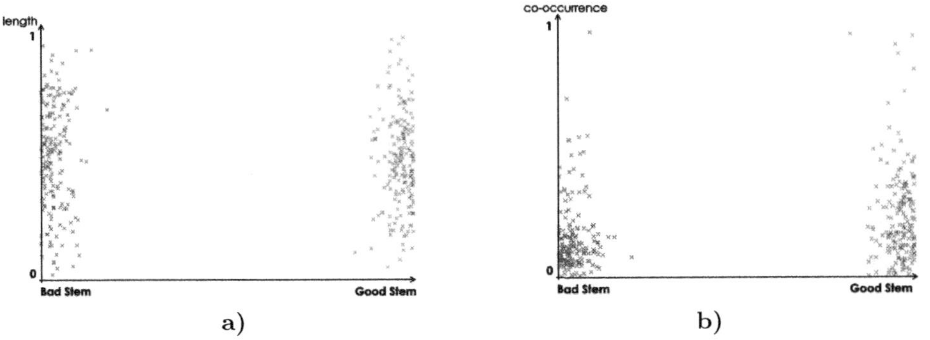

Fig. 4. Distribution (with jitter) of the features *length* (**a**) and *co-occurrence* (**b**)

Table 1 presents the results for the classification task with only two features.

Table 1. Results of the classifier using the *length* and *co-occurrences* criteria

	TP	FP	Precision	Recall	F-Measure
Bad Stem	66.7	31.1	68.2	66.7	67.4
Good Stem	68.9	**33.3**	67.4	68.9	68.1
AVG	67.8	32.2	67.8	67.8	**67.8**

With a *f-measure* of 67.8%, the classifier behaves better than simple chance. However, the percentage of false positives is higher than it would be desirable. 60 sentences were classified as good stems when they were bad stems. Nevertheless, this is still a good result since one is interested in the best few good sentences, instead of complete precision.

Table 2 presents the results when all the features were used, and Table 3 shows the confusion matrices comparison between the first and second approaches. With an increase of 7.7% of the *f-measure*, the inclusion of the new features improved the classifier. The rate of false positives decreased to 21.7% (11.6% lower than in the first attempt).

While analyzing the *information gain* of the features, the *co-occurrence* criterion proved to be the most relevant with a score of 0.2379, followed by *length* with 0.0364. These two features were the ones that were considered in the first experiment, and were pointed out in previous work as good estimates. The only criterion that the classifier dismissed was the *level* feature.

Another interesting result was that 23% of the good stems of the gold standard had a probability estimate greater than 90%. The sentence that ranked higher was *"They stole my laptop* **computer***, full of my company's files, personal texts and personal e-mail"*. The majority of the false positives were caused by enumerations undetected by the STRING chain and by the fact that the level criterion was discarded, since these two aspects were considered while building the gold standard.

Table 2. Results of the classifier using all criteria

	TP	FP	Precision	Recall	F-Measure
Bad Stem	78.3	27.2	74.2	78.3	76.2
Good Stem	72.8	**21.7**	77.1	72.8	74.9
AVG	75.6	24.4	75.6	75.6	**75.5**

Table 3. Confusion matrix of the classifier using only the *length* and *co-occurrence* features **(a)** and using all criteria **(b)**

a)

Actual Value	*Prediction Outcome*: Bad Stem	*Prediction Outcome*: Good Stem
Bad Stem	120	**60**
Good Stem	56	124

b)

Actual Value	*Prediction Outcome*: Bad Stem	*Prediction Outcome*: Good Stem
Bad Stem	141	**39**
Good Stem	49	131

In order to conclude on the utility of the classifier, it was tested for the noun *office* and for the verb *release*. For the first ten higher ranked sentences, 7 were good stems. The top result for the word *office* was *"Instead of opening a private* ***office****, Maria worked in the Military Hospital but was dismissed for political reasons"* and for the word *release* was *"One of his first acts, already in power, was to* ***release*** *three dozen political prisoners, including former President Olusegun Obasanjo"*.

4.2 Classifying Linguists' Stems

When the stems developed by linguists were submitted to the classifier only 13% of them were classified positively. When looking at the data, the reason why several of those sentences were rejected became clear, and two conclusions could be drawn.

The lowest ranked stems (the ones with high probability of being negative examples) did not respect the set of criteria that was used by the classifier. Some sentences had the target word in the end of the sentence, some had many numbers and acronyms and some were too short or too long. Some examples of these phenomena are *"Allotting,* ***proportionally****, the potential voters of the AD between PSD and PP, the votes in both parties is 32.3 and 4.1% respectively"* or *"Santana Lopes* ***assists*** *TVI"*. This confirmed the motivation of this work: experts' time should be applied on this task on sentences that are already filtered and flagged as potential good stems, reducing errors generated by the exhaustive process of selecting sentences from scratch.

However, the rejected 87% of the stems are far from being composed only of mistakes. The way we applied the classifier turned out to be inappropriate. The classifier should be used to classify stems one target word at a time, instead of classifying stems from different words in one passage. The manually generated set of stems is composed of sentences that aim to test a great variety of words (the ones in the P-AWL set), containing two or three examples for each inflected form of the target word. These inflected forms will have lower co-occurrence scores than words that are more common, as *computer*, and, when normalized, will be hindered by higher occurring words.

5 Conclusions and Future Work

This work presented a method to generate stems for *cloze* questions, using a supervised machine learning technique. The features developed here proved to be word-independent, and able to select good stems for words not represented in the train set.

Despite being able to reduce the number of false positives as the number of features increased, some criteria could have been used as a filter instead of being a parameter to the SVM (such as the presence of an enumeration). Additionally, the classifier could benefit if the computation of the *co-occurrence* feature took into consideration the lemma of the word (instead of considering the inflected forms themselves).

This work also identified some problems with the set of sentences developed by experts, contributing for the motivation of having automatic techniques for stem generation.

Being only interested in 3 to 5 sentences per word, this resource can help teachers build *cloze* question exercises in an expedite way, using their expertise to focus on a few sentences that are potentially good to be used as stems. In this setup, there is also the advantage of bootstrapping the model as teachers identify good suggestions while REAP.PT is being used.

Acknowledgments. This work was partially supported by FCT (INESC-ID multiannual funding) through the PIDDAC Program funds and by FCT project CMU-PT/HuMach/0053/2008. The authors would also like to thank Professors Bruno Martins and Pável Calado for the discussion throughout the execution of this work.

References

[1] Ait-Mokhtar, S., Chanod, J., Roux, C.: A Multi-Input Dependency Parser. In: Proceedings of the Seventh International Workshop on Parsing Technologies, pp. 17–19 (2001)

[2] Baptista, J., Costa, N., Guerra, J., Zampieri, M., de Lurdes Cabral, M., Mamede, N.: P-AWL: Academic Word List for Portuguese. In: Computational Processing of the Portuguese Language, pp. 120–123 (2010)

[3] Correia, R., Baptista, J., Mamede, N., Trancoso, I., Eskenazi, M.: Automatic Generation of Cloze Question Distractors. In: Proceedings of the Workshop on Second Language Studies: Acquisition, Learning, Education and Technology, Tokyo, Japan (September 2010)

[4] Diniz, C.: Um Conversor Baseado em Regras de Transformação Declarativas. Master's thesis, Instituto Superior Técnico–Universidade Técnica de Lisboa, Lisbon, Portugal (2011)
[5] Fellbaum, C.: WordNet: An electronic lexical database. The MIT press (1998)
[6] Hatcher, E., Gospodnetic, O.: Lucene in action (2004)
[7] Heilman, M., Collins-Thompson, K., Callan, J., Eskenazi, M.: Classroom success of an Intelligent Tutoring System for lexical practice and reading comprehension. In: Ninth International Conference on Spoken Language Processing, Citeseer (2006)
[8] Hoshino, A., Nakagawa, H.: A Real-Time Multiple-Choice Question Generation for Language Testing: A Preliminary Study. In: Proceedings of the Second Workshop on Building Educational Applications Using NLP, pp. 17–20. Association for Computational Linguistics (2005)
[9] Marujo, L., Mamede, N., Lopes, J., Trancoso, I., Pino, J., Eskenazi, M., Baptista, J., Viana, C.: Porting REAP to European Portuguese. In: International Workshop on Speech and Language Technology in Education, ISCA, Citeseer, Wroxall Abbey Estate, Warwickshire, UK (September 2009)
[10] Medeiros, J.: Processamento Morfológico e Correcção Ortográfica do Português. Master's thesis, Instituto Superior Técnico–Universidade Técnica de Lisboa, Lisbon, Portugal (1995)
[11] Mihalcea, R., Tarau, P.: TextRank: Bringing Order Into Texts. In: Proceedings of the Empirical Methods on Natural Language Processing Conference, pp. 404–411. Association for Computational Linguistics, Barcelona (2004)
[12] Paulo, J.: PAsMo-Pós Analisador Morfológico. Ph.D. thesis, Instituto Superior Técnico–Universidade Técnica de Lisboa, Lisbon, Portugal (2001)
[13] Pino, J., Heilman, M., Eskenazi, M.: A Selection Strategy to Improve Cloze Question Quality. In: Proceedings of the Workshop on Intelligent Tutoring Systems for Ill-Defined Domains. 9th International Conference on Intelligent Tutoring Systems, Citeseer, Montreal, Canada, pp. 22–32 (2008)
[14] Ribeiro, R., Oliveira, L., Trancoso, I.: Using Morphossyntactic Information in TTS Systems: Comparing Strategies for European Portuguese. In: Computational Processing of the Portuguese Language, p.195 (2003)
[15] Santos, D., Rocha, P.: Evaluating CETEMPúblico, a Free Resource for Portuguese. In: Proceedings of the 39th Annual Meeting on Association for Computational Linguistics, pp. 450–457. Association for Computational Linguistics (2001)
[16] Skory, A., Eskenazi, M.: Predicting Cloze Task Quality for Vocabulary Training. In: Proceedings of the NAACL HLT 2010 Fifth Workshop on Innovative Use of NLP for Building Educational Applications, pp. 49–56. Association for Computational Linguistics (2010)

Toponym Disambiguation Using Ontology-Based Semantic Similarity

David S. Batista[1], João D. Ferreira[2], Francisco M. Couto[2], and Mário J. Silva[1]

[1] IST/INESC-ID Lisbon, Portugal
{dsbatista,msilva}@inesc-id.pt
[2] University of Lisbon, LaSIGE

Abstract. We propose a new heuristic for toponym sense disambiguation, to be used when mapping toponyms in text to ontology concepts, using techniques based on semantic similarity measures. We evaluated the proposed approach using a collection of Portuguese news articles from which the geographic entity names were extracted and then manually mapped to concepts in a geospatial ontology covering the territory of Portugal. The results suggest that using semantic similarity to disambiguate toponyms in text produces good results, in comparison with a baseline method.

Keywords: semantic similarity, ontologies, toponym sense disambiguation, geographic information retrieval.

1 Introduction

Word-sense-disambiguation deals with the problem of selecting the correct semantic meaning of ambiguous words mentioned in an unstructured text [11]. A particular case of ambiguous words are toponyms (place names), whose geographic meaning, that is the geographic concept identified by a unique place on the Earth to which the word is referring to, needs to be identified. For instance, the place name *Lisboa* can represent up to 41 different locations in the territory of Portugal alone, from streets to a municipality, a city or a region.

In this work we address one specific type of ambiguity, namely *referent ambiguity*, when the same toponym can represent more than one geographic concept (for other types of ambiguity see the work of Martins et. all [10]). Given a toponym in a text and a geospatial ontology, we assign, from the set of concepts having that toponym, the one representing the toponym's context.

A common approach to resolve *referent ambiguity* uses heuristics, usually based on hierarchy constraints derived from administrative subdivisions encoded in the ontology [12]. For example, we expect that a large city is more likely to be referred than a small village with the same name. Another method uses discourse interpretation heuristics. For instance, if the same toponym is used multiple times in the same text or section, it is assumed that it is always referring to the same location rather than different locations that share the same name [5].

H. Caseli et al. (Eds.): PROPOR 2012, LNAI 7243, pp. 179–185, 2012.

Yet another method, based on using geospatial information associated with each concept, is to minimize the bounding polygon that contains all candidate referents, using the geographic bounding boxes associated with each concept [7].

In this paper we introduce a new heuristic. It is reasonable to expect that in a section of a text, close geospatial concepts share a higher degree of closeness also in semantics. For instance, if a news article mentions *Lisboa* and *Porto*, than its expected that *Lisboa* refers to the city and not to one of the streets with that name, since both these names can refer to cities. Based on this idea, we developed two mapping techniques based on semantic similarity measures to solve *referent ambiguity*: Global-Mapping and Sequential-Mapping. In the evaluation we used a geospatial ontology of the Portuguese territory and a set of geographic annotated news articles. Experiences show that Sequential-Mapping based on the Jiang-Conrath measure gives the best results. The rest of this paper is organized as follows: in section 2 we present the semantic similarity measures used, section 3 describes the two mapping techniques, section 4 the assessment and results, and finally in section 5 we present the conclusions.

2 Semantic Similarity

According to Budanitsky and Hirst, the most effective semantic similarity measures are the ones based on the information content (IC) that two concepts share [3].

If the ontology is structured as a directed acyclic graph (DAG) the IC of an ontology concept is inversely proportional to its frequency in a given corpus. The frequency is propagated to its ancestors, making the IC of a concept roughly proportional to its depth in the DAG. Thus, if $f(c)$ is the frequency of concept c, including its descendants, IC is defined as $\text{IC}(c) = -\log \frac{f(c)}{\max_c f(c)}$, where $\max_c f(c)$ is the maximum frequency of all concepts, i.e. the frequency of the root concept. The shared information between two concepts is normally proportional to the IC of the Most Informative Common Ancestor (MICA) in the ontology:

$$\text{IC}_{\text{MICA}}(c_1, c_2) = \max\{\text{IC}(a) : a \in \text{Anc}(c_1) \cap \text{Anc}(c_2)\}$$

where $\text{Anc}(c_x)$ represents the ancestors of c_x.

Resnik defined similarity between two concepts as the amount of information content they share, given by the information content of their MICA [13]. Jiang and Conrath defined a distance measure as the difference between the IC of both concepts and the IC of their MICA [6]; assuming that the IC is normalized for values between 0 and 1, the distance can be converted to similarity. Lin defined similarity as the IC of their MICA over the IC of both concepts [8]. Every one of these measures is based on Resnik's definition of shared information (they use a single common ancestor), and are summarized in Table 1.

Table 1. Semantic Similarity Measures

Measure	Formula
Jiang-Conrath	$sim_{\mathrm{JC}}(c_1, c_2) = 1 - (\mathrm{IC}(c_1) + \mathrm{IC}(c_2) - 2 \times \mathrm{IC}_{\mathrm{MICA}}(c_1, c_2))$
Lin	$sim_{\mathrm{Lin}}(c_1, c_2) = 2 \times \mathrm{IC}_{\mathrm{MICA}}(c_1, c_2) \div (\mathrm{IC}(c_1) + \mathrm{IC}(c_2))$
Resnik	$sim_{\mathrm{Resnik}}(c_1, c_2) = \mathrm{IC}_{\mathrm{MICA}}(c_1, c_2)$

3 Toponym Disambiguation

Having as input a sequence of toponyms (extracted, for instance, from a text) $T = \{t_1, \ldots, t_n\}$, we define for each toponym, the set of geographic concepts labeled with the toponym as:

$$GeoConcepts(t_x) = \{g_1, \ldots, g_n\}$$

the goal is to define a function that maps each toponym to the geographic concept it is intended to represent in the input sequence:

$$GeoMap(t_x) = g_x : g_x \in GeoConcepts(t_x)$$

Global-Mapping identifies for each toponym the concept that maximizes its semantic similarity with the concepts for all the other toponyms. One-sense-per-word is assumed, that is, if the same toponym occurs in the text more than once we always assume that it is referring to the same geographic location [5]. For every toponym t_x a geographic concept is assigned:

$$GeoMap_{\mathrm{global}}(t_x) = \arg\max_{g_x} (\max_{g_y} sim(g_x, g_y))$$

where $g_x \in GeoConcepts(t_x)$ and $g_y \in GeoConcepts(T \backslash \{t_x\})$. At the end, each toponym t_x is mapped to the unique geographic concept that has the highest similarity score among all pairs of geographic concepts. This technique explores all the possible combinations between the different geographic meanings that each toponym can have, which represents a high computational complexity.

Sequential-Mapping takes into consideration the order of the toponyms in the text. First, it calculates the semantic similarity between the pairs of concepts for the first pair of toponyms, t_1 and t_2. From the set of possible pairs, the one with the highest semantic similarity is chosen, and the two geographic concepts are mapped to the corresponding toponyms:

$$GeoMap_{\mathrm{seq}}(t_1, t_2) = \arg\max_{g_1, g_2} sim(g_1, g_2)$$

Then, the next toponym in the text, t_3, is disambiguated. For all the pairs, composed by the geographic concept that gave the highest similarity score to toponym t_2 and all the possible geographic concepts for t_3, the pair with the highest semantic similarity is chosen. This pattern is applied sequentially, until the last toponym is reached. This technique always selects the geographic concept assigned to the last toponym and uses it to calculate the semantic similarity with all the possible geographic concepts for the next toponym in the text. This ensures that the geographic concept that yields the maximum similarity is always propagated to the next pair:

$$GeoMap_{\text{seq}}(t_x : 3 \leq x \leq n) = \underset{g_x}{\arg\max}\ sim(GeoMap_{\text{seq}}(t_{x-1}), g_x)$$

4 Assessment

To evaluate our techniques we used three public resources, described below.

Geo-Net-PT is a public geographic ontology covering the territory of Portugal. It is divided in two domains: administrative and physical. The administrative domain contains the administrative divisions of the territory and the physical domain includes physical geography features, such as natural regions and man-made spots [9].

The Information Content (IC) for any given concept was calculated with basis on the number of occurrences of the capitalized version of the name of a concept in a Portuguese n-grams collection [2].

CHAVE [14] is a Portuguese collection of news articles, with toponyms recognized by REMBRANDT [4]. The articles were scanned to map each identified toponym to the geographic concepts it might represent in Geo-Net-PT. A total of 195 news articles were selected for manual mapping. From the set of possible geographic concepts associated to each toponym, human mappers discarded all but the one representing the correct geographic concept associated to the toponym's context. Only toponyms for parts of the Portuguese territory having a geographic concept in Geo-Net-PT were considered. The result is Geo-CHAVE-PT, a subset of news articles from CHAVE with the toponyms linked to Geo-Net-PT concepts. Geo-Chave-PT is available for download[1], along with a detailed description of the corpus and mapping guidelines.

As baseline for assessing the effect of the proposed semantic similarity measures, we applied a naïve disambiguation technique that simply selects the geographic concept with the highest IC: $GeoMap_{\text{baseline}}(t_x) = \arg\max_{g_x} IC(g_x)$

4.1 Assessing Geographic Similarity

We applied the three techniques to automatically map the toponyms to geographic concepts. To evaluate the mappings we adapted a previously proposed formula to measure the geographic similarity between two concepts [1]. For a given pair (g_1, g_2), where g_1 represents the geographic concept manually mapped and g_2 the concept automatically disambiguated, the following characteristics were calculated:

[1] http://dmir.inesc-id.pt/reaction/Geo-Net-PT_02_in_English

Table 2. Average *GeoSimilarity* and processing time for the Geo-CHAVE-PT articles

Technique	Similarity Measure	Average *GeoSimilarity*	CPU time
$GeoMap_{\text{seq}}$	Jiang-Conrath	0.54	01:02:20
	Lin	0.45	01:03:45
	Resnik	0.43	03:04:57
$GeoMap_{\text{global}}$	Jiang-Conrath	0.51	02:17:27
	Lin	0.43	02:18:45
	Resnik	0.43	02:18:47
$GeoMap_{\text{baseline}}$		0.28	00:01:12

$$\text{closeness}(g_1, g_2) = (1 + \text{shortestpath}(g_1, g_2))^{-1}$$

$$\text{relatedness}(g_1, g_2) = \begin{cases} \text{desc}(g_1) \div \text{desc}(g_2) & \text{if } g_1 \subseteq g_2 \\ \text{desc}(g_2) \div \text{desc}(g_1) & \text{if } g_2 \subseteq g_1 \\ 0 & \text{otherwise} \end{cases}$$

$$\text{sibilings}(g_1, g_2) = \begin{cases} 1 & \text{if parent}(g_1) = \text{parent}(g_2) \\ 0 & \text{otherwise} \end{cases}$$

where $\text{desc}(g_x)$ the number of descendants of g_x in the graph, $\text{shortestpath}(g_x, g_y)$ defines the minimum distance between g_x and g_y, measured in number of edges and $\text{parent}(g_x)$ is the concept in the ontology hierarchy immediately above g_x. Those concepts are then combined by a sum and normalized to $[0, 1]$, yielding the metric adopted for this study:

$$GeoSimilarity(g_1, g_2) = \frac{1}{3}(\text{closeness}(g_1, g_2) + \text{relatedness}(g_1, g_2) + \text{sibilings}(g_1, g_2))$$

4.2 Results

The semantic similarity measures were implemented in Java, querying a relational database representation of Geo-Net-PT. Each mapping technique, implemented in Python, processed the articles from Geo-CHAVE-PT as a batch job. The processing time and average *GeoSimilarity* for the whole collection are shown in Table 2.

The Jiang-Conrath measure applied to the Sequential-Mapping achieves the best results. It also takes less time to process, because it does not explore all the possibilities. It simply chooses the best one locally as it sequentially maps the toponyms to geographic concepts. The Resnik measure took three times more to process in the Sequential-Mapping, because the highest similarity score was too often the same for different pairs, probably because it only uses the IC of a single common ancestor, the most informative one, the MICA. This increased the number of disambiguation possibilities exponentially.

This result, for a geospatial ontology, is in line with the work of Budanitsky and Hirst, where the Jiang-Conrath measure also outperformed the other tested measures on WordNet [3].

5 Conclusions

The Jiang-Conrath semantic similarity measure yields the best results and both mapping techniques have comparable results. The Global-Mapping technique, however, has high computational costs and assumes One-sense-per-word, the Sequential-Mapping is faster, and allows repeated toponyms in the same text to be correctly mapped to different geographic concepts.

The extraction of toponyms did not take into consideration linguistic features such as sentence boundaries or paragraphs. Geographic features usually associated to a toponym, such as municipality (*concelho*), street (*rua*), were not taken into consideration. Such geographic features alone can disambiguate the toponyms. Combining this new heuristic with others can also improve the geographic mapping process.

Acknowledgements. This work was supported by FCT, through the project UTA-Est/MAI/0006/2009 (REACTION), the scholarship SFRH/BD/70478/2010 and the Multiannual Funding Program.

References

1. Andrade, L., Silva, M.J.: Relevance Ranking for Geographic IR. In: Purves, R., Jones, C. (eds.) GIR. Department of Geography, University of Zurich (2006)
2. Batista, D., Silva, M.J.: A Statistical Study of the WPT05 Crawl of the Portuguese Web. In: FALA 2010 VI Jornadas en Tecnología del Habla and II Iberian SLTech Workshop, Vigo, Spain (2010)
3. Butanitsky, A., Hirst, G.: Semantic Distance in WordNet: An Experimental, Application-Oriented Evaluation of Five Measures. In: Proceedings of WordNet and Other Lexical Resources Workshop (2001)
4. Cardoso, N.: REMBRANDT - Reconhecimento de Entidades Mencionadas Baseado em Relações e ANálise Detalhada do Texto. In: Encontro do Segundo HAREM, PROPOR 2008, Aveiro, Portugal (2008)
5. Gale, W.A., Church, K.W., Yarowsky, D.: One Sense per Discourse. In: Proceedings of the Workshop on Speech and Natural Language, HLT 1991 (1992)
6. Jiang, J.J., Conrath, D.W.: Semantic Similarity Based on Corpus Statistics and Lexical Taxonomy. In: Proc. of the Int'l. Conf. on Research in Computational Linguistics, pp. 19–33 (1997)
7. Leidner, J.L., Sinclair, G., Webber, B.: Grounding Spatial Named Entities for Information Extraction and Question Answering. In: Proceedings of the HLT-NAACL 2003 Workshop on Analysis of Geographic References, vol. 1 (2003)
8. Lin, D.: An Information-Theoretic Definition of Similarity. In: ICML 1998: Proceedings of the Fifteenth International Conference on Machine Learning, pp. 296–304. Morgan Kaufmann Publishers Inc., San Francisco (1998)

9. Lopez-Pellicer, F.J., Chaves, M., Rodrigues, C., Silva, M.J.: Geographic Ontologies Production in GREASE-II. Tech. Rep. TR 09-18, University of Lisbon, Faculty of Sciences, LASIGE (November 2009)
10. Martins, B., Anastácio, I., Calado, P.: A Machine Learning Approach for Resolving Place References in Text. In: Proceedings of the 13th AGILE International Conference on Geographic Information Science. Association of Geographic Information Laboratories for Europe. Springer, Guimarães (2010)
11. Navigli, R.: Word Sense Disambiguation: A Survey. ACM Comput. Surv. 41 (February 2009)
12. Rauch, E., Bukatin, M., Baker, K.: A Confidence-Based Framework for Disambiguating Geographic Terms. In: Proceedings of the HLT-NAACL 2003 Workshop on Analysis of Geographic References. Association for Computational Linguistics (2003)
13. Resnik, P.: Using Information Content to Evaluate Semantic Similarity in a Taxonomy. In: Proceedings of the 14th International Joint Conference on Artificial Intelligence, vol. 1, pp. 448–453. Morgan Kaufmann Publishers Inc, San Francisco (1995)
14. Santos, D., Rocha, P.: The Key to the First CLEF with Portuguese: Topics, Questions and Answers in CHAVE. In: Multilingual Information Access for Text, Speech and Images (2005)

Automatic Hyponymy Identification from Brazilian Portuguese Texts

Leonardo Sameshima Taba and Helena de Medeiros Caseli

Department of Computer Science, LaLiC/NILC,
Federal University of São Carlos (UFSCar), São Carlos/SP, Brazil
{leonardo_taba,helenacaseli}@dc.ufscar.br

Abstract. In the natural language processing (NLP) scenario, Brazilian Portuguese (and Portuguese in general) still suffers from the lack of good quality base tools (e.g. parsers) and resources (e.g. annotated corpora). Corpora annotated with semantic information is particularly scarce and is a very costly resource to be produced manually. In order to provide some help to mend that situation, this paper presents an automatic hyponymy identification method for Brazilian Portuguese texts. The proposed method uses lexical and syntactic data alongside common sense information obtained from the Brazilian Open Mind Common Sense project (OMCS-Br). The results obtained so far are compatible with previous work and encourage other directions for further research.

Keywords: hyponymy identification, information extraction, lexical semantics, common sense, natural language processing.

1 Introduction

The usage and importance of semantic information in Natural Language Processing (NLP) tasks is growing by the minute, as can be seen in areas such as web search[1], information retrieval [3] and automatic translation[2]. However, the rate at which semantic information can be produced by human annotators is much less than that which is needed by NLP applications. As one of the resources that will hopefully help bridge that gap, this paper presents an automatic hyponymy identification method using lexical and syntactic data alongside common sense information.

Semantic relation identification is the task of finding semantic relations between terms in texts. There's not a single formal definition for "term" and "semantic relation", therefore, in this paper, "term" will refer to any entity that is mentioned in a text, and "semantic relation" stands for any relation, explicit or implicit, between terms on a semantic level. The semantic relation focused in this work is the hyponymy relation, also called is-a relation. Hyponymy was selected because (i) it is one of the most frequently occurring and studied semantic

[1] http://www.google.com/

[2] http://translate.google.com/

H. Caseli et al. (Eds.): PROPOR 2012, LNAI 7243, pp. 186–192, 2012.

relations and (ii) we intend to extend previous work [5] carried out on Brazilian Portuguese by using common sense knowledge. Therefore, this paper brings the first results of an ongoing work on automatic identification of semantic relations for Brazilian Portuguese. To do so, the investigated methodology tries to extract semantic relations from texts using lexical and syntactic patterns and also a novel feature regarding previous related studies: common sense information.

Common sense can be defined as a set of facts and beliefs that are shared by a group of people at a certain time and space [12]. Sentences like "The sky is blue" and "Balls are used to play football" could be categorized as common sense as they are common conceptions to many people in the world. The Open Mind Common Sense Project[3], from the Massachusetts Institute of Technology (MIT), aims to collect such knowledge via the internet, amassing data provided by textual templates filled by volunteers. There's also the Open Mind Common Sense Brazil Project[4] (OMCS-Br), the Brazilian branch of the MIT project, which is the one that this work is related to. Combining common sense information with NLP is still an under-researched area, and that combination could have a positive impact on many NLP tasks. Some possibilities are using common sense to customize texts in order to make them more understandable or to translate slangs and regionalisms [13].

The OMCS project's facts database is based on Minsky's [10] theory of how knowledge is organized inside the human mind. Minsky hypothesized that facts are stored as relations between terms; therefore, the OMCS project defined a set of semantic relations henceforth called "Minksy relations". Some examples of Minsky relations are hyponymy, meronymy, locality, made-of (something is made of some material) and effect-of (action and consequence). It is important to state that this paper reports only experiments carried out to identify hypomyny relations. Other Minsky relations will be investigated in a near future.

The rest of this paper is structured as follows: the next section gives an overview of related works on semantic relation identification, Section 3 describes the methods used in this work, Section 4 presents some of the preliminary results of this work and Section 5 provides a discussion of the preliminary results, the direction in which this work is heading and future developments.

2 Related Work

There has been extensive work on the subject of semantic relation identification and some of it will be overviewed in this section. Hearst [8] was the first to point some textual constructs as strong indicators of semantic relations in a text. In her paper, Hearst describes 6 textual patterns that indicate, with high reliability, the presence of a hyponym relation between two noun phrases. She proposed an algorithm to find patterns that imply a semantic relation R, which will be described in more details in the next section. Hearst's method showed good results: by applying her patterns on encyclopedic and journalistic corpora

[3] `http://openmind.media.mit.edu/`

[4] `http://www.sensocomum.ufscar.br/`

composed of around 28 million words, 63% of a random sample of all the relations found were considered of good quality. The evaluation was done manually and thus is highly subjective; aware of that fact, Hearst tried to be as "conservative" as possible, using WordNet [4] as a reference to measure the quality of the relations.

Berland and Charniak [1] follow Hearst's algorithm but for searching meronym relations. The authors found two textual patterns that were prolific in finding the desired relation. Their results, obtained by applying the patterns on a 100 million words journalistic corpus, show that, on average, 55% of the relations found were correct. Girju and Moldovan [7] also follow Hearst's algorithm, looking for causality relations on a 3 gigabytes journalistic corpus and reporting a 65% accuracy.

Freitas and Quental's [5] work is one of the few that focuses on the Portuguese language. They adapted Hearst's patterns to Portuguese, creating 4 patterns that indicate hyponym, and applied them to a corpus composed of around 2 million words of the public health domain. The results are compatible with Hearst's, showing that 73% of the relations found were of high quality.

3 Methodology

As the first step to the automatic identification of semantic relations, we implemented and evaluated two approaches based on Hearst's [8] method for hyponymy relations. In the following, we describe the data resources used in the experiments (Sect. 3.1) and the investigated approaches (Sect. 3.2).

3.1 Resources

The main linguistic resource used in this work is a scientific-journalistic corpus comprised of around 870000 words from articles of the scientific dissemination magazine *Pesquisa FAPESP*[5]. The corpus was morphologically tagged by the PALAVRAS parser [2] and it's noun phrases were identified using a noun phrase (NP) identifier described in [11].

The other resource used is the OMCS-Br common sense facts database. Among all the semantic relations that it contains, there are around 12000 hyponymy relations that were used in the second approach evaluated in this paper.

3.2 Approaches

The first approach consists in applying manually constructed lexico-syntactic patterns on a text to identify hyponymy relations. Particularly, the work of Freitas and Quental [5], who adapted Hearst's patterns to Brazilian Portuguese, was the main base for this paper. Freitas and Quental defined 4 lexico-syntactic patterns that are strong indicators of hyponymy relations, which are very

[5] `http://revistapesquisa.fapesp.br/`

1: NP_Hyper (tais como|como) NP {, NP}* (e|ou) NP
2: NP {, NP}* ,? (e|ou) outros NP_Hyper
3: tipos de NP_Hyper: NP {, NP}* (e|ou) NP
4: NP_Hyper chamad(o|a|os|as) de? NP

Fig. 1. Freitas and Quental's [5] patterns

similar to Hearst's "such as" ("*tais como*" or "*como*") and "or/and others" ("(*e*|*ou*) *outros*") patterns. Figure 1 shows these patterns as regular expressions.[6] After the tagging and NP identifying process described on Sect. 3.1, Freitas's patterns were applied to the corpus in order to find hyponym relations.

The second method implemented was the algorithm (Fig. 2) defined by Hearst to find new patterns based on already known related terms and the context of occurrence of these terms in texts. The common sense knowledge from the OMCS-Br project was applied in the second step of the algorithm: the list of seed terms used was extracted from the OMCS-Br facts database as the terms of already known hyponymy relations (see Sect. 3.1).

1. A semantic relation R of interest is chosen (e.g. hyponymy, meronymy, etc.);
2. A list of pairs of terms which are known to be related by R is constructed (e.g. for the hyponymy relation, "Brazil-country", "Portugal-country", and "dog-animal"). That list can be obtained by applying hand-made textual patterns on the corpus or from a pre-existing lexical or knowledge database;
3. Sentences of the corpus in which these terms occur syntactically close to each other are searched, and the contexts (words around which they appear) are stored (e.g. "Brazil is an emerging country").
4. The stored contexts are analyzed, taking into account that common contexts may be indicators of the relation of interest;
5. One or more contexts are converted into patterns which are used to find more instances of the relation R and the process is repeated from step 2.

Fig. 2. Hearst's [8] algorithm

Thus, following Hearst's algorithm in Fig. 2, the list of common sense seed terms was used to obtain a set of 76 distinct contexts that may indicate the relation of interest, in this case hyponymy. After one iteration of the algorithm, the automatically extracted contexts were manually reviewed to generate new patterns, described in Fig. 3. These patterns, in turn, were used to find new relation instances. In this first experiment only one iteration of the algorithm was performed.

[6] In the patterns, NP stands for a noun phrase and NP_Hyper denotes the hyperonym NP in the pattern.

```
5: NP {, NP}* ,? (e | ou) (qualquer | quaisquer) outro{s}? NP Hiper
6: NP é (o|a|um|uma) NP_Hyper
7: NP são NP_Hyper
```

Fig. 3. New patterns found with the application of Hearst's [8] algorithm described in Fig. 2

4 Results

The results obtained by applying Freitas and Quental's patterns (Fig. 1) to the corpus are akin to the results obtained by them [5] and Hearst [8]: of the 1309 relations found, 816 (62.34%) were considered correct by manual review. As in these previous works, the automatically identified relations were manually analyzed and so the evaluation is subjective; even so, the analysis was conducted as conservatively and carefully as possible.

It's also worth mentioning that other 317 relations (24.22%) were partially correct, with the error being in the identification of the noun phrases because of prepositional phrases – some noun phrases encompass more or less terms than they should. For example, in the sentence *"(...) [centros urbanos] em [crescimento contínuo] como [Uberaba] (...)"* ([urban centers] in [continuous development] like [Uberaba]), the correct NP[7] for the relation should encompass both identified NPs, starting at "centros" and only ending at "contínuo". That incorrect NP identification results in the partially correct relation *is-a(Uberaba, crescimento contínuo)*, when it should be *is-a(Uberaba, centros urbanos em crescimento contínuo)*.

When applying only the new patterns (Fig. 3) found using Hearst's algorithm and the list of common sense related terms, 854 new relations were found, of which 605 (70.84%) were considered correct and 163 (19.09%) were partially correct. Table 1 summarizes all the obtained results and Table 2 presents some examples of correctly identified relations.

Table 1. Number of correct, partially correct and incorrect relations identified using the patterns on Fig. 1 and Fig. 3

Patterns applied	**Correct**	**Partially correct**	**Incorrect**	**Total**
Freitas and Quental's (Fig. 1)	816 (62.34%)	317 (24.22%)	176 (13.44%)	1309 (100%)
Our new patterns (Fig. 3)	605 (70.84%)	163 (19.09%)	86 (10.07%)	854 (100%)
Total	1421 (65.70%)	480 (22.19%)	262 (12.11%)	2163 (100%)

Although the precision shown in Table 1 alone doesn't indicate the real quality of the proposed approach, it was not possible to calculate the recall values due to the corpus used in the experiments not being annotated with the correct semantic relations that should be identified. Therefore, the total amount of relations that

[7] The noun phrases are delimited by square brackets ("[" and "]").

Table 2. Examples of correctly identified relations, the contexts around which they occurred and the patterns that were used to identify them

Relation	Original context	Pattern
is-a(Rio Branco, capital) *is-a(Rio Branco, capital)*	...aeroportos de capitais como Rio Branco... *...airports of capitals such as Rio branco...*	1
is-a(formigas, insetos) *is-a(ants, insects)*	...formigas e outros insetos acabam grudados... *...ants and other insects end up stuck...*	2
is-a(macrófagos, células) *is-a(macrophages, cells)*	...ativa células chamadas macrófagos, que... *...activates cells called macrophages, that...*	4
is-a(cerveja, bebida) *is-a(beer, beverage)*	...tomar cerveja, vinho ou qualquer outra bebida... *...drinking beer, wine or any other beverage...*	5
is-a(dióxido de estanho, semicondutor) *is-a(tin dioxide, semiconductor)*	...o dióxido de estanho é um semicondutor... *...tin dioxide is a semiconductor...*	6

are codified in the corpus is not known. Even so, the feeling is that the recall is very low, since from around 36000 sentences only 2163 hyponymy relations were found. Creating an annotated corpus is one of the future tasks that are going to be done in order to allow further experiments.

5 Discussion and Future Work

As noted by Hearst [9] and Freitas and Quental [5] and discussed in the last section, one of the difficulties when extracting relations is the correct identification of the noun phrases. Also, some of the extracted relations, although correct, are very general, like *is-a(engenharia mecânica, área)* (is-a(mechanical engineering, area)).

So far, only one type of Minsky relation – hyponymy – was identified. The remaining relations are still untested. Some of them, like meronymy, were already studied in past works [1], but others weren't. More experiments must be made in order to evaluate the possibility of identifying them with textual patterns, and the quality of such identification.

This work and all the works mentioned in Sect. 2 are based on the textual patterns paradigm. One of the problems of this approach is that it has high precision but low recall. In order to improve these results, machine learning (ML) methods have been advocated.

Works such as [14] and [6] have shown that ML can be applied successfully to the semantic relation identification task. Combining machine learning techniques with common sense knowledge seems to be a promising direction, and is the one that will be taken in a near future.

Acknowledgements. The authors acknowledge CNPq and FAPESP for the financial support.

References

1. Berland, M., Charniak, E.: Finding parts in very large corpora. In: Proceedings of the 37th Annual Meeting of the ACL, pp. 57–64. ACL, College park (1999)
2. Bick, E.: The Parsing System" Palavras": Automatic Grammatical Analysis of Portuguese in a Constraint Grammar Framework. Aarhus University Press (2000)
3. Carlson, A., Betteridge, J., Kisiel, B., Settles, B., Hruschka Jr., E.R., Mitchell, T.M.: Toward an architecture for never-ending language learning. In: Proceedings of the AAAI (2010)
4. Fellbaum, C.: WordNet: An electronic lexical database. The MIT press, Cambridge (1998)
5. de Freitas, M.C., Quental, V.: Subsídios para a elaboração automática de taxonomias. In: Anais do XXVII Congresso da SBC, pp. 1585–1594. V TIL, Rio de Janeiro (2007)
6. Girju, R., Beamer, B., Rozovskaya, A., Fister, A., Bhat, S.: A knowledge-rich approach to identifying semantic relations between nominals. Information Processing and Management 46(5), 589–610 (2010)
7. Girju, R., Moldovan, D.: Text mining for causal relations. In: Proceedings of the FLAIRS 2002, pp. 360–364. AAAI Press, Pensacola (2002)
8. Hearst, M.A.: Automatic acquisition of hyponyms from large text corpora. In: Proceedings of the 14th International Conference on Computational Linguistics, vol. 2, pp. 539–545. ACL, Nantes (1992)
9. Hearst, M.A.: Automated discovery of wordnet relations. In: WordNet: An Electronic Lexical Database, ch.5, pp. 131–151. The MIT press (1998)
10. Minsky, M.: The Society of Mind. Simon and Schuster (1986)
11. dos Santos, C.N., Oliveira, C.: Aplicação de aprendizado baseado em transformaçoes na identificação de sintagmas nominais. In: Anais do XXV Congresso da SBC, pp. 2138–2147. III TIL, Brasil (2005)
12. Singh, P., Lin, T., Mueller, E.T., Lim, G., Perkins, T., Zhu, W.L.: Open Mind Common Sense: Knowledge Acquisition from the General Public. In: Meersman, R., Tari, Z. (eds.) CoopIS/DOA/ODBASE 2002. LNCS, vol. 2519, pp. 1223–1237. Springer, Heidelberg (2002)
13. Sugiyama, B.A., Anacleto, J.C., de Medeiros Caseli, H.: Assisting users in a cross-cultural communication by providing culturally contextualized translations. In: Proceedings of the 29th ACM International Conference on Design of Communication, SIGDOC 2011, October 03-05, pp. 189–194. ACM, Pisa (2011)
14. Zhang, M., Zhang, J., Su, J., Zhou, G.: A composite kernel to extract relations between entities with both flat and structured features. In: Proceedings of the 21st COLING and the 44th Annual Meeting of the ACL, ACL-44, pp. 825–832. ACL, Stroudsburg (2006)

Semantic Role Labeling for Portuguese – A Preliminary Approach –

João Sequeira, Teresa Gonçalves, and Paulo Quaresma

Universidade de Évora
m5071@alunos.uevora.pt, {tcg,pq}@uevora.pt

Abstract. Currently there are increasingly more private and academic publications in the form of digital content on the Internet making extremely difficult to extract and maintain the content information manually. Normally, these tasks follow approximations based on natural language processing. This paper presents a preliminary approach for obtaining a semantic role labeler for Portuguese, a little explored aspect of natural language processing for this language. The approach was evaluated for the 3 most frequent semantic roles (relation, subject and object) with a subset of Bosque 8.0 corpus. The same approach was applied to an English corpus – the CONLL'2004 one and its results were compared to the ones obtained on the CONLL'2004 shared task. At the same time it presents BosqueUE, a Portuguese corpus for semantic role labeling that can be the basis material for future research in the area. This corpus has the same format as the CONLL'2004 one, facilitating multi-language evaluations.

1 Introduction

Currently there is a large amount of digital content (academic, personal,news and other) available on the internet. The task of extracting information content from these different kind of sources became practically impossible [22]. With the increase of digital content published there was also an increase in research applications able to automatically analyze and extract information from them [7].

The semantic role labeling, portraying the semantic relationships between the different constituents of the sentence, has been an area of increasing interest due to their importance in applications of information extraction, question-answering, document summarization and others that require semantic information [6]. This aspect of natural language processing already has several available resources for the English language, product of several projects under international conferences [6], but there is still much material to be explored in other languages, such as the Portuguese one.

This paper describes the construction of a Portuguese corpus for the Semantic Role Labeling task and the use of the MinorThird tool [9] as a preliminary work for this NLP task. To have a means of comparison, MinorThird is also used with

H. Caseli et al. (Eds.): PROPOR 2012, LNAI 7243, pp. 193–203, 2012.

the English corpus built for the CONLL'2004 Conference[1] and its results are compared with the ones obtained there [6].

It is organized as follows: Section 2 introduces the Semantic Role Labeling task and some systems built to perform it, Section 3 describes the construction of a Portuguese corpus for the SRL task and Section 4 presents a preliminary semantic role labeler built with MinorThird. Finally Section 5 discusses the obtained results and enumerates future work.

2 Semantic Role Labeling

This section introduces the semantic role labeling task and presents some work done in the field.

2.1 The Task

Semantic Role Labeling is currently one of the most active subgroups in the area of natural language processing. It intends to identify the verbs in sentences and its syntactic arguments [7], such as the subject of the action and the object of the action among others.

Figure 1 shows the semantic roles for a Portuguese sentence present in Bosque 8.0 [1], a Portuguese corpus parsed by the Palavras tool [3] and manually revised by linguists. The sentence has a subject ("`Vera`"), a verb that makes up the predicate ("`apagou`") and an object ("`a luz`"). The predicate is tagged as the *Relation*, the subject as *Arg0* and the object as *Arg1* (*Arg0* and *Arg1* are numbered arguments of the predicate used by the Propbank annotation).

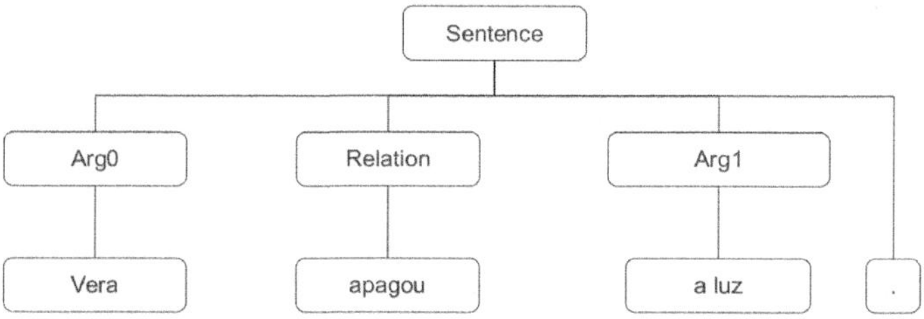

Fig. 1. A Portuguese sentence and the semantic roles present in it

Gildea and Jurafsky [13], pioneers in the semantic role labeling task, listed two prominent methods to perform the analysis of texts: the grammar based systems and the data-driven ones. The process of creating grammars is very time consuming as they are created by hand and need to include a description for each existing case of the language. On the other hand, data-driven systems

[1] Conference on Computational Natural Language Learning.

need a classified corpus, and an application to create a model from them. This model is then used to classify new texts. Examples of these applications are the participants of CONLL shared tasks, specifically for the years of 2004 and 2005 [6,7].

Palmer *el. al* [24] introduces SRL and discusses the most important and discriminating features used by this task. Next sub-section presents some semantic role labeling systems.

2.2 Related Work

Bick [5] describes a grammar based semantic-role annotator for the Portuguese language: it uses a constraint grammar to map and disambiguate 40 different semantic roles. The grammar has 500 mapping rules and a small number of disambiguation rules. It is reported to have an average f-score of 88.6%.

Amancio *et. al.* [2] describe a similar task to SRL, which is the assigning of Wh-Questions to Verbal Arguments via machine learning, using the same features used in SRL.

Gildea and Hockenmaier [14] present a systems that uses a combinatory categorial grammar to identify the semantic roles for the English language and is reported to have a f-score of 80.4 on PropBank [16,23].

CONLL'2004 and CONLL'2005 shared tasks aimed at developing data-driven systems for semantic role labeling. They used the Penn Treebank [21,26] with predicate-argument annotations from the PropBank. On CONLL'2005 the Brown corpus [12] was also used to realize tests.

On CONLL'2004 the systems with better performance were the ones developed by Hacioglu et al. [15] with an f-score of 69.49 for the test set and Punyakanok et al. [28] with an f-score of 66.39.

The Hacioglu system was developed using TinySVM [17], a Support Vector Machine implementation, with a polynomial kernel of degree 2 using features from three categories [15]:

- **Base** Features. These can be inferred directly from the simplest data: words, verb lemmas, part-of-speech, BP positions using IOB, clause labels present in the sentence and named entities (using person, local, organization and others);
- **Token** Features. These are those that can be inferred at the level above the basic features (the level of BP) such as: token position with respect to the predicate, path between the token and the predicate, clause patterns, distances between the token and the predicate and the number of words in a the token
- **Sentence** level features such as part-of-speech of the predicate and the words that precede and follow it, frequent/rare predicate, context window of the predicate, semantic frames of the predicates present in PropBank and number of predicates in the sentence.

The Punyakanok et al. system [28] is composed by a set of classifiers and an inference procedure used both to clean the classification results and to ensure structural integrity of the final role labeling. The learning algorithm used is a variation of the Winnow update rule incorporated in SNoW [31,32], a multi-class classifier that is specifically tailored for large scale learning tasks.

The system consists of three phases [28]:

1. Find **Argument Candidates.** The system tries to filter out unlikely candidates with two classifiers: one to detect beginning phrase locations and other to detect end phrase locations. Both use the following features: word, POS tag, IOB tags for chunks, lemma and POS tag of the active predicate, active/passive voice of the current predicate, word position with respect to the predicate, the boundary of clauses, the sequence of chunks from the current word to the predicate, the path formed from a semi-parsed tree containing clauses and chunk and the position of the target word relative to the predicate;
2. **Phrase Classification.** This phase is accomplished with a multi-class classifier used to supply confidence scores of how likely individual phrases have specific argument types and The most likely solution is chosen using the matrix of confidences and linguistic information. It uses a multitude of linguistic, position and features.
3. **Filter Function.** This phase applies global constraints derived from linguistic information and structural considerations.

In CONLL'2005, the systems that obtained the best results were [27] and [25] with f-scores of 77.92 and 77.30 respectively for test set combining the Penn Treebank and the Brown corpus.

The Punyakanok et al. system [27] uses the same learning algorithm as the system presented in CONLL'2004. It has four stages:

1. **Pruning.** Very unlikely constituents are filtered by means of an heuristic presented in [36];
2. **Argument Identification.** Uses binary classification to identify whether a candidate is an argument or not;
3. **Argument Classification.** Uses a multi-class classifier trained to differentiate the types of the arguments supplied by the previous stage;
4. **Inference.** It incorporates linguistic and structural knowledge to resolve any inconsistencies of argument classification. The process is formulated as an integer linear programming (ILP) problem that takes as inputs the confidences over each type of the arguments supplied by the argument classifier using several constraints.

The Pradhan et al. system [25] uses TinySVM to train one-vs-all classifiers with Support Vector Machines developing a binary classifier to each semantic class plus a "NULL" class. It uses two systems: one chunk based that are very efficient and robust and other based on full syntactic parses that normally are more accurate; the goal was to preserve the robustness and flexibility of the segmentation of the phrase-based chunker, and take advantage of features from full syntactic parses. For an input sentence, syntactic constituent structure parses are generated by a Charniak [8] parser and a Collins [10] parser. Semantic role labels are assigned to the constituents of each parse using Support Vector Machine classifiers. The resulting semantic role labels are converted to an IOB representation. These IOB representations are used as additional features, along with flat syntactic chunks, by a chunking SVM classifier that produces the final SRL output [25].

3 BosqueUE – A Portuguese Corpus for SRL

As already mention, to build a data-driven SRL system for the Portuguese language it is necessary to have a classified corpus. Besides including the semantic roles of each sentence's chunk, it helps to have the morphologic, syntactic and other kind of features.

This enriched corpus is primally based on Bosque 8.0[2], a corpus incorporated in Forest Sintá(c)tica project [1]. This project consists of plain text with syntactically analyzed sentences (tree structures) by the PALAVRAS parser [4]. Figure 2 shows the information retained in Bosque 8.0 for the sentence '`Vera apagou a luz.`'.

```
'source' => 'CP429-7 Vera apagou a luz.',
'number' => 1,
'cod' => 'CETEMPúblico n=429 sec=clt sem=96a',
't' => [
  'fcl||STA',
  [
    'np||SUBJ',
    'prop(\'Vera\' F S)||H::Vera'
  ],
  [
    'vp||P',
    'v-fin(\'apagar\' PS 3S IND)||MV::apagou'
  ],
  [
    'np||ACC',
    'art(\'o\' <artd> F S)||>N::a',
    'n(\'luz\' <np-def> F S)||H::luz'
  ],
  'jjpunct(-.-)'
]
```

Fig. 2. Bosque 8.0 representation for the sentence '`Vera apagou a luz.`'

Bosque 8.0 consists of 9368 sentences from the first 1000 extracts from the CETEMPúblico and CETEMFolha, prioritizing quality over quantity [20]. CETEMPúblico uses news taken from the Público newspaper and CETEMFolha uses news excerpts taken from the Folha de S. Paulo newspaper.

A subset of 4416 sentences from Bosque 8.0 was used to build BosqueUE; this subset was obtained by selecting the CETEMPublico sentences that ended with a punctuation mark.

[2] `http://www.linguateca.pt/Floresta/corpus.html`

BosqueUE corpus was built using the same format as the corpus used in CONLL'2004; Figure 3 presents the extract for same sentence.

Vera	Vera	PROP	B-np	B-SUBJ	B-PER	(S*
apagou	apagar	V	B-vp	B-p	O	*
a	o	DET	B-np	B-ACC	O	*
luz	luz	N	I-np	I-ACC	O	*
.	.	PU	O	O	O	*S)

Fig. 3. BosqueUE representation for the sentence '`Vera apagou a luz.`'

This format has a word per line and contains the following 7 features: word, lemma, part-of-speech tag, chunks with IOB, semantic roles with IOB, named entities with IOB and clauses. Words, lemmas, chunks and clauses were extracted from the Bosque8.0 corpus; the part-of-speech column uses LABEL-LEX [18] tags having performed a manual review in ambiguous situations; the named entities were obtained with the application presented in [22].

A similar annotated corpus for SRL for Portuguese language, is the Propbank-Br [11], a Brazilian treebank annotated with semantic role labels. This corpus consists of 6142 instances for SRL annotation, with 1068 different predicates. This corpus follows the Propbank guidelines and uses the syntactic trees of the Brazilian portion of Bosque.

4 A Preliminary Semantic Role Labeler

This section presents a preliminary approach of using a Portuguese corpus for the SRL task: it starts by introducing MinorThird tool and then presents corpora and the experimental setup. It ends by displaying the results obtained.

4.1 MinorThird

MinorThird is an open source set of Java classes to perform tasks over texts, such as text classification and named entity extraction. It was created by Professor William W. Cohen of the Carnegie Mellon University and is currently maintained Frank Lin [9].

It uses a collection of documents to create a database called *TextBase* and logical statements over text chunks are stored in *TextLabels* objects. Since the annotations on *TextLabels* are independent of the contents of the documents it is possible to have different annotations on same set of documents. The annotations in *TextLabels* describe syntactic or semantic properties for words, documents or spans and can be created manually or automatically through an application [9].

It implements sequential learning methods such as Conditional Random Fields [35,19] and semi, conditional, maximum entropy and hidden Markov Models [33].

4.2 Corpora

The BosqueUE corpus was transformed into XML documents with syntactic annotations. For example, the sentence '`Vera apagou a luz`' is represented as showed in Figure 4 (tag`<P>` stands for *Predicate*, `<Arg0>` for *Subject* and `<Arg1>` for *Object*).

```
<Arg0>Vera</Arg0> <P>apagou</P> <Arg1>a luz</Arg1>.
```

Fig. 4. SRL tags for the sentence '`Vera apagou a luz.`'

In order to compare the outcome of the BosqueUE Portuguese corpus with an English one, a similar process was carried out with the corpus used for the CONLL'2004 shared task [6][3]. This corpus consists of six sections of the Wall Street Journal part of the Penn Treebank [21] adding information from predicate-argument syntactic structures [16,23].

4.3 Experimental Setting

In order to obtain the best results several MinorThird algorithms with the default values were tried with a context window of size three. The best results were obtained for:

- SVMCMM: Conditional Markov Models [30,29] trained with Support Vector Machines [34];
- CRF: Conditional Random Fields [35,19]

As already said, BosqueUE is composed of 4416 sentences; from all its syntactic tags, models were built for the ones with statistic validity, namely, *Predicate*, *Subject* and *Object*.

CONLL'2004 corpus is divided into train, development and test sets composed respectively of 8936 (sections 15–18 of Penn TreeBank), 1671 (section 20) and 2012 sentences (section 21). Only train and test sets were used, since MinorThird algorithms were tried with default values and models were built for the same tags.

Table 1 shows the syntactic tags with the number of times they appear in each corpus.

Table 1. Main tags, semantic roles and count for BosqueUE and CONLL'2004 corpora

		BosqueUE	CONLL'2004	
tag	semantic role	#	# train	#test
`P`	Predicate	7268	19098	3627
`Arg0`	Subject	4673	12709	1671
`Arg1`	Object	3802	18046	3429

[3] Retrieved from: `http://www.lsi.upc.edu/srlconll/st04/st04.html`

A 10-fold cross-validation procedure was applied over the BosqueUE corpus and a train/test procedure was applied to CONLL'2004 one. Model's performance was analyzed through precision (π), recall (ρ) and F_1 (f_1) measures.

4.4 Results

Table 2 shows the results obtained with SVMCMM and CRF algorithms for the BosqueUE corpus. In it, it's possible to observe that CRF algorithm consistently presents better precision values while SVMCMM presents better recall ones.

For both algorithms precision values are at least 0.1 above recall ones for all tags (for CRF algorithm and `Arg0` and `Arg1` tags precision is 0.2 higher than recall). As expected the *Predicate* tag presents the best results with f_1 values above 52%, while the *Object* tag has f_1 values below 20%.

Table 2. BosqueUE: Precision, recall and F_1 for SVMCMM and CRF algorithms

	SVMCMM			CRF		
tag	π	ρ	f_1	π	ρ	f_1
`P`	.588	.477	.527	.648	.477	.549
`Arg0`	.388	.259	.311	.434	.237	.306
`Arg1`	.269	.147	.190	.317	.117	.171

Table 3 shows the results for the CONLL'2004 corpus obtained with SVM-CMM and CRF algorithms.

As opposed to BosqueUE, in CONLL'2004 corpus both algorithms present similar precision and recall values (except for `Arg0` tag where CRF has a 0.1 higher precision value). Once again, the *Predicate* tag presents the best results with f_1 values above 82%, while the *Object* tag has f_1 values below 24%.

Table 3. CONLL'2004: Precision, recall and F_1 for SVMCMM and CRF algorithms

	SVMCMM			CRF		
tag	π	ρ	f_1	π	ρ	f_1
`P`	.850	.823	.836	.842	.805	.823
`Arg0`	.599	.464	.523	.699	.463	.557
`Arg1`	.372	.170	.234	.414	.151	.221

Comparing Table 2 and Table 3, one can conclude that the ones obtained for the Portuguese corpus are below the corresponding ones for English corpus. This gap could be explained by the different sizes of the datasets: CONLL'2004 is around 3 times bigger than BosqueUE.

Table 4 compares F_1 values for `Arg0` and `Arg1` tags obtained by SVMCMM Minorthird algorithm with the best and worst ones from CONLL'2004 shared task as reported on [6] (*Predicate* values are not shown since they were not reported on the shared task).

Table 4. F_1 values obtained by SVMCMM and the best algorithm from CONLL'2004 shared task

		CONLL'2004	
tag	SVMCMM	best	worst
`Arg0`	.523	.814	.562
`Arg1`	.234	.716	.490

From the table one can see that the use of linguistic information such as words, part-of-speech and chunk labels, clauses and named entities are useful for the semantic role labeling problem and the sequential learning methods alone are not enough. While these features improve both labelers (`Arg0` and `Arg1`) the increase is greater for *Object* role than for the *Subject* one.

5 Conclusions and Future Work

This work attempts to apply a line of research still little explored for the Portuguese language. It was found that the preliminary results obtained with a Portuguese corpus are below those obtained with an English corpus. As already mentioned this difference could be explained by the different corpus size. Another possible explanation is the use of more complex syntactic structures and many word flexions that exists in the Portuguese language when compared with the English one.

On the other hand it is possible to conclude that the use of linguistic information such as words, part-of-speech and chunk labels, clauses and named entities are useful for the semantic role labeling problem and the sequential learning methods alone does not produce good results.

As future work we intend to increase the size of the Portuguese corpus and develop a classifier that makes use of all the linguistic information that the built BosqueUE corpus provides. Only then a comparison between both languages will be fair.

References

1. Afonso, S., Bick, E., Haber, R., Santos, D.: Florestra sintá(c)tica: A treebank for portuguese. In: LREC 2002, the Third International Conference on Language Resources and Evaluation, pp. 1698–1703 (2002)
2. Amancio, M.A., Duran, M.S., Aluisio, S.M.: Automatic question categorization: a new approach for text elaboration. Procesamiento del Lenguaje Natural (46), 43–50 (March 2011)
3. Bick, E.: The Parsing System "Palavras": Automatic Grammatical Analysis of Portuguese in a Constraint Grammar Framework. Ph.D. thesis, Aarhus University, Aarhus, Denmark (November 2000)
4. Bick, E.: The Parsing System "PALAVRAS": Automatic Grammatical Analysis of Portuguese in a Constraint Grammar Framework. Aarhus University Press (2000)

5. Bick, E.: Automatic semantic-role annotation for portuguese. In: Anais do XXVII Congresso de SBC (2007)
6. Carreras, X., Màrquez, L.: Introduction to the conll-2004 shared task: Semantic role labeling. In: Proceedings of CoNLL 2004 (2004)
7. Carreras, X., Màrquez, L.: Introduction to the conll-2005 shared task: Semantic role labeling. In: Proceedings of the Ninth Conference on Computational Natural Language Learning, CoNLL 2005 (2005)
8. Charniak, E.: A maximum-entropy inspired parser. In: Proceedings of NAACL 2000 (2000)
9. Cohen, W.: Minorthird: methods for identifying names and ontological relations in text using heuristics for inducing regularities from data (2004), `http://minorthird.sourceforge.net`
10. Collins, M.: Head-driven statistical models for natural language parsing. Computational Linguistics 29(4), 589–637 (2003)
11. Duran, M.S., Aluisio, S.M.: Propbank-br: a brazilian portuguese corpus annotated with semantic role labels. In: STIL 2011 – 8th Symposium in Information and Human Language Technology (October 2011)
12. Francis, W., Kucera, H.: Brown corpus manual (1997), `http://icame.uib.no/brown/bcm.html`
13. Gildea, D., Jurafsky, D.: Automatic labeling of semantic roles. Computational Linguistics 28, 245–288 (2002)
14. Gildea, D., Hockenmaier, J.: Identifying semantic roles using combinatory categorial grammar. In: Proceedings of the 2003 conference on Empirical Methods in Natural Language Processing, EMNLP 2003, pp. 57–64. Association for Computational Linguistics, Stroudsburg (2003)
15. Hacioglu, K., Pradhan, S., Ward, W., Martin, J., Jurafsky, D.: Semantic role labeling by tagging syntactic chunks. In: Proceedings of CoNLL 2004 Shared Task, pp. 110–113 (2004)
16. Kingsbury, P., Palmer, M.: From treebank to propbank (2002), `http://citeseerx.ist.psu.edu/viewdoc/summary?doi=10.1.1.13.7566`
17. Kudo, T.: Tinysvm: Support vector machines (2002), `http://chasen.org/~taku/software/TinySVM`
18. Laboratório de Engenharia da Linguagem: Label-lex (1995), `http://label.ist.utl.pt/pt/apresentacao.php`
19. Lafferty, J., McCallum, A., Pereira, F.: Conditional random fields: Probabilistic models for segmenting and labeling sequence data. In: Proceedings of 18th International Conference on Machine Learning, pp. 282–289 (2001)
20. Linguateca: Florestra sintá(c)tica (2009), `http://www.linguateca.pt/floresta/corpus.html`
21. Marcus, M., Santorini, B., Marcinkiewicz, M.: Building a large annotated corpus of english: The penn treebank. Computational Linguistics 19(2), 313–330 (1993)
22. Miranda, N., Raminhos, R., Seabra, P., Sequeira, J., Gonçalves, T., Quaresma, P.: Named entity recognition using machine learning techniques. In: EPIA 2011, 15th Portuguese Conference on Artificial Intelligence, Lisbon, PT (October 2011)
23. Palmer, M., Gildea, D., Kingsbury, P.: The preposition bank: An annotated corpus of semantic roles. Computational Linguistics 31 (2005)
24. Palmer, M., Gildea, D., Xue, N.: Semantic Role Labeling. Synthesis Lectures on Human Language Technologies. Morgan & Claypool Publishers (2010)
25. Pradhan, S., Hacioglu, K., Ward, W., Martin, J., Jurafsky, D.: Semantic role chunking combining complementary syntactic views. In: Proceedings of the Ninth Conference on Computational Natural Language Learning, CoNLL 2005 (2005)

26. Project, T.P.T.: The penn treebank project (1999), http://www.cis.upenn.edu/~treebank/
27. Punyakanok, V., Koomen, P., Roth, D., Yih, W.: Generalized inference with multiple semantic role labeling systems. In: Proceedings of the Ninth Conference on Computational Natural Language Learning (CoNLL 2005), pp. 181–184 (2005)
28. Punyakanok, V., Roth, D., Yih, W., Zimak, D., Tu, Y.: Semantic role labeling via generalized inference over classifiers. In: Proceedings of CoNLL 2004 Shared Task (2004)
29. Rabiner, L.: A tutorial on hidden markov models and selected applications in speech recognition. Proceedings of the IEEE, 257–286 (1989)
30. Rabiner, L., Juang, B.: An introduction to hidden markov models. IEEE ASSP Magazine (Janeiro 1986)
31. Roth, D.: Learning to resolve natural language ambiguities: A unified approach. In: Proc. of AAAI, pp. 806–813 (1998)
32. Roth, D., Yih, W.: Probabilistic reasoning for entity & relation recognition. In: The 19th International Conference on Computational Linguistics, COLING 2002, pp. 835–841 (2002)
33. Stamp, M.: A revealing introduction to hidden markov models (2004), http://citeseerx.ist.psu.edu/viewdoc/summary?doi=10.1.1.136.137&rank=1
34. Vapnik, V.: Statistical Learning Theory. Wiley-Interscience (Setembro 1998)
35. Wallach, H.: Conditional random fields: An introduction (2004), http://citeseer.ist.psu.edu/viewdoc/summary?doi=10.1.1.124.6711
36. Xue, N., Palmer, M.: Calibrating features for semantic role labeling. In: Proc. of the EMNLP 2004, pp. 88–94 (2004)

An Architecture for Semantic Role Labeling on Portuguese

Erick Rocha Fonseca and João Luís G. Rosa

Instituto de Ciências Matemáticas e de Computação,
Universidade de São Paulo, São Carlos, Brazil
{erickrf,joaoluis}@icmc.usp.br

Abstract. We present an adaptation of the architeture of the system SENNA, which performs various NLP tasks, to Portuguese, considering the richly inflected morphology of the language. We propose to separate words in lemmas and their flexional attributes. We point out the major problems that could arise from this approach as well as their potential solutions. This architecture can greatly benefit from the use of unlabeled data, which is especially good considering the small amounts of labeled resources in Portuguese.

1 Introduction

Most Natural Language Processing (NLP) end-user applications need one or more types of linguistic annotation on the pieces of text they are working with, among which some of the most common are syntactic trees, part-of-speech tags, and, more recently, semantic role labels. In order to obtain these annotations in real time, specialized tools for each kind of annotation are usually employed.

The internals of these tools are often focused on a single task, and provide no way for interacting with different ones except for having their output used as input to other tools. Thus, when training machine learning based systems for each task, instead of some kind of shared knowledge being produced, what happens is that the error from the first systems propagates to the next ones.

This article presents an implementation idea for a multipurpose NLP system for Portuguese, inspired on the architecture of SENNA [3, 4]. The system, based on machine learning, would ultimately be capable of performing the tasks of chunk parsing, complete syntactic parsing, language modeling, named entity recognition (NER) and semantic role labeling (SRL). The last one is of special interest to us, since it has been scarcely explored in Portuguese and, according to benchmark results, is the hardest of all.

Especially appealing to us is that the architecture can benefit from using unlabeled data in its training, since there is a shortage of labeled resources in Portuguese. For SRL, the only resource available is the PropBank-Br [5], built with the same style of the original PropBank [7] on a corpus of Brazilian Portuguese newswire text. However, while the original project counts more than

H. Caseli et al. (Eds.): PROPOR 2012, LNAI 7243, pp. 204–209, 2012.

100 thousand predicate instances, its Brazilian counterpart has around 7 thousand. This huge difference in the quantity of labeled data may seriously affect statistical approaches for Portuguese SRL.

The rest of the paper is structured as follows. Section 2 explains the architecture proposed by Collobert et al. in [3, 4]. In section 3, we show the problems in adapting the architecture to Portuguese and our solutions to them. Section 4 presents our conclusions.

2 Architecture Overview

Machine learning based systems for NLP usually determine beforehand which attributes should be observed in the input data. In the SRL task, for example, the seminal work of Gildea and Jurafsky [6] introduced attributes that became widely used in the area, such as phrase type, head word and parse tree path. Other works included verb subcategorization [10], part-of-speech (POS) tag of the head word [8], among others, including variations in existing attributes to deal with data sparsity [9].

In contrast with this practice, the rationale of the architecture presented here is to automatically extract features from the words presented to the system, given the number of attributes to be observed. The extraction is achieved through word representations in the form of numeric vectors. The values of these vectors are adjusted in an artificial neural network and can be shared by the various NLP tasks.

The authors of [3, 4] show that training the system for language modeling, which uses unlabeled text, improves performance on other tasks, as very large amounts of data can be used to adjust the feature vectors.

In the tasks of SRL, NER and parsing, the system input consists of a complete sentence and its output refers to one word at a time. Specifically in the SRL task, for each predicate in a sentence, all words are tagged as the predicate itself, part of an argument (and which type) or unrelated. For the language modeling task, which also interests us, the system tags a given sequence of words as positive (i.e., acceptable) or negative.

2.1 Word Representations

Each input sentence can be viewed as a sequence of words $(w_1, w_2, \ldots w_n)$. Furthermore, each word can be treated as an index in a finite dictionary $\mathcal{D}$. The first layer of the network thus maps each word to its corresponding feature vector using the lookup table LT_W. So, for the i-th word index in $\mathcal{D}$, we have:

$$LT_W(i) = W_i,$$

where $W_i \in \Re^d$ is a vector corresponding to the i-th column of the matrix of parameteres $W \in \Re^{d \times |\mathcal{D}|}$, and d is the number of features per word, which is a system parameter. Therefore, the input sentence is converted to a sequence of numeric vectors $(W_{w_1}, W_{w_2}, \ldots W_{w_n})$.

It should be noted that, as happens with neural networks parameters, the content of the feature vectors is no more than numbers, hardly understandable by a human being (which may not be clear from the name *feature*). We do not aim to extract any meaning thereof, but rather to leave them as internal representations.

It is also possible to include additional information about discrete attributes to a word's representation, such as the presence of capitalization or morphological tags. For each new attribute k, the process is analogous to the raw word representation: an associated dictionary $\mathcal{D}^k$ lists all its possible values and a matrix $W^k \in \Re^{d^k \times |\mathcal{D}^k|}$ contains the parameters to be adjusted, where each column W_i^k corresponds to the feature vector for the i-th value of attribute k and d^k is the number of features for that attribute, also a system parameter.

The resulting representation for each word in a input sentence is then the concatenation of all its feature vectors, i.e., the vector for the raw word itself and those for each of its attributes. This corresponds to a vector with dimension $d = \sum_k d^k$.

2.2 Including Neighboring Words

Simple tasks like POS tagging and chunking require knowledge about a word's neighbors, while more complex ones like SRL and full syntactic parsing require knowledge about the whole sentence. The architecture can provide that in the following way.

For the simpler tasks where a fixed size text window provides enough information, each word fed to the network has its representation concatenated with those of k words to its left and k to its right, where k is a system parameter. Further layers of the network perform standard weighted summation interleaved with a nonlinear activation function.

To bring knowledge about the whole sentence, a slight modification is required. First, if all words from the sentence are to be fed to provide information to tag a single target word, the relative distance between each input and the target must be provided in the form of an additional attribute. Second, the architecture needs a mechanism to deal with size varying sequences, as is the case of natural language sentences. This is achieved through a convolutional and a max layer.

If we view the input sentence as a function over time, at each instant t the t-th word is fed to the network, along with its k neighbors to the left and to the right. The convolutional layer will perform standard weighted summation on this input and store the result. When all the words have been fed, the max layer extracts the maximum value for each feature from the convolutional layer, representing the most relevant value for each one. This yields a fixed size vector that can go to the further layer of the network, as in the simpler tasks.

3 The Case of Portuguese

The architecture shown in the previous section presents a considerable problem if applied unchanged for Portuguese, as it includes inflected words in the dictionary $\mathcal{D}$. The aim of its creators was to deliberately avoid almost any kind of pre-processing, and while it is feasible for English, the richly inflected morphology of Portuguese would cause much data sparsity, a major problem for machine learning.

A straightforward solution to this problem would be to lemmatise every word found, and then to store lemmas in $\mathcal{D}$ and associated morphological information in separate feature dictionaries. This solution has some shortcomings, which we list below along with our proposed ideas for remedying them.

Differing from the Original Philosophy. The aim of the system SENNA is to build up knowledge about words with almost no pre-processing at all. By introducing a previous lemmatisation process, we are aware to be breaking that philosophy in some sense.

However, since our main interest is the SRL task, and given the difficulties of working with the Portuguese language, we believe this is still a valid approach. It is also worth noting that lemmatisation is a very well defined problem, and unlike many other NLP tasks, feasible to be implemented without machine learning methods. It is enough that a lemmatiser know the language's lexicon, inflection rules, irregular inflected forms, and preferentially heuristics to deal with neologisms. Anyway, we plan to compare the results of our proposed architecture with the original approach (raw words and presence of capitalisation only).

Feature Dictionaries. Word classes do not share all inflection types. For instance, only verbs can be inflected for tense and mood, while only nominal classes (adjectives, determiners, nouns and pronouns) have gender.

We propose to approach this problem by defining dictionaries for tense, mood, person (all of which only apply to verbs), number (for verbs and nominal classes) and gender (only for nominal classes). We choose to ignore degree inflection (present in adjectives and some adverbs) since it is relatively uncommon in Portuguese. Besides their usual values, each dictionary would also have a NA value for cases when that attribute is non applicable. Contractions such as *do*, for *de* + *o*, should be treated as a single word with nominal attributes.

Lemmatisation Errors. Lemmatisation errors, in the sense that the correct word form was not found even if listed among incorrect ones, are very rare. The morphological module of the parser PALAVRAS [1] achieves more than 99.5% success rate on it, the best performance for Portuguese we are aware of.

Non-recognized words would be treated as if their lemma were their surface form, while its flexional information would all be NA, which is a deterministic solution. We know that when it happens, we are bringing in some error to the system training, however small it is. On the other hand, the very low error rate combined with the capacity of neural networks to ignore small portions of noisy data would minimize this problem.

Ambiguity. This is by far our greatest concern, as in most NLP research. As stated before, a single inflected word may correspond to more than one lemma. There are a couple of measures we can take to circumvent this problem. First, we can train the network in a semissupervised fashion, using the manually corrected corpus Bosque (a part of the Floresta Sintá(c)tica [2]). The initial lookup table adjustments would then be done with correct lemmas. Second, when dealing with unlabeled data, since we are training a language model, we may reject lemmas tagged as negative according to our (still in training) system. This measure is obviously susceptible to errors, and we plan to evaluate if it actually improves the system performance.

There are still cases of *real* ambiguity, where more than one interpretation is plausible and should be classified as positive by the language model. For example, in the sentence *Disse o que queria* (I/he/she said what I/he/she wanted), both verbs (*disse* and *queria*) could be in the first or third person, yielding a total of four different sentence interpretations. Context often disambiguates such cases, but examining it is out of the scope of our architecture. Our proposed solution is to treat all of them as if they were separate inputs, since any could (at least theoretically) occur in a text.

This approach could at an extreme lead us to a point where, upon seeing many examples like the one given above, the system would not distinguish first and third person verb forms, even when they are conjugated differently. Although not desirable, this would be a result of the training data. If a machine learning system is given a massive number of example sentences where first and third person verbs are indistiguishable, it *is expected* not to learn to distinguish them.

4 Conclusions

We have presented an adaptation to Portuguese of a promising architeture for multiple NLP tasks. We have listed the major difficulties inherent to this adaptation, and pointed out how they could be solved. We expect to obtain good performance on the tasks of NER, surface and deep parsing, and SRL in Portuguese, and we also want to evaluate the impact of our proposed changes when compared to the original architecture.

Acknowledgments. João Luís G. Rosa thanks Fapesp - Fundação de Amparo à Pesquisa do Estado de São Paulo, Brazil, for the research support under project number 2008/08245-4, with which this paper is associated. Also, the authors would like to thank the anonymous reviewers for their constructive criticism and useful suggestions.

References

[1] Bick, E.: The parsing system PALAVRAS: automatic grammatical analysis of Portuguese in a constraint grammar framework. Ph.D. thesis, Aarhus University (2000)

[2] Bick, E., Santos, D., Afonso, S., Marchi, R.: Floresta sintá(c)tica: Ficção ou realidade? In: Avaliação Conjunta: Um Novo Paradigma no Processamento Computacional Da Língua Portuguesa, pp. 291–300. IST Press (2007)

[3] Collobert, R.: Deep learning for efficient discriminative parsing. In: AISTATS (2011)

[4] Collobert, R., Weston, J., Bottou, L., Karlen, M., Kavukcuoglu, K., Kuksa, P.P.: Natural language processing (almost) from scratch. Journal of Machine Learning Research 12, 2493–2537 (2011)

[5] Duran, M.S., Aluísio, S.M.: PropBank-Br: a Brazilian Portuguese corpus annotated with semantic role labels. In: Proceedings of STIL 2011 8th Brazilian Symposium in Information and Human Language Technology (2011)

[6] Gildea, D., Jurafsky, D.: Automatic labeling of semantic roles. Computational Linguistics, 245–288 (2002)

[7] Palmer, M., Kingsbury, P., Gildea, D.: The Proposition Bank: An annotated corpus of semantic roles. Computational Linguistics, 71–105 (2005)

[8] Pradhan, S., Ward, W., Hacioglu, K., Martin, J.H.: Shallow semantic parsing using support vector machines. In: Proceedings of HLT/NAACL 2004 (2004)

[9] Pradhan, S.S., Ward, W., Martin, J.H.: Towards robust semantic role labeling. Computational Linguistics 34(2), 289–310 (2008)

[10] Xue, N., Palmer, M.: Calibrating features for semantic role labeling. In: Proceedings of EMNLP 2004, pp. 88–94 (2004)

Towards Semi-supervised Brazilian Portuguese Semantic Role Labeling: Building a Benchmark

Fernando Emilio Alva-Manchego and João Luís G. Rosa

Núcleo Interinstitucional de Linguística Computacional (NILC)
Instituto de Ciências Matemáticas e de Computação, Universidade de São Paulo
Av. Trabalhador São-carlense, 400, Centro
13560-979 – São Carlos, SP, Brasil
{falva,joaoluis}@icmc.usp.br

Abstract. One of the main research challenges in semantic role labeling (SRL) is the development of applications for languages other than English. For Brazilian Portuguese, recent projects in lexical semantics are about to provide the necessary computational resources for research in this area. However, the amount of annotated data provided is not significant enough for successful supervised learning. Hence, we propose to use a semi-supervised approach capable of taking advantage of both annotated and unannotated data available. In this paper, we outline the methodology for the development of this SRL system, the same as the benchmark to be used to test its performance.

Keywords: Semantic Role Labeling, Semi-supervised Learning, Natural Language Processing.

1 Introduction

Semantic Role Labeling (SRL) is the natural language processing (NLP) task of identifying the basic structures of events in sentences, such as *who* did *what* to *whom*, *when* and *where* [21]. This has proven useful in a variety of NLP applications, such us: information extraction, question answering systems, automatic summarization and machine translation [21].

Although English SRL has been researched the most, at least one work has been done for European Portuguese using constraint grammar rules [4]. However, writing these rules is time-consuming and they, typically, have limited coverage, since they must anticipate each possible way in which semantic roles can be syntactically realized [14].

Currently, supervised learning methods are commonly used for automatic SRL. For Brazilian Portuguese, lexical resources which provide the necessary annotated data are under construction: PropBank.Br [10] and FrameNet.Br [25]. However, their dimensions are still not adequate for successful supervised learning in this language. Hence, we suggest to explore *semi-supervised* learning methods, which are capable of extracting relevant information from both annotated and unannotated data.

H. Caseli et al. (Eds.): PROPOR 2012, LNAI 7243, pp. 210–217, 2012.

We propose training a classifier on the annotated corpus of the PropBank.Br project, using the self-training strategy with maximum entropy as basic classification algorithm. The system's performance will be evaluated using precision, recall and F_1 measures. In order to perform the necessary tests, a benchmark is being built based on the methodology of the CoNLL Shared Tasks (STs) [5,6,28,15].

This paper is organized as follows: related work is described in section 2, the proposal is detailed in section 3, section 4 describes the benchmark under construction and section 5 presents some final remarks.

2 Related Work

For languages other than English, most of the research started after a corpus with sufficient annotated data was made available. Then, supervised methods used for English SRL were adapted for the specific characteristics of the other languages. For example, in [29], the author works with supervised learning and maximum entropy, using data from the Chinese PropBank and NomeBank. Also, in [8], they use Kernels with the Arabic PropBank made available during the SemEval-2007 Task [7]. An exception is [19], which explores bootstrapping using the Dutch PropBank.

To solve the problem of scarcity of annotated data, several approaches can be explored. One of them proposes to reuse a resource already with semantic annotation in one language, and project it to a paralell corpus in another language [22,23,24]. Other approach consists on using semi-supervised methods. For example, [11,12,13] propose and algorithm to increase a small number of manually annotated instances with unannotated examples, which roles were infered automatically through projection. Finally, unsupervised methods have also been proposed for the different stages of the SRL process: argument identification [1] and argument classification [16,17,18].

3 Proposal

In this section, the proposal of using a semi-supervised approach for Brazilian Portuguese SRL is detailed.

3.1 Learning Corpus

We plan to use the PropBank.Br corpus. It is being created based on the annotation of the Brazilian Portuguese section (CETENFolha) of the corpus Bosque from the Floresta Sintá(c)tica[1], which is a corpus annotated by the parser Palavras [3] and manually corrected by linguistics [26]. The PropBank.Br corpus is composed of 4213 sentences, which results in 7107 instances, and a total of 1068 target verbs, whose distribution in the corpus can be seen in Fig. 1.

[1] `http://www.linguateca.pt/Floresta/principal.html`

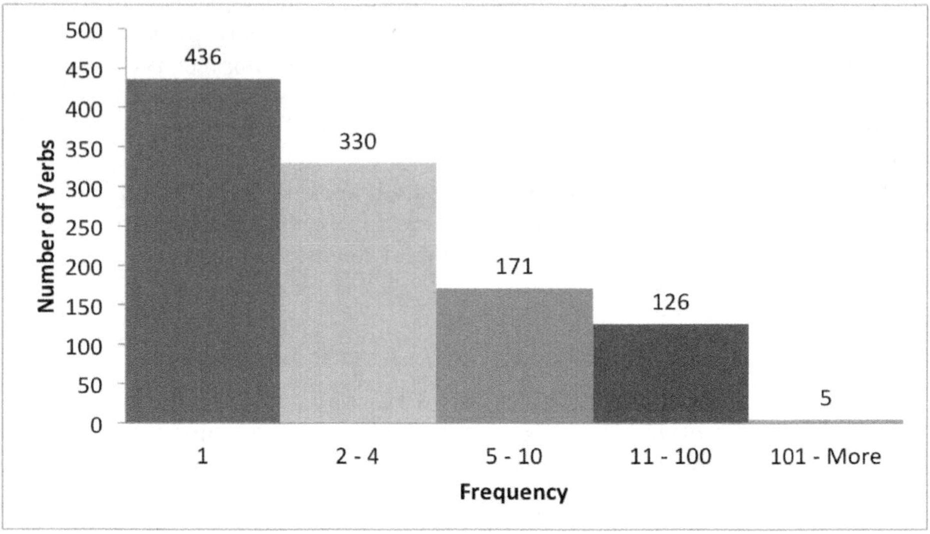

Fig. 1. Target verbs distribution in the PropBank.Br corpus

3.2 System Architecture

We chose to use a three-phase system: pruning, argument identification and argument classification.

For *pruning*, we will use the strategy of [30], which helps eliminating most of the constituents that are definitely not arguments. Then, for *argument identification*, a binary NULL-NON NULL classifier will be trained. For these first stages, a list of compound verbs in Portuguese [9] will probably be used to help finding the appropriate target verb. Finally, for *argument classification*, there are two main approaches: to use a classifier for all the possible verbs and their arguments, and to use a specific classifier for the arguments of each verb sense. Considering the data available in the corpus (Fig. 1), the former will be used.

3.3 Classification Algorithms

From Fig. 1, we can see that around 70% of the target verbs appear at most in 4 instances in the corpus. Using a supervised method with this data would result in poor learning. Considering this, we decided to use a *semi-supervised* approach.

Specifically, the self-training strategy of [31] will be explored, since it has been recently adapted and optimized for SRL in [32]. Maximum entropy models will serve as the basic classifier, considering their multi-class nature and that, in general, discriminative approaches are more appropiate for exploring a big number of attributes than models based on frequency (like decision trees).

3.4 Sentences' Features

Three sets of features are to be explored:

1. The standard set of [14] used in most SRL systems, which includes: phrase type, governing category, parse tree path, position, voice, head word and subcategorization.
2. The set of [2]. Since it proved to be useful in the automatic labeling of questions for Brazilian Portuguese, it is expected that those features also benefit SRL, considering how similar both tasks are. Some of the features considered are: argument order, specific syntactic function, number of arguments, principal verb token, semantic values of the argument tokens, simple or multiword verb, among others.
3. The set used in [20], a SRL system for Spanish, considering that this language has a stronger relation with Portuguese than with English. These features are grouped in classes: features of the verb (VForm, VLemma, etc.), features on the sibling of the verb in focus (syntactic category, syntactic function, prepostion, etc.), features that describe the properties of the content word of the focus sibling (word, lemma, POS, POS type, gender, etc.) and features on the clause (number of siblings with function circumstancial complement, relative positions of siblings with functions SUBJ, CAG, etc).

The features described above may be modified in order to take full advantange of them for the task in hand. Also, eventually, features extracted from other lexical resources, such as VerbNet.Br [27], could be explored.

It is worth mentioning that, at this point, we can not determine all the levels of annotation that will be required from the annotated and the unannotated data. The experiments to be performed with the features listed in this subsection will help us establish which type of morphological, syntactic and/or semantic information is needed for the SRL system to obtain a good performance.

4 Building a Benchmark

Unfortunately, the labeler to be implemented can not be compared directly with the one from [4] since the set of semantic roles and the corpus used are different. Therefore, we decided to create our own benchmark for comparison and testing. The STs have been widely used as a benchmark for English SRL. Hence, we decided to build our own based on their methodology for data format and evaluation, which are described in the following subsections.

For the benchmark to be completed, a baseline system for comparison is also required. Although not available right now, a minimal supervised system will be built, using similar characteristics to the ones of the semi-supervised system described previously.

4.1 Formatting the Data

All data was extracted from the PropBank.Br corpus. In Table 1 we show the type of information that is currently available (levels of annotation). At the same time, in Fig.2, an example of an annotated sentence in the "new" corpus is given.

Table 1. Column format. The fields below the line will not be available in the test set.

Number	Name	Description
1	ID	Token counter, starting at 1 for each new sentence
2	FORM	Word form or punctuation symbol
3	LEMMA	Gold-standard lemma of FORM
4	GPOS	Gold-standard part-of-speech tag
5	FEAT	Gold-standard morphological features
6	CLAUSE	Clauses in Start-End format
7	FCLAUSE	Full clause information in Start-End format
8	SYNT	Full syntactic tree
9	PRED	The semantic predicates in this sentence.
10...	ARG	Columns with argument labels for each semantic predicate following textual order.

```
1  Sob            sob            PRP    -            (S* (FCL* (FCL(PP*  -        (AM-PRD*
2  o              o              ART    M|S          *   *     (NP*        -        *
3  comando        comando        N      M|S          *   *     *           -        *
4  de             de             PRP    -            *   *     (PP*        -        *
5  Ronaldo_Rosas  Ronaldo_Rosas  PROP   M|S          *   *     (NP*))))    -        *)
6  ,              ,              PU     -            *   *     *           -        *
7  o              o              ART    M|S          *   *     (NP*        -        (A0*
8  programa       programa       N      M|S          *   *     *)          -        *)
9  mostrará       mostrar        V-FIN  FUT|3S|IND   *   *     (VP*)       mostrar  (V*)
10 reportagens    reportagem     N      F|P          *   *     (NP*        -        (A1*
11 especiais      especial       ADJ    F|P          *   *     (ADJP*)     -        *
12 de             de             PRP    -            *   *     (PP*        -        *
13 Sônia_Pompeu   Sônia_Pompeu   PROP   F|S          *   *     (NP*)))     -        *)
14 .              .              PU     -            *)  *)    *)          -        *
```

Fig. 2. An example of an annotated sentence in the training and development sets

4.2 Evaluation

The system will be evaluated in three tasks: **argument identification** (given a correct parse tree, label each node as being an argument or not), **argument classification** (given the set of nodes in the tree that are, in fact, arguments, label each one with the corresponding semantic role tag) and **combination of identification and classification** (the system must classify the nodes as being a specific argument or as not being an argument).

The standard measures used in the STs will be calculated: *precision* (percentage of the labels output by the system which are correct), *recall* (percentage of the true labels correctly identified by the system) and F_1 (harmonic mean of precision and recall).

An advantage of having the data in the STs format is that the *srl-eval.pl* script (the STs official evaluation program) can be used to evaluate the system's performance in our corpus. As a consequence, their evaluation rules will also be applied. This means that, for an argument to be correctly recognized, the words spanning the argument as well as its semantic role have to be correct. In addition, the verb argument of each proposition is excluded from the evaluation. This is because, most of the time, the verb corresponds to the target verb of the proposition (which is provided as input) and it is fairly easy to identify. Since there is one verb with each proposition, evaluating its recognition over-estimates the overall performance of a system.

5 Final Remarks

In this paper, we first outlined the proposal for implementing a SRL system for Brazilian Portuguese using a three-phase architecture, the self-training strategy with maximum entropy models, and different sets of features. Experiments are still to be performed in order to validate this proposal.

Many unexplored areas in SRL for Portuguese could be considered as future work, such as: using other semi-supervised algorithms (like *co-training*), studying and determining more Portuguese specific sentences' features, performing an extrinsic evaluation of the labeler as part of a more complex NLP system, etc.

As a second part, we described a benchmark that is being built based on the STs methodology, which will be used to test the performance of our system. Although it already has many useful characteristics, others can be added to improve both the corpus and the evaluation process.

For example, the training and development sets could be enriched with other information also present in the STs corpora, such as: chunks and named entities. Also, the test set could contain sentences from a corpus different to the one used for training, which will allow a cross-corpora evaluation (like in CoNLL-2005). Additionally, the evaluation script could also show results of the intermediate steps performed by the system (identification and classification) individually, and not just the final one.

By the end of this project, we expect to determine a combination of features extracted from the sentences' constituents in the corpus which provides a greater benefit for Brazilian Portuguese SRL, the same as implementing a tool for automatic SRL in this language. We expect that the labeler to be developed contributes to increase the interest in the development of applications for semantic analysis and benefits many other areas of NLP for Brazilian Portuguese.

Acknowledgments. The authors are grateful to FAPESP (Fundação de Amparo à Pesquisa do Estado de São Paulo, Brazil) for the research support under project numbers 2010/04647-0 and 2008/08245-4, respectively, which this paper is associated with. Also, the authors would like to thank the anonymous reviewers for their constructive criticism and useful suggestions.

References

1. Abend, O., Reichart, R., Rappoport, A.: Unsupervised argument identification for Semantic Role Labeling. In: 47th Annual Meeting of the ACL and the 4th International Joint Conference on Natural Language Processing of the AFNLP, Singapore, pp. 28–36 (2009)
2. Amancio, M.A., Duran, M.S., Aluisio, S.M.: Automatic Question Categorization: a New Approach for Text Elaboration. In: 2010 Workshop on Natural Language Processing and Web-based Technologies – IBERAMIA 2010, pp. 21–30. Bahía Blanca, Argentina (2010)
3. Bick, E.: The parsing system PALAVRAS: automatic grammatical analysis of Portuguese in a constraint grammar framework. Aarhus University Press (2000)
4. Bick, E.: Automatic Semantic Role Annotation for Portuguese. In: V Workshop em Tecnologia da Informação e da Linguagem Humana – XXVII Congresso da Sociedade Brasileira de Computação, RJ, Brazil, pp. 1713–1716 (2007)
5. Carreras, X., Màrquez, L.: Introduction to the CoNLL-2004 Shared Task: Semantic Role Labeling. In: 8th Conference on Computational Natural Language Learning: Shared Task, pp. 89–97. ACL, Boston (2004)
6. Carreras, X., Màrquez, L.: Introduction to the CoNLL-2005 Shared Task: Semantic Role Labeling. In: 9th Conference on Computational Natural Language Learning: Shared Task, pp. 152–164. ACL, Ann Arbor (2005)
7. Diab, M., Alkhalifa, M., ElKateb, S., Fellbaum, C., Mansouri, A., Palmer, M.: SemEval-2007 Task 18: Arabic Semantic Labeling. In: 4th International Workshop on Semantic Evaluations, pp. 93–98. ACL, Prague (2007)
8. Diab, M., Moschitti, A., Pighin, D.: Semantic Role Labeling Systems for Arabic using Kernel Methods. In: 46th Annual Meeting of the ACL: Human Language Technologies, pp. 798–806. ACL, Columbus (2008)
9. Duran, M.S.: Tratamento Automático de Verbos Auxiliares no Português do Brasil. Seminário do GEL, 58 (2010), `http://www.gel.org.br/?resumo=6572-10`
10. Duran, M.S., Aluísio, S.M.: Propbank-Br: a Brazilian Portuguese Corpus Annotated with Semantic Role Labels. In: 8th Brazilian Symposium in Information and Human Language Technology, pp. 164–168. Cuiabá, Mato Grosso (2011)
11. Fürstenau, H., Lapata, M.: Graph Alignment for Semi-Supervised Semantic Role Labeling. In: 2009 Conference on Empirical Methods in Natural Language Processing, pp. 11–20. ACL and AFNLP, Singapore (2009)
12. Fürstenau, H., Lapata, M.: Semi-supervised Semantic Role Labeling. In: 12th Conference of the European Chapter of the ACL, pp. 220–228. ACL, Athens (2009)
13. Fürstenau, H., Lapata, M.: Semi-supervised Semantic Role Labeling via Structural Alignment. Computational Linguistics 38(1), 135–171 (2012)
14. Gildea, D., Jurafsky, D.: Automatic labeling of semantic roles. Computational Linguistics 28(3), 245–288 (2002)

15. Hajic, J., Ciaramita, M., Johansson, R., Kawahara, D., Martí, M.A., Màrquez, L., Mayers, A., Nivre, J., Padó, S., Stepánek, J., Stranák, P., Surdeanu, M., Xue, N., Zhang, Y.: The CoNLL-2009 Shared Task: Syntactic and Semantic Dependencies in Multiple Languages. In: 13th Conference on Computational Natural Language Learning: Shared Task, pp. 1–18. ACL, Boulder (2009)
16. Lang, J., Lapata, M.: Unsupervised Induction of Semantic Roles. In: Human Language Technologies: The 2010 Annual Conference of the North American Chapter of the ACL, pp. 939–947. ACL, Los Angeles (2010)
17. Lang, J., Lapata, M.: Unsupervised Semantic Role Induction via Split-Merge Clustering. In: 49th Annual Meeting of the Association for Computational Linguistics, pp. 1117–1126. ACL, Portland (2011)
18. Lang, J., Lapata, M.: Unsupervised Semantic Role Induction with Graph Partitioning. In: 2011 Conference on Empirical Methods in Natural Language Processing, pp. 1320–1331. ACL, Edinburgh (2011)
19. Monachesi, P., Stevens, G., Trapman, J.: Adding semantic role annotation to a corpus of written Dutch. In: 1st Linguistic Annotation Workshop, pp. 77–84. ACL, Prague (2007)
20. Morante, R., Busser, B.: ILK2: Semantic Role Labelling for Catalan and Spanish using TiMBL. In: 4th International Workshop on Semantic Evaluations, pp. 183–186. ACL, Prague (2007)
21. Màrquez, L.: Semantic Role Labeling: Past, Present and Future. In: Tutorial Abstracts of ACL-IJCNLP 2009, p. 3. ACL, Singapore (2009)
22. Padó, S., Lapata, M.: Cross-linguistic Projection of Role-Semantic Information. In: Human Language Technology Conference and Conference on Empirical Methods in Natural Language Processing 2005, pp. 859–866. ACL, Vancouver (2005)
23. Padó, S., Lapata, M.: Optimal Constituent Alignment with Edge Covers for Semantic Projection. In: 21st International Conference on Computational Linguistics and the 44th Annual Meeting of the ACL, pp. 1161–1168. ACL, Sydney (2006)
24. Padó, S., Lapata, M.: Cross-lingual Annotation Projection of Semantic Roles. Journal of Artificial Intelligence Research 36, 307–340 (2009)
25. Salomão, M.M.M.: Framenet brasil: um trabalho em progresso. Calidoscópio 7(3), 171–182 (2009)
26. Santos, D., Bick, E., Afonso, S.: Floresta sintá(c)tica: apresentação e história do projecto. Encontro Um passeio pela Floresta Sintá(c)tica (Setembro 2007)
27. Scarton, C.E.: VerbNet.Br: construção semiautomática de um léxico computacional de verbos para o português do Brasil. In: 8th Brazilian Symposium in Information and Human Language Technology, pp. 20–29. Cuiabá, Mato Grosso (2011)
28. Surdeanu, M., Johansson, R., Meyers, A., Màrquez, L., Nivre, J.: The CoNLL 2008 Shared Task on Joint Parsing of Syntactic and Semantic Dependencies. In: 12th Conference on Computational Natural Language Learning, pp. 159–177. ACL, Manchester (2008)
29. Xue, N.: Labeling Chinese Predicates with Semantic Roles. Computational Linguistics 34, 225–255 (2008)
30. Xue, N., Palmer, M.: Calibrating Features for Semantic Role Labeling. In: 2004 Conference on Empirical Methods in Natural Language Processing, pp. 88–94. ACL, Barcelona (2004)
31. Yarowsky, D.: Unsupervised word sense disambiguation rivaling supervised methods. In: 33rd Annual Meeting of the Association for Computational Linguistics, pp. 189–196. ACL, Cambridge (1995)
32. Kaljahi, R.S.Z.: Adapting Self-training for Semantic Role Labeling. In: ACL 2010 Student Research Workshop, pp. 91–96. ACL, Uppsala (2010)

Building a Sentiment Lexicon for Social Judgement Mining

Mário J. Silva[1], Paula Carvalho[2], and Luís Sarmento[3]

[1] IST/INESC-ID
mjs@inesc-id.pt
[2] INESC-ID
pcc@inesc-id.pt
[3] SAPO/Portugal Telecom
las@co.sapo.pt

Abstract. We present a methodology for automatically enlarging a Portuguese sentiment lexicon for mining social judgments from text, i.e., detecting opinions on human entities. Starting from publicly-availabe language resources, the identification of human adjectives is performed through the combination of a linguistic-based strategy, for extracting human adjective candidates from corpora, and machine learning for filtering the human adjectives from the candidate list. We then create a graph of the synonymic relations among the human adjectives, which is built from multiple open thesauri. The graph provides distance features for training a model for polarity assignment. Our initial evaluation shows that this method produces results at least as good as the best that have been reported for this task.

1 Introduction

Synonyms have identical semantic orientation when used in the same context, which has motivated different authors to explore synonymy in language resources for automatically enlarging their sentiment lexicons. In general, lexicon-based approaches start from a confined list of manually annotated polar words, which are then used as seeds for finding new polar items. Classical methods typically explore WordNet [7] and other lexical resources publicly available to determine the polarity of the words with which known polar words are semantically related.

Regarding synonymy, the basic solution consists in propagating the polarity information of known polar words to the elements belonging to the same synset. However, such procedure is fallible because identical lexical forms (or homographs) can make part of different synsets presenting a diversity of meanings and polarities. Additionally, the majority of publicly available lexical resources describing synonymy only provide an inventory of potential synonyms for a given word, without defining the context where they have an identical interpretation. For example, the adjective *fresh* can be used as modifier of a human noun, and it can be replaced by adjectives such as *impertinent* or *impudent*, which have a negative semantic orientation. Also, it can modify non-human nouns, exhibiting

H. Caseli et al. (Eds.): PROPOR 2012, LNAI 7243, pp. 218–228, 2012.

in this case an opposite polarity. For example, when combined with an abstract noun such as *portrayal*, *fresh* is interpretable as *new* or *novel*, conveying a positive semantic orientation.

The polarity of predicates can change according to the syntactic-semantic nature of the nouns they relates to. This means that the polarity assignment will only be successful if such combination constraints will be taken into account.

We present a methodology for automatically enlarging a sentiment lexicon for mining social judgments from text, which takes into account the syntactic-semantic nature of predicate's arguments. We use a two-step approach to accomplish such goal: first, we find which adjectives can be used as human modifiers, and, next, assign them a polarity attribute. The identification of human adjectives is performed through the combination of a linguistic-based strategy, for extracting human adjective candidates from corpora, and a machine learning strategy, for automatically selecting the best human candidates. Like Rao and Ravichandran [15], we treat polarity labeling as a propagation problem in a graph, built from several open thesauri, but we restrict the polarity propagation to synonyms sharing similar syntactic-semantic properties.

We address human adjective predicates (i.e. adjectives modifying human nouns), because we are interested in extracting opinions targeting human entities, but the proposed methodology is language independent and could be applied to a wider range of syntactic-semantic predicates. In fact, we have used the methodology to create a sentiment lexicon that has been applied in opinion mining tasks involving social judgements of human targets, based on public linguistic resources available for Portuguese.

The remainder of the paper is organized as follows: Section 2 presents some related work; Section 3 describes the linguistic resources used in our experiments and for generating the sentiment lexicon; Sections 4 and 5 detail the methodology we used for identifying human adjectives and classifying their polarity, respectively; evaluation results are discussed in Section 6; some concluding remarks are, finally, presented in Section 7.

2 Related Work

First approaches on sentiment lexicon construction aim at identifying the subjective lexical units in general language and determining their semantic orientation (or polarity). The pioneer work of Hatzivassiloglou and McKeown [9] tackles the problem of determining the semantic orientation of adjectives by exploring the co-occurrence of positive and negative adjectives with other expressions in corpora, namely in the scope of copulative constructions (whose constituents tend to be coherent in terms of polarity) and adversative constructions (whose constituents tend to exhibit different polarity). After combining these constraints across many adjectives, the authors use a clustering algorithm that separates the adjectives into different groups, which are then labeled as positive or negative.

Turney and Littman [18] propose a bootstrapping method for inferring the semantic orientation of new polar words by computing the pointwise mutual information (PMI) between each target word and a few set of positive and negative paradigm words (or seeds) previously classified. On the other hand, [16] use two bootstrapping algorithms that exploit extraction patterns to learn sets of subjective nouns. The authors show that *Meta-Bootstrapping* and *Basilisk* algorithms, typically used for automatically generating extraction patterns to identify words belonging to a semantic class, can also be effective in identifying subjective words. This approach is based on the principle that words of the same semantic class or category tend to occur in similar contexts.

Significant research on sentiment lexicon construction has been also exploring WordNet (e.g. [10,6,8]) and other lexical resources for languages where WordNet-like resources are not available [15], to acquire new polar words. For example, Kamps et al. [10] try to determine sentiments of adjectives in WordNet by measuring the relative distance of each adjective to the reference positive and negative words "good" and "bad", respectively. Kim and Hovy [12] built a sentiment classifier that also uses WordNet, but performing a tri-polarity classification (positive, negative and neutral), based on a manually annotated dataset composed of verbs and adjectives.

Rao and Ravichandran [15] treat polarity detection as a semi-supervised label propagation problem in a graph. Takamura et al. [17] exploit the gloss information associated to words in dictionaries, for determining their semantic orientation. They construct a lexical network by linking two words whenever one of them appears in the gloss of the other word. Semantic orientations are regarded as spins of electrons, and the mean field approximation is used to compute the approximate probability function of the system.

Despite the availability of a number of sentiment lexicons, especially for English, it has being argued that their use is frequently unsatisfactory, because they do not reflect domain-specific lexical usage [5]. Hence, different approaches have been proposed to create domain-dependent polarity lexicons, instead of general-purpose lexicons. To face this problem, Fahrni and Klenner propose a two-stage method for determining the sentiment of adjectives in a given domain [1]. First, they use the Wikipedia for automatically detecting candidate targets associated to adjectives in a given domain, followed by a bootstrapping approach to determine the target-specific adjective polarity. Choi and Cardie propose integer linear programming to adapt an existing lexicon into a new specific one [5]. They consider the relations among words to derive the most likely polarity of each lexical item for a given domain. Kanayama and Nasukawa use manually-crafted syntactic patterns to identify polar atoms in a given domain [11].

In this paper, we propose a methodology for developing a fine-grained sentiment lexicon, which can be seen as an alternative to general-purpose and domain-specific lexicons. Polarities in this lexicon are assigned based on the syntactic-semantic category of targets of sentiment, applying to texts from any domain. It only handles human adjectives presently, but it can be adapted to other syntactic and semantic categories. As Riloff et al. [16] and Kanayama and Nasukawa [11], we make use of manually-crafted lexico-syntactic patterns, but they do not require any POS tagging or parsing. Similar to Kim and Hovy [12], we use a manually annotated lexicon for assigning positive, negative and neutral polarity to adjectives linked to these ones by a synonymy relationship. As suggested by Rao and Ravichandran [15], we used information aggregated from several publicly-available thesauri, since Portuguese, like many other languages, does not have comprehensive and publicly available WordNets.

3 Input Linguistic Resources

The identification of human adjective candidates relies on a small set of lexical resources, namely a lexicon of adjectives (Section 3.1), a list of names (Section 3.2), and a dictionary of profession and position names (Section 3.3). These are applied to a large n-gram collection (Section 3.5). The lexicon of adjectives is also used, together with a dictionary of synonyms (Section 3.4), in the polarity assignment stage.

3.1 Lexicon of Adjectives

We started with a lexicon of 24,792 adjectival lemmas, which are partially annotated with their semantic category and polarity. In detail, 4,034 adjectives were manually assigned to the human attribute and the remaining 511 lemmas to the non-human attribute. Human adjectives are characterized as co-occurring with a human subject (e.g. *the prime-minister is popular*). On the contrary, this type of subject is interdicted with non-human adjectives (e.g. *the prime-minister is sporadic*).

Human adjectives were also manually labeled with their prior polarities, which may be positive (1), negative (-1) or neutral(0). In terms of polarity distribution, 56% of the entries were labeled as negative (2,242 lemmas), 19% as positive (785 lemmas) and the remaining 25% as neutral (1,009 lemmas).

Polar adjectives were mostly collected from a lexico-syntactic database of human intransitive adjectives available in contemporary European Portuguese [3].

3.2 Dictionary of Names

In addition to this lexicon, the patterns described below make also use of a dictionary of names and surnames. These were collected from the public lists of placed secondary teacher names in the 2009 recruitment, available from the Portuguese Ministry of Education website. We obtained a list of 562 proper names and 1,388 surnames. First names correspond to the first element in name combinations. After removing all possible prepositions and conjunctions, we extracted all the tokens from name combinations after the first two (in Portuguese countries, many people have two given names), and classified them as surnames.

3.3 Dictionary of Profession and Position Names

Finally, we use a dictionary composed by 383 lemmas ($\sim$ 1200 inflected forms) denoting a professional or official position. Dictionary entries were semi-automatically compiled from news corpora, by exploring syntactic structures where such type of nouns typically occurs (e.g. in apposition to a human named entity).

3.4 Synonym Dictionaries

To expand the original polarity lexicon of human adjectives, we explored synonymy among adjectives in different publicly-available thesauri for Portuguese. We specifically used PAPEL 2.0 [14], TeP [13] and DicSin[1]. The previously mentioned resources comprise 87,327 different lemmas; distributed in 136,913 pairs of synonyms, 36,326 involving adjectives.

3.5 N-gram Corpus

In our experiments, we explore WPT05, a collection of over 10 million documents from the Portuguese web [2]. We used the n-grams (and their frequencies) generated from the documents in the collection with language automatically identified as Portuguese ($\sim$ 7 million documents, 26 Gb of text). We filtered the tokens with $length > 32$, but did not exclude n-grams from the set with low frequency. Given that we will be looking for the occurrence of multiple patterns with a given lemma in our classification process, low frequency n-grams can combine to produce high frequency patterns. In this research, we explored 8 million unigrams, 501 million trigrams, 984 million tetragrams and 1,321 million pentagrams. This corpus contains a large and representative sample of the Portuguese documents available on the Web, including a comprehensive range of types and genres of texts.

4 Identification of Human Adjectives

To identify human adjective candidates, we create a library of hand-crafted lexico-syntactic patterns representing elementary copular and adnominal constructions where such predicates can be found. These are then applied to WPT05, to gather evidence about adjective behavior. The adjective and pattern frequencies in the n-gram corpus are then used as input features to a binary classifier that we have trained and tested using the manually labeled adjectives.

4.1 Pattern Recognition

A total of 29 distinct lexico-syntactic patterns were formalized and applied to WPT05 trigrams, quadrigrams and pentagrams (Table 1). These patterns apply only to masculine and feminine singular forms, and some of them differ very slightly from each other, depending, for example, on the nature of the subject involved.

[1] `http://www.dicsin.com.br/`

Table 1.

ID	Pattern	Matches
301	N COP ADJ	14,474
302	IHREF COP ADJ	27,655
303	HREF COP ADJ	9,157
304	ERGO COP ADJ	23,337
305	tu COP ADJ	1,412
306	IND HREF ADJ	11,489
307	IND ERGO ADJ	31,933
308	IND ADJ HREF	1,206
309	IND ADJ ERGO	13,497
401	N S COP ADJ	7,240
402	S S COP ADJ	4,466
403	N COP MODIF ADJ	2,755
404	N COP IND ADJ	2,218
405	ART ADJ DO N	2,659
406	ART ADJ DO ERGO	5,846
407	tu COP MODIF ADJ	616
408	HREF COP MODIF ADJ	2,238
409	ERGO COP MODIF ADJ	4,170
410	IREF COP MODIF ADJ	11,994
411	IREF COP IND ADJ	3,957
412	HREF COP IND ADJ	1,102
413	ERGO COP IND ADJ	2,146
414	IND HREF MODIF ADJ	4,822
415	IND ERGO MODIF ADJ	4,030
501	N S COP MODIF ADJ	805
502	S S COP MODIF ADJ	465
503	N S COP IND ADJ	1,088
504	S S COP IND ADJ	667
505	ART ADJ DO S S	698
	Total	191,782

In these patterns, we match the subject position against dictionaries of Portuguese person names, and profession and position names (ERGO), such as *primeiro-ministro* (prime-minister) and *professor* (professor). Regarding names, we explored first names (N) and combinations of *first-name surname* (N S) and *surname surname* (S S). The subject position may also be filled by a human generic noun (HREF), such as *pessoa* (person) or *indivíduo* (individual) and by the person singular pronouns *ele* (he), *ela* (she), *você* (you), and *tu* (you), coded as IREF.

Within these constructions, adjectives can relate to their subject through the elementary copulative verbs (COP) *ser* and/or *estar* (both translatable by to be) and other aspectual variants (namely, *andar, continuar, ficar, permancer, encontrar-se, mostrar-se, reveler-se, tornar-se* and *viver*). We confine the tense in predicative constructions to simple present, past and future (third-person singular). The constructions invoking the second-person singular (tu, you) only consider simple present verb forms, the most representative when using direct address pronouns.

Furthermore, adjectives can be found in adnominal position, post-modifying or pre-modifying the noun they are correlated, which may be preceded by an indefinite article (IND). We also account for the possibility of adjectives filling the head of a cross-construction, linking to the noun they modify by the preposition de (of). Pre-modification of adjectives is also considered. We use of a small list of quantifier and intensifier adverbs (MODIF) that usually co-occur with human adjectives (e.g. *very, particularly, truly*).

Together, the patterns used in our experiments matched 191,782 different sequences, containing 8,579 different adjectival lemmas.

4.2 Classification

To refine the results provided by the lexico-patterns and filter out potential erroneous cases, we first explored, for each candidate human adjective, (i) the number of matches in the corpus, and (ii) the number and type of instantiated patterns.

For example, the adjective *ventilado* (*ventilated*) only matches once, while *impotente* (*impotent*) has a total of 97 matches, instantiating 9 different patterns. It is reasonable to infer that the adjective impotent is more prone to be considered as a valid human adjective than *ventilado*. It can also be assumed that the adjectives not matching any pattern in the entire collection are either rare in language or they have not a human behavior.

We train a statistical classifier to automatically distinguish high-human evidence (HHE) adjectives from low-human evidence (LHE) adjectives. The automatic classification is performed based on the following identified features: (i) frequency of each pattern, (ii) total number of instantiated patterns, (iii) frequency of matches, and (iv) lemma frequency in the n-gram collection. These attributes are associated to each recognized adjective from our original lexicon, including those whose semantic category is already known.

In our experiments, the training set was composed by 4,042 entries, distributed in the following proportion: 2,580 HHE adjectives, and 1,462 LHE adjectives. The HHE adjectives correspond to the lemmas in the corpus labeled as human in the original lexicon. LHE adjectives include both the adjectives recognized in the corpus that are assigned to the not-human attribute in the original lexicon, and the adjectives that do not occur in the n-gram collection, regardless of their prior semantic classification in the original lexicon.

5 Polarity Assignment

Once the human adjectives are identified in the lexicon, we focus on assigning polarities to the adjectives with high human evidence (HHE). The procedure starts by deriving a synonym graph, called a *syngraph*, where the nodes are the previously identified human lemmas and the edges represent synonymy relationships between lemmas. Each node in the syngraph is named as the concatenation of a lemma, its grammatical category and semantic class. This combination, which we designate henceforth as a qualified lemma, *qualiflemma*, was previously applied to the normalization of entries in dictionaries. By having a network of qualiflemmas, instead of just lemmas, we can prevent the propagation of synonymy relations between lemmas of distinct syntactic-semantic categories (homographs), and enables the assignment of different polarities to such lemmas.

To automatically assign polarities to the unlabelled qualiflemmas, we train a statistical classifier, which explores a feature vector extracted from the syngraph. We used 80% of the polar qualiflemmas for generating the syngraph, and saved the remaining 20% for learning a model to assign polarities to lemmas.

In the syngraph we have nodes with polarities, $-1, 0, 1, null$, where null designates unassigned polarity. The goal is to learn a model that predicts the polarity of a node with null polarity given the polarity information of its neighborhood. A qualiflemma in the syngraph with unassigned polarity may possibly have adjacent nodes exhibiting the four distinct polarities, making the decision complex. A situation where all the adjacent nodes have null polarity is quite common. However, we can attempt to observe across the adjacent nodes and assign polarities based on the polarities of the cloud of the connected qualiflemmas in the synonym graph. We capture that information by computing the shortest-paths and distances to the nearest nodes with assigned polarity.

The distances are computed using Dijkstra's shortest-path algorithm on a modified syngraph, to which we added three start nodes, labeled "1", "-1" and "0", each representing a polarity value. These are directly connected to all the nodes representing qualiflemmas with the same assigned polarity. The distances from each qualiflemma q, to each of these three start nodes correspond to the $dpos_q$, $dzer_q$ and $dneg_q$ features used in the subsequent statistical classification.

Besides these features, we also calculate three polarity weights ($wpos_q$, $wzer_q$, $wneg_q$) as the sum of the inverses of the distances of each node to the corresponding start node. For $wpos_q$, we have:

$$wpos_q = \sum_i \frac{1}{1 + dpos_i}$$

where the i represent the nodes adjacent to q.

6 Evaluation

We use in all experiments the C4.5 (J48) classifier implementation of the Weka toolkit with default parameters and 5-fold cross validation [19]. There is no particular reason for picking this algorithm other than it enables the identification of the features that contribute to polarity assignment in a learning algorithm. To reduce the impact of unbalanced data in automatic classification, we use the SMOTE filter implemented in Weka, which creates new minority class examples by interpolating between existing minority instances. In the polarity assignment experiments, we used 36,326 pairs of synonyms, which were obtained from the publicly-available resources previously described.

The derived syngraph contains 5,063 nodes, of which 1,989 have a prior polarity (500 positive, 380 neutral, 1,109 negative). An inspection of this graph of human adjectives shows that deciding on the polarities based on the synonyms is not trivial. The graph is highly connected: its order is 7.15 and the counts of nodes directly connected to a node of positive/neutral/negative polarity are 4,340, 4,460 and 3,731, respectively.

6.1 Human Classifier

The generated models are able to correctly predict the semantic category of adjectives in 94% of the cases, with a recall also of 94%. These evaluation results reinforce both the adequacy of methodology proposed, and the pertinence of the previously identified features for classifying adjectives as having LHE or HHE.

6.2 Lexical Expansion

We started with a large set of human adjective lemmas with initial polarities and restricted the expansion to those lemmas with high evidence of usage in social judgments in our corpus. However, we were still able to uniformly expand the lexicon by 70% for all polarity classes with good accuracy.

6.3 Polarity Classifier

The learned model has an accuracy of 87%. The Chi-square test of independence for this contingency table indicates a significance level of 0.1% ($p - value < 0,001$). The effect of the SMOTE filter in the training of the model is a higher precision for the classification of positive and negative adjectives (88 and 89%, respectively). The most problematic cases involve, as expected, the neutral polarity, which is, in average, correctly assigned only in 82% of the cases. The highest recall is obtained for positive polarities (93%), and the lowest for neutral polarities (70%). For negatives, recall is 92%. We intentionally trained the model to improve prediction performance with the positive lemmas because we have found that positive opinions represent the best predictor of politicians' popularity [4].

The quality of our classification appears to be at least as good as the best recently reported results. Rao and Ravichandran [15] report F-measure values of 93.00% with their label-propagation method on Hindi (relations among lemmas extracted from WordNet) and 82.46% on French (relations extracted from the OpenOffice French Thesaurus). In their evaluation, however, they classified all the adjective lemmas instead of just the human lemmas. In addition, their evaluation only assigned two polarity values (positive and negative). To compare the performance of our method we reassigned the results of our classifier into two classes, by dividing the neutral lemmas between the two other classes proportionally to their frequency. The resulting average F-measure of our method is higher: 97.03%.

Our method appears to perform better, despite the differences between the two evaluation settings, namely different languages and different resources for synonymy extraction. Given that performance is lower when the synonymic relationships are not as well accurate (French OpenOffice thesaurus), we conjecture that the performance of our methods may too be highly sensitive to the accuracy of the relationships present in the *syngraph* upon which the polarity classification algorithm is based. The strategy that we adopted of restricting the classification to human lemmas and discarding the relationships between non-human lemmas in the graph may be the reason for the observed improved F-measure.

7 Conclusions and Future Work

The experiments show that the automatic identification and classification of human adjectives in a lexicon can be carried out successfully by obtaining evidence about their use in large corpora. This can be performed using specific sets of handcrafted lexico-syntactic patterns describing the contexts where such categories are expected to occur. The frequency of matches recognized by those patterns, together with the adjective frequency in corpora, proved to be important and a distinctive feature for automatically distinguishing high-human evidence adjectives from low-human evidence adjectives.

Other analyses need to be conducted with the proposed method in order to assess its robustness, such as measuring how accuracy is affected by the size of the initial lexicon or by the percentage of lemmas in that lexicon reserved for the synonyms graph construction. An updated version of the lexicon described in this paper is available from our website (`http://dmir.inesc-id.pt/reaction/SentiLex-PT_02_in_English`).

Acknowledgements. We are grateful to Carlos Costa for programming and validating part of the software. This work was partially supported by FCT (Portuguese re- search funding agency) under grant UTA-Est/MAI/0006/2009 (REACTION project), and scholarship SFRH/BPD/45416/2008. We also thank FCT for its INESC-ID multi-annual support.

References

1. Fahrni, A., Klenner, M.: Old wine or warm beer: Target-specific sentiment analysis of adjectives. In: Symposion on Affective Language in Human and Machine, AISB Convention (2008)
2. Batista, D., Silva, M.J.: A statistical study of the wpt05 crawl of the portuguese web. In: FALA 2010: VI Jornadas en Tecnología del Habla and II Iberian SLTech Workshop, Vigo, Spain (2010)
3. Carvalho, P.: Análise e Representaço de Construçes Adjectivais para Processamento Automático de Texto: Adjectivos Intransitivos Humanos. PhD thesis, University of Lisbon (2007)

4. Carvalho, P., Sarmento, L., Teixeira, J., Silva, M.J.: Liars and saviors in a sentiment annotated corpus of comments to political debates. In: Proceedings of the 49th Annual Meeting of the Association for Computational Linguistics: Human Language Technologies, pp. 564–568. Association for Computational Linguistics, Portland (2011)
5. Choi, Y., Cardie, C.: Adapting a polarity lexicon using integer linear programming for domain-specific sentiment classification. In: EMNLP (2009)
6. Esuli, A., Sebastiani, F.: Determining term subjectivity and term orientation for opinion mining. In: E-ACL, Trento, Italy (2006)
7. Fellbaum, C.: WordNet: An Electronic Lexical Database (Language, Speech, and Communication). The MIT Press (1998)
8. Godbole, N., Srinivasaiah, M., Skiena, S.: Large-scale sentiment analysis for news and blogs. In: International Conference on WebLogs and Social Media (ICWSM 2007), Colorado, USA (2007)
9. Hatzivassiloglou, V., McKeown, K.: Predicting the semantic orientation of adjectives. In: ACL (1997)
10. Kamps, J., Marx, M., Mokken, R.J., de Rijke, M.: Using wordnet to measure semantic orienta-tion of adjectives. In: LREC (2004)
11. Kanayama, H., Nasukawa, T.: Fully auto-matic lexicon expansion for domain-oriented sen-timent analysis. In: ACL (2006)
12. Kim, S.M., Hovy, E.: Extracting opinions, opinion holders, and topics expressed in online news media text. In: ACL/COLING Workshop on Sentiment and Subjectivity in Text (2006)
13. Maziero, E., Pardo, T., Felippo, A., da Silva, B.D.: A base de dados lexical e a interface web do tep 2.0 - thesaurus eletrônico para o português do brasil. In: TIL (2008)
14. Oliveira, H.G., Santos, D., Gomes, P.: Extracção de relações semânticas entre palavras a partir de um dicionário: o papel e sua avaliação. Linguamática 2(1) (2010)
15. Rao, D., Ravichandran, D.: Semi-supervised polarity lexicon induction. In: EACL (2009)
16. Riloff, E., Wiebe, J.: Learning extraction patterns for subjective expressions. In: EMNLP (2003)
17. Takamura, H., Inui, T., Okumura, M.: Extracting semantic orientation of words using spin model. In: ACL (2005)
18. Turney, P., Littman, M.: Measuring praise and criticism: Inference of semantic orientation from association. ACM Transactions on Information Systems 21(4) (2003)
19. Witten, I.H., Frank, E.: Data Mining: Practical machine learning tools and techniques, 2nd edn. Morgan Kaufman, San Francisco (2005)

A Bootstrapping Algorithm for Learning the Polarity of Words

António Paulo Santos[1], Hugo Gonçalo Oliveira[2], Carlos Ramos[1], and Nuno C. Marques[3]

[1] GECAD, Institute of Engineering, Polytechnic of Porto, Portugal
[2] CISUC, University of Coimbra, Portugal
[3] DI-FCT, Universidade Nova de Lisboa, Monte da Caparica, Portugal
{pgsa,csr}@isep.ipp.pt, hroliv@dei.uc.pt, nmm@di.fct.unl.pt

Abstract. Polarity lexicons are lists of words (or meanings) where each entry is labelled as *positive*, *negative* or *neutral*. These lists are not available for different languages and specific domains. This work proposes and evaluates a new algorithm to classify words as *positive*, *negative* or *neutral*, relying on a small seed set of words, a common dictionary and a propagation algorithm. We evaluate the *positive* and *negative* polarity propagation of words, as well as the *neutral* polarity. The propagation is evaluated with different settings and lexical resources.

Keywords: polarity of words, polarity lexicon, sentiment analysis, lexicon expansion, opinion mining.

1 Introduction

Sentiment analysis, also known as opinion mining, often relies on polarity lexicons, also known as sentiment lexicons (e.g. [13], [3]). These lexicons are lists of words (e.g. [10], [11], [12]) or lists of meanings (e.g. [1], [8]) where each entry is labelled with their **a priori polarity**. For instance, even without knowing the context, words like *love*, *peace*, *fun* and *success* are usually labelled with a *positive prior polarity*, while words like *hate*, *war*, *bore*, *failure* are labelled with a *negative prior polarity*, and words like *people*, *table*, and *tree* are labelled as *neutral* in lexicons that also consider the *neutral* polarity. In a subsequent step this a priori polarity can be then used to determine the **contextual polarity** [13].

Many works (e.g. [10], [11]) classify words as *positive* or *negative*. However besides *positive* and *negative* connotations, words might have no connotation at all, which means they have *neutral* polarity. In this work, we extend the algorithm originally presented in [14], by considering the *neutral* class and representing the dictionary as an undirected graph. The method relies on a small seed set of words, a common dictionary represented as a graph, and a propagation algorithm.

The paper is organized as follows. In section 2 we review the related work and in section 3 we present the proposed propagation algorithm. In section 4 we evaluate the algorithm and finally, we conclude and discuss on further work, in section 5.

H. Caseli et al. (Eds.): PROPOR 2012, LNAI 7243, pp. 229–234, 2012.

2 Background

In [11], 9,107 words are extracted from a common online dictionary and represented as a *directed graph.* On that graph, nodes represent words and edges represent the semantic relation between words. With 5 *positive* and 5 *negative* seed words, with manually labelled polarity, the authors applied a simple propagation and voting algorithm. Starting from the seed nodes, the algorithm visits every word in the graph by a breadth-first traversal and their polarity is iteratively propagated to the unlabelled words.

The representation of a dictionary varies. Some authors have chosen, to represent it as a *directed graph* (e.g. [2], [7], [11]), while others as an *undirected graph* (e.g. [10]). Particularly in [11], the dictionary is represented as a *directed graph* in order to preserve the original structure of the dictionary. In this work, we considered the property of symmetry of some semantic relations (e.g. synonymy, antonymy) and therefore we used an undirected representation. This representation relies on the fact that on a binary relation R between two words, if *word1 is related to word2* then *word2 is related to word1*, at least in some contexts. For instance, if *word A is a synonym of B*, *B* is as well a *synonym of A*.

3 The Polarity Propagation Algorithm

In this section we present the polarity propagation algorithm for classifying words as *positive*, *negative* and *neutral.* Assuming that we have a dictionary represented as an undirected graph (Fig. 1, at left), in which, each number represents a word, we get the list of classified words (graph on the right), by applying the following steps:

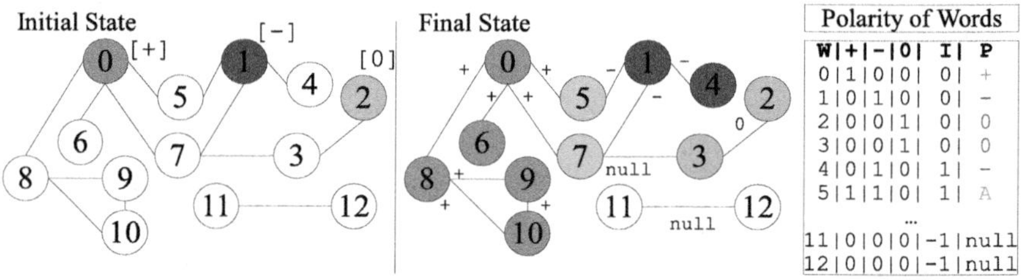

Fig. 1. Polarity propagation (*+, -, 0, A, null = positive, negative, neutral, ambiguous, and no polarity; w = word; I. = Iteration; P = final polarity*). Close to each node and edge is the polarity propagated to its neighbour.

1. Associate a *positive* (C_+), a *negative* (C_-), a *neutral* (C_0) and *iteration* (I) counter to each word. Initialize the first three counters to 0 and the fourth with a *negative* value. The *iteration* counter will record the shortest distance between each word and the closest seed word (the number of edges between them). This distance is useful to re-run the algorithm and apply a weighted propagation, as we will discuss in step 3.

2. Label a set of words $W = \{w_1, \ldots, w_n\}$ that should ideally contain the same number of *positive*, *negative* and *neutral* words. This is done by incrementing the respective counter with a *positive* value. Set the *iteration* counter of each word to 0 (this means that they are seed words). Finally, add all these words in a queue Q.
3. Retrieve the first word w_1 from the queue Q and get all their unvisited neighbours $Nb_{w1} = \{nb_1, \ldots, nb_m\}$. If $Nb_{w1} = \{\}$ go to step 4, else:
   (a) Propagate the *positive*, *negative* or *neutral* polarity of word w_1 to each one of its neighbours nb_i ($i = 1$ to m), by incrementing the counter of each neighbour, according to the following rules:

   (1) **If** $w_1 > 0 \wedge w_1 \longrightarrow nb_i$ **Then** $pos. \Leftarrow pos. + 1 * Weight$
   (2) **If** $w_1 < 0 \wedge w_1 \longrightarrow nb_i$ **Then** $neg. \Leftarrow neg. + 1 * Weight$
   (3) **If** $w_1 = 0 \wedge w_1 \longrightarrow nb_i$ **Then** $neu. \Leftarrow neu. + 1 * Weight$
   (4) **If** $w_1 > 0 \wedge w_1 \dashrightarrow nb_i$ **Then** $neg. \Leftarrow neg. + 1 * Weight$
   (5) **If** $w_1 < 0 \wedge w_1 \dashrightarrow nb_i$ **Then** $pos. \Leftarrow pos. + 1 * Weight$
   (6) **If** $w_1 = 0 \wedge w_1 \dashrightarrow nb_i$ **Then** $neu. \Leftarrow neu. + 1 * Weight$
   (7) **If** $w_1 = ambigous$ **Then** *do nothing*

   Where:
   - Word w_1 is *positive* ($w_1 > 0$), *negative* ($w_1 < 0$), or *neutral* ($w_1 = 0$) if the respective counter is a unique maximum. Otherwise, the polarity of word w_1 is ambiguous and therefore the last rule is applied.
   - $w_1 \longrightarrow n_i$, represents any semantic relation between word w_1 and its neighbour n_i, where the polarity should be maintained (e.g. synonymy relation). $w_1 \dashrightarrow w_1$, represents any semantic relation that inverts the polarity (e.g. antonymy relation).
   - The $Weight$ variable should be set to 1 if we want to apply an unweighted propagation approach. If we want to apply a weighted propagation approach the $Weight$ should decrease as iteration increases, as discussed in [11].

   (b) For each neighbour nb_i of w_1 that does not exist on queue Q set the iteration counter to *iteration counter of* $w_1 + 1$ and add it to the end of Q. If n_i already exists on Q don't do nothing.
4. Mark word w_1 as visited.
5. If the Q is not empty, go to step 3, otherwise the algorithm ends (step 6).
6. At this step we have a list of words, each classified as:

`null`	If $C_+ = C_- = C_0 = 0$
`+`	If $C_+ > C_-$ AND $C_+ > C_0$
`-`	If $C_- > C_+$ AND $C_- > C_0$
`0`	If $C_0 > C_+$ AND $C_0 > C_-$
`Ambiguous`	Otherwise

In addition to the polarity, the values of the counters can be used to compute the polarity strength. For instance, the polarity strength of a positive word can be computed as the ratio of $C_+ \ / \ (C_+ + C_- + C_0)$.

4 Evaluation

4.1 Lexical Resources Used

In our experiments, we used a synonymy graph established by the semantic relations extracted from three dictionaries, namely *PAPEL 2.0*[1] [6], *Wikcionário.PT*[2] and *Dicionário Aberto*[3]. All the semantic relations were extracted in the scope of the Onto.PT project[4] [5] [4] and are available as triples in the form **lexical-item1 SEMANTIC-RELATION lexical-item2**, in which lexical-item is a lemmatized word or a multiword expression. In PAPEL 2.0 there are 79,161 synonymy triples containing 43,375 distinct lexical items, and, from Wikcionrio.PT and DA, there are 24,542 and 47,387 triples, containing 19,839 and 40,583 distinct lexical items respectively.

After building the graph and applying the propagation algorithm we evaluated the classified words using the following datasets:

- **SentiLex-PT01**[5] [12]: a publicly available sentiment lexicon for Portuguese. It contains 6,321 adjective lemmas classified in one of three classes: *positive* (+), *negative* (-), *neutral* (0). From those, 3,585 are manually classified and 2,736 are automatically classified. We used the entries manually classified.
- **Human 1:** a list of 460 Portuguese words (148 nouns, 107 verbs, 200 adjectives and 5 adverbs) manually classified by a native speaker. Each word was classified in one of three classes: *positive* (+), *negative* (-), and *neutral* (0).
- **Human 2:** the same previous dataset annotated by a different person.

The words in the datasets *Human 1* and *Human 2* were classified out of context and in a domain independent way. The annotator was instructed to classify each word according its primary thought. The goal was to try to capture the polarity of the words in most contexts, whenever possible (e.g. words such as *beautiful*, *peace* tend to be *positive* in most contexts, words such as *war* tend to be *negative*, and words such as *car*, *people* tend to be *neutral*).

The inter-annotator agreement for *Human 1* and *Human 2* datasets was 76.30% (Cohen's Kappa = 0.61). According to a commonly cited scale this value of the kappa is in the substantial agreement range (0.61-0.80) [14]. The obtained agreement can be further increased, for instance, by providing to the annotators words within sentences. However, for this study the obtained value is sufficient to measure the agreement of the algorithm in respect to those two annotators.

4.2 Results and Discussion

The goal of this experiment was to evaluate the classification performance of the propagation algorithm using: *PAPEL 2.0*, *Wikcionário.PT* and *Dicionário*

[1] Available at `http://www.linguateca.pt/PAPEL/`
[2] `http://pt.wiktionary.org/`
[3] `http://www.dicionario-aberto.net`
[4] Available at `http://ontopt.dei.uc.pt/index.php?sec=recursos`
[5] Available at `http://xldb.fc.ul.pt/wiki/SentiLex-PT01_in_English`

Aberto. Experiment settings: *graph type* = undirected; *semantic relations* = synonymy; *num. of seed words* = 4(+), 4(-), 4(0); *evaluated classes* = +, -, 0.

Table 1 shows the results for 9 experiments. Each experiment was performed for 10 runs, varying the seed words. The seed words were retrieved from a list of about 150 seed words (50 (+), 50 (-), 50 (0)), most of them from the manually labelled entries in SentiLex-PT01.

Table 1. Average results for 9 experiments, 10 runs each, varying the seed words ($\overline{Class}$ = average number of classified words; $\overline{Eval}$ = average number of evaluated words; $\overline{Acc}$ ± SD = average accuracy ± standard deviation)

Eval.Dataset	PAPEL			D.Aberto			Wik.PT		
	$\overline{Class}$	$\overline{Eval}$	$\overline{Acc}$ ±SD	$\overline{Class}$	$\overline{Eval}$	$\overline{Acc}$ ±SD	$\overline{Class}$	$\overline{Eval}$	$\overline{Acc}$ ±SD
SentiLex	30,044	2,338	**66 ±5.89**	20,391	1,188	**57 ±9.47**	11,069	1,034	**52 ±11.56**
Human 1	30,045	393	54 ±4.20	20,391	252	44 ±9.32	11,069	226	41 ±10.17
Human 2	30,045	393	60 ±9.24	20,391	252	46 ±21.30	11,069	226	41 ±18.03

According to the resource used, the best accuracy was obtained with *PAPEL*. These results seem to be related with the number of synonymy relations because the best results were obtained with the resource with more synonymy relations (79,161 relations) and the worst results with the resource with less synonymy relations (Wik.PT with 24,541 relations).

According to the evaluation dataset, the higher accuracy was obtained with *SentiLex*. Part of the reason may be because when we are using *SentiLex*, we are evaluating just adjectives. The accuracy of 66.03% is very close from the accuracy of 67% reported in [12] for the same task and evaluation dataset.

In conclusion we would like to draw attention to some aspects. First, since the inter-human agreement between *Human 1* and *Human 2* was 76.30%, the 0-76% scale seems more proper than the 0-100% scale, when analysing these results. Second, these are results considering words classified by the algorithm of all the iterations. Considering only words from the first iterations (e.g. until no further than 4th iteration) improve the results. Third, we just used 12 seed words. It is expected that increasing the number of words would improve the results.

5 Conclusion and Future Work

As any other approach that relies only on words to build or expand general purpose lexicons are limited. In order to reduce this limitation, we intend to adapt the current method to classify words in a domain or topic-specific way. Relying in word senses such as in *SentiWordNet* [1] for English, it may be also a possible future direction. As future work, we can also determine the contextual polarity in larger units of text like sentences, taking advantage of models similar to those used in part-of-speech tagging [9].

Acknowledgements. António Paulo Santos is supported by the FCT grant SFRH/BD/47551/2008.

Hugo Gonçalo Oliveira is supported by the FCT grant SFRH/BD/44955/2008 co-funded by FSE. Nuno C. Marques wishes to thank InspirennovIT for sponsoring this work in the context of Best Supplier project[6].

References

1. Baccianella, S., Esuli, A., Sebastiani, F.: SentiWordNet 3.0: An enhanced lexical resource for sentiment analysis and opinion mining. In: P. of the 7th Conf. on Lang. Resources and Evaluation (LREC 2010), Valletta, MT, vol. 25, pp. 2200–2204 (2010)
2. Blair-Goldensohn, S., Hannan, K., Mcdonald, R., Neylon, T., Reis, G.A., Reynar, J.: Building a Sentiment Summarizer for Local Service Reviews. Electrical Engineering (2008)
3. Ding, X., Liu, B., Yu, P.: A holistic lexicon-based approach to opinion mining. In: Proceedings of the International Conference on Web Search and Web Data Mining, pp. 231–240. ACM (2008)
4. Gonçalo Oliveira, H., Antón Pérez, L., Costa, H., Gomes, P.: Uma rede léxico-semântica de grandes dimensões para o português, extraída a partir de dicionários electrónicos. Linguamática 3(2), 23–38 (2011)
5. Gonçalo Oliveira, H., Gomes, P.: Onto.PT: Automatic Construction of a Lexical Ontology for Portuguese. In: Proceedings of 5th European Starting AI Researcher Symposium (STAIRS 2010), pp. 199–211. IOS Press (2010)
6. Gonçalo Oliveira, H., Santos, D., Gomes, P.: Extracção de relações semânticas entre palavras a partir de um dicionário: o PAPEL e a sua avaliação. Linguamática 2(1), 77–93 (2010)
7. Jijkoun, V., Hofmann, K.: Generating a Non-English Subjectivity Lexicon: Relations That Matter. Computational Linguistics, 398–405 (2009)
8. Maks, I., Vossen, P.: Different Approaches to Automatic Polarity Annotation at Synset Level. In: Proceedings of the First International Workshop on Lexical Resources, WoLeR 2011, pp. 62–69 (2011)
9. Marques, N.C., Pereira Lopes, G.: Tagging with Small Training Corpora. In: Hoffmann, F., Adams, N., Fisher, D., Guimarães, G., Hand, D.J. (eds.) IDA 2001. LNCS, vol. 2189, pp. 63–72. Springer, Heidelberg (2001)
10. Rao, D., Ravichandran, D.: Semi-supervised polarity lexicon induction. In: Proceedings of the 12th Conference of the European Chapter of the Association for Computational Linguistics on EACL 2009, pp. 675–682 (2009)
11. Paulo-Santos, A., Ramos, C., Marques, N.C.: Determining the Polarity of Words through a Common Online Dictionary. In: Antunes, L., Pinto, H.S. (eds.) EPIA 2011. LNCS, vol. 7026, pp. 649–663. Springer, Heidelberg (2011)
12. Silva, M.J., Carvalho, P., Costa, C., Sarmento, L.: Automatic Expansion of a Social Judgment Lexicon for Sentiment Analysis. Technical Report. TR 10-08. University of Lisbon, Faculty of Sciences, LASIGE (2010)
13. Taboada, M., Brooke, J., Tofiloski, M., Voll, K., Stede, M.: Lexicon-based methods for sentiment analysis. Computational Linguistics, 1–41 (2011)
14. Viera, A.J., Garrett, J.M.: Understanding interobserver agreement: the kappa statistic. Family Medicine 37(5), 360–363 (2005)

[6] Portuguese National Strategic Reference Framework — *QREN/Programa Operacional de Factores de Competitividade*, proposal n. 18627 — 06/2010 SI I&DT supported by European Regional Development Fund.

The Role of Language Registers in Polarity Propagation

António Paulo Santos[1], Hugo Gonçalo Oliveira[2], Carlos Ramos[1], and Nuno C. Marques[3]

[1] GECAD, Institute of Engineering - Polytechnic of Porto, Portugal
[2] CISUC, University of Coimbra, Portugal
[3] DI-FCT, Universidade Nova de Lisboa, Monte da Caparica, Portugal
{pgsa,csr}@isep.ipp.pt, hroliv@dei.uc.pt, nmm@di.fct.unl.pt

Abstract. This paper presents work on the automatic creation of a polarity lexicon based on a lexical-semantic network. During this work, we noticed that the language registers of a relation should be considered in polarity propagation. After analysing the possible registers and performing some experiments, our intuition was confirmed – there are registers that should invert the transmitted polarity (irony), while others always transmit negative polarity (pejorative, disparaging).

Keywords: sentiment analysis, opinion mining, polarity, lexical-semantic resources, language registers.

1 Introduction

Polarity lexicons (e.g. SentiWordNet [1], for English, or SentiLex [10], for Portuguese) are useful resources for sentiment analysis, and their creation is one of the main research lines in this area. They consist of a set of lexical items (words or expressions) and the typical sentiment they transmit, usually referred to as polarity. Common values for polarity are positive, neutral and negative.

Most approaches on the automatic construction of sentiment lexicons fall in one of two categories: corpora-based approaches (e.g. [3][11][4][6]), that explore the co-occurrence of words in large collections of texts; and dictionary or wordnet-based approaches (e.g. [5][7][9][8]), that exploit information provided by lexical-semantic resources.

In this work, while applying a polarity propagation algorithm [8] to PAPEL [2], a lexical-semantic network for Portuguese, we noticed that there were "erroneous" synonymy relations that were originating propagation errors. We analysed these relations and verified that some of them had additional language information fields, typically found in dictionaries, namely: domain, register and variant. Also, we observed that some of the registers could change the simple polarity propagation. After performing several experiments, we confirmed that these registers provide valuable information for polarity propagation. Therefore, when available, they should be handled properly.

H. Caseli et al. (Eds.): PROPOR 2012, LNAI 7243, pp. 235–240, 2012.

We start this paper by describing, briefly, the polarity propagation algorithm used. Then, we present PAPEL, with special focus to the information fields its relations contain, and to the synonymy relations, which were the ones exploited in this work. Before concluding, we present the performed experiments, which confirmed our assumptions regarding the correct treatment of the registers.

2 Polarity Propagation

The polarity propagation algorithm used in this work [8] sees a dictionary (or a lexical-semantic network) as a graph, where lexical items are the nodes, and the edges connecting the items represent the semantic relations. It starts with a small set of seed words, for instance, five positive and five negative, manually labelled. Then, the algorithm visits every lexical item in the graph by a breadth-first traversal. The polarity of the lexical items is iteratively propagated to the unlabelled items. While most relations tend to preserve polarity (e.g. synonymy) others invert polarity (e.g. antonymy).

3 PAPEL

PAPEL [2] is a public domain lexical-semantic network, automatically extracted from a proprietary dictionary. PAPEL 2.0 contains about 97,000 lexical items – 55,900 nouns, 24,000 verbs, 21,000 adjectives and 1,400 adverbs – and about 198,000 connections between them. The latter denote semantic relations and are represented as triples, with the following structure:

`arg1 RELATION_NAME arg2    domain;register;variant`
(e.g. `divertimento SINONIMO_N_DE alegria`)

A triple indicates that one sense of the lexical item in the first argument (`arg1`) is related to one sense of the lexical item in the second argument (`arg2`) by means of a relation identified by `RELATION_NAME`. In PAPEL, the name of the semantic relation defines the part-of-speech of its arguments. Furthermore, some of the triples have additional information fields, typically found in dictionaries.

3.1 Additional Fields

In this work, we used PAPEL 2.0[1]. In this version, some of the triples have the following additional language information fields, also obtained from the dictionary definitions, that provide information about the way words are, or may be used:

- **Register:** the situation or context where the definition holds.
- **Domain:** the specific sphere of knowledge where the definition is common or valid.
- **Variant:** the Portuguese variant where the definition applies.

[1] Available through `http://www.linguateca.pt/PAPEL/`

Table 1 shows some of the possible values for these fields, their meaning and the number of triples of PAPEL 2.0 in which they occur (Occurrences). Whereas domain and variant do not seem relevant for polarity propagation, some of the registers might be, as we will show in the next section.

Table 1. Information fields, some of their possible values and number of occurrences

Field	Value	Meaning	Occurrences
Domain	bot.	botany	6,397
	zool.	zoology	2,515
	medic.	medicine	2,256
	quim.	chemistry	1,001
	mus.	music	669
Register	fig.	figurative	7,357
	pop.	popular	2,072
	coloq.	informal	747
	pej.	pejorative	431
	depr.	disparaging	251
	vulg.	vulgarism	75
	cal.	slang	31
	irón.	ironic	17
Variant	Bras.	Brazil	2,092
	reg.	regionalism	1,900
	Ang.	Angola	386
	Moçamb.	Mozambique	247

3.2 Synonymy in PAPEL

PAPEL contains several semantic relations. In this work we exploited the synonymy relation. In PAPEL 2.0, there are 79,161 relational triples denoting synonymy – 37,452 between nouns, 21,465 between verbs, 19,073 between adjectives and 1,171 between adverbs.

On the context of polarity propagation, synonymous lexical items tend to have the same polarity. However, some of the registers presented in the previous section might change the typical polarity of two lexical items connected by synonymy. After analysing different triples with different registers, we divided the register values into two types, according to their ability of changing polarity transmitted in synonymy relations:

- **Polarity Keepers:** registers that preserve the regular behaviour of the synonymy relation, in terms of transmitted polarity.
- **Polarity Modifiers:** registers that have the ability of changing the regular behaviour of the synonymy relation, in terms of polarity transmitted.

While registers like figurative or informal (see examples 1 and 2 below) belong to the first type, based on observation, we assume that irony, pejorative and disparaging registers belong to the second (see examples 3, 4 and 5 below). In section 4, the former assumption is the main goal of our experimentation.

(1) `caro SINONIMO_ADJ_DE salgado ;fig;`

(2) `excelente SINONIMO_ADJ_DE porreiro ;coloq;`

(3) `punição SINONIMO_N_DE recompensa ;irón;`

(4) `concubina SINONIMO_N_DE fêmea ;pej;`

(5) `enfraquecer SINONIMO_V_DE efeminar ;depr;`

Still, irony behaves differently than pejorative and disparaging registers. The presence of irony in a synonymy relation means that the connected lexical items are only synonymous in an ironic context, whereas in most typical contexts they have an opposite meaning. Therefore, in order to propagate the typical polarity, we can handle ironic synonymy relations as antonymy. On the other hand, pejorative and disparaging synonymy relations always transmit negative polarity. So, for the former registers, if the propagated polarity is positive or neutral, it becomes negative. Otherwise, the negative polarity is preserved.

There are other interesting registers in PAPEL which we thought about exploiting for polarity propagation. For instance, vulgarism (see examples 6 and 7 below) and slang (see examples 8 and 9 below) are usually associated with negative polarities (as 6 and 8), which suggests that they belong to the polarity modifiers group. However, even though less common, there are as well positive relations with these registers (as 7 and 9).

(6) `insignificância SINONIMO_DE merdice ;vulg;`

(7) `erecção SINONIMO_DE tesão ;vulg;`

(8) `excremento SINONIMO_N_DE trampa ;cal;`

(9) `força SINONIMO_N_DE tusa ;cal;`

4 Experiments

So far, our assumptions regarding the impact of language registers in polarity propagation are the following:

1. Ironic relations invert the propagated polarity
2. Pejorative relations always propagate negative polarity
3. Disparaging relations always propagate negative polarity

The goal of our experimentation is thus to confirm the former assumptions. We assess the performance of polarity classification when, in propagation, the target registers in the synonymy triples are handled differently. Therefore, we ran the propagation algorithm with different numbers of seeds, obtained from the manually annotated SentiLex-PT01 [10], a public sentiment lexicon for Portuguese with 6,321 adjectives, 3,585 of which have their polarity manually classified. The algorithm only stops when there are no more nodes to visit.

The results shown in table 2 were computed after comparing the automatically labelled adjectives in PAPEL with the manually labelled adjectives in SentiLex-PT01. Also, the results were measured following three different criteria:

1. Ignoring all triples with registers that would be exploited (iron, pej, depr);
2. Considering all the triples of the target register in the graph construction but ignoring their extra information during the polarity propagation. The triples of all the other exploited registers are ignored;
3. Considering all the triples, and handling the target register properly. The triples of all the other exploited registers are ignored.

Table 2 shows the average number of classified, evaluated words, and accuracy for each criteria, obtained while handling the target registers differently. Each combination "target register(s)+seeds" was performed for 10 runs varying the seed words. We present the accuracies for three seeds (one positive, one negative and one neutral), and 12 (four positive, four negative and four neutral), which confirms that, as expected, the number of seeds is proportional to accuracy.

As for the target registers, using only irony (iron) addresses our first assumption, only pejorative (pej) addresses the second assumption, only disparaging (depr) our third assumption, and, finally, using the three exploited registers (All) intends to show the overall performance improvement.

Table 2. Average results, according to the handled registers (Register = target registers; $\overline{Class}$ = average number of classified words; $\overline{Eval}$ = average number of evaluated adjectives; $\overline{Acc}$ ± SD = average accuracy ± standard deviation)

Registers	**Seeds**	Criteria 1			Criteria 2			Criteria 3		
		$\overline{Class}$	$\overline{Eval}$	$\overline{Acc}$ $\pm SD$	$\overline{Class}$	$\overline{Eval}$	$\overline{Acc}$ $\pm SD$	$\overline{Class}$	$\overline{Eval}$	$\overline{Acc}$ $\pm SD$
iron	3	30,783	2,308	57.67 ±11.0	30,781	2,307	57.63 ±11.0	30,781	2,306	**57.73** ±10.9
	12	29,857	2,315	64.64 ±7.9	29,854	2,312	64.39 ±7.8	29,855	2,315	**64.73** ±7.9
pej	3	”	”	”	30,851	2,311	57.60 ±10.9	30,852	2,310	**58.73** ±11.0
	12	”	”	”	29,978	2,322	64.80 ±8.0	30,049	2,331	**65.14** ±7.3
depr	3	”	”	”	30,829	2,305	62.31 ±04.2	30,841	2,307	**62.81** ±3.9
	12	”	”	”	29,865	2,319	65.20 ±06.6	29,871	2,323	**65.48** ±6.4
All	3	30,783	2,308	57.67 ±11.0	30,898	2,308	62.12 ±4.2	30,917	2,312	**63.77** ±3.7
	12	29,857	2,315	64.64 ±7.9	29,977	2,322	65.13 ±6.5	30,045	2,338	**66.03** ±5.9

All our assumptions were confirmed by these experiments, as the proper handling of each and all the registers lead to higher accuracies. When the three exploited registers are handled according to our assumptions, there is an improvement on accuracy: about 1.5% when using three seeds and 0.9% when using 12 seeds. Even though the improvements might not look significant, we must have in mind that the number of target registers in synonymy triples is almost residual – 11 irony registers, 159 pejorative, and 88 disparaging, in a total of 79,000 triples. Also, as the current version of SentiLex only contains adjectives, these were the only evaluated words, which are always about 7.5% of the classified words.

5 Concluding Remarks

We confirmed that information about the language register of semantic relations might be valuable for polarity propagation. Since synonymy relations connect lexical items with the same meaning, it is expectable that both items have the same (typical) polarity. However, we noticed that some registers might change the transmitted polarity and should thus be handled properly.

After identifying the registers that modify polarity, we did some experiments with a polarity propagation algorithm in a lexical-semantic network and confirmed that our intuition was true. Therefore, once available, information about language registries, more precisely irony, pejorative and disparaging markers, should be exploited in the automatic construction of polarity lexicons.

Acknowledgements. António Paulo Santos is supported by the FCT grant SFRH/BD/47551/2008. Hugo Gonçalo Oliveira is supported by the FCT grant SFRH/BD/44955/2008 co-funded by FSE.

References

1. Esuli, A., Sebastiani, F.: Sentiwordnet: A publicly available lexical resource for opinion mining. In: Proceedings of the 5th Conference on Language Resources and Evaluation LREC 2006, pp. 417–422 (2006)
2. Gonçalo Oliveira, H., Santos, D., Gomes, P.: Extracção de relações semânticas entre palavras a partir de um dicionário: o PAPEL e sua avaliação. Linguamática 2(1), 77–93 (2010)
3. Hatzivassiloglou, V., Mckeown, K.R.: Predicting the semantic orientation of adjectives. In: Proceedings of 35th Annual Meeting of the Association for Computational Linguistics (ACL 1999), pp. 174–181. Association for Computational Linguistics, Madrid (1997)
4. Kaji, N., Kitsuregawa, M.: Building lexicon for sentiment analysis from massive collection of HTML documents. In: Proceedings of the Joint Conference on Empirical Methods in Natural Language Processing and Computational Natural Language Learning (EMNLP-CoNLL), pp. 1075–1083 (2007)
5. Kamps, J., Mokken, R.J., Marx, M., de Rijke, M.: Using WordNet to measure semantic orientation of adjectives. In: Proceedings of the 4th Intl. Conference on Language Resources and Evaluation, LREC 2004, vol. IV, pp. 1115–1118. ELRA, Paris (2004)
6. Kanayama, H., Nasukawa, T.: Fully automatic lexicon expansion for domain-oriented sentiment analysis. In: Proceedings of the 2006 Conference on Empirical Methods in Natural Language Processing, EMNLP 2006, pp. 355–363. Association for Computational Linguistics, Stroudsburg (2006)
7. Kim, S.M., Hovy, E.: Determining the sentiment of opinions. In: Proceedings of the 20th Intl. conference on Computational Linguistics, COLING 2004, pp. 1267–1373. Association for Computational Linguistics, Stroudsburg (2004)
8. Paulo-Santos, A., Ramos, C., Marques, N.: Determining the Polarity of Words through a Common Online Dictionary. In: Antunes, L., Pinto, H.S. (eds.) EPIA 2011. LNCS, vol. 7026, pp. 649–663. Springer, Heidelberg (2011)
9. Rao, D., Ravichandran, D.: Semi-supervised polarity lexicon induction. In: Proceedings of 12th Conference of the European Chapter of the Association for Computational Linguistics (EACL 2009), Athens, Greece, pp. 675–682 (2009)
10. Silva, M., Carvalho, P., Costa, C., Sarmento, L.: Automatic expansion of a social judgment lexicon for sentiment analysis. Tech. rep., Faculdade de Ciências da Universidade de Lisboa (2010)
11. Turney, P.D.: Thumbs up or thumbs down?: semantic orientation applied to unsupervised classification of reviews. In: Proceedings of the 40th Annual Meeting on Association for Computational Linguistics, ACL 2002, pp. 417–424. Association for Computational Linguistics, Stroudsburg (2002)

Sentiment Analysis on Twitter Data for Portuguese Language

Marlo Souza and Renata Vieira

Pontifícia Universidade Católica do Rio Grande do Sul, Porto Alegre RS, Brasil
marlo.souza@acad.pucrs.br, renata.vieira@pucrs.br

Abstract. This work presents an study on Sentiment Analysis on Twitter data for the Portuguese language. It evaluates the impact of different preprocessing techniques, Portuguese polarity lexicons and negation models showing low impact of preprocessing and negation modelling in classification of tweets.

Keywords: Sentiment Analysis, Twitter, Portuguese language.

1 Introduction

Twitter is a microblog system in which users publish limited length messages. The importance of this system is due to its worldwide and fast growth - reaching 200 millions users and around 150 millions user publishings daily in 2011[1]. Twitter has been shown as an important source of information in a different range of areas - from social sciences [9], marketing and economy [12] and others-, but Twitter data poses difficulties for automatic linguistic analysis - due to a great deal of language variation and slangs.

Sentiment Analysis, or Opinion Mining, corresponds to the problem of identifying or extracting emotions, opinions or points of view expressed in text. This area has received a great deal of attention in the last years due to its potential applicability, according to Wilson et al. [26], among others.

Sentiment analysis techniques have been applied to several tasks in the literature (e.g. [10,18]) improving the treatment of non-factual information in text. More recently, Twitter data has been used in Sentiment Analysis studies [8,12,16,21]. Most of these works, however, focus on a rather shallow content analysis of the text due to the difficulties involved in processing Twitter data.

This work presents an study on sentiment analysis techniques on Twitter messages to improve Sentiment Analysis performance on this medium. We evaluate the impact of different models of negation and different sentiment lexicons for the Portuguese language and the effect of pre-processing techniques for the task. The remainder of this work is structured as follows: Section 2 presents the related work on sentiment analysis for Twitter; Section 3 presents our method for sentiment analysis on Twitter data, discussing some of our decisions and limitations; Section 4 presents the experiment performed using the described method and its results and, in Section 5, we conclude this work.

[1] Source: http://business.twitter.com/basics/what-is-twitter. Statistics of july, 2011

H. Caseli et al. (Eds.): PROPOR 2012, LNAI 7243, pp. 241–247, 2012.

2 Related Work

Work on sentiment analysis is expanding on the last years, applying the proposed solutions to a myriad of problems, such as opinion summarization [10], Information Extraction [18] among others.

Despite most work focus on a document-level analysis to identify and extract sentiment from text, a deeper level of analysis has been explored in the literature. It is the case of works like [17] that explore subjectivity in a sentence-level, or [26,27] that use machine learning techniques to identify phrase-level sentiment, [13,24] exploring statistics and semantic relations to access sentiment in a word-level case.

Works as [3,15] show that in the sub-sentential treatment of sentiment, the semantic structure of the sentence plays an important role in determination of polarity, specially the treatment of negation.

Work on Sentiment Analysis with Twitter data has only recently begin to flourish in literature, due to its potential applicability to market analysis and other areas. Some of these focus on a bag-of-word and n-gram approach to classify tweets according to their subjective content.

Early papers on Twitter processing as [8,16] use n-gram and POS features, and simple models of negation - as a valence shifter with limited scope - to feed different classifiers to predict the polarity of a tweet.

Davidov et al.[4] and Kouloumpis et al. [14] use similar approach based on traditional features for sentiment analysis and Twitter specific feature - such as hastags, links and emoticons - to predict the tweet polarity.

For the Portuguese language, the Twittómetro, a tool for analyzing opinion about candidates for the Portugal's presidential campaign, is discussed in [21]. The authors execute the sentiment analysis using a lexicon-based approach combined with a lexical-syntactic patterns to identify and compose sentiments expressed in tweets.

Other work for sentiment analysis in Twitter setting for Portuguese language are [2] - which relies on an association discovery over lexical features for classifying tweets - and [20] - which relies on detecting bias for a user towards a particular topic.

The automatic construction of training data is also an interesting topic, when dealing with Twitter. Many works rely on information specific to microblogging texts, such as emoticons and hashtags. For example, [8,16] use emoticons to provide polarity information for the tweets - an approach that has proven to be unreliable, by our analysis. [4] use the hashtags as source of polarity, but in their case the hashtags were the sentiment class attributed to the tweet, not a polarity value. Other approach is to use Amazon Mechanical Turk[2] - or similar services - to annotate tweets about polarity [5].

Our work is related to [21], since we intend to perform sentiment analysis in Twitter data for the Portuguese language by a lexicon-based approach. Our work, however, does not rely on a defined domain - and thus on domain-specific resources as that one.

[2] `https://www.mturk.com/mturk/welcome`

3 Sentiment Analysis for Twitter Data

When dealing with micro-blogging data an extra effort must be made to preprocess it, since well-established techniques do not perform well on this new source of data. Gimpel et al. [7], for example, discuss that in this context, problems like tokenization and POS tagging, are much more difficult to deal with than in normal English texts.

This may have a major impact on the method used as, for example, Part-of-Speech features, which are important for Sentiment Analysis in traditional data, but has proven to have little impact on Twitter texts [14].

Most of the work on Sentiment Analysis deal with this problem in a rather ad-hoc way, not paying much attention to it. We propose the use of heuristics to perform a lexical normalization on the data. To define our heuristics, a set of 500 Portuguese Twitter messages were collected and analyzed. Those texts have no intersection with the ones we use in the experiment described further in the work.

The most common case of variation found was the repetition of vowels to denote intensity, which can be easily treated. Another common case of lexical variation in the texts is misspelling of words based on phonetic similarity. To solve this, we applied a variation of the metaphone algorithm for Brazilian Portuguese. More difficult cases, however, are the ad-hoc abbreviations commonly made by the users. A regular abbreviation pattern is the omission of vowels. Thus, words as "saudade" (to miss something/someone) are written as "sdd". It can be corrected by comparison of words by consonants. This heuristic, however, must be applied only in specific cases, as when no vowels are found in the word, since many errors can be performed by applying it.

We used a lexicon-based sentiment analysis method and evaluate the use of the heuristics and of different models to the scope of negation in our experiments. The sentiment lexicons used were the Sentilex lexicon [22] of adjectives composed of around 6000 annotated human predicatives and a expanded version of the OpLexicon [23], a domain-independent sentiment lexicon for the Portuguese language.

The negation modeling, as discussed by [3,15,25], is of major importance when dealing with sentential and sub-sentential sentiment analysis. In our method, we model negation by a set of negation-indicators which are words and expressions - such as "não" (no, not), "nunca" (never), "ninguém" (no one) - indicating intrasentential negation and a defined scope in the sentence.

The negation-indicators act as polarity shifters of expressions within their scope. An important fact to remember is that the Portuguese language allows three kinds of verbal negation - pre-verbal, post-verbal and a double negation both pre- and post-verbal [19].

Lastly, the polarity of a tweet is identified by the algebraic sum of the polarities, after applying the negation treatment. The internal structure of the sentence may provide more useful information for sentiment analysis, based on the discursive relations it encompass such as discussed by [1]. We do intend to analyze those features in the future in a heuristic-based manner, similar to [6].

4 Experiment and Results

In this section we present our experiments on twitter sentiment analysis. First we describe corpus and lexicon, then we evaluate two negation models, our preprocessing technique and differences based on the choice of lexicon.

4.1 The Corpus

The corpus used as test set for our experiments is composed of 1700 tweets balanced between two classes - positive and negative - and was automatically built using the Twitter API. To collect solely tweets written in Portuguese we used the language parameter choosing the 'pt' value - set language to Portuguese.

As query and to classify the tweets according to their polarity, we used the Twitter hashtags #win and #fail - for positive and negative polarities, respectively. Our approach relies on two special hashtags - the #win and #fail - which denote a positive and a negative sentiment, respectively.

Tweets containing both #win and #fail hashtags were discarded. From the total pool collected on a 3-day process, we randomly selected 1700 tweets - divided by polarity - 540 of which presents negation of terms.

4.2 The Lexicon

We use in this work the OpLexicon - a sentiment lexicon for the Portuguese language built using multiple sources of information [23]. The lexicon is constituted of around 15,000 polarized words classified by their morphological category annotated with polarities positive, negative and neutral. It has been enlarged by re-applying the corpus-based method using a bigger corpus and extracting polar verbs using the thesaurus-based method. Also, the adjectives present in the SentiLex which were not annotated in the OpLexicon were included in the later.

4.3 Results

We implemented two models of negation scope: one based on a pre-fixed window - we used a 5-word window (NegWind), and one with the negator's scope over the entire sentence(NegSent). One method that disregards the effects of the negation (NoNeg) was also implemented aiming to evaluate the impact of the negation on this type of texts.

The Table 1 summarizes the results achieved by our method presenting the accuracy and F-measures ($\beta = 1$, F1) values for the positive and negative classes.

Note that, the Sentential model of negation seems the most adequate, which indicates that the 5 word window is constraint too rigid. Taking negation - in the sentential scope - into consideration has increased the precision which indicates a better modeling of opinion identification. The drop in the recall, on the other hand, may indicate that - since the scope of a negator is too broad - negation may have been applied on polarized terms not syntactically related to the negator.

Table 1. Results using different models for negation

	pos			neg		
	Prec	Rec	F1	Prec	Rec	F1
NoNeg	0.62	0.50	0.55	0.67	0.30	0.42
NegSent	0.66	0.46	0.54	0.74	0.33	0.45
NegWind	0.61	0.5	0.55	0.67	0.29	0.40

We also decided to evaluate the impact of our pre-processing technique in the results by applying the same method without lexical normalization and also we evaluated the performance of our proposed polarity lexicon, OpLexicon, against other available lexicon for Portuguese - the Sentilex lexicon. The results may be seen in Table 2.

Table 2. Results without pre-processing and using the Sentilex lexicon

	pos			neg		
	Prec	Rec	F1	Prec	Rec	F1
Without PP	0.66	0.46	0.54	0.74	0.33	0.45
Sentilex	0.52	0.22	0.31	0.58	0.19	0.29

We can see that the pre-processing technique had minor or no effect on the performance. The OpLexicon, however, yielded higher rates than SentiLex. This is not surprising at all since the SentiLex is built for specific domain analysis and was used to enrich the OpLexicon.

5 Discussion and Conclusions

The better performance obtained by the Oplexicon (Table 1) is due to the fact that, unlike the SentiLex, it is composed by different types of words and not just adjectives as that one. Besides, the SentiLex was constructed for a specific domain and, for that reason, it may yield a low performance when used in other domains. This is not surprising since [23] reports similar results about Sentilex and Oplexicon and since Sentilex was used for the enlargement of our version of the Oplexicon.

The pre-processing techniques seem to have no impact on the results. We believe that the set of heuristics is yet quite small and simple, which is probably the main reason for such a result. Other methods of lexical normalization, e.g. [11], may be applied to improve the performance. Our work is, however, the first work on Sentiment Analysis for Twitter, in our knowledge, that evaluates the impact of the preprocessing techniques applied.

It is important to note that the language detection of the Twitter API is not perfect. From the texts which the system could not classify either as positive nor negative - due to lack of opinion-bearing words from the lexicon - for the

OpLexicon using the NegSent strategy, around 15% where texts written in the Spanish language. Besides, about 5% from those which could not be classified, were spam using the #fail hashtag in its text.

The low impact of the negation may be explained firstly by the high coverage of the lexicon and by the specificities of the data, as discussed. We intend to explore further different models of negation, the inclusion of other negators and the best scope of negation for this source of data. Our model of negation, nevertheless has increased the precision of the classification - indication a better modeling of opinion.

Finally, note that the accuracy of our system is yet quite below others, as the Twittómetro. This is because the polarity identification method relies solely on the algebraic sum of the polarities and a simple model of negation. We intend to further explore methods for tweet classification including pseudo-syntactic and discursive information. We also plan to propose a sub-sentential entity-centered method for sentiment analysis for Twitter messages.

References

1. Asher, N., Benamara, F., Mathieu, Y.: Appraisal of opinion expressions in discourse. Lingvisticæ Investigationes 31.2, 279–292 (2009)
2. Calais Guerra, P.H., Veloso, A., Meira Jr., W., Almeida, V.: From bias to opinion: a transfer-learning approach to real-time sentiment analysis. In: Proceedings of the 17th ACM SIGKDD International Conference on Knowledge Discovery and Data Mining, KDD 2011, pp. 150–158. ACM, New York (2011), `http://doi.acm.org/10.1145/2020408.2020438`
3. Choi, Y., Cardie, C.: Learning with compositional semantics as structural inference for subsentential sentiment analysis. In: EMNLP 2008, pp. 793–801. ACL, Stroudsburg (2008)
4. Davidov, D., Tsur, O., Rappoport, A.: Enhanced sentiment learning using twitter hashtags and smileys. In: COLING 2010, pp. 241–249. ACL, Stroudsburg (2010)
5. Diakopoulos, N.A., Shamma, D.A.: Characterizing debate performance via aggregated twitter sentiment. In: CHI 2010, pp. 1195–1198. ACM, New York (2010)
6. Ding, X., Liu, B., Zhang, L.: Entity discovery and assignment for opinion mining applications. In: KDD 2009, pp. 1125–1134. ACM, New York (2009)
7. Gimpel, K., Schneider, N., O'Connor, B., Das, D., Mills, D., Eisenstein, J., Heilman, M., Yogatama, D., Flanigan, J., Smith, N.A.: Part-of-speech tagging for twitter: Annotation, features, and experiments. In: ACL 2011 (Short Papers), pp. 42–47. ACL (2011)
8. Go, A., Huang, L., Bhayani, R.: Twitter sentiment analysis. Entropy, 17 (2009)
9. Golder, S.A., Macy, M.W.: Diurnal and Seasonal Mood Vary with Work, Sleep, and Daylength Across Diverse Cultures. Science 333(6051), 1878–1881 (2011)
10. Grefenstette, G., Qu, Y., Shanahan, J.G., Evans, D.A.: Coupling niche browsers and affect analysis for an opinion mining application. In: RIAO 2004, pp. 186–194. CID (2004)
11. Han, B., Baldwin, T.: Lexical normalisation of short text messages: Makn sens a #twitter. In: ACL-HLT 2011 (2011)
12. Jansen, B.J., Zhang, M., Sobel, K., Chowdury, A.: Twitter power: Tweets as electronic word of mouth. Journal of the American Society for Information Science and Technology 60(11), 2169–2188 (2009)

13. Kamps, J., Marx, M., Mokken, R.J., de Rijke, M.: Using WordNet to measure semantic orientation of adjectives. In: LREC 2004 (2004)
14. Kouloumpis, E., Wilson, T., Moore, J.: Twitter sentiment analysis: The good the bad and the omg! In: Artificial Intelligence, pp. 538–541 (2011)
15. Moilanen, K., Pulman, S.: Sentiment composition. In: RANLP 2007, Borovets, Bulgaria, pp. 378–382 (2007)
16. Pak, A., Paroubek, P.: Twitter as a corpus for sentiment analysis and opinion mining. Computer, 1320–1326 (2010)
17. Pang, B., Lee, L.: A sentimental education: Sentiment analysis using subjectivity summarization based on minimum cuts. In: Proceedings of the ACL, pp. 271–278 (2004)
18. Riloff, E., Wiebe, J., Phillips, W.: Exploiting subjectivity classification to improve information extraction. In: AAAI 2005 (2005)
19. Schwenter, S.A.: The pragmatics of negation in Brazilian Portuguese. Lingua 115(10), 1427–1456 (2005)
20. Silva, I.S., Gomide, J., Veloso, A., Meira Jr., W., Ferreira, R.: Effective sentiment stream analysis with self-augmenting training and demand-driven projection. In: Proceedings of the 34th International ACM SIGIR Conference on Research and Development in Information, SIGIR 2011, pp. 475–484. ACM, New York (2011), `http://doi.acm.org/10.1145/2009916.2009981`
21. Silva, M.J., Team, R.: Notas sobre a realizao e qualidade do twitómetro. Tech. rep., University of Lisbon, Faculty of Sciences, LASIGE (May 2011)
22. Silva, M.J., Carvalho, P., Costa, C., Sarmento, L.: Automatic expansion of a social judgment lexicon for sentiment analysis. Technical Report TR 1008 University of Lisbon Faculty of Sciences LASIGE (2010)
23. Souza, M., Vieira, R., Busetti, D., Chishman, R., Alves, I.M.: Construction of a portuguese opinion lexicon from multiple resources. In: STIL 2011, Cuiabá, Brazil (2011)
24. Turney, P.D.: Thumbs up or thumbs down?: semantic orientation applied to unsupervised classification of reviews. In: ACL 2002, pp. 417–424. Association for Computational Linguistics, Morristown (2002)
25. Wiegand, M., Balahur, A., Roth, B., Klakow, D.: A survey on the role of negation in sentiment analysis. Imagine, 60–68 (July 2010)
26. Wilson, T., Wiebe, J., Hwa, R.: Recognizing strong and weak opinion clauses. Computational Intelligence 22, 73–99 (2006)
27. Wu, Y., Zhang, Q., Huang, X., Wu, L.: Phrase dependency parsing for opinion mining. In: Proceedings of the 2009 Conference on Empirical Methods in Natural Language Processing, pp. 1533–1541. Association for Computational Linguistics, Singapore (2009), `http://www.aclweb.org/anthology/D/D09/D09-1159`

REAP.PT
Serious Games for Learning Portuguese

André Silva[1,2], Cristiano Marques[1,2],
Jorge Baptista[2,3], Alfredo Ferreira Jr.[1,2], and Nuno Mamede[1,2]

[1] Instituto Superior Técnico, Universidade Técnica de Lisboa, Lisboa, Portugal
[2] Spoken Language Lab., INESC-ID Lisboa, Lisboa, Portugal
andre.silva@l2f.inesc-id.pt, cristiano.jlm@gmail.com,
alfredo.ferreira@ist.utl.pt, Nuno.Mamede@inesc-id.pt
[3] Universidade do Algarve, Faro, Portugal
jbaptis@ualg.pt

Abstract. Language learning resources are constantly evolving alongside technology. Computer-Aided Language Learning (CALL) is an area of research that focuses on developing tools to improve the process of learning a language. REAP.PT is a system that aims to teach Portuguese in an appealing way, addressing issues that the user is interested in. Initially conceived for vocabulary learning, this paper presents new trends in the REAP.PT development. For text-based exercises, it focus on automatic generation of syntactic and vocabulary questions. These exercises are set in a gaming context, to better motivate students. The paper also introduces a new evolution of REAP.PT, a 3D gaming environment for the learning of expressions denoting spacial relations between objects and object manipulation. These gaming aspects increase students motivation and help promote language learning.

Keywords: Computer Assisted Language Learning, Serious Games, Pictorial Exercises, Vocabulary Acquisition, Syntactic-Semantic Exercises, Automatic Exercise Generation, Portuguese.

1 Introduction

With the advancement of technologies for information systems, the use of Computer Aided Language Learning (CALL) has emerged as a tempting alternative to traditional modes of supplementing or replacing direct student-teacher interaction, such as the language laboratory or audio-tape-based self-study. Currently, CALL can be seen as an approach to language teaching and learning in which computer technology is used as an aid to the presentation, reinforcement and assessment of material to be learned, usually including a substantial interactive element. It also includes the search and the investigation of applications in language teaching and learning [1]. Gamper and Knapp (2002) define CALL as "a research field which explores the use of computational methods and techniques as well as new media for language learning and teaching" [2].

H. Caseli et al. (Eds.): PROPOR 2012, LNAI 7243, pp. 248–259, 2012.

Nowadays, people have come to expect more from language learning tools. It is known that video games have an intrinsic motivation appeal that makes them a valid tool for learning [3][4][5]. Serious Games emerged as digital games and equipment with an agenda of educational design and beyond entertainment. As Kurt Squire said, "e-Learning designers struggle to compel users who have paid thousands of dollars to complete an online course. Yet, game players routinely spend dozens, if not hundreds and thousands of hours mastering complex skills in digital worlds that are time-consuming, challenging, and difficult to master" [6]. Video games also allow players to be placed in rich environments, otherwise inaccessible, giving them increased motivation. A study involving 100 students showed that the right combination of both interactivity and media-richness results in an increase in knowledge acquisition, sustainability and topic interest [7], making video games a trustworthy environment for learning.

Although Serious Games can have a broad range of purposes and areas of application – such as healthcare, military and education [8] – we will focus on language learning. Recent projects show that most of the time, Serious Games are used to learn specific parts of a language, or to prepare someone for a certain situation, be it a person in a vacation trip or a soldier going to war. For example, *Mingoville*[1] is an online learning environment featuring English lessons for children, and has currently more than one million users [9]; *Polyglot Cubed*[2] is an educational game designed to aid in foreign language learning, currently available for Mandarin Chinese and Cape Verdean Creole [10]; *Global Conflicts*[3] is a series of role-playing, educational games used for teaching history, citizenship, geography and media courses; finally, Tactical Language & Culture Training Systems[4] (TLTS) are courses that use virtual-world simulations to help people acquire communicative skills in foreign languages and cultures; these courses are in widespread use by U.S. marines and soldiers, and increasingly by military service members in other countries [11].

In short, the use of Serious Games for language learning has been increasing in recent years, and there are already some successful systems in widespread use. The scarcity of (good quality) resources for Portuguese is a strong motivation for research in this area. Thus, these systems served as inspiration for some of the aspects of our approach.

2 REAP.PT

REAP[5] [12] (REAder-specific Practice) project, is a tutoring system for second language learning taking advantage of CALL technologies and based on Natural Language Processing. The system focuses on vocabulary learning by providing the students real documents featuring target vocabulary words [13] in context.

---

[1] `http://www.mingoville.com` (last access: June 2011).
[2] `http://www.polyglotgame.com` (last access: June 2011).
[3] `http://www.globalconflicts.eu` (last access: June 2011).
[4] `http://www.alelo.com/tactical_language.html` (last access: June 2011).
[5] `http://reap.cs.cmu.edu` (last accessed in June 2011).

The REAP.PT[6] results from porting the REAP system, originally built for English, to Portuguese [14][15]. It presents to the students rich and authentic study material (texts, exercises, etc.) that is deemed interesting by the students and adequate to their learning needs and current skills, thus being able to advance their learning process. In order to offer an interactive and individualized experience to the students, these have the possibility to define their topics of interest, which allows the system to present the most suitable documents for a specific student. The documents are extracted from the web, and because of this, students have access to both recent and varied readings.

This paper introduces two new trends in the REAP.PT development. The first is a set of exercises, presented in a gaming context, and focusing on aspects of syntax that are especially problematic for students of Portuguese as a Second Language (PSL). Because the exercises are generated automatically, careful design of the generation process is necessary in order to assure the linguistic adequacy and the relevance of the question. Above all, the exercises automatically generated by the system should not present ambiguous solutions. The second trend is the introduction of a 3D gaming environment for the learning of expressions denoting spacial relations between objects and object manipulation. Even if REAP.PT has been constantly evolving since inception, no step had yet been taken away from text-based exercises. The addition of a 3D environment opens up many possibilities both in terms of the exercises that can be made and in the ways that they can be presented to the student. The paper is structured as follows: Section 3 present the four written-based games: the *Lexical Mahjong* (3.1), *The Right Mood* (3.2), *Nominal Determinants* and *Collective Nouns* (3.3). Section 4 presents *The Office*, the 3D game developed for REAP.PT. Section 5 describes the evaluation and presents the results. Finally, section 6 concludes the paper pointing new directions for future research.

3 Written-Based Games

3.1 *Lexical Mahjong*: Word-Definition Association

In these exercises the student has to make a correspondence between the lemma and the definition of a word. Words and their definitions are taken from the Infopédia[7]. Then, a set of filters performs the following selection procedures: (i) using the Levenshtein distance algorithm as an approximation to the linguistic concept of cognate words (e.g. *escrito*/*escrever*), definitions containing cognates of the target word are discarded, since they are an obvious cue to the student; (ii) only definitions of more than one word (to avoid similarities with synonyms' exercises) and less than 150 words (to avoid very long definitions) are considered; (iii) characters that hinder the understanding of a given definition are removed (e.g. numbering of definitions, semicolon, cardinal, etc.); (iv) the learning level

[6] Available at: `http://call.l2f.inesc-id.pt/` (last acess: January 2012).

[7] `http://www.infopedia.pt/` (last access: December 2012).

of the words in the definition must be equal or less than the level of the exercise and that of the target word it corresponds to.

As REAP.PT is student-oriented, the word-definition pairs are chosen according to the student profile. A set of classifiers is applied to the definitions in order to determine the appropriate level of the exercise. Three levels were considered: beginners, intermediate and advanced. The level of a given word is determined by a classifier trained on a corpus of secondary school textbooks and national exams, structured by grade, from the 5th to the 12th [15]. All these preprocessing steps are performed once for each word, and the results are stored in the REAP.PT database.

Additionally, a scoring mechanism was added. From an initial set of points, more points are awarded or taken depending on correct/wrong answer, the time elapsed, or the student's hesitations. The evaluation module, apart from the scoring' system, checks if the student hits/misses a given word-definition correspondence and if the student has finished the exercise. The implementation of this scoring' system tries to accomplish several objectives: to provide the student some feedback on his/her performance; to keep the student motivated in the exercise; to prevent the student from solving the exercise by repeating his/her tries; and to provide a "summary" of the student's performance to the teacher. The result of this evaluation is stored in the REAP.PT database and it will be used later to analyze the student's learning progress.

3.2 *The Right Mood*: Choice of Mood in Subordinate Clauses

Learning the vocabulary of subordinating conjunctions and conjunctive phrases implies the acquisition of syntactic restrictions imposed to the mood of the subordinate clause they introduce. This game consists of cloze questions allowing students to practice reading comprehension while enhancing their syntactic knowledge of the language. Thus, for example, *até* (until) impose the infinitive or the subjunctive mood (1), but does not accept the indicative mood (2):

(1) *O Pedro fez isso até o João chegar / até que o João chegasse*
(Peter did that until John arrive_`inf/subj`)
(2) **O Pedro fez isso até que o João chega / chegava / chegou / chegará*

For the automatic generation of the exercise, the CETEMPúblico Corpus [16] is used, after being processed by the STRING processing chain [17]. A large set of conjunctions and conjunctive phrases have been listed in the system's lexicon. A set of general chunking rules creates a `SC` (subclause) chunk that links these conjunctions to the first verb of the subordinate clause, e.g.

O Pedro fez isso `SC[até o João chegar]`
O Pedro fez isso `SC[até que o João chegasse]`
(Peter did that until John arrive_`inf`/until that John arrive_`subj`)

To run the filters that generate the potential stems for this exercise, the distributed computing platform Hadoop is used to process this large corpus. To generate the distractors (wrong answers, e.g. indicative/subjunctive/infinitive mood alternation), a verb generator was used with a set or filtering rules to avoid ambiguity, i.e. verb homographs.

3.3 Nominal Determinants

Two more new games were built with the purpose of help learning the subtle distributional constraints observed between a determinative noun and the noun it determines. This syntactic-semantic relations may also serve to teach the classifying relationship between collective nouns and common nouns, since collectives often function as determinants on the common nouns they classify. Thus, while based on similar syntactic-semantic relations, two distinct exercises could be produced. For example:

Exercise: *O Pedro bebeu um ... de vinho* (Peter drank a ... of wine)
Possible answers: *copo* (glass), *saco* (bag), *cesto* (basket), *molho* (bunch)
Exercise: *O Pedro encontrou uma ... de ovelhas* (Peter found a ... of sheep)
Possible answers: *rebanho* (herd (sheep)), *cardume* (school (fish)), *exame* (swarm (bees)), *bando* (flock (birds))

These exercises are generated from real sentences, taken from the corpus. In these sentences, a quantifying dependency (`QUANTD`) has been extracted by the syntactic parser [17]. This dependency holds between a nominal determinant functioning as the head of a nominal or prepositional phrase (`NP` or `PP`) and the head of the immediately subsequent `PP` introduced by preposition *de* (of). The same relation holds for collective nouns. In the exercises above, this dependency links *copo* (glass) with *vinho* (wine) and *rebanho* (herd) with *ovelhas* (sheep), respectively. Again, the distributed computing platform HADOOP is used to retrieve from the large-sized corpus the sentences that can be potential stems for this exercise.

A new layer of lexical information was added to the lexicon in order to generate adequate distractors for the exercises. To do that, a list of determinative and collective nouns was added and a set of semantic features was defined and accorded to these words. These semantic features are based on categories such as: *Human*, *Animal*, *Food*, *Organization/Institution*, *Object*, *Nature*, *Military* and *Local*. By using these features, potential distractors that share the same traits as the target word are never select, thus avoiding to generate unwanted correct solutions as foils. Another set of constraints prevents general-purpose nominal determinants, such as *conjunto* (set) or *grupo* (group), from being selected as foils. An additional set of constraints was also added to prevent the generation of distractors that, in a figurative use, would make for possible solutions. For example, while normally nominal determinants associated to the *Animal* feature (e.g. *alcateia*, "pack", speaking of wolves) are not used for human nouns,

Fig. 1. *The Office* 3D game: One of the rooms available

this combination might, however, be used in a ironic turn (v.g. *Uma alcateia de políticos* "a pack of politicians"). To make the exercise more interesting and adequate, the distractors' generation keeps the same number and gender of the target word.

For these last two games, a feedback mechanism has been put in place: when a wrong word is chosen, the systems presents a real sentence taken from the corpus and showing an appropriate use of that wrong word. If available, the system also shows an image of the wrong word. In this way, positive feedback is provided.

4 *The Office*: Learning Spatial Relations in a 3D Game

This game also takes advantage of a Serious Games approach to language learning in order to make both the interface and the exercises more appealing to the student. It provides a 3D environment filled with objects that the student can interact with. In this environment, students perform exercises that focus on the verbs and prepositions used to describe the spatial relations between objects. Exercises consist in asking the student – represented by an avatar on screen – to perform different actions in an office scenario (Fig. 1). These actions include rearranging the position of objects so that certain spatial conditions are fulfilled, e.g. *Coloque o objecto A* ***em cima de*** *o objecto B* (Put the object A **on top of** the object B). The office scenario allows for the student's avatar to move around different rooms, unlocked by successful completion of a set of challenges put to him/her, and thus getting points.

The list of action verbs denoting the avatar movements and object manipulation, as well as the locative prepositions denoting spatial relations has been taken from lexical resources specifically built for and integrated in the STRING NLP chain[17]. This list has been validated by teachers of the Portuguese as Second Language (PSL) from University of Algarve, who also provided information regarding the appropriate level of each item.

Some accessibility utilities are also available to the student. One of them is the possibility of clicking on a certain object to check its definition in a dictionary. Another is the integration with the text-to-speech synthesizer already in use in REAP.PT, so that the student is able to hear the instructions, as well as any words he/she selects. In fact, although of a very diverse nature, the game is fully integrated as any other module of REAP.PT, namely, the system databases are used to retrieve information on the student using the application and also for storing her/his results and progress. Using the same databases allows the teachers to check their students' progress.

The Transformative, Adaptive, Responsive and enGaging EnvironmenT (TARGET) Platform[8] was the framework chosen to implement the game [18].

Much care has been taken in the a game plan developing in order for it to be successful regarding the user's enjoyment. Even more so when that game is intended for learning purposes. [19] discusses some heuristics that make things fun to learn, in particular when applied to instructional games. Those heuristics, along with the Serious Games review, were used as a base during the creation of this game plan and helped define many of its aspects, such as the importance of goals and of progression in keeping the user engaged; the need for appropriate performance and informative, positive feedback.

5 Evaluation

5.1 Evaluation of Written-Base Games

To assess the performance of the REAP.PT, two groups of students tested it and comment on its use. This evaluation consisted in a session involving: filling-in an initial form (tracing the students' profile); solving the exercises; and answering a final questionnaire, aimed at a qualitative assessment of the system. For each game, three exercises were given, one for each level: beginners, intermediate and advanced.

The exercises generated by the REAP.PT for the games presented in this paper already require some knowledge of Portuguese as they call upon more advanced language contents. If they are non-native Portuguese speakers, they should already have an elementary knowledge of the language. Due to the impossibility of gathering in time a group of subjects with these features, the tests were conducted with a group with relatively similar characteristics - Portuguese native speakers in the 3rd and 4th grade (Group 1). The choice of Group 1 can be justified by their knowledge of Portuguese as mother language but their

[8] `http://www.reachyourtarget.org/` (last access: June 2011).

still limited vocabulary - this being one of the REAP.PT main learning targets. In order be able to contrast the performance of this group , the same test was also performed by another group of native speakers, with at least a College degree (Group 2). Thus, 45 subjects performed this exercise, 18 in Group 1 and 31 in Group 2. Naturally, while the age of the subjects from Group 1 is quite homogeneous, age range of Group 2 varied from 19 to 70, the average age being 24.

The testing environment was also different for each group. Children from Group 1 did the exercises one at time in the classroom computer, and a team member of REAP.PT project helped them to access the starting page and occasionally had to explain the exercises or some unknown word since some children had not yet fully mastered their reading skills. Each child took about 26 minutes to complete the exercises and the questionnaire. Users from Group 2 did the test at their homes, without any aid from the REAP.PT team.

Based on the questionnaire, more than 77% of the users found the system easy to use, while only 39% needed to use the "Help" button. In this way, we can say that one of the goals of this project was achieved, namely, to create a system with a simple and easy to use interface, so that all users (even some inexperienced users) were able to use the REAP.PT. The pre-processing of the exercises is made in advance and stored in the system database. Thus, the generation stage, which is the most time-consuming step of the processing, is already done when the user access the system. This is why the system is so quick when responding to requests from users as it is reflected in their answers.

Users from both groups reported that the exercise they liked the most (44%) was the *Lexical Mahjong*, followed by the *Collective Nouns* (27%), and *Nominal Determinants* (17%), and, finally, by *The Right Mood* (12%). However, it is remarkable that the 2 Groups had clearly different impressions about these last three exercises. While Group 1 liked *Collective Nouns* as 2nd best (40%), Group 2 shows similar preference for each of these three (16%, 19% and 21%). This may have to do with the fact that collective nouns are an explicit grammatical subject at "Escola Primária" (Elementary School) thus contributing to a higher familiarity with the topic and the exercise, which entailed this preference trend. In the *Lexical Mahjong* exercises, Group 2 obtained better results with a performance of 84% (standard deviation = 6,6%), while the users from Group 1 made more errors and obtained a performance of 54% (standard deviation = 12,4%). Another interesting aspect to note is the error rate progression for each exercise. In both groups, as the difficulty level of exercises increases, so the students do more mistakes, which confirms the adequacy of the strategy here followed and its implementation, namely to distinguish the level of different definitions inside the dictionary entry of each target word. The results of the *Collective Nouns* and *Nominal Determinants* exercises were very similar. In Group 1, on average, students missed one in every three submitted exercises while Group 3 misses less, which is naturally understandable given the age difference and educational level of the subjects from each group. This information confirms the students self-assessment regarding the difficulty of the exercises.

In the overall assessment of the system, 50% of the subjects classify it as "very good", 17% as "good", 9% as "acceptable" and the remaining 18% classified it as "bad" or "very bad". In this way a global positive appreciation was achieved (67% good or very good). Group 2 is more critical about the system: only 41% found it very good against 55% from Group 1. The most critical opinion relates to the high repetition of the infinitive in *The Right Mood* exercises. This is due to the fact that most of the stems found in the CETEMPúblico corpus show the verb in the infinitive. Because of this skewed distribution of mood, a high proportion of stems has also the target verb form in the infinitive. This can lead the users to sense that this exercise is the easiest, due to repetition, but also the less interesting one. Finally, we verified that Group 1 takes on average 5 minutes more than the Group 2 to finish the exercises, the *Lexical Mahjong* being the game that takes the longest time to complete.

5.2 Evaluation of *The Office* 3D Game

In a preliminary evaluation of *The Office* 3D game, a total of 14 students from the Portuguese as Second Language (PSL) course of University of Algarve have played with the application. 32 exercises were scattered throughout five different rooms. When starting the game, the students were encouraged to complete a brief tutorial that explained the various mechanics of the game, including how to control the avatar, start exercises and interact with objects. When this tutorial was finished or skipped, the first level was unlocked and the students were on their own. When the student finished playing, a questionnaire was presented focusing on the interaction with the application and its interface; and on the student's learning experience. From the group of players, 8 students, 3 female and 5 male, also answered this questionnaire. Students' age varied from 23 to 29, average age being 27. Contact with Portuguese averages at 6 months. Students native languages included Spanish (the larger group), Russian, Romanian, Bengali, Nepali, Catalan, and Ukrainian.

62,5% of the students considered the tutorial helpful, and only 12,5% did not find it helpful. Some students did not notice the option to change the interface to English, and this may have impacted negatively in their opinion on the tutorial, seeing as it was text-heavy. Moving the objects was easy for most students, only 12,5% found it difficult. Controlling the avatar was less positively assessed, as 25% did not find it easy. The two results are related as those who found it very easy to move objects also had less trouble controlling the avatar and vice-versa. Future work should focus on polishing these controls in order to improve the results. Most students did not think that learning how to play the game interfered with their learning experience. When asked this question, only 10% said it had interfered at all.

The questionnaire also asked for the students' perception of how much they had learned, the difficulty level of the exercises and their general satisfaction with the knowledge provided by the game. 25% answered that they thought they had learned more with the game than with a traditional class, while everyone else thought they had learned the same. No one thought they had learned less with the game, which is good. Also, every student considered the difficulty to be at their own level, neither being too high nor too low and 62.5% answered they had noticed an increase in the difficulty level of the exercises as they progressed. This means that the progression system may still need some tuning, even if it is already sufficiently well implemented to become noticeable. In general, students were satisfied (50%) or very satisfied (25%) with the game. In their general comments, students seem to prefer knowing exactly where to go and what exercise to do next, thus showing an attitude different from the games' expectations, where an inquisitive attitude towards exploration of the scenario had been envisioned. Some students wished to exit the office setting and explore different scenarios. This is certainly something to be done as future work.

Using the interaction logs, it was possible to compare, for every exercise, its expected time for completion with the average time taken in the evaluation, and the maximum points obtainable with the average points awarded. For lack of space, only the conclusions are presented here. Three levels of difficulty were defined, and for each one an expected time for completion was set to 60 seconds (easy), 90 seconds (intermediate) and 120 seconds (difficult), respectively. In average, students concluded the exercises in 40% of the expected time, but standard deviation is high (35%). The forfeit rate of the exercises is very close to zero. There was only a single case in which a student tried the same exercise 3 times unsuccessfully and then decided to move to another exercise without returning. Every other failure resulted in the student trying again – either right away of after solving other exercises in between – and then finishing correctly. This may be interpreted as a result of the successful development of the game plan and, in particular, of constructive/positive feedback the system provided for wrong answers.

This first evaluation shows that this learning resource has the potential to be well received by PSL students and may contribute to improve learning these particular expressions of spatial relations an place actions in a motivating and appealing way.

6 Conclusion and Future Work

CALL resources, based on NLP technology, can contribute to the development of a new learning paradigm, where the student has a more active role, and the motivation for learning can be enhanced by presenting learning materials and game-like exercises more appealing and adequate to the student profile, language skills and topics of interest.

This paper presented REAP.PT, a system that aims to teach Portuguese in an appealing way, addressing issues that the user is interested in. While initially conceived for vocabulary learning by Portuguese as Second Language (PSL) students, this paper presented new trends in the REAP.PT current development, now focusing on automatic generation of text-based exercises for syntactic properties and vocabulary, also relevant to native speakers. These exercises were set in a gaming context, to better motivate students. A series of NLP techniques were used to automatically generate the stems and distractors from real texts, producing sets of exercises adequate to the student model (his language skills and level of proficiency), and with gradually increasing difficulty. The paper also introduced a new evolution of REAP.PT, a 3D gaming environment for the learning of expressions denoting spacial relations between objects and object manipulation. This is one of the topics covered by PSL courses where the 3D gaming environment can be most helpful in providing a motivating CALL setting.

These REAP.PT games were evaluated with real PSL students (and some with native children) and results showed a (very) positive feedback, while demonstrating the adequacy of the CALL approach. Particularly relevant is also the fact that all NLP tools, resources and scripts can now be reused to create new scenarios, and to pursue other venues in other Portuguese learning topics and in brand new gaming settings. For new text-based exercises, already being developed, REAP.PT will include topics such as clitic pronoun placement, active-passive and direct/indirect speech sentence transformation, idiomatic and formulaic expressions – topics particularly relevant to, but not exclusive of, PSL courses. For 3D games, extension to new scenarios (exploring the City, the School, the Mall, the Amusement Park) will allow the learning of specific vocabulary, and will include a more active role from the user, by having him/her producing small sentences and interacting with other gaming partners (real people or artificial agents). In the near future, attention shall be given to improve some features of REAP.PT: the interface needs improving in the design and conviviality, making more appealing and adequate to the users' age. The gradual difficulty of exercises makes REAP.PT particularly apt to be used in assessment of language skills, once extensive testing has been done. Field tries are already in place at University of Algarve to prepare these new developments in the near future. The web interface, tested on all major browsers, makes REAP.PT freely accessible all around the world to any registered user.

Acknowledgments. This work was partially supported by FCT (INESC-ID multi-annual funding), through the PIDDAC Program, the FCT project REAP.PT (proj. ref. CMU-PT/HuMach/0053/2008).

References

1. Hacken, P.T.: Computer-assisted language learning and the revolution in Computational Linguistics. Journal Linguistik online 17, 23–39 (2003)
2. Gamper, J., Knapp, J.: A review of intelligent CALL systems. Computer Assisted Language Learning 15, 329–342 (2002)
3. Kirriemuir, J., Mcfarlane, A.: Literature review in games and learning. Technical report, NESTA Futurelab (2004)
4. Malone, T., Lepper, M.: Making learning fun: A taxonomy of intrinsic motivations for learning. In: Snow, R., Farr, M. (eds.) Aptitude, Learning, and Instruction. Cognitive and Affective Process Analyses, LEA, Hillsdale, vol. 3, pp. 223–253 (1987)

5. Gee, J.: What Video Games Have to Teach Us About Learning and Literacy. St. Martin's Press (2008)
6. Squire, K.: Game-based learning: Present and future state of the field (2005)
7. Wong, W.L., Shen, C., Nocera, L., Carriazo, E., Tang, F., Bugga, S., Narayanan, H., Wang, H., Ritterfeld, U.: Serious video game effectiveness. In: Proceedings of the International Conference on Advances in Computer Entertainment Technology, ACE 2007, pp. 49–55. ACM, New York (2007)
8. Michael, D.R., Chen, S.L.: Serious Games: Games That Educate, Train, and Inform. Muska & Lipman/Premier-Trade (2005)
9. Srensen, H.B., Meyer, B.: Serious Games in language learning and teaching - a theoretical perspective. In: Akira, B. (ed.) Situated Play: Proceedings of the Digital Games Research Association Conference, Tokyo University, pp. 559–566 (2007)
10. Grace, L.D.: Polyglot cubed: the design of a multi-language learning game. In: Proceedings of the International Conference on Advances in Computer Entertainment Technology, ACE 2009, pp. 421–422. ACM (2009)
11. Johnson, W., Valente, A.: Tactical language and culture training systems: Using artificial intelligence to teach foreign languages and cultures. In: Proceedings of the 20th National Conference on Innovative Applications of Artificial Intelligence, pp. 1632–1639. AAAI Press (2008)
12. Heilman, M., Eskenazi, M.: Language learning: Challenges for intelligent tutoring systems. In: Aleven, V., Pinkwart, N., Ashley, K., Lynch, C. (eds.) Proceedings of the Workshop of Intelligent Tutoring Systems for Ill-Defined Domains. 8th International Conference on Intelligent Tutoring Systems, pp. 20–28 (2006)
13. Levy, M.: Computer Assisted Language Learning: Context and Conceptualization. Clarendon Press (1997)
14. Marujo, L., Lopes, J., Mamede, N.J., Trancoso, I., Pino, J., Eskenazi, M., Baptista, J., Viana, C.: Porting REAP to European Portuguese. ISCA International Workshop on Speech and Language Technology in Education, SLaTE 2009 (2009)
15. Correia, R.: MSc Thesis, Automatic Question Generation for REAP.PT Tutoring System, Instituto Superior Técnico - Universidade Técnica de Lisboa (2010)
16. Santos, D., Rocha, P.: Evaluating CETEMPúblico, a free resource for Portuguese. In: Proceedings of the 39th Annual Meeting on Association for Computational Linguistics, pp. 450–457 (2001)
17. Mamede, N.: STRING - A Cadeia de Processamento de Lngua Natural do L2F. Technical Report, L2F/INESC-ID, Lisboa (2011)
18. Ribeiro, C., Jepp, P., Pereira, J., Fradinho, M.: Lessons learnt in building serious games and virtual worlds for competence development. In: Taisch, M., Cassina, J., Smeds, R. (eds.) Experimental Learning on Sustainable Management, Economics and Industrial Engineering, pp. 52–62 (2010)
19. Malone, T.: What makes things fun to learn? Heuristics for designing instructional computer games. In: Proceedings of the 3rd ACM SIGSMALL Symposium and the first SIGPC Symposium on Small Systems (SIGSMALL 1980), pp. 162–169. ACM (1980)

Graph-Based Methods for Multi-document Summarization: Exploring Relationship Maps, Complex Networks and Discourse Information

Rafael Ribaldo[1], Ademar Takeo Akabane[1], Lucia Helena Machado Rino[2], and Thiago Alexandre Salgueiro Pardo[1]

[1] Instituto de Ciências Matemáticas e de Computação, Universidade de São Paulo
[2] Departamento de Computação, Universidade Federal de São Carlos
{ribaldo,takeusp}@grad.icmc.usp.br, lucia@dc.ufscar.br, taspardo@icmc.usp.br

Abstract. In this work we investigate the use of graphs for multi-document summarization. We adapt the traditional Relationship Map approach to the multi-document scenario and, in a hybrid approach, we consider adding CST (Cross-document Structure Theory) relations to this adapted model. We also investigate some measures derived from graphs and complex networks for sentence selection. We show that the superficial graph-based methods are promising for the task. More importantly, some of them perform almost as good as a deep approach.

Keywords: summarization, graphs, discourse.

1 Introduction

Nowadays, with the huge and growing amount of information available in the Web and the sparse time to read and grasp it, manual analysis of the conveyed documents becomes almost impossible. According to a research conducted by the International Data Corporation [12], textual information in digital format currently amounts to c.a. 1.8 zettabytes, a number nine times bigger than five years ago. This makes evident that automatic treatment of texts is necessary, including information retrieval and extraction, topic detection and tracking, and text summarization, which is the focus of this paper.

Multi-Document Summarization (MDS) is defined as the task of automatically producing a unique summary of a set of documents on the same topic [17]. This task differs from single-document summarization in that it must detect and treat many phenomena that are typical of relating information on an inter-document basis. Important multi-document phenomena comprise dealing with redundant, complementary, and contradictory information, and both retrieving adequate links provided by referring expressions and ordering events and facts in time. More profound challenges refer to making writing styles uniform, fusing information, and balancing different perspectives of the same stories, among others. Practical applications for MDS include, e.g., making a web search engine produce summaries

H. Caseli et al. (Eds.): PROPOR 2012, LNAI 7243, pp. 260–271, 2012.

of groups of news texts based on users queries, instead of just hits as normally do Google or Google News; indexing complex work, such as collections of scientific articles on a specific theme; producing biographies or synthesizing diverse opinions (being them in agreement or not) about a topic.

Traditional approaches to summarization comprise the well-known surface (shallow/superficial) and deep-based ones. The former uses relatively shallow linguistic information or no linguistic information at all, whilst the latter makes significant use of linguistic information during processing, usually including world knowledge and discourse models. In this paper, we address surface and hybrid approaches by dealing with graphs and discourse information processing.

Modeling texts as graphs implies normally having as their vertices (or nodes) text segments and as their links information on how these nodes relate to each other. Usually, text segments may be words, sentences, or paragraphs. At this level, having a surface-based or deep-based approach refers basically to how graph links depict information relationships. For summarization purposes, graph metrics signal the importance of a text segment. Based on those, the segment may be chosen to compose a summary or not. [3], [4], [13], [22] and [31] are examples of works that follow such approach. In particular, [3], [4] and [13] use a special type of graph, the complex networks. These differ from simple graphs in that they follow complex principles of organization and usually have a large number of nodes and show special topographic features [2].

In summarization, adding discourse features for deep processing implies having the system making decisions based on linguistic knowledge. In this case, it is possible to keep the granularity of the text segments similar to those used for a graph organization. However, weighting the importance of segments is now usually dependent on deeper text analysis. As a result, linguistically motivated relationships between those text spans may be depicted. The Cross-document Structure Theory (CST) [26] is an exemplar of the discourse models or theories that are used in MDS. It consists in a finite set of relations that are used to relate information from different texts, including relations for equivalent and contradictory segments, elaborated topics, citation of sources and authorship identification, etc. [1], [7], [9], [27], [28], [29] and [34] are good representatives of this research line. While some of these works explicitly use discourse models, others try to approach such relationships only indirectly.

In this paper we explore some graph-based summarization methods, aiming at (i) adapting to MDS the traditional Relationship Map [31], which has been originally proposed for single-document summarization, (ii) evaluating the impact of incorporating knowledge on the CST relations in the adapted model, and (iii) investigating how good content selection may be by taking into account some graph and complex network metrics. We focus on the production of generic and informative summaries. Experiments using a corpus of Brazilian Portuguese news texts were carried out to evaluate the contribution of each MDS method. Only summary informativeness was considered. Graph-based methods turned out to be well suited for the envisioned task. Actually, the results were close to the best ones obtained for Portuguese when a deep CST-based approach was considered. Since the latter is very effortful, MDS based on surface methods seem more promising and scalable. We also show that adding CST relations to the graph metrics does not alter the quality of content selection. Instead, it only reinforces that graph-based methods are adequate for MDS so far.

In the next section, we briefly review the main initiatives related to the work presented in this paper. In Section 3, we present the graph-based methods that we explore. MDS assessment is described in Section 4, followed by some final remarks in Section 5.

2 Related Work

Graphs have shown to be applicable to many Natural Language Processing applications [21] and there are several graph-based approaches for both single and multi-document summarization (see, for example, [3], [4], [7], [11], [13], [18], [22], [31] and [32]). In this section we briefly introduce the ones that served as the basis for our work.

The work of Salton et al. [31] probably introduced the first widespread graph-based approach to single-document summarization. In the proposed relationship map method, the authors model a text as a graph/map in the following way: each paragraph is represented as a node, and weighted links are established only among paragraphs that have some lexical similarity. This may be pinpointed through lexical similarity metrics. The choice for representing paragraphs (and not words, clauses, or sentences) as nodes is due to the assumption that paragraphs provide more information surrounding their main topics and, thus, may be used for more coherent and cohesive summaries. For summarization purposes, only the highly weighted links are considered: given a graph with N nodes, only the 1.5 * N best links provide the means to select paragraphs to include in a summary. Once established such a threshold, three different ways of traversing a graph are proposed, namely, the Bushy Path, the Depth-first Path, and the Segmented Bushy Path. In the Bushy Path, the density, or *bushiness,* of a node is defined as the number of connections it has to other nodes in the graph. So, a highly linked node has a large overlapping vocabulary with several paragraphs, representing an important topic of the text. For this reason, it is a candidate for inclusion in the summary. Selection of highly connected nodes is done until compression rate is satisfied in the Bushy Path. In this way, the coverage of the main topics of the text is very likely to be good. However, the summary may be non-coherent, since relationships between every two nodes are not properly tackled. To overcame that, instead of simply selecting the most connected nodes, the Depth-first Path starts with some important node (usually the one weighted the highest) and continues the selection with the nodes (i) that are connected to the previous selected one and (ii) that come after it in the text, also considering selecting the most connected one among these, trying to avoid sudden topic changes. This procedure is followed until the summary is fully built. Its advantage over the Bushy Path is that more legible summaries may be built due to choosing sequential paragraphs. However, topic coverage may be damaged. The Segmented Bushy Path aims at overcoming the bottlenecks introduced by the other two methods: it tackles the topic representation problem by first segmenting the graph in portions that may correspond to the topics of the text. Then, it reproduces the Bushy Path method in each subgraph. It is guaranteed that at least one paragraph of each topic will be selected to compose the summary. In their evaluation, Salton et al. show that the methods produce good results for a corpus of encyclopedic texts, with the Bushy Path being the best method.

Antiqueira et al. [3] [4] use complex networks to model texts for single-document summarization. In their networks, each sentence is represented as a node and links are established among sentences that share at least one noun. Once the network is built, sentence ranking is performed by using graph and complex network measures. The best-ranked sentences are thus selected to compose the summary, as usual. From several measures explored by Antiqueira et al., we selected only three for our work: degree, clustering coefficient and shortest path. The well-known degree measure indicates how many connections one node has to the others. It accounts for the density measure in Salton et al.'s model. It is assumed that the bigger the degree of a node, the more important the corresponding sentence is. Notice that this selection strategy is also similar to Salton et al.'s Bushy Path (although graph construction is different). The clustering coefficient measure signals how nodes tend to cluster together. It was introduced in [33] and, for summarization, it may indicate central topics to the summary. Also well-known, the shortest path measure indicates the length of the shortest path between 2 nodes. Antiqueira et al. use the average of the lengths of the shortest paths from a node to every other node in the network as an indication of its importance: the nearer a node is (in average) to the other nodes, the better the sentence is to compose the summary. The authors evaluated their work with news texts in Brazilian Portuguese using the TeMário corpus [23]. Good results were achieved, but they did not outperform the best single-document summarization system for Portuguese - SuPor [13], which is based on machine learning over a rich set of features.

Turning to MDS, Castro Jorge and Pardo [7] model several texts as just one graph, with nodes representing sentences and links representing discourse relations among the sentences. Discourse relations are CST relations [26], in particular, the ones refined by Maziero et al. [19]. Such relations pinpoint similar and different sentence content, as well as different writing styles and decisions among the texts. For sentence selection, sentences that have more relations/links to others are preferred, similarly to those approaches for single-document summarization. A further step here is verifying whether a selected sentence is redundant, i.e., if it embeds information that has already been conveyed by other previously selected sentences. By analyzing the CST relations, it is possible to detect redundancy of a candidate sentence. In this case, such sentence is ignored and the next candidate sentence in the graph is considered. Castro Jorge and Pardo evaluated their approach on the CSTNews corpus of news texts in Brazilian Portuguese [5]. This corpus is manually annotated with CST and it also conveys manual summaries. The results were the best produced so far for this language, for the MDS task.

3 Graph-Based Methods for MDS

In this section we report on graph-based methods adapted for the MDS task and compare them with previous results for Portuguese. We considered three distinct models, namely: the classical Relationship Map one [31], now conveying a multi-document representation; its enrichment with CST relations [19] [26]; and a proposal using graphs and complex networks metrics, following the work of Antiqueira et al. [3] [4]. Annotating relationship maps with CST relations aimed at verifying their impact in the

Salton et al.'s model. The hypothesis here was that since the best results for Portuguese were obtained by a CST-based approach, as reported by Castro Jorge and Pardo [7], an MDS model based upon relationship maps could benefit from that enrichment.

As far as we know, this is the first time that the classical Relationship Map approach is used for MDS purposes. Complex networks had already been explored in this context, providing the means for a machine learning solution [9] along with other surface and CST-based features. It is also interesting to notice that the approaches in this paper are complementary to current work in MDS for the Portuguese language, which has mainly focused on deep-based approaches (e.g., see [6], [7], [8], [9] and [10]).

For adapting the Relationship Map model, sentences (instead of paragraphs) are represented in nodes. To our view, this may allow for more refined summarizing decisions, although it may endanger coherence and cohesion of the final summaries. A multi-document graph is built in the following way: each sentence from a group of texts is represented by a node in the graph; links between nodes signal how similar the related sentences are. The cosine similarity measure [30] is used to compute lexical similarity among sentences, after removing stopwords and stemming the remaining words. A graph conveys as many nodes as sentences in the focused set of texts, and duplication is not verified beforehand. It turns out that even identical sentences, which are likely to occur in a set of texts on the same topic, are represented in the graph as different nodes. This is the reason for treating redundancy afterwards. As suggested by Salton et al., if there are N nodes in a graph, only the 1.5 * N best weighted links are considered for MDS. After modeling the texts in a graph, sentence selection proceeds as described in Section 2, for two paths only: the Bushy Path and the Depth-first Path. The Segmented Bushy Path was not adopted because it requires more sophisticated reasoning, which must be investigated in future work. Another modification took place in adapting the Depth-first Path for MDS: in the original single-document summarization model, paragraphs selection is subordinated to previous paragraphs choices, in order to observe paragraph ordering for cohesiveness. In MDS, it only makes sense to observe sentence ordering if sentences are selected from the same text. For this reason, this restriction is relaxed if the sentences under analysis come from different texts.

Finally, the summary is built with the most prominent sentences of all texts. As expected, there is a high degree of redundancy in texts, which is natural in multi-document analysis tasks. Such redundancy is frequently indicated by links with very high degree of similarity. Therefore, in order to avoid having redundant sentences in the summary, we stipulated a redundancy limit that a new selected sentence may have in relation to any of the previously selected sentences. If this limit is reached, this new sentence is considered redundant and ignored, and the summarization process goes to the next candidate sentence; otherwise, the sentence is included in the summary. The limit was defined as the sum of the highest and the lowest cosine values in the graph divided by 2, resulting in an intermediate value between them. So, the redundancy limit is not fixed; it depends on the graph produced for the input texts. We believe that this flexibility allows a better redundancy treatment for varied domains and text types and genres.

The number of selected sentences to compose the summary is also limited by the specified compression rate, which gives the proportion between the length of the summary and the length of the texts. In this work, we consider the length of the longest text, in number of words. Therefore, a 70% compression rate indicates that the summary may be 30% long, in relation to the length of the longest text in the group.

To illustrate the process under focus here, Figures 1, 2 and 3 show 2 short source texts (from CSTNews corpus [5]) and the corresponding automatic summary produced by the previous method, using a 70% compression rate. The original language of the texts is Brazilian Portuguese, and so is the summary one. For clarity, English translation is also provided. Both Bushy and Depth-first Paths produced the same summary for the text set shown. One may see that the summary is good.

A ginasta Jade Barbosa, que obteve três medalhas nos Jogos Pan-Americanos do Rio, em julho, venceu votação na internet e será a representante brasileira no revezamento da tocha olímpica para Pequim-2008.
A tocha passará por vinte países, mas o Brasil não estará no percurso olímpico. Por isso, Jade participará do evento em Buenos Aires, na Argentina, única cidade da América do Sul a receber o símbolo dos Jogos.
O revezamento terminará em 8 de agosto, primeiro dia das Olimpíadas de Pequim.

=== English translation ===
The gymnast Jade Barbosa, who won three medals in the Pan-American Games in Rio in July, won an Internet poll and will be the Brazilian representative in the Olympic torch relay to Beijing in 2008.
The torch will pass through twenty countries, but Brazil is not in the Olympic route. Therefore, Jade will participate in the event in Buenos Aires, Argentina, the only city in South America to receive the symbol of the Games.
The relay will end on August 8, the first day of the Beijing Olympics.

Fig. 1. First Text

Um dos destaques desta temporada do esporte brasileiro, a ginasta Jade Barbosa foi escolhida, na noite desta terça-feira, para ser a representante do Brasil no revezamento da tocha dos Jogos Olímpicos de Pequim.
Em votação pela internet, a ginasta recebeu mais de 100 mil votos e superou o nadador Thiago Pereira, que ganhou seis ouros nos Jogos Pan-Americanos.
O Brasil não faz parte do trajeto da tocha olímpica. Na América do Sul, a chama passará por Buenos Aires, onde Jade participará do revezamento, no dia 11 de abril.
Aos 16 anos, Jade conquistou três medalhas no Pan: ouro na disputa dos saltos, prata na apresentação por equipes e bronze no solo.
Ao todo, a chama olímpica percorrerá 20 países antes de chegar a Pequim para a abertura da competição, no dia 8 de agosto.

=== English translation ===
One of the highlights of this season in Brazilian sports, the gymnast Jade Barbosa was chosen Tuesday night to be the representative of Brazil in the torch relay in Beijing Olympic Games.
In an Internet poll, the gymnast has received over 100,000 votes and has beaten the swimmer Thiago Pereira, who won six gold medals in the Pan-American Games.
Brazil is not part of the Olympic torch route. In South America, the flame will pass through Buenos Aires, where Jade will participate in the relay on April 11.
At the age of 16, Jade won three medals in the Pan: gold in the Vault, silver in the Team competition, and bronze in the Floor.
In general, the Olympic flame will travel through 20 countries before arriving in Beijing for the competition opening on August 8.

Fig. 2. Second Text

A ginasta Jade Barbosa, que obteve três medalhas nos Jogos Pan-Americanos do Rio, em julho, venceu votação na internet e será a representante brasileira no revezamento da tocha olímpica para Pequim-2008. Na América do Sul, a chama passará por Buenos Aires, onde Jade participará do revezamento, no dia 11 de abril.

=== English translation ===
The gymnast Jade Barbosa, who won three medals in the Pan-American Games in Rio in July, won an Internet poll and will be the Brazilian representative in the Olympic torch relay to Beijing in 2008. In South America, the flame will pass through Buenos Aires, where Jade will participate in the relay on April 11.

Fig. 3. Summary

In a variation of the above method, we add CST relations to the graph in order to verify the impact of discourse information in the summary informativeness. We expect that the use of such refined knowledge may improve sentence selection. Two strategies are followed in order to consider CST relations in the graph: (i) simply summing the number of relations that each sentence presents to its number of links, without considering the relation type itself, and (ii) assigning scores to every CST relation that appears in each sentence and, then, summing up those scores to the number of links of that sentence. Relations that account for content matters are considered more important and receive values from 0.5 to 1, with relations that indicate redundancy getting higher values, since redundancy usually indicates importance in MDS [17]; relations that deal mainly with different writing styles and decisions receive values below 0.5. After considering CST in the graph, sentence selection is performed in the same way it is before.

Finally, our last method consists in using some graph and complex network measures to rank the sentences in the graph, in order to select the candidates to a summary. A graph is built in the same way as before: sentences are represented in the nodes and links are weighted according to their lexical similarity with other sentences through the cosine similarity measure. However, differently from Salton et al. method, in this approach we keep all the links, as Antiqueira et al. [3] [4] do. We use degree, clustering coefficient and shortest path measures to rank sentences, and the best-ranked ones are selected for inclusion in the summary. While degree and shortest path measures are also usual graph metrics, the clustering coefficient one is typical of complex networks.

Independently from the chosen measure to score the importance of the candidate sentences, redundancy is treated in the same way as that by Salton et al. adapted method, i.e., by applying a redundancy threshold to select sentences.

4 Evaluation of the Proposed Methods

The evaluation of the proposed MDS strategies was carried out over the CSTNews corpus, which amounts to 140 news texts in Brazilian Portuguese divided into 50 groups. Each group has from 2 to 3 texts on a same topic, having in average 49 sentences and 945 words. This corpus includes manual multi-document summaries (one per group) with 70% compression rate (in relation to the longest text). The texts are manually annotated with CST, with satisfactory agreement values among the annotators [5]. That indicates that the annotation is reliable and, for this reason, was used for evaluation in our work.

We used ROUGE (Recall-Oriented Understudy for Gisting Evaluation) [15], a tool created to enable direct comparison among an automatically generated summary and the corresponding human summaries. It provides precision, recall and f-measure values based on the number of common n-grams conveyed by the summaries. As its authors showed, ROUGE is as good as humans in ranking summaries according to their informativeness, being a good indicator of the quality of the summarization methods.

In this work, we report ROUGE results just for 1-gram comparisons, hence using the so-called ROUGE-1. This metric has been shown to be enough for comparing summary informativeness. We also produced ROUGE results for other n-gram comparisons, but do not show them here, since they corroborated the ROUGE-1 results.

We also evaluated summaries produced by other systems that can tackle texts on Brazilian Portuguese, namely: GistSumm [24] [25], Castro Jorge and Pardo's CSTSumm [7] [8], MEAD [28], and a baseline method that randomly select sentences. GistSumm was the first MDS system produced for Portuguese and follows a very naive approach, by simply juxtaposing all the texts and selecting the sentences according to the frequency of their words. CSTSumm has been graded so far the best MDS system for Portuguese and follows the purely CST-based method presented in Section 2. MEAD is one of the most famous multi-lingual MDS systems and it is based on centroids, sentence position and simple lexical features extracted from the sentences.

Table 1 shows the results of our assessment, ordered according to the systems f-measures. As expected, the best system is still CSTSumm. This is certainly due to its deep approach, which usually provides better results. Similarly to Antiqueira et al.'s findings for single-document summarization, the degree measure yields very good results for MDS, following CSTSumm in the rank. Amongst Salton et al.'s methods, the Bushy Path was slightly better. This has also been reported by Salton et al. before. It is noticeable that MEAD f-measure is smaller than most of those provided by our proposed summarization strategies. It is also curious that some systems were worse than the random baseline. The MDS system based on the clustering coefficient measure was the worst one.

Table 1. Evaluation results

System/Method	Precision	Recall	F-measure
CSTSumm	0.5761	0.5065	0.5297
Degree	0.5328	0.5037	0.5155
Shortest Path	0.5306	0.5009	0.5131
Bushy Path	0.4844	0.5397	0.5083
Bushy Path with CST	0.4844	0.5397	0.5083
Depth-first Path	0.4811	0.5340	0.5040
Depth-first Path with CST	0.4811	0.5340	0.5040
MEAD	0.5242	0.4602	0.4869
Random Baseline	0.4494	0.4864	0.4652
GistSumm	0.3599	0.6643	0.4599
Clustering coefficient	0.4671	0.4476	0.4560

The system with the highest precision was also CSTSumm. Although being one of the worst systems, GistSumm obtained the highest recall value, much better than all other values. It is followed by Salton et al. methods.

While some results were expected, others were surprising. Except the clustering coefficient method, all the others were good, performing closely to CSTSumm, possibly indicating that it is worthy following a superficial approach, mainly if we consider that such methods are more scalable and do not depend on sophisticated resources or models based on discourse annotation. If we consider that the results of CSTSumm evaluation were obtained with a manually CST-annotated corpus and that an automatic CST annotation would produce worse summarization results (as it usually happens in the area), it would not be a surprise if some our proposed graph-based methods would outperform CSTSumm with automatic annotation. This is in line with [16], which had already indicated that discourse information could be replaced by superficial measures without great loss. However, to investigate that further, in future work we intend to use a CST parser available for Portuguese [20] in order to pursue automatic annotation of the same CSTNews corpus and reproduce the above evaluation setting.

It was also surprising that enriching Salton et al. paths with CST did not alter any of the results. In fact, CST was only useful to reinforce the results obtained by using the original path models. This observation also corroborates the conclusion that graphs are promising for MDS.

The machine learning solution to MDS presented in [9], which uses complex networks and other superficial and CST-based features to induce decision trees for classifying important and non-important sentences for composing summaries, produced very good summaries for Portuguese. The CSTNews corpus was also used in their evaluation. However, their results may not be directly compared to ours, since they used another evaluation strategy, based on training and test sets (not using the full corpus for evaluation, therefore). In spite of that, according to the results in [9], which also evaluated CSTSumm and MEAD summaries, the machine learning approach ranked right after CSTSumm, which was the best system in the corresponding evaluation. Another interesting issue in that work is that the clustering coefficient feature showed to be a relevant feature when used together with other CST-based features. In this work such measure was the worst one when used alone, probably indicating that it is better used as complementary information rather than a unique indicator of sentence importance.

Based on the results in [9], we may expect that the same machine learning approach would rank after CSTSumm in this paper and would also suffer from using automatically CST-annotated texts, probably being outperformed by our graph-based methods.

5 Final Remarks

We showed in this paper that graph-based methods for MDS of texts in Brazilian Portuguese provide very good results, being close to the best system available for that language. Next steps consist in developing the Segmented Bushy Path to MDS, as well as to evaluate the methods not only for informativeness, but for other quality criteria, as coherence and cohesion. For such evaluation, it will also be important to incorporate in the methods a sentence ordering method, since the sentences are currently juxtaposed in a summary in the order they are selected. One of the sentence ordering strategies investigated in [14] might be used. Finally, it may also be worthy to explore other graph-based MDS strategies, as the ones proposed in [22].

Acknowledgements. The authors are grateful to FAPESP, CAPES and CNPq for supporting this work.

References

1. Afantenos, S.D., Doura, I., Kapellou, E., Karkaletsis, V.: Exploiting Cross-Document Relations for Multi-document Evolving Summarization. In: Proceedings of the 3rd Hellenic Conference on Artificial Intelligence, Samos Island, Greece, May 5-8, pp. 410–419 (2004)
2. Albert, R., Barabási, A.L.: Statistical mechanics of complex networks. Reviews of Modern Physics 74(1), 47–97 (2002)
3. Antiqueira, L.: Desenvolvimento de Técnicas Baseadas em Redes Complexas para Sumarização Extrativa de Textos. MSc Dissertation. Instituto de Ciências Matemáticas e de Computação, Universidade de São Paulo. São Carlos/SP, Brazil, p. 124 (March 2007)
4. Antiqueira, L., Oliveira Jr., O.N., Costa, L.F., Nunes, M.G.V.: A Complex Network Approach to Text Summarization. Information Sciences 179(5), 584–599 (2009)
5. Cardoso, P.C.F., Maziero, E.G., Castro Jorge, M.L.R., Seno, E.M.R., Di Felippo, A., Rino, L.H.M., Nunes, M.G.V., Pardo, T.A.S.: CSTNews - A Discourse-Annotated Corpus for Single and Multi-Document Summarization of News Texts in Brazilian Portuguese. In: Proceedings of the 3rd RST Brazilian Meeting, Cuiabá/MT, Brazil, October 26, pp. 88–105 (2011)
6. Cardoso, P.C.F., Pardo, T.A.S., Nunes, M.G.V.: Métodos para Sumarização Automática Multidocumento Usando Modelos Semântico-Discursivos. In: Proceedings of the 3rd RST Brazilian Meeting, Cuiabá/MT, Brazil, October 26, pp. 59–74 (2011)
7. Castro Jorge, M.L.R., Pardo, T.A.S.: Experiments with CST-based Multidocument Summarization. In: Proceedings of the ACL Workshop TextGraphs-5: Graph-based Methods for Natural Language Processing, Uppsala, Sweden, July 16, pp. 74–82 (2010)
8. Castro Jorge, M.L.R.: Sumarização automática multidocumento: seleção de conteúdo com base no Modelo CST (Cross-document Structure Theory). MSc Dissertation. Instituto de Ciências Matemáticas e de Computação, Universidade de São Paulo. São Carlos/SP, Brazil, p. 86 (April 2010)
9. Castro Jorge, M.L.R., Agostini, V., Pardo, T.A.S.: Multi-document Summarization Using Complex and Rich Features. In: Anais do VIII Encontro Nacional de Inteligência Artificial, Natal/RN, Brazil, July 19-22, pp. 1–12 (2011)
10. Castro Jorge, M.L.R., Pardo, T.A.S.: A Generative Approach for Multi-Document Summarization using the Noisy Channel Model. In: Proceedings of the 3rd RST Brazilian Meeting, Cuiabá/MT, Brazil, October 26, pp. 75–87 (2011)
11. Erkan, G., Radev, D.: LexRank: Graph-based Lexical Centrality as Salience in Text Summarization. Journal of Artificial Intelligence Research 22(1), 457–479 (2004)
12. Gantz, J., Reinsel, D.: Extracting Values from Chaos. IDC IView (June 2011)
13. Leite, D.S.: Um Estudo Comparativo de Modelos Baseados em Estatísticas Textuais, Grafos e Aprendizado de Máquina para Sumarização Automática de Textos em Português. MSc Dissertation. Departamento de Computação, Universidade Federal de São Carlos. São Carlos/SP, Brazil, p. 231 (December 2010)

14. Lima, J.B.P., Pardo, T.A.S.: Ordenação de Sentenças em Sumários Multidocumento: Uma Abordagem Utilizando Relações CST. In: Proceedings of the 2nd STIL Student Workshop on Information and Human Language Technology, Cuiabá/MT, Brazil, October 24-25, pp. 1–3 (2011)
15. Lin, C.Y., Hovy, E.: Automatic Evaluation of Summaries Using N-gram Co-occurrence Statistics. In: Proceedings of the Conference of the North American Chapter of the Association for Computational Linguistics on Human Language Technology, Edmonton, Canada, May 27 - June 1, pp. 71–78 (2003)
16. Louis, A., Joshi, A., Nenkova, A.: Discourse indicators for content selection in summarization. In: Proceedings of the 11th Annual Meeting of the Special Interest Group on Discourse and Dialog, Tokyo, Japan, September 24-25, pp. 147–156 (2010)
17. Mani, I.: Automatic Summarization. John Benjamins Publishing Co., Amsterdam (2001)
18. Mani, I., Bloedorn, E.: Summarizing Similarities and Differences Among Related Documents. Information Retrieval 1(1-2), 35–67 (1997)
19. Maziero, E.G., Castro Jorge, M.L.R., Pardo, T.A.S.: Identifying Multidocument Relations. In: Proceedings of the 7th International Workshop on Natural Language Processing and Cognitive Science, Funchal/Madeira, Portugal, June 8-12, pp. 60–69 (2010)
20. Maziero, E.G., Pardo, T.A.S.: Multi-Document Discourse Parsing Using Traditional and Hierarchical Machine Learning. In: Proceedings of the 8th Brazilian Symposium in Information and Human Language Technology, Cuiabá/MT, Brazil, October 24-26, pp. 1–10 (2011)
21. Mihalcea, R., Radev, D.: Graph-based Natural Language Processing and Information Retrieval. Cambridge University Press (2011)
22. Mihalcea, R., Tarau, P.: An Algorithm for Language Independent Single and Multiple Document Summarization. In: Proceedings of the 2nd International Joint Conference on Natural Language Processing, Jeju Island, Korea, October 11-13 (2005)
23. Pardo, T.A.S., Rino, L.H.M.: TeMário: Um Corpus para Sumarização Automática de Textos. Technical Report NILC-TR-03-09. Instituto de Ciências Matemáticas e de Computação, Universidade de São Paulo. São Carlos/SP, Brazil, p. 13 (October 2003)
24. Pardo, T.A.S., Rino, L.H.M., Nunes, M.G.V.: GistSumm: A Summarization Tool Based on a New Extractive Method. In: Proceedings of the 6th Workshop on Computational Processing of the Portuguese Language - Written and Spoken, Faro, Portugal, June 26-27, pp. 210–218 (2003)
25. Pardo, T.A.S.: GistSumm - GIST SUMMarizer: Extensões e Novas Funcionalidades. Technical Report NILC-TR-05-05. Instituto de Ciências Matemáticas e de Computação, Universidade de São Paulo. São Carlos/SP, Brazil, p. 8 (February 2005)
26. Radev, D.R.: A common theory of information fusion from multiple text sources, step one: Cross-document structure. In: Proceedings of the 1st ACL SIGDIAL Workshop on Discourse and Dialogue, Hong Kong, China, October 7-8 (2000)
27. Radev, D.R., Jung, H., Budzikowska, M.: Centroid-based summarization of multiple documents: sentence extraction, utility-based evaluation and user studies. In: Proceedings of the ANLP/NAACL Workshop on Automatic Summarization, Seattle, USA, April 30, pp. 21–30 (2000)
28. Radev, D.R., Blair-Goldensohn, S., Zhang, Z.: Experiments in single and multidocument summarization using MEAD. In: Proceedings of the 1st DUC Workshop on Text Summarization, New Orleans, USA, September 13-14 (2001)
29. Radev, D.R., Blair-Goldensohn, S., Zhang, Z., Raghavan, R.S.: NewsInEssence: A system for domain-independent, real-time news clustering and multi-document summarization. In: Proceedings of the 1st International Conference on Human Language Technology Research, San Diego, USA, March 18-21 (2001)

30. Salton, G.: Automatic text processing. Addison-Wesley Longman Publishing Co., Inc., Boston (1988)
31. Salton, G., Singhal, A., Mitra, M., Buckley, C.: Automatic Text Structuring And Summarization. Information Processing & Management 33(2), 193–207 (1997)
32. Wan, X.: An Exploration of Document Impact on Graph-Based Multi-Document Summarization. In: Proceedings of the Conference on Empirical Methods in Natural Language Processing, Waikiki, USA, October 25-27, pp. 755–762 (2008)
33. Watts, D.J., Strogatz, S.H.: Collective dynamics of 'small-world' networks. Nature 393, 440–442 (1998)
34. Zhang, Z., Blair-Goldensohn, S., Radev, D.R.: Towards CST-enhanced summarization. In: Proceedings of the 18th National Conference on Artificial Intelligence, Edmonton, Canada, July 28 - August 1, pp. 439–446 (2002)

Clustering and Categorization of Brazilian Portuguese Legal Documents

Luis Otávio de Colla Furquim* and Vera Lúcia Strube de Lima

Pontifícia Universidade Católica do Rio Grande do Sul,
Av Ipiranga 6681, Porto Alegre, Brazil
luisfurquim@gmail.com
vera.strube@pucrs.br
http://www3.pucrs.br/portal/page/portal/facinppg/ppgcc

Abstract. This study explores the use of machine learning in case law search in electronic trials. We clustered case law documents, automatically generating classes to a categorizer. These classes are used when a user uploads new documents to an electronic trial. We selected the algorithm TClus, created by Aggarwal, Gates and Yu, removing its document/group discarding features and adding a cluster division feature. We introduced a new paradigm "bag of terms and law references" instead of "bag of words" by generating attributes using a law domain thesaurus to detect legal terms and using regular expressions to detect law references. We clustered a case law corpus. The results were evaluated with the Relative Hardness Measure (RH) and the $\bar{\rho}$-Measure (RHO). The results were tested both with Wilcoxon's Signed-ranks Test and Count of Wins and Losses Test to determine their significance. The categorization results were evaluated by human specialists. We compared true/false positives against document similarity with the centroid, cluster size, quantity and type of the attributes in the centroids and cluster cohesion. The article also discusses attribute generation and its implications to the classification results.

Keywords: categorization, clustering, hard clustering, bag of terms and law references, document retrieval, Relative Hardness Measure, $\bar{\rho}$-Measure, Wilcoxon's Signed-ranks test, Count of Wins and Losses Test, law, judicial decisions, case law.

1 Introduction

Today, in Brazil, people spend too much time searching case law documents due to limitations on available tools:

1. **Search Scope:** The format of case law documents includes an abstract with keywords followed by a case report and the judge's decision. The available search tools limit the search to the abstract and the keywords. Due

* We thank HP that sponsored this research with a 24 month scholarship.

H. Caseli et al. (Eds.): PROPOR 2012, LNAI 7243, pp. 272–283, 2012.

to well known lack of precision when writing these sections, we commonly find incomplete abstracts and keyword lists. Sometimes these sections show keywords referring to themes not discussed in the subsequent text;
2. **Search Arguments:** Currently available search tools offer indexed search. Sometimes they allow the use of boolean operators. Although the search may be refined, many people are unused to boolean operators. Documents that lack one or more of the arguments are hard to be found, even when discussing themes similar to those in the documents returned by the query.

Since January 2010, Brazilian courts started to virtualize their trials, even without solving previous limitations of case law search. To deal with these problems, we propose to classify the petitions uploaded to the system using a taxonomy previously generated by clustering the case law documents. The query result is, thus, the list of case law documents in the class the petitions were classified. The attribute vectors used in the clustering phase are generated by the entire case law document, instead of only by the abstract and keyword section. A benefit of this approach is retrieving documents discussing related themes, even in absence of some of the query terms [19]. Other benefit is achieving more specific results even when the user does not have the skills to make complex boolean queries.

Section 2 brings a review on methods combining categorization and clustering. In Section 3 we present our proposal of categorization aided by clustering to search case law documents. Section 4 describes the methods of evaluation we used. The results obtained indicate that our algorithm outperforms the one in [1], successfully reducing attribute dimensionality. It also shows that groups with many documents, higher cohesion, higher dimensionality in the centroid or predominance of one type of attribute (term/law reference) lead to more true positives. Finally, we conclude (Section 5) commenting on the results.

2 Related Work

Research on document categorization rarely uses clustering techniques. For example, in [12] Portuguese legal documents are classified using SVM and a set of manually generated classes. Despite not being so popular, experiments combining the benefits of clustering techniques with those of categorization are not new. Methods, however, differ acording to their purpose. In [9,10,6] modified versions of the Expectation Maximization algorithm (EM) [4] are used to change bayesian networks by discovering hidden variables and inserting or removing graph edges, improving the performance of the algorithm. A later work [5] uses agglomerative clustering and bayesian model fusion techniques to discover hidden variables. Naïve Bayes parameters are determined in [26] using a pre-defined hierarchical topic mixture model and the EM algorithm. The term position in the hierarchy influences the probability of classification of a document in a given class. In [3], the assumption of a one-to-one relation between semi-supervised generated groups and classes from labeled data is rejected. Clustering is used to partition data so each partition achieves this one-to-one relationship. In [23] attribute vector disjunction is provided to a co-training algorithm using clustering to generate new attributes to the vectors. The original attributes, extracted from the text documents are used to train the first classifier. The cluster generated attributes are used to train the second classifier. A new categorization technique, proposed in [27], may be used when labeled data is scarce, e.g. fewer than 10 labeled documents in each class. To achieve this, all documents are clustered and then labels are applied according to the labeled documents in the same cluster. The clustering and the categorization algorithms alternate their iterations: the clustering labels data near the centroids and the classifier labels data with greater SVM margin. To solve SVM scalability issues, [18] reduced training data by clustering them and using the centroids as the new training data. A class hierarchy is generated in [14] using clustering. Documents can be then classified in any node, not just the leaves. A binary SVM classifier is used.

In [7] classes are generated using clustering. A one-to-one relationship between groups and classes is assumed. In order to face scalability issues, semisupervised clustering is used by attaching keywords to specific clusters (Keyword Based Clustering). The categorization just tested if the new documents were from the domain or not. Legal documents of the Administrative Supreme Court of Austria were used. We took inspiration from this technique in our proposal, as we have two domain thesauri available.

Aggarwal, Gates and Yu [1] propose a semisupervised hard clustering algorithm to generate classes based on a one-to-one relationship with the groups. These classes are then used to classify documents. The starting data in [1] is a labeled dataset. The initial groups are populated with the documents from the related class. The initial centroids' dimensionality is reduced pre-discarding infrequent attributes and those with higher Normalized Gini index [22]. Successive clustering iterations may change document/group association, join groups or discard small groups or documents too distant from any centroid. Each iteration also discards infrequent attributes from the centroids, reducing their dimensionality. The clustering phase ends when the centroids reach a minimum predefined dimensionality.

The categorizer in [1] is based on the similarity function already used in the clustering phase, comparing the unlabeled documents to the centroids of the groups that generated the classes. It employs an additional step, based on a modified version of the Chakrabarti *et al.* [2] algorithm, to improve precision when the document is too close to more than one centroid. In this step the algorithm recalculates the similarity to the nearest centroids using only those non null attributes of the centroids which are null in the other centroids.

3 Legal Document Categorization Using Semisupervised Clustering Generated Classes

This section presents our proposal for the use of machine learning methods to search case law documents. We classify petitions uploaded to electronic trials using cluster generated classes.

3.1 Clustering and Categorization Process

Case law databases do not include precise abstracts and keywords, so it is not possible to provide well labeled documents to train a classifier here. As in [1,7], we provide classes using a clustering algorithm. Given the magnitude and the constant growth of case law databases, it is not possible to figure out the exact number of classes. So, the clustering algorithm should not depend on previous configurations of the number of groups. The algorithm shown in [1], starts from an original predefined taxonomy, changes it and then presents a new taxonomy which is still related to the original one. It permits to initialize the groups based on the abstract and the keywords of the case law documents and cluster them according to the entire document, obtaining a new improved taxonomy. The chosen algorithm discards documents and groups. In order to find isolated previous cases, we decided to disable these features. We added a cluster division phase, which is not present in the original algorithm. Such new phase should bring us more refined clusters. Notice that the algorithm in [1] assumes a one-to-one relationship between groups and classes. Although [3] argues that this may be a wrong assumption and provides a solution to it, this solution demands trusted labeled data and, as already stated, this is not the case.

As shown in Figure 1, we propose a two phase process in which:

Phase "A": (a) **We preprocess** the j_i documents from the $\mathcal{J}$ case law corpus, as described in Section 3.2. From each j_i document we extract an attribute vector $j_i = (a_{i_0}, ..., a_{i_n})$, where a_{i_n} is the n_{th} attribute of j_i. Thus we get an attribute vector set $\mathbb{J} = \{j_i \in \mathbb{J}\}$; (b) **We select** from the j_i attributes, the first a_0 attributes of the document abstract as the C_{a_0} initial class label. We assume the S_{a_0} class

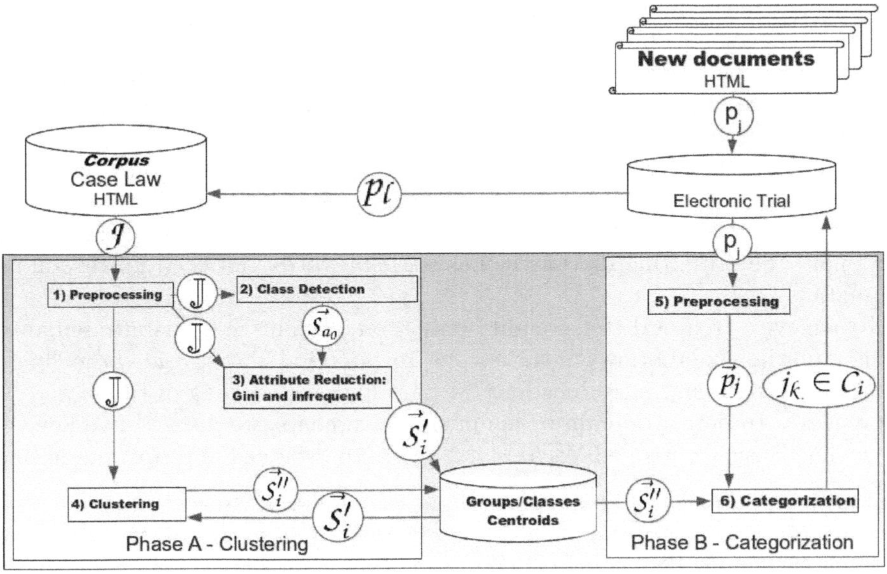

Fig. 1. Process of Clustering and Categorization

as the initial group; (c) **We discard** from $\mathbb{J}$ the a_i attributes with the higher Normalized Gini Index and those that appear in only one document; with the remaining a_i attributes in the $j_i \in \mathbb{J}$ documents, we compute the $\mathcal{S}'_i$ centroids; (d) **We group** the $\mathbb{J}$ documents, using the $\mathcal{S}'_i$ centroids as initial seed, changing the document/group relationship, obtaining new groups and new $\mathcal{S}''_i$ centroid set (Section 3.2).

Phase "B": (a) **We preprocess** the p_j documents uploaded to the $\mathcal{P}$ electronic trials, as described in Section 3.2. From each p_j document, we get $p_j = (a_{j_0}, ..., a_{j_n})$ attribute vector and a_{j_n} is the n_{th} attribute extracted from p_j. Thus we get $\mathbb{P} = \{p_j \in \mathbb{P}\}$ attribute vector set; (b) **We categorize** the p_j attribute vectors in one of the C_i classes defined by the $\mathcal{S}''_i$ groups generated in phase "A", using the algorithm in Section 3.2. The $j_k \in C_i$ case law document references from the group related to the class obtained in this categorization are shown in the p_j electronic trial.

As trials are judged, new p_ℓ documents are added to the case law corpus. So, new executions of the clustering phase ("A") are required.

3.2 Example of Use

Corpus, Dictionary and Thesaurus Construction. We built a corpus using case law documents downloaded from the 4th Region Brazilian Federal Tribunal[1]. We checked the fields "Acórdãos", "Súmulas" and "Decisões Monocráticas a partir de 08/2006" and selected the date range Jan-09-06 to May-27-09. We discarded documents covering multiple themes. The remaining 1,192 documents were single themed, thus satisfying the hard clustering assumption. A consequence of this strategy was that many discussion themes appeared only in one document, determining cluster fragmentation.

As in [17] with Finnish texts, in [11] with Portuguese texts and in [13] with Portuguese legal texts, we used the lemmatization method to normalize the words. To help the preprocessing phase, we merged 3 dictionaries: the Brazilian Portuguese Unitex Dictionary (Unitex-PB[2]) [20] and the Portuguese[3] and Latin[4] versions of *Wiktionary*, a project of Wikimedia Foundation[5]. The Latin version was necessary because of the frequent use of Latin expressions in case law documents. We added words present in the corpus but absent in this merged dictionary. The resulting dictionary has a list of words and their corresponding lemmata.

To achieve a greater dimensionality reduction, keeping the attribute semantics focused on the domain (law) of the corpus, inspired in [7], we avoided the "bag of words" paradigm and introduced a new paradigm called "bag of terms and law references". To detect domain terms in the documents, we merged two law thesauri. The Basic Controlled Vocabulary, VCB [16] with 3,400 terms, maintained

[1] `http://www.trf4.jus.br/trf4/jurisjud/pesquisa.php?tipo=2`

[2] `http://www.nilc.icmc.usp.br/nilc/projects/unitex-pb/web/dicionarios.html`

[3] `http://dumps.wikimedia.org/ptwiktionary/`

[4] `http://dumps.wikimedia.org/lawiktionary/`

[5] `http://www.wikimedia.org/`

by the Brazilian Federal Senate[6], and the Brazilian Federal Justice Thesaurus[7], TJF [24]. Both contain single and multiword terms organized in term sets in an hierarchical semantic structure. The merge process was semi-automatic: 1) we automatically merged two synsets if both had, at least, one common term; 2) a specialist manually indicated equivalence relations between synsets of the thesauri. So, the 8,840 terms from the VCB were merged to the 6,310 terms from the TJF and 13,354 synsets were obtained.

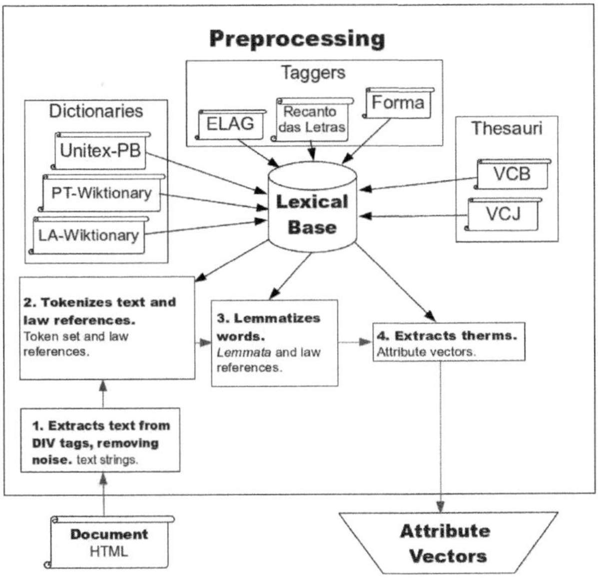

Fig. 2. Document Preprocessing

Document Preprocessing. In Figure 2, we show the preprocessing to obtain the attribute vectors from the documents. We developed a parser to preprocess the corpus' documents extracting words and law references. We first converted the downloaded HTML case law documents to plain text. We selected only the paragraphs with discussion content. After the regular expression token extraction, the parser identifies law references and normalizes this sequence to a single token, as shown in Table 1. When the law reference indicates just a section of a law, the parser emits multiple tokens, one for the whole law, and one for each rule-section referred in the text. This procedure prevents generating equal attributes on documents about distinct laws.

After obtaining the sequence of words and law references, we normalize the words with a lemmatizer that iterates between applying grammatical disambiguation rules based on unambiguous neighbors [20] and suffix based neighboring probabilistic disambiguation [11]. We then scan the lemmata extracting

[6] `http://www.senado.gov.br/`

[7] `http://www2.jf.jus.br/jspui/bitstream/1234/5509/3/tesauro_juridico.pdf`

the occurrences of terms and law references. These attributes are then weighted. Lower weights are given to term attributes near the root of the thesaurus and higher weights to those near the leaves. Law references are given weight 6 for any kind of law if there is a reference to the law section. Otherwise, it is given weight 1 if it is a constitution, 2 if it is a generic code or 4 for other types of law.

Table 1. Law References Normalization

"artigo 34 do decreto-lei n° 8192 de fevereiro de 1972" "article #34 of the Decree-Law #8,192 of february 1972"	"dl 8192/1972" "dl 8192/1972 art. 34"
"decreto-lei n° 8192 de fevereiro de 72, por meio dos artigos 34, 35 e 36" "Decree-Law #8,192 of february 72, by articles 34, 35 and 36"	"dl 8192/1972" "dl 8192/1972 art. 34" "dl 8192/1972 art. 35" "dl 8192/1972 art. 36"

Finally, attribute dimensionality is reduced by discarding the infrequent attributes and also those with the higher Normalized Gini Index.

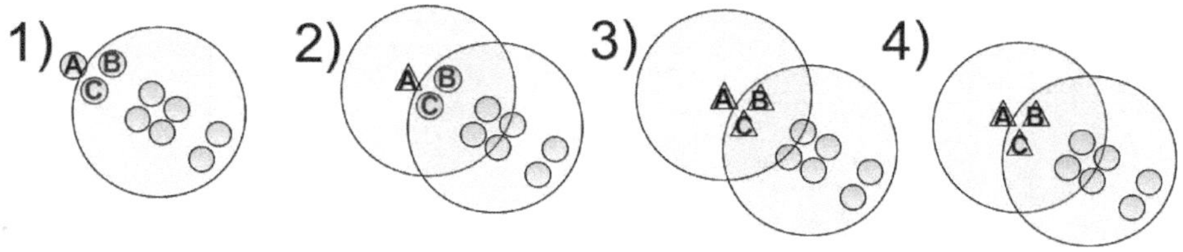

Fig. 3. Successive iterations may attract documents to the newly created cluster

Clustering and Categorization Algorithm. In addition to disabling the document/cluster discards, we introduced a cluster division feature, to achieve more refined clusters. Inspired by the TOD [8] algorithm, we changed the document attribution pass of the algorithm TClus [1]. The documents that should be discarded are turned into a single document cluster. Successive iterations may attract more documents from near clusters. In Figure 3, we show that if 1) the document "A" is beyond the similarity threshold of the cluster and 2) we create a new cluster with a centroid given by "A", 3) in the next iteration, the documents "B" and "C", although inside the current cluster threshold, are closer to the new centroid thus being attached to its cluster, and at last, 4) we recompute the centroids.

Categorization. The categorization follows the algorithm described in [1]. The similarity function used in the clustering phase finds the most similar centroid. If the similarity is higher than a certain threshold, the document is considered to belong to the class originated from the corresponding cluster. Otherwise, a second similarity function is computed between the document and the two most similar centroids, using only the non null attributes of the centroids which are null in the other centroid.

4 Evaluation

The case law corpus built for this study, was divided in 3 sets: 716 training documents, 238 test documents and 238 documents left to the operation phase. For the evaluation of this study, due to the cost of specialists manual evaluation, we worked with first 105 documents from the operation partition. In the preprocessing step we extracted 1,255,266 tokens from the training corpus, 1,753 per document, on average. The extraction of terms and law references generated 63 distinct attributes per document.

4.1 Clustering Evaluation

We ran both clustering algorithms: the original version from [1] as the baseline and the one disabling group and document discards and enabling group division. We set the similarity discard threshold to 40% and the group discard threshold to 4 documents. Values higher than these led to a 100% document/group discarding. Iterations started with 70 non null attributes in the centroids and stopped with 15. Starting with more than 70 attributes took too much time to process. Stopping between 15 and 7 attributes almost did not changed results at all and below 7 attributes the groups degenerated and different themes started to be merged in big groups. This process generated 453 clusters: 362 with 1 document, 50 with 2 documents, 11 with 3 documents, 10 with 4 documents and 20 with more than 4 documents. The largest cluster has 25 documents. The average is 1.98 documents per cluster.

After running clustering we computed the *Relative Hardness Measure* (RH) [21] and the $\bar{\rho}$-*Measure* (RHO) [25] internal quality indexes for each algorithm result. RH is a sum of intercluster similarities, so lower values are better. RHO normalizes the clusters' density with the total density, higher values are better. We choose these two indexes based on [15], that shows they are the most similar to what a human evaluation should indicate and are applicable to short documents. Even if the documents we use are significantly bigger than theirs, our bag of terms and law references model determines significant dimensionality reduction, resulting in similar sized attribute vectors. The baseline algorithm [1] scored RH=0.071 and RHO=1.01. Our algorithm scored RH=0.035, 50.19% better than the baseline and RHO=0.89, 12.2% poorer than the baseline. We established the null hypothesis $H0_{RH} : RH_{BL} = RH'$ and $H0_{\bar{\rho}} : \bar{\rho}_{BL} = \bar{\rho}'$. The Sign Test for the RH measure indicates that we can reject the null hypothesis $H0_{RH} : RH_{BL} = RH'$. We concluded that our algorithm version's performance is better than the original one regarding to the RH measure. On the other side, both Sign and Wilcoxon Tests indicate that we can not reject the null hypothesis $H0_{\bar{\rho}} : \bar{\rho}_{BL} = \bar{\rho}'$ and conclude that the original algorithm's performance is better than our version with respect to the $\bar{\rho}$-Measure.

4.2 Categorization Evaluation

The 105 documents were categorized in 74 of the 453 generated classes. For evaluation purpose, we split them into 3 subsets. The subsets A and B contained

50 documents and the subset C contained 5 documents. The results then underwent a human specialist evaluation. The first specialist evaluated the results from subsets A and C. The second specialist evaluated the results from subsets B and C. They were given the original texts from the classified documents and the original texts from the documents grouped in the cluster that originated the related classes. We then asked them to consider the document categorized as the description of a trial and the documents in the group that originated the class as a result of a case law search. A positive classification should be given, if the texts in the returned documents provided enough arguments for them to argue in a court when acting in trial represented by the classified document. For the C subset, their evaluation agreed in 4 of the 5 documents. The result with no agreement on classification was counted as a false positive. Figure 4 shows the true positive (TP) and false positive (FP) counts, vertical axis, splitted into similarity ranges, horizontal axis, starting with those classified with more than 40% of similarity with the centroid, followed by ranges of $\Delta 5\%$ until 5% minimum similarity. With higher similarities we have more chances of getting true positives. There was no false positive categorized with more than 30% of similarity.

We noticed that 26 of the 53 TP (almost 50%) occurred in classes with less than 4 documents, 18 (34%) of them were groups with just one document. We consider this a key indication that disabling the document/group discard in the algorithm is a right decision. We remark that more TP where found when one of the following conditions was present in the group that originated the selected class: 1) big group, 2) higher cohesion, 3) higher attribute dimensionality in its centroid, 4) predominance of attributes originated from terms and 5) predominance of attributes originated from law references. When none of these conditions were met, there were no predominance between TP/FP. We didn't find the simultaneous incidence of more than one of these conditions. We also noticed that conditions 4 and 5 were mutually excluding. When there is a predominance of attributes originated from terms, there are a too few law reference attributes, and vice-versa. This means that both kinds of attributes are useful, except when mixed. We finally remark that, comparing the incidence of TP in the presence of these conditions, the least effective was condition 4 and the most effective was condition 5. This reveals the importance of the law references in attribute

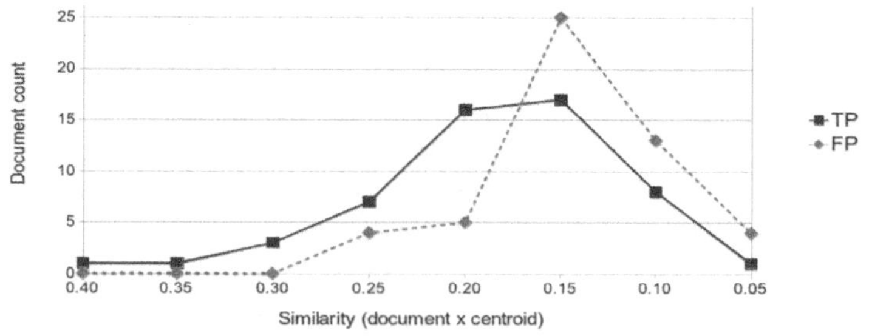

Fig. 4. Similarity between documents and the classes

generation. We compared the attributes of the original centroids with those in the final centroids and found that the law reference represented about 4.9% of the attributes in the initial centroids, rising to about 15.17% in the final centroids.

5 Conclusion

In this study we present a proposal to case law documents search using machine learning. Its architecture was organized in two phases, the first one clustering the case law documents to generate the classes and the other classifying lawyers' petitions uploaded to electronic trials, retrieving the documents from their respective classes. We conclude that this proposal can be successfully used mainly because of its good results and its performance viability. We add the following contributions: a) evolving of the algorithm proposed in [1] by disabling document/group discard and creating a group division method; b) creating a new paradigm we call "bag of terms and law references" which reduces dimensionality and bounds the vector features to the domain thesaurus and to related texts (law references), c) finding that TP may be raised when we have: 1) a big group, 2) higher cohesion, 3) higher dimensionality in its centroid, 4) predominance of attributes originated from terms and 5) predominance of attributes originated from law references.

We notice that attributes originated from law references played a key role in categorization results. In future work we'll focus on the semi-supervision limitations of the clustering algorithm by using only these kind of attributes when in clustering phase. The attributes originated from terms, however, should be kept in the centroids and used together with the law reference attributes in the categorization phases. We believe this could be useful when the petitions uploaded to the electronic trials lack some key law references raising their similarity with the centroids through the term generated attributes.

References

1. Aggarwal, C.C., Gates, S.C., Yu, P.S.: On using partial supervision for text categorization. IEEE Transactions on Knowledge and data Engineering, 245–255 (2004)
2. Chakrabarti, S., Dom, B., Agrawal, R., Raghavan, P.: Scalable feature selection, classification and signature generation for organizing large text databases into hierarchical topic taxonomies. The VLDB Journal The International Journal on Very Large Data Bases 7(3), 163–178 (1998)
3. Cong, G., Lee, W.S., Wu, H., Liu, B.: Semi-supervised Text Classification Using Partitioned EM. In: Lee, Y., Li, J., Whang, K.-Y., Lee, D. (eds.) DASFAA 2004. LNCS, vol. 2973, pp. 482–493. Springer, Heidelberg (2004)
4. Dempster, A.P., Laird, N.M., Rubin, D.B.: Maximum likelihood from incomplete data via the em algorithm. Journal of the Royal Statistical Society. Series B (Methodological), 1–38 (1977)
5. Elidan, G., Friedman, N.: Learning the dimensionality of hidden variables. In: Proceedings of the 17th Conference in Uncertainty in Artificial Intelligence, Citeseer, pp. 144–151 (2001)

6. Elidan, G., Lotner, N., Friedman, N., Koller, D.: Discovering hidden variables: A structure-based approach. Pattern Recognition Letters 21, 779–786 (2000)
7. Feinerer, I., Hornik, K.: Text mining of supreme administrative court jurisdictions. Data Analysis, Machine Learning and Applications, 569–576 (2008)
8. Friedman, M., Kandel, A.: Introduction to pattern recognition: Statistical, structural, neural and fuzzy logic approaches (1999)
9. Friedman, N.: Learning belief networks in the presence of missing values and hidden variables. In: ICML, pp. 125–133 (1997)
10. Friedman, N.: The bayesian structural em algorithm. In: Proc. UAI, Citeseer, vol. 98 (1998)
11. Gonzalez, M.: Termos e relacionamentos em evidência na recuperao de informação. Ph.D. thesis, Universidade Federal do Rio Grande do Sul, Porto Alegre, RS (2005)
12. Gonçalves, T., Quaresma, P.: A Preliminary Approach to the Multilabel Classification Problem of Portuguese Juridical Documents. In: Pires, F.M., Abreu, S.P. (eds.) EPIA 2003. LNCS (LNAI), vol. 2902, pp. 435–444. Springer, Heidelberg (2003)
13. Gonçalves, T., Quaresma, P.: Is linguistic information relevant for the classification of legal texts? In: ICAIL 2005: Proceedings of the 10th International Conference on Artificial Intelligence and Law, pp. 168–176. ACM, New York (2005)
14. Hao, P.Y., Chiang, J.H., Tu, Y.K.: Hierarchically svm classification based on support vector clustering method and its application to document categorization. Expert Systems with applications 33(3), 627–635 (2007)
15. Ingaramo, D., Pinto, D., Rosso, P., Errecalde, M.: Evaluation of internal validity measures in short-text corpora. Computational Linguistics and Intelligent Text Processing, 555–567 (2008)
16. Jaegger, F., et al.: Vocabulário controlado básico. Serviço de Gerência da Rede Virtual de Bibliotecas, Congresso Nacional, RVBI. Brasília, DF (junho 2007)
17. Korenius, T., Laurikkala, J., Järvelin, K., Juhola, M.: Stemming and lemmatization in the clustering of finnish text documents. In: Proceedings of the Thirteenth ACM International Conference on Information and Knowledge Management, pp. 625–633. ACM (2004)
18. Li, B., Chi, M., Fan, J., Xue, X.: Support cluster machine. In: Proceedings of the 24th International Conference on Machine Learning, pp. 505–512. ACM (2007)
19. Maarek, Y., Fagin, R., Ben-Shaul, I., Pelleg, D.: Ephemeral document clustering for web applications. Tech. Rep. RJ 10186, IBM Research (2000)
20. Muniz, M., Nunes, M.: A Construoção de Recursos Linguístico-computacionais para o Português do Brasil: o Projeto de Unitex-PB. Master's thesis, Universidade de So Paulo. Instituto de Ciências Matemáticas e de Computção, São Carlos, SP (2004)
21. Pinto, D., Rosso, P.: On the Relative Hardness of Clustering Corpora. In: Matoušek, V., Mautner, P. (eds.) TSD 2007. LNCS (LNAI), vol. 4629, pp. 155–161. Springer, Heidelberg (2007)
22. Porath, E.B., Gilboa, I.: Linear measures, the gini index, and the income-equality trade-off. Journal of Economic Theory 64, 443–467 (1994)
23. Raskutti, B., Ferrá, H., Kowalczyk, A.: Combining clustering and co-training to enhance text classification using unlabelled data. In: Proceedings of the Eighth ACM SIGKDD International Conference on Knowledge Discovery and Data Mining, pp. 620–625. ACM (2002)

24. Sordi, N., et. al.: Tesauro jurídico da justiça federal. Conselho da Justiça Federal (Fev 2007)
25. Stein, B., zu Eissen, S.M., Potthast, M.: Syntax versus semantics. In: 3rd International Workshop on Text-Based Information Retrieval (TIR 2006), Citeseer, p. 47 (2006)
26. Toutanova, K., Chen, F., Popat, K., Hofmann, T.: Text classification in a hierarchical mixture model for small training sets. In: Proceedings of the Tenth International Conference on Information and Knowledge Management, pp. 105–113. ACM (2001)
27. Zeng, H.J., Wang, X.H., Chen, Z., Lu, H., Ma, W.Y.: Cbc: Clustering based text classification requiring minimal labeled data. In: Third IEEE International Conference on Data Mining, ICDM 2003, pp. 443–450. IEEE (2003)

SIGA, a System to Manage Information Retrieval Evaluations

Luis Costa, Cristina Mota, and Diana Santos

Linguateca/FCCN and University of Oslo
luis.f.kosta@gmail.com, cmota@ist.utl.pt, d.s.m.santos@ilos.uio.no

Abstract. This paper provides an overview of the current version of SIGA, a system that supports the organization of information retrieval (IR) evaluations. SIGA was recently used in Págico, an evaluation contest where both automatic and human participants competed to find answers to 150 topics in the Portuguese Wikipedia, and we describe its new capabilities in this context as well as provide preliminary results from Págico.

Keywords: Information extraction, information retrieval, evaluation, question answering, usability, wikipedia.

1 Introduction

SIGA is a web-based management and evaluation system supporting the organization of Information Retrieval (IR) evaluations, distributed by Linguateca, and included in the GIRA package[1]. Its source code is open, so anyone can improve and extend it to the particular requirements of a specific IR evaluation.

The need for this computational environment arose during the organization of GikiCLEF [1,2], because there was a considerable number of people creating and assessing topics in geographically distinct sites, dealing with large amounts of data (the Wikipedia collections for the several languages involved and many systems' submissions). SIGA has a similar structure to other systems such as DIRECT [3] or the system used in INEX to back the evaluation [4], and supports multiple user roles for different tasks. Different choices and privileges are thus available, namely topic creation, run submission and validation, document pool generation, (cooperative) assessment, system scoring and display of results. Compared to these two systems, SIGA offers an important additional capability introduced in the context of GikiCLEF: the support for topic assessment overlap (several judgements for the same answer) and the semi-automation of the subsequent conflict resolution process.

More recently, in 2011, SIGA was adapted and extended to support the organization and participation in Págico[2] [5], an evaluation contest in information retrieval in Portuguese organized by Linguateca [6] whose goal was to evaluate systems aiming to find non-trivial answers to complex information needs in Portuguese, and is a follow-up of GikiCLEF that builds on Linguateca's previous experience but focuses on a specific

[1] http://www.linguateca.pt/GikiCLEF/GIRA/

[2] http://www.linguateca.pt/Pagico/

H. Caseli et al. (Eds.): PROPOR 2012, LNAI 7243, pp. 284–290, 2012.

cultural sphere (the Portuguese-speaking one) instead of cross-linguality or geographical subjects. Three main capabilities were added: automatic assessment of answers and justifications, an interface for human participants, and navigation inside a static version of Wikipedia. The current paper, although also describing SIGA, focuses on the new features required by this evaluation. For details about earlier versions and uses of SIGA, please check [1,2] or the GikiCLEF site.[3]

2 Technologies Used and Functionalities

SIGA is developed mainly in PHP and JavaScript, with data stored in a MySQL database. The choice of these technologies was driven by the following requirements:

Easy Installation. As users of the system would range from hard-core software developers to computer users with just basic knowledge, a web application was chosen, with no need for local installation.

Intuitive Interface. Catering for the broad spectrum of users, the interface should be satisfactory for the different types of users, especialy when human participants are concerned.

Ability to Deal with Large Amounts of Data, due to the large size of the document collections, of the results created by the participants and of the evaluation data created by the evaluators.

Topic Creation. SIGA supports the creation of topic sets for IR evaluations, helping topic managers look for answers in titles of Wikipedia documents included in the collections.

From the point of view of system evaluation, SIGA had a major shortcoming: it did not support adding justifications during topic creation, which entailed that the pre-assessment was only based on the comparison of answers, putting the burden of assessing the justifications on the human assessors. We improved SIGA so that it now allows the addition of justifications while creating the topics, so that the automatic pre-assessment can also be based on the justifications.

SIGA was also modified to allow a two level categorization of topics. The topic creators can associate one or more classes to each topic, and group those classes into major thematic subjects, which are then presented to the participants during the evaluation context proper.

Support for the Participation of Automatic Systems. The interface for system participation allows: (i) the download of different topic sets (evaluation, examples, testing, training), (ii) the validation and submission of runs, (iii) the inspection of the system scores and (iv) the comparison of results with the other systems. These tasks are somewhat similar to those performed by SAHARA [7], which provides a comparison between the scores of the new submitted runs and the runs officially submitted to HAREM [8].

[3] http://www.linguateca.pt/GikiCLEF/

Interface for Human Participation. Due to the considerable number of topics (three times more topics than in GikiCLEF), it was not expected that all participants would answer all questions. Therefore, aiming for a higher coverage of answered topics, the interface presented to each participant the topics in a different order.

Still, the participants had the option of altering the order in which they navigated in two ways, namely:

- Direct navigation to the particular topic they were interested in answering, when-viewing the list of all topics and the list of topics previously answered;
- Choice of the subject of the next topic.

The interface provides a keyword based search on a static version of the Wikipedia. Participants can use this functionality to find documents, which can then be selected either as answers to the current topic, or as justifications for a particular answer of the current topic. As for justifications, in addition to providing a list of justification documents, participants have the option of providing a textual description of how the list of documents constitutes a justification to the given answer, if they feel that just listing the documents still does not make the justification obvious.

The result of a search is an ordered list of documents whose titles match the keywords provided by the participants. Each document can be visualized, allowing the participant to decide whether it is an appropriate answer or justification. If this is the case, it should be simple to add it as answer or justification using the buttons on top of the page being visualized.

The current system logs the participants' actions in the background: visited topics, keyword searches, documents viewed and the answers and justifications given. This allows the study of the time used for each topic and answer, as well as investigate the strategies used by the participants to find answers and justifications.

Support for Assessment. SIGA provides extensive support in the assessment phase of IR evaluations. The first task consists in pooling all answers returned by the participants. Answers and justifications provided by automatic and human participants are pooled together and treated almost the same way in the assessment process. The only difference is that the human participants have the possibility of providing a textual explanation on how the justification documents support the answers, which is displayed to the assessors in the assessment interface.

The assessment interface of SIGA allows assessors are able to judge the candidate answers, and check the correctness of their justifications. Prior to this, there is an automatic process where answer and justification documents are marked as correct if they had already been listed as answers and justifications by the topic manager during the topic preparation period.

Assessment using SIGA is thus performed in three steps:

1. All answers and justifications provided by the participants which were listed as answers and justifications during the topic preparation period are automatically classified as correct.
2. The remaining answers are then assessed by the assessors, and classified either as Incorrect, Correct, or Dubious. If the answer is classified as Correct, assessors

must also indicate whether it is Justified or Not Justified (by looking at the answer document and possibly at the chain of justification documents).

3. Finally, for answers for which different assessments exist, conflict resolution is performed. This process allows the assessors to discuss and become aware of complications and/or mistakes or mistaken assumptions. The administrator can choose to question the diverging assessors, or decide in straightforward cases.

Scoring, and Display of Comparative Results. The final scores are automatically computed after the completion of the assessment task and made available to participants, who have access to several different measures (precision, pseudo-recall, originality, tolerant-precision, etc.[4]) and to the detailed assessment of their answers, which they can contest or comment upon. (See [9] for an overview of the assessment problems.)

3 Towards Better Portuguese IR Systems with SIGA

SIGA was used in Págico where both automatic and human participants competed to find (justified) answers to 150 topics in the Portuguese Wikipedia.

Altogether the (6 human and 2 automatic) participants in Págico submitted a total of 32,488 unique answers, see Tables 1 and 2, which still reflect preliminary numbers.

Table 1. Statistics about answers and justifications

	Topic owners	Participants
Auto-justified answers	635	31,714
Answers that include justification	72	774
(with one document)	(67)	(678)
(with more than one document)	(5)	(96)
Total	707	32,488

This version of SIGA allowed us to significantly enlarge the set of topic answers and justifications, given that human participants provide more reliable answers than automatic systems[5], hence resulting in a better evaluation contest.

The human participation brought in the challenging topic of non-topical factors in information access (see [10]). With this amount of data we have produced a large base of information for subsequent statistical processing of data in Portuguese, as opposed to the case of GikiCLEF, where most of the answers were in English, and very few were only found in Portuguese.

[4] Pseudo-recall is computed by considering the set of all correct answers jointly returned or prestored, as if they were all correct answers in Wikipedia. Originality measures the capacity of finding answers that others have not found. Tolerant-precision is more lenient in that it also accepts correct but not justified answers.

[5] Table 3 shows that human precision is always above .5, while automatic participation was not higher than .1173.

Table 2. Answers per participant: AA=assessed automatically; HA=assessed by humans; C+J=correct and justified; C+nJ=correct but not justified; nC=incorrect

Participant	Sent	AA C+J	AA C+nJ	AA C+nJ but HA C+J	AA nC but HA C+J	AA nC but HA C+nJ
ludIT	1387	263	135	18	683	22
GLNISTT	1016	180	64	1	419	52
Ângela Mota	157	46	0	0	42	3
João Miranda	101	25	26	4	29	3
Bruno Nascimento	34	19	1	0	4	2
RAPPORTAGICO	5184	226	7	0	355	38
RENOIR	45000	305	12	0	856	81

Table 3. Participant scores: score=C*C/N, where C=correct answers, N=number of answers given by participant

Participant	Run	Topics	Answers	Correct	Precision	Pseudo-recall	Score
ludIT	1	150	1387	1067	0.7693	0.5105	820.8284
GLNISTT	1	148	1016	660	0.6496	0.3158	428.7402
João Miranda	1	40	101	80	0.7921	0.0383	63.3663
Ângela Mota	1	50	157	88	0.5605	0.0421	49.3248
RAPPORTAGICO	3	116	1730	207	0.1197	0.0990	24.7682
RAPPORTAGICO	2	115	1736	202	0.1164	0.0967	23.5046
RAPPORTAGICO	1	114	1718	180	0.1048	0.0861	18.8591
Bruno Nascimento	1	18	34	24	0.7059	0.0115	16.9412
RENOIR	1	150	15000	437	0.0291	0.2091	12.7313
RENOIR	3	150	15000	399	0.0266	0.1909	10.6134
RENOIR	2	150	15000	330	0.0220	0.1579	7.2600

Table 4. Analysis of correct and justified answers: SHA=answers given both by systems and humans; HA=answers given only by humans; SA=answers given only by systems

	SHA	HA	SA	Total
Assessed automatically	133	254	49	436
Assessed by judges	195	886	302	1383
Total	328	1140	351	1819

Furthermore, by logging the human participants navigation through the topics while they are trying to find the answers and justifications, we can provide important information for IR system developers and interface designers that can help them to identify some of the strategies used by people when seeking information in Wikipedia, as the following figures illustrate.

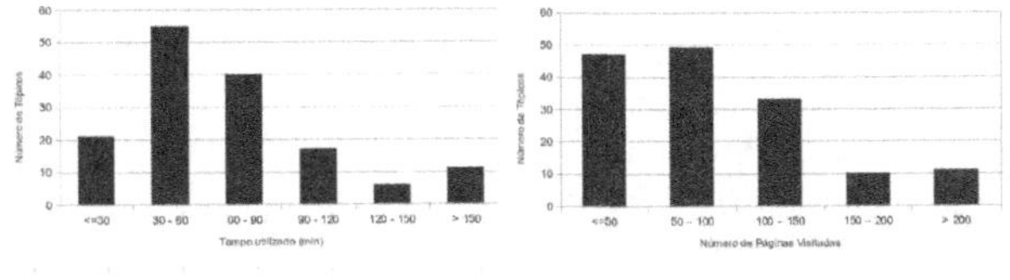

Fig. 1. A bird eye's view of human participation in Págico

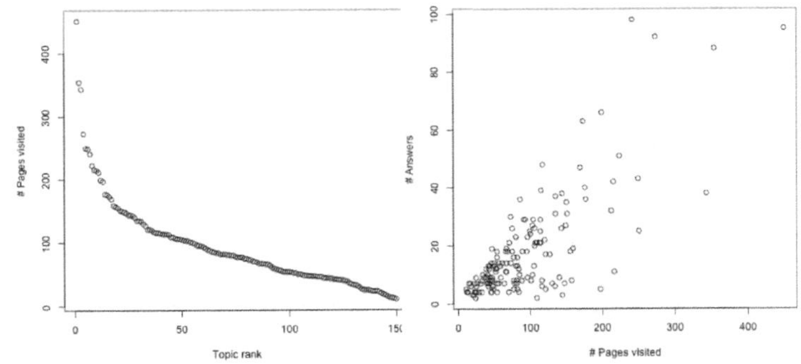

Fig. 2. Interplay between topics and number of different pages scrutinized

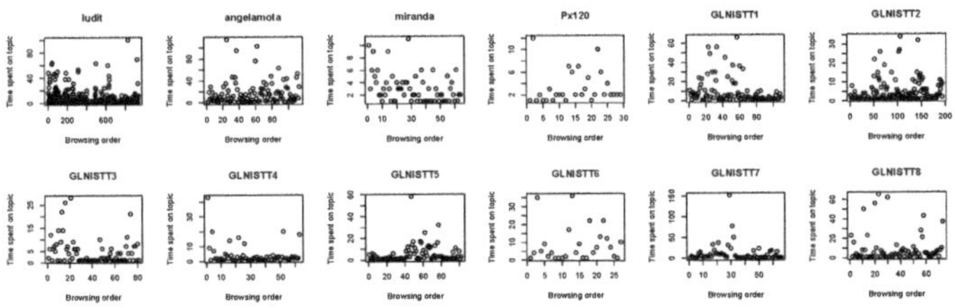

Fig. 3. Comparison between time spent on answers and browsing order

Acknowledgments. Linguateca has throughout the years been jointly funded by the Portuguese Government, the European Union (FEDER and FSE), UMIC, FCCN and FCT. Págico is also supported by the Universities of Oslo, PUC-Rio, and Coimbra. We are most grateful to the rest of the Págico team and to the Págico participants.

References

1. Santos, D., Cabral, L.M.: Summing GikiCLEF up: expectations and lessons learned. In: Peters, C., Nunzio, G.D., Kurimo, M., Mandl, T., Mostefa, D., Peñas, A., Roda, G., Borri, F., Nardi, A., Peters, C. (eds.) Multilingual Information Access Evaluation. Text Retrieval Experiments, vol. I, pp. 212–222. Springer (2009)
2. Santos, D., Cabral, L.M., Forascu, C., Forner, P., Gey, F., Lamm, K., Mandl, T., Osenova, P., Peñas, A., Rodrigo, Á., Schulz, J., Skalban, Y., Sang, E.T.K.: GikiCLEF: Crosscultural Issues in Multilingual Information Access. In: Calzolari, N., Choukri, K., Maegaard, B., Mariani, J., Odijk, J., Piperidis, S., Rosner, M., Tapias, D. (eds.) Proceedings of the 7th conference on International Language Resources and Evaluation (LREC 2010), Valletta, Malta, European Language Resources Association, ELRA (May 2010)
3. Dussin, M., Ferro, N.: Direct: applying the DIKW hierarchy to large-scale evaluation campaigns. In: Larsen, R.L., Paepcke, A., Borbinha, J.L., Naaman, M. (eds.) Proceedings of the 8th ACM/IEEE-CS Joint Conference on Digital Libraries, pp. 424–424. ACM, New York (2008)
4. Lalmas, M., Piwowarski, B.: INEX 2006 relevance assessment guide. In: INEX 2006 Workshop Pre-Proceedings, pp. 389–395 (2006)
5. Santos, D.: Porquê o Págico? Linguamática 4 (2012)
6. Santos, D.: Caminhos percorridos no mapa da portuguesificação: A Linguateca em perspectiva. Linguamática 1(1), 25–59 (2009)
7. Gonçalo Oliveira, H., Cardoso, N.: SAHARA: an online service for HAREM Named Entity Recognition Evaluation. In: The 7th Brazilian Symposium in Information and Human Language Technology, STIL 2009 (2009)
8. Mota, C., Santos, D. (eds.): Desafios na avaliação conjunta do reconhecimento de entidades mencionadas: O Segundo HAREM. Linguateca (2008)
9. Mota, C., Freitas, C., Costa, L., Rocha, P.: O que é uma resposta? notas de uns avaliadores estafados. Linguamática 4 (2012)
10. Karlgren, J.: Stylistic Experiments for Information Retrieval. PhD thesis, Stockholm University (2000)

E-commerce Market Analysis from a Graph-Based Product Classifier

Andréa Britto Mattos, Marcelo Van Kampen, Camila Carriço, André Ricardo Dias, and Alexandre Crivellaro

Instituto Brasileiro de Opinião Pública e Estatística (IBOPE), São Paulo, Brazil
{andrea.mattos,marcelo.kampen,camila.carrico,andre.dias, acrivellaro}@ibope.com

Abstract. This work describes a system for analyzing e-commerce markets, by tracking the consumers' habits in a set of webstores. In order to retrieve the types of products that are mostly purchased or viewed by a sample of consumers, we extract product titles displayed at the pages browsed by the monitored sample. Then, the collected titles are used as input to an automatic classification process that aims to assign each product to a suitable category and brand. We introduce a graph-based classifier that explores the relation between the categories and brands to be determined and is built from a supervised training set. We will show that the classifier was able to obtain good results, improving a Bayesian technique and allowing a further analysis of the considered market.

Keywords: Product classification, e-commerce analysis.

1 Introduction

The electronic commerce market has grown widely in the last decade affecting the habits of people from different demographic backgrounds and achieving a global reach [10]. Once the Internet represents a powerful sales channel, advertisements companies are increasingly turning their attention to the analysis of the consumers' preferences on the web, and recent works have focused on studying this behavior [4,8,5]. Additionally, webstores are interested in monitoring the consumers' purchasing decisions, so they can select the most appropriate products to be advertised or recommended on their home page.

Although webstores can monitor the activities that users are executing in their webpages, they are unable to determine their socioeconomic profiles or to know what is happening in other webstores. Therefore, this works proposes an approach to analyze how users interact with the e-commerce market, by monitoring a sample of consumers with known profiles and tracking their activities on a set of webstores. Our tests were applied considering the Brazilian market.

We use a specialized tool to inform us every browsed webpage from the monitored sample, and extract the product titles displayed in the entries related to webstores. Our goal is to analyze how many people have executed a certain activity considering these products (e.g. viewing in detail, purchasing, etc.). Due to

H. Caseli et al. (Eds.): PROPOR 2012, LNAI 7243, pp. 291–297, 2012.

the large products' variety and the lack of standardization of their descriptions in different stores, only queries related to categories and brands are allowed. According to this restriction, we introduce an automatic procedure for product classification. Figure 1 shows a flowchart of the proposed system.

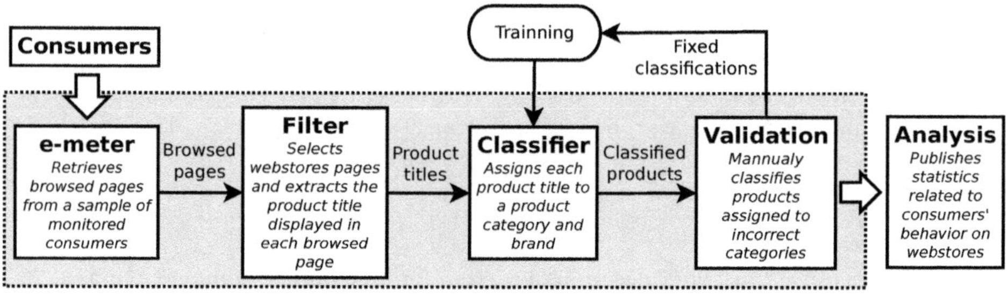

Fig. 1. Scheme for the proposed approach on e-commerce market analysis

1.1 Related Work on Product Classification

The goal of e-commerce product classification is to assign each product of an electronic catalog to a class of a given classification system, so that all items of the same class have similar attributes [11]. Automatic and semi-automatic approaches for this task have been addressed and the most common techniques are based on Bayesian classifiers, Support Vector Machines and K-Nearest neighbors [6,7,9,2,3]. Most studies focus on classifying structured product descriptions, e.g., inputs containing separate fields for name, description, brand etc. Our inputs are given by the captured content of different pages, therefore, we deal with unstructured titles that describe the full information related to each product.

Also, all the previous approaches are only concerned in assigning each product to a single category, while we are interested in obtaining mappings to pairwise classes of category and brand. This research is the first to deal with product classification in Portuguese and to offer an application for the proposed approach, describing a full system that is able to offer a comprehensive analysis about e-commerce market and consumers' online behavior.

This paper is organized as follows. In section 2 we describe the main characteristics of the system and section 3 focus on our product classification method. The results are shown in section 4 and the paper is concluded in section 5.

2 System Overview

The proposed system for analyzing e-commerce operations relies in the navigation log of a sample of consumers. We monitored 25 of the main Brazilian webstores and 2.000 households with an average number of three residents each, which gives us about 6.000 customers who agreed to provide their browsing data and demographic profiles for this research. We developed an application called e-meter, that generates reports of a user's browser navigation, when installed in his machine. Every two seconds, it gathers the currently open browser windows and tabs via Automation (Internet Explorer) or via Plug-ins (Firefox and Chrome) and evaluates each page's URL. If it matches against a list of webstores' URLs, the full source of the viewed page is retrieved and sent to our server.

The filtered sources from the browsed webstores are analyzed, using regular expressions that extract every product description displayed in these pages. Then, a category and brand are assigned to each description, and finally, all these information yields rankings that contain broad information about online shopping behavior (for example, reporting top product categories or webstores).

The product categories are defined according to the deepest level of an hierarchical market structure divided in *Class*, *Sector* and *Category*[1]. Our structure contains 23 classes, 195 sectors and 1113 categories, and was designed to encompass all product's types of the considered webstores. For product classification, we propose an automatic method based on supervised training, and include a validation interface for fixing incorrect results, that may be added to the training database in order to prevent the misclassification of similar entries.

3 Product Classification

This section describes the proposed method for assigning a product description to its most likely pair of brand and category. Table 1 shows some examples of collected entries[2] that must be classified and analyzed by the system.

Table 1. Examples retrieved from the monitored navigation (first column).

Product description	Category	Brand
SMARTPHONE LG GM750 WI-FI 3G GPS WINDOWS MOBILE MP3 BLUETOOTH 2GB	Smartphone	LG
NOTEBOOK VAIO S110 CORE I3 2.13GHZ 4GB 500GB 13.3" DVDRW PRETO WINDOWS 7 HOME PROFESSIONAL	Notebook	Sony
CÂMERA DIGITAL CYBER-SHOT DSC-W310 12MP LCD 2,7" PRETA SONY	Digital camera	Sony
GUARDA-ROUPA DITALIA, 3 PORTAS, 3 GAVETAS, BRANCO	Bedroom furniture	Ditalia
TÊNIS NIKE AIR MAX TURBULENCE+ 16	Tênis	Nike

The inputs are pre-processed according to some general rules and specific characteristics of products. For example, besides prepositions and conjunctions, we remove words that can occur in any type of products, like colors. Common units or acronyms (e.g. GB, MB, MP, HD, CM, etc.) are separated from the values that precedes them, and numbers are discarded. Also, common adjectives are standardized to the masculine gender and singular form, spellings that may appear in different forms are standardized (e.g. {*smart phone*, *smart fone*} → *smartphone*) and abbreviations are expanded (e.g. *cam* → *camera*).

[1] For example, the *Electronics* class can be refined in *Televisions* sector, which in turn can contain categories related to *Plasma Televisions*, *LCD Televisions*, and so forth.

[2] *Câmera, guarda-roupa, portas, gavetas, tênis, branco* and *preto(a)* [pt] stands for *camera, wardrobe, doors, drawers, tennis shoes, white* and *black*, respectively.

3.1 Bayesian Approach

Bayesian classifiers are commonly used in text classification, as mentioned in subsection 1.1. In [6,7], Kim *et. al.* proposed an weighted Bayesian classifier that assigns categories for structured products, based on the structure fields and the frequency in which words and categories were assigned in the training process. Words are weighted according to how discriminant they are for the classification.

Once we deal with unstructured titles and seek for a pairwise classification, the previous work could be adapted to our problem as follows. Given an item w to be classified, composed of n words, and sets of candidates categories C and candidate brands B, the probability that the item w belongs to category C_j and the probability that it belongs to brand B_k are computed as:

$$P(C_j|w = \{w_1, \ldots, w_n\}) = P(C_j) \times \prod_{i=0}^{n} \left(\frac{P(w_i|C_j)}{P(w_i)} \right)^{\frac{1}{N_C(w_i)}}$$

$$P(B_k|w = \{w_1, \ldots, w_n\}) = P(B_k) \times \prod_{i=0}^{n} \left(\frac{P(w_i|B_k)}{P(w_i)} \right)^{\frac{1}{N_B(w_i)}}$$

where, regarding the training step, $P(C_j)$ denotes the number of times category C_j appeared, $P(w_i|C_j)$ denotes the number of times w_i was associated with C_j and $P(w_i)$ denotes the number of times w_i appeared (brand case is analogous).

We denote by $N_C(w_i)$ the function that counts in how many different categories the word w_i has occurred. Likewise, $N_B(w_i)$ stores the number of different brands associated with w_i in the training set. Then, if a given word is associated to many different categories, it should not represent a discriminative attribute for classification and its weight must be small (brand case is analogous)[3].

3.2 Graph-Based Approach

The application of the previous approach to our problem has the limitation that category and brand are inferred independently. In order to explore the relation between them, we proposed a graph-based approach as an alternative for the Bayesian method. A graph-based method for text classification was introduced in [1], being successfully applied together with different approaches.

In our method, in order to classify an item $w = \{w_1, \ldots, w_n\}$, we build a directed graph $G(V, E)$ with vertices for all w_i and for every C_j and B_k that were assigned to any word of the item during the training step. Then, we add directed weighted edges from the brands vertices to the words vertices that are related in the training set. Also, we define edges from words to categories, and finally, from categories back to brands, as shown in the scheme of Figure 2.

Then, we evaluate which pair (B_m, C_m) yields the best classification result for w using the graph G. This is made by assigning costs to the graph's edges

[3] If w_i was not associated to C_j (or B_k), instead of assigning $P(w_i|C_j)$ to zero, the authors define a mismatching probability of 10^{-5}.

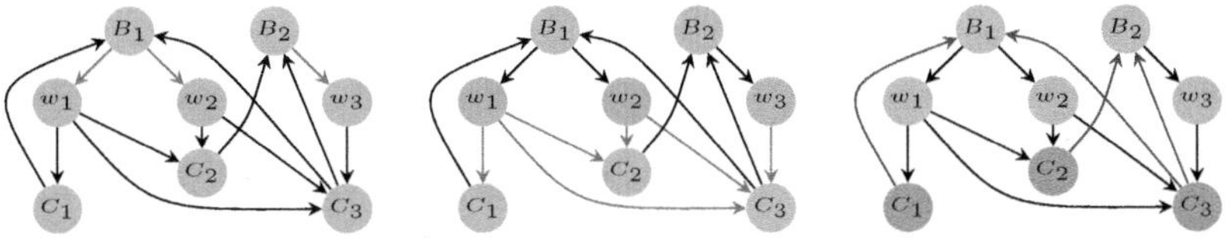

(a) $edge(B_k, w_i)$ (b) $edge(w_i, C_j)$ (c) $edge(C_j, B_k)$

Fig. 2. An example graph with two brands, three words and three categories

and computing maximum cost paths[4] between each combination of B_k and C_j. The edge costs are given according to the vertices that delimiters them:

$$edge(B_k, w_i) = \left(\frac{P(w_i|B_k)}{P(w_i)}\right)^{\frac{1}{deg^-(w_i)}} \times deg^+(B_k) \quad \text{(i)}$$

$$edge(w_i, C_j) = \left(\frac{P(w_i|C_j)}{P(w_i)}\right)^{\frac{1}{deg^+(w_i)}} \times deg^-(C_j) \quad \text{(ii)}$$

$$edge(C_j, B_k) = \left(\frac{P(C_j|B_k)}{P(C_j)}\right)^{\frac{1}{deg^+(C_j)}} \quad \text{(iii)}$$

were $P(C_j|B_k)$ denotes the number of times category C_j appeared in the training step associated to brand B_k, $deg^-(v)$ denotes the *indegree* of a vertex v and $deg^+(v)$ denotes its *outdegree*. Notice that $N_C(w_i)$ is equivalent to $deg^+(w_i)$, and so are $N_B(w_i)$ and $deg^-(w_i)$.

The proposed edge cost defines that words appearing in many categories (high $deg^+(w_i)$) have small weights for classification but also states that categories related to many words of the input (high $deg^-(C_j)$) must have higher weights (i). The same holds for brand classification, with high $deg^-(w_i)$ and $deg^+(B_k)$, respectively (ii). The relation between categories and brands is taken in account, and weighted according to number of connections within the graph (iii).

When computing all possible paths between candidates B_k and C_j, considering each w_i, we sum the costs of the edges that are part of each path, and also the cost related to $edge(C_j, B_k)$ (cost values are summed to avoid the definition of a mismatching probability value). In our work, we included the constraint to check whether B_k and C_j are adjacent in the graph, only considering paths when $edge(C_j, B_k)$ exists. Formally, our goal is to find the pair B_m and C_m such that:

$$(B_m, C_m) = \underset{C_j \in C, B_k \in B}{\arg\max}\ cost(B_k, C_j)$$

$$cost(B_k, C_j) = \begin{cases} edge(B_k, C_j) + \sum_{i=1}^{n}(edge(B_k, w_i) + edge(w_i, C_j)) & \textbf{if } \exists \text{ edges } (B_k, C_j), (B_k, w_i) \text{ and } (w_i, C_j) \\ 0 & \textbf{otherwise} \end{cases}$$

[4] A path is a sequence of vertices with edges connecting each vertex to the next one. In our work, the edges could have been defined in the opposite direction, once graph theory proves that the sum of the cost values for the paths would be the same.

4 Results

We used 118.000 products extracted from the monitored webstores that were manually classified, and performed 10 different tests. In each, we randomly extracted 8.000 items of the sample, and reclassified the products automatically, using the remaining 100.000 items as a training set. We evaluated the Bayesian approach of Kim *et. al.* and our graph-based method, as shown in Table 2.

Table 2. Mean error comparison between the two approaches in 10 different tests

	Average success rate (%)		**Error (%)**	
	Bayesian classifier	Graph-based classifier	Average	Maximum
Category	85.84	91.59	5.74	6.30
Brand	87.70	93.80	6.09	6.60
Both	80.80	89.51	8.71	9.28

The results show that the graph-based method was able to achieve better results, increasing individual brand and category classification accuracy in $\approx$ 6%, and joint accuracy in $\approx$ 8%. Considering these results, using the graph-based classifier, only $\approx$ 11% of the input items would have to be reclassified manually, while this rate would increase to $\approx$ 19% in the Bayesian approach.

Finally, Table 3 shows a ranking with the 10 most frequent categories and brands out of the 118.000 products, which represents the final step of our system.

Table 3. The most popular categories and brands regarding the monitored consumers

Top categories		Top brands	
1. Books	6. Tennis shoes	1. Samsung	6. Nokia
2. Movie DVDs	7. Notebooks	2. LG	7. Electrolux
3. Mobile phones	8. Music CDs	3. Nike	8. Brastemp
4. Wrist Watches	9. Blu-ray movies	4. Adidas	9. Technos
5. TV Series DVDs	10. Bedding	5. HP	10. Tramontina

5 Concluding Remarks

This paper proposed a novel approach for classifying products described as unstructured data, and shown how this task was used in a system for analyzing e-commerce market. The system is based on a sample of monitored consumers, whose navigation logs feed the classifier. Then, the classified products are used to formulate rankings that express the monitored shopping behavior.

Product descriptions are classified in pairs of brands and categories, using a graph that is built from a supervised training set. This is made by computing maximum weighted paths between each candidate pair, through the words from the description. The weights are defined according to the frequency in which brands, categories and words are related in the training step, and the edge connectivity given by the graph. Our graph-based method could improve the Bayesian approach and can be adapted to any type of pairwise classification.

References

1. Angelova, R., Weikum, G.: Graph-based text classification: learn from your neighbors. In: SIGIR 2006: Proceedings of the 29th Annual International ACM SIGIR Conference on Research and Development in Information Retrieval, pp. 485–492. ACM, New York (2006)
2. Bergamaschi, S., Guerra, F., Vincini, M.: Product classification integration for e-commerce. In: Proceedings of the 13th International Workshop on Database and Expert Systems Applications, DEXA 2002, pp. 861–867. IEEE Computer Society, Washington, DC (2002)
3. Bratko, A., Filipič, B.: Exploiting structural information for semi-structured document categorization. Inf. Process. Manage. 42, 679–694 (2006)
4. Cheema, A., Papatla, P.: Relative importance of online versus offline information for internet purchases: Product category and internet experience effects. Journal of Business Research 63(9-10), 979–985 (2010)
5. Kiang, M.Y., Ye, Q., Hao, Y., Chen, M., Li, Y.: A service-oriented analysis of online product classification methods. Decision Support Systems (2011)
6. Kim, Y.-G., Lee, T., Chun, J.H., Lee, S.-G.: Modified Naïve Bayes Classifier for E-Catalog Classification. In: Lee, J., Shim, J., Lee, S.-g., Bussler, C.J., Shim, S. (eds.) DEECS 2006. LNCS, vol. 4055, pp. 246–257. Springer, Heidelberg (2006)
7. Kim, Y.-G., Lee, T., Lee, S.-G., Park, J.-H.: Exploiting Attribute-Wise Distribution of Keywords and Category Dependent Attributes for E-Catalog Classification. In: Huang, D.-S., Wunsch II, D.C., Levine, D.S., Jo, K.-H. (eds.) ICIC 2008. LNCS, vol. 5226, pp. 985–992. Springer, Heidelberg (2008)
8. Korgaonkar, P., Becerra, E., O'Leary, B., Goldring, D.: Product classifications, consumer characteristics, and patronage preference for online auction. Journal of Retailing and Consumer Services (March 2010)
9. Omelayenko, D.K., Ding, Y., Korotkiy, M., Omelayenko, B., Kartseva, V., Zykov, V., Klein, M., Schulten, E., Fensel, D.: Goldenbullet: Automated classification of product data in e-commerce. In: Proceedings of Business Information Systems Conference BIS 2002 (2002)
10. Poong, Y., Zaman, K.-U., Talha, M.: E-commerce today and tomorrow: a truly generalized and active framework for the definition of electronic commerce. In: Proceedings of the 8th International Conference on Electronic Commerce: The New E-commerce: Innovations for Conquering Current Barriers, Obstacles and Limitations to Conducting Successful Business on the Internet, ICEC 2006, pp. 553–557. ACM, New York (2006)
11. Volker, J.L., Schmitz, V., Dorloff, F.D.: A modeling approach for product classification systems. In: Proc. of the 13th International Workshop on Database and Expert Systems Applications (DEXA 2002), Aix-en-Provence, pp. 868–874 (2002)

A Description Logic for InferenceNet.Br

Wellington Franco, Thiago Alves, Henrique Viana, and João Alcântara

Departamento de Computação, Universidade Federal do Ceará, P.O. Box 12166, Fortaleza, CE, Brasil 60455-760
{jwellingtonfranco,thiagoalves,henriqueviana,jnando}@lia.ufc.br

Abstract. The InferenceNet.Br is a new linguistic resource for Portuguese language knowledge bases, which follows some principles of the WordNet, ConceptNet and FrameNet. Unlike these other linguistic resources, InferenceNet.Br allows representation of more expressive relationships between concepts, permitting to characterize premises or conclusions of these relationships. However, the integration between InferenceNet.Br with other linguistic resources is not possible directly. In order to settle this problem, in this paper we present a mapping of the knowledge bases of InferenceNet.Br to the Description Logic *DL-Lite*$_{\mathcal{A}}$, which is a fragment of OWL. As we know, OWL is a standard language designed to facilitate machine interpretability of Web content, through explicitly representation of the meaning of concepts in vocabularies and relationships between them. By doing this, we provide a connection between InferenceNet.Br and resources as WordNet and ConceptNet, since there are some efforts to link them to OWL.

Keywords: Linguistic Resource, InferenceNet.Br, Description Logic, OWL.

1 Introduction

Natural Language Processing (NLP) is a field of computer science and linguistics concerned with technics to construct computational systems which treat different levels of meanings and uses of natural languages. Roughly speaking, NLP systems are conceived to enable computers to manipulate linguistic signs reasonably in order to take decisions, extract and retrieve information, summarize texts, and solve other tasks involving the understanding of sentences and texts in natural language.

In Mitkov [21], it was stated that the reasoners of current NLP systems employ predominantly syntactic approaches, facing difficulties in capturing certain kind of know- ledge, mainly because of the lack of linguistic resources that support a complete understanding of concepts and sentences. The usual reasoners define taxonomies or ontologies through objects of determined domain, classes of objects and relationships between objects, but do not represent their practical content. For instance, when we take into consideration the word *chair*, we can either conclude that it is made of a specific material, as wood for example, or we can conclude that it has some practical utilities, as a seat or even a weapon.

An alternative applied to improve the quality of inferences in NLP systems is to resort to lexical-semantic base systems, whose aim is to add contextual elements that enhance the ability of these systems. The three principal bases used in semantic systems and applications of NLP are the WordNet [19], the FrameNet [6] and the ConceptNet [15]. WordNet provides lexical-semantic resources for the Portuguese language;

H. Caseli et al. (Eds.): PROPOR 2012, LNAI 7243, pp. 298–305, 2012.

however it expresses only basic semantic hierarchy relationships (hyperonymy and hyponymy), inclusions (holonymy and meronymy), equivalences (synonymy) or oppositions (antonymy), and it is not available for large-scale use. FrameNet is based on the theory of frames proposed by Minsky [20]. A frame is a conceptual hierarchical structure defining a situation, object or event by participants and their relationships. ConceptNet is a commonsense knowledge base in Portuguese language, which creates a net of relationships between concepts, such that these relationships may express causality, functionality, location, time, etc.

The Semantic Inferencialism Model (SIM) [22,23] is another lexical-semantic base with good results. It consists of a new model for treatment of pragmatic-semantic level of natural languages, adding new information in relationships and allowing richer inferences. SIM comprises four components: Conceptual Base, Sentence Patterns Base, Rule Base for Practical Reasoning and an algorithm to deduce information from these knowledge bases. To represent Conceptual and Sentence Patterns Bases it is used the Portuguese linguistic resource InferenceNet.Br [24] and the Rule Base for Practical Reasoning is represented as a knowledge base in logic programming [16].

ConceptNet, the InferencentNet.Br is a commonsense knowledge base which can represent many different types of knowledge, as relationships between concepts, expressed as simple phrases of natural language. This resource can be accessed by the portal www.inferencenet.org, where its Conceptual Base contains around 180.000 relationships between concepts and 700.000 relationships of pragmatical content, and the Sentence Patterns Base contains 5963 sentence patterns through 1061 relationships.

However, the InferenceNet.Br is not directly related to any other resources known, as WordNet, FrameNet and ConceptNet. Following the works [12,14], which converts WordNet and ConceptNet databases in Web Ontology Language (OWL) [18], the standard language to represent ontologies in the Semantic Web, in this paper we will translate InferenceNet.Br into a Description Logic namely $DL\text{-}Lite_{\mathcal{A}}$ [7]: a logic of the *DL-Lite* family that can be expressed in OWL-Lite [13,18] and that is expressive enough to represent InferenceNet.Br knowledge bases. By doing this translation, we can establish a connection between InferenceNet.Br with databases of WordNet and ConceptNet.

This paper is organized as follows: Section 2 describes the general aspects of InferenceNet.Br and explains the component and main characteristics of the resource. In Section 3 we describe the Description Logic $DL\text{-}Lite_{\mathcal{A}}$ and we show how to represent the content of Conceptual and Sentence Patterns Bases using this logic. Finally, we conclude the paper in Section 4.

2 InferenceNet.Br

One motivation for introducing a new linguistic resource is the lack of linguistic resources with large scale semantic knowledge for the Portuguese language. Lexical-semantic bases in Portuguese as WordNet.Pt [10], WordNet.Br [9], TeP 2.0 [17], PAPEL [11], FrameNet Brasil [25] or OMCS-Br [1] are restricted to specific domains, i.e., they depend of a text corpus, a dictionary or a thesaurus [26]. Another motivation is the non-existence of a linguistic resource with inferentialist semantic knowledge, either for Portuguese or other languages [22,23]. InferenceNet.Br contains two knowledge bases included in the SIM (Semantic Inferentialism Model): the Conceptual Base and the Sentence Patterns Base. Thus, SIM permits to realize inferences through three knowledge bases (Conceptual Base + Sentence Patterns Base + Rule Base for Practical Reasoning), an input text generated by a parser and an inference algorithm.

2.1 Conceptual Base

The Conceptual base is represented by a set of relations R_C, which is denoted by a tuple $R_C =$ (relationship, c_i, c_j, type), where the relationship denotes a relation between two concepts: c_i and c_j. A concept can represent simple or composite words from the noun, verb, adjective, or adverb classes, e.g., *crime* (crime), *morte* (death), *viver* (to live), *prova de inglês* (english test) or *anteontem* (day before yesterday). Words which belong to the preposition, pronoun or conjunction classes do not have semantic value, and therefore are not expressed by concepts. Table 1 shows the types of relationships presented in the Conceptual Base, whose relationships were inherited from the ConceptNet knowledge bases. These relationships may express physical, functional or even causal characteristics of the concepts. Each relationship defines uniquely a pre-condition or post-condition of origin of the concept. Pre-conditions or premises of use are what gives someone the right to use the concept and what could exclude such right, serving as premises for utterances and reasoning. Post-conditions or conclusions of use are what follows or what are the consequences of using the concept, which let one know what someone is committed to by using a particular concept, serving as conclusions from the utterance and as premises for future utterances and reasoning. We use type = Pre or Pos, to denote pre-condition or post-condition. For instance, a partial vision of the relations of the concept *crime* can be:

1. *(capazDe, crime, ter vítima, Pre).*
2. *(efeitoDe, crime, culpa, Pos).*
3. *(usadoPara, crime, vingança, Pre).*

1. (capableOf,crime,to have victim, Pre).
2. (effectOf, crime, guilt, Pos).
3. (usedFor, crime, vengeance, Pre).

Table 1. Types of relationships expressed by ConceptNet

Category	Type of relationship (ConceptNet)	Type of inferential relation (InferenceNet.Br)
Things	PropertyOf, PartOf, MadeOf, IsA, DefinedAs	Pre-condition
Events	LastSubEventOf, PreRequirementEventOf, FirstSubEventOf, SubEventOf	Post-condition
Causal	EffectOf, DesirousEffectOf	Post-condition
Affective	DesireOf, MotivationOf	Pre-condition
Functional	UsedFor, CapableOfReceivingAction	Post-condition
Agents	CapableOf	Pre-condition
Spatial	LocationOf	Pre-condition

2.2 Sentence Patterns Base

The Sentence Patterns base is represented by a set of relations R_P, which is denoted by a tuple $R_P =$ (isA, t_p, c_i, type), where t_p can be a noun phrase (NP) or a prepositional phrase (PP) of a sentence, c_i is a concept and type = Pre or Pos. Note that in the Sentence Patterns Base, there are only relationships of the kind *isA* (in Portuguese: *éUm*). For instance, a partial vision of the relations of sentence $p =$ *X ser assassinado por Y* (*X* to be killed by *Y*) can be (where sn(p) and sp(p) are respectively a noun phrase (NP) and prepositional phrase (PP) of p, i.e., sn(p) = *X* and sp(p) = *Y*):

1. *(éUm, sn(p), pessoa, Pre).*
2. *(éUm, sn(p), vítima, Pos).*
3. *(éUm, sp(p), assassino, Pos).*

1. (isA, sn(p), person, Pre).
2. (isA, sn(p), victim, Pos).
3. (isA, sp(p), assassin, Pos).

Unlike the Conceptual Base, the inferential content of the relationship *isA* is not defined only as a pre-condition. The sentence patterns permit either create pre-conditions (generated from the the Conceptual Base), as well post-conditions, generated according to the circumstance expressed by the adverbial complement of main verbs and auxiliary verbs of a phrase [24].

3 The *DL-Lite*$_{\mathcal{A}}$ Logic

3.1 Syntax and Semantics

The Semantic Web [2,5] aims for machine-understandable Web resources, whose information can be shared and processed both by automated tools, such as search engines, and by human users. To make sure that the resources will be understandable, one needs ontologies in which the knowledge is described. Basically, an ontology is a collection of definitions of concepts.

Description Logics (DLs) [4] are a family of knowledge representation languages that can be used to represent ontologies of an application domain in a structured and formally well-understood way. Besides, these logics are more expressive than propositional logic and have a decision problem more efficient than first order logic. In general, they have a good relation between expressivity and complexity (most of them are decidable).

In this section we describe *DL-Lite*$_{\mathcal{A}}$ [7], a logic of the family of Description Logics *DL-Lite* [3], which will represent the content of the knowledge bases in InferenceNet.Br. *DL-Lite* are logics with low expressivity and good results of complexity. In general, these logics have a polynomial-time algorithm for its satisfiability problem [3]. This DL was chosen as the base of our work because its expressivity is sufficient to represent knowledge bases of InferenceNet.Br. Indeed, *DL-Lite*$_{\mathcal{A}}$ is more expressive than InferenceNet.Br, but a good result obtained is that this logic maintains the polynomial-time algorithm for its satisfiability problem [8].

The logic *DL-Lite*$_{\mathcal{A}}$ works with three components: individuals, concepts and roles. An individual represents an element of the universe of discourse (domain). Concept represent unary predicates related to individuals. For instance, let *john* be an individual and *Person* be a concept name; we can say that *john* is an instance of *Person*, denoting that John is a person. In this work, we will employ concept names to express words of the noun, adjective, article or adverb classes. Roles represent binary predicates between individuals. If *john* and *mary* are individuals and *IsMarried* is a role name, by *IsMarried(john, mary)*, we mean that the individual *john* is married to the individual *mary*. Elementary descriptions as *Person* and *IsMarried* are called, respectively, *atomic concepts* and *atomic roles*. In *DL-Lite*$_{\mathcal{A}}$, general descriptions can be inductively built from them with concept constructors as follows, where B (basic concept) and L (light concept) are concepts, A is an atomic concept, R is an atomic role and R^- is inverse of an atomic role:

$$B \longrightarrow A \mid \bot \mid \exists R \mid \exists R^-$$
$$L \longrightarrow B \mid \neg B \mid \exists R.L \mid \exists R^-.L$$

The basic concept B may denote an atomic concept (A), which can be a noun, adjective, article or adverb, e.g., *Crime, Person, Death, Assassin*; or a bottom concept ($\bot$) which denotes the empty set; or an unqualified existential concept ($\exists R$). Indeed, this last concept can be used to represent the noun phrases of a sentence pattern with a verbal phrase (R), e.g., the concept $\exists$*KilledBy*, denoting someone that was killed by something/someone. The inverse unqualified existential concept ($\exists R^-$) represents the prepositional phrase of a sentence pattern, e.g., the concept $\exists$*KilledBy*$^-$, which denotes someone that killed someone. The light concept L may denote a basic concept B or a complement of basic a concept ($\neg B$), as example $\neg$*Man*, which denotes the concept *Woman*; or a qualified existential concept ($\exists R.L$), which can be used to represent the noun phrase of a sentence pattern with a verbal phrase + prepositional phrase of a sentence. For example, $\exists$*Kill.Man* denotes someone that killed a man. Finally, for the inverse qualified existential concept ($\exists R^-.L$), we can have the concept $\exists$*Kill*$^-$*.Man* as a example, which denotes someone that was killed by man.

The semantics of *DL-Lite*$_{\mathcal{A}}$ concepts is given by an interpretation $I = (\Delta^I, \cdot^I)$, where the domain Δ^I is a non-empty set of elements and $\cdot^I$ is a mapping function defined by: each individual a is mapped to $a \in \Delta^I$; each atomic concept A is mapped to $A^I \subseteq \Delta^I$; each atomic role R is mapped to $R^I \subseteq \Delta^I \times \Delta^I$. Complex concepts can be inductively interpreted as follows: $\bot^I = \emptyset$; $\neg B^I = \Delta^I \backslash B^I$; $(\exists R)^I = \{a \in \Delta^I \mid \exists b \in \Delta^I, (a,b) \in R^I\}$; $(\exists R^-)^I = \{a \in \Delta^I \mid \exists b \in \Delta^I, (b,a) \in R^I\}$; $(\exists R.L)^I = \{a \in \Delta^I \mid \exists b \in \Delta^I, (a,b) \in R^I \wedge b \in L^I\}$; $(\exists R^-.L)^I = \{a \in \Delta^I \mid \exists b \in \Delta^I, (b,a) \in R^I \wedge b \in L^I\}$.

The knowledge representation on Description Logics can be divided into two components: the TBox and the ABox. TBox introduces the *terminology*, i.e., the vocabulary of an application domain, while the ABox contains assertions about individuals in terms of this vocabulary. The vocabulary consists of concepts and roles. The TBox can be used to assign names to complex concept and role descriptions. Syntactically, the terminology is represented by a finite set of *terminological axioms*, which have the form $B \sqsubseteq L$. Axioms of these kinds are called inclusions. The inclusion axiom $B \sqsubseteq L$ means that each individual of B is an individual of L. For instance, we can define *Crime* $\sqsubseteq$ *ViolaçãoDeLei* (*Crime* $\sqsubseteq$ *LawViolation*), denoting that every crime is a law violation; or we can define *Crime* $\sqsubseteq$ $\exists$*motivaçãoDe.Vingança* (*Crime* $\sqsubseteq$ $\exists$*motivationOf.Vengeance*), to denote that every crime is a motivation of vengeance; or

Assassino $\sqsubseteq \exists$*AssassinadoPor*$^-$ (*Assassin* $\sqsubseteq \exists$*KilledBy*$^-$), denoting that every assassin is an individual that killed someone.

An ABox (Assertional Box) consists of a finite set of *assertion axioms* of the form $L(a)$, $R(a, b)$ and $R^-(a, b)$, where a and b are individuals. These assertions denote that an individual a is an instance of the concept L and that the pair of individuals (a, b) is an instance of R or R^-, respectively. For instance, the assertions *Assassino(john), AssassinadoPor(mary, john)* (*Assassin(john), KilledBy(mary, john)*) denotes that John is an assassin, that he killed Mary. The semantics of inclusion and assertion axioms is given by: $B \sqsubseteq L$ iff $B^I \subseteq L^I$; $L(a)$ iff $a \in L^I$; $R(a, b)$ iff $(a, b) \in R^I$; $R^-(a, b)$ iff $(b, a) \in R^I$.

3.2 Mapping InferenceNet.Br to *DL-Lite*$_{\mathcal{A}}$

In this section, we will show that we can convert any knowledge base in InferenceNet.Br into *DL-Lite*$_{\mathcal{A}}$, and consequently we can reason about InferenceNet.Br using some Semantic Web tools, specifically those one using the OWL-Lite language. We will show how the tuples of InferenceNet.Br can be translated into inclusion axioms in *DL-Lite*$_{\mathcal{A}}$. We regard that we are only considering knowledge bases in *DL-Lite*$_{\mathcal{A}}$ with an empty ABox, since the InferenceNet.Br represents only terminologies. First we will show that all tuples of the Sentence Patterns Base correspond to inclusion axioms in the TBox:

- For all tuple of the form (*éUm*, sn(p), c, type) in the Sentence Patterns Base, we can transform it in an axiom inclusion in which
  - If type = Pos, then the axiom inclusion will be of the form $\exists$sv(p) $\sqsubseteq A$, where sv denotes the verbal phrase (VP) of a sentence in Portuguese and A is the atomic concept created from the concept c;
  - If type = Pre, then the axiom inclusion will be of the form $A \sqsubseteq \exists$sv(p).

Suppose the following tuple *(éUm, sn(X ser assassinar por Y), vítima, Pos)* in the Sentence Patterns Base . The translated axiom inclusion will be of the form $\exists$sv(*X* ser assassinar por *Y*) $\sqsubseteq$ *Vítima* which is equivalent to the axiom $\exists$*serAssassinarPor* $\sqsubseteq$ *Vítima*.

- For all tuple of the form (éUm, sp(p), c, type) in the Sentence Patterns Base
  - If type = Pos, then the axiom inclusion will be of the form $\exists$sv(p)$^-$ $\sqsubseteq A$;
  - If type = Pre, then the axiom inclusion will be of the form $A \sqsubseteq \exists$sv(p)$^-$.

Let *(éUm, sp(X ser assassinar por Y), assassino, Pos)* be a tuple in the sentence Patterns Base. The translated axiom will be $\exists$*serAssassinarPor*$^-$ $\sqsubseteq$ *Assassino*. Finally, we will show the mapping from the tuples of Conceptual Base to inclusion axioms in *DL-Lite*$_{\mathcal{A}}$.

- For all tuples of the form (r, c_1, c_2, type) of the Conceptual Base, we can transform them into inclusions in $\mathcal{T}$ of the form $A_1 \sqsubseteq \exists r.A_2$, where r is a relationship name and A_1 and A_2 are atomic concepts created from the concepts c_1 and c_2, respectively.

Suppose that we have the tuple *(usadoPara, crime, vingança, Pre)* in the Conceptual Base; thus the axiom inclusion in $\mathcal{T}$ will be *Crime* $\sqsubseteq \exists$*usadoPara.Vingança*.

4 Conclusion

In this paper, we used the Description Logic $DL\text{-}Lite_{\mathcal{A}}$ to represent the knowledge bases of the web resource InferenceNet.Br. We showed a mapping of InferenceNet.Br to $DL\text{-}Lite_{\mathcal{A}}$, which enables the integration of InferenceNet.Br with Semantic Web tools, in special, tools based on the OWL-Lite language. Consequently we can make inferences using some related Description Logics reasoners.

As a future work, we can consider to investigate a mapping for the case in which the ABox is non-empty. In this case, we are treating not only with the InferenceNet.Br resource, but with the whole SIM (Semantic Inferentialism Model), where the ABox can be generated by an input text and a morphosyntactical parser. Another line of research is to extend the expressivity of InferenceNet.Br or SIM and consequently extend the logic introduced in this paper in order to represent more expressive concepts and relationships between concepts. This extension will enable us to express more complex concepts rather than only the atomic ones or even to represent other kinds of relationships, e.g., similarities between concepts.

References

1. Anacleto, J., Lieberman, H., Tsutsumi, M., Neris, V., Carvalho, A., Espinosa, J., Godoi, M., Zem-Mascarenhas, S.: Can Common Sense Uncover Cultural Differences in Computer Applications? Artificial Intelligence in Theory and Practice, 1–10 (2006)
2. Antoniou, G., van Harmelen, F.: A Semantic Web Primer (Cooperative Information Systems). The MIT Press (April 2004)
3. Artale, A., Calvanese, D., Kontchakov, R., Zakharyaschev, M.: The DL-Lite Family and Relations. J. Artif. Int. Res. 36, 1–69 (2009)
4. Baader, F., Calvanese, D., McGuinness, D.L., Nardi, D., Patel-Schneider, P.F.: The Description Logic Handbook: Theory, Implementation, and Applications. Cambridge University Press, New York (2003)
5. Baader, F., Horrocks, I., Sattler, U.: Description Logics as Ontology Languages for the Semantic Web. In: Hutter, D., Stephan, W. (eds.) Mechanizing Mathematical Reasoning. LNCS (LNAI), vol. 2605, pp. 228–248. Springer, Heidelberg (2005)
6. Baker, C.F., Fillmore, C.J., Lowe, J.B.: The Berkeley Framenet Project. In: Proceedings of the 36th Annual Meeting of the Association for Computational Linguistics and 17th International Conference on Computational Linguistics, vol. 1, pp. 86–90. Association for Computational Linguistics (1998)
7. Botoeva, E., Calvanese, D., Rodriguez-Muro, M.: Expressive Approximations in *DL-Lite* Ontologies. In: Dicheva, D., Dochev, D. (eds.) AIMSA 2010. LNCS, vol. 6304, pp. 21–31. Springer, Heidelberg (2010)
8. Calvanese, D., De Giacomo, G., Lembo, D., Lenzerini, M., Poggi, A., Rodriguez-Muro, M., Rosati, R.: Ontologies and Databases: The DL-Lite Approach. In: Reasoning Web, pp. 255–356 (2009)
9. Dias da Silva, B.C., Di Felippo, A., Hasegawa, R.: Methods and Tools for Encoding the WordNet.Br Sentences, Concept Glosses, and Conceptual-Semantic Relations. In: Vieira, R., Quaresma, P., Nunes, M.d.G.V., Mamede, N.J., Oliveira, C., Dias, M.C. (eds.) PROPOR 2006. LNCS (LNAI), vol. 3960, pp. 120–130. Springer, Heidelberg (2006)
10. Marrafa, P., et al.: WordNet.Pt - Uma Rede Léxico-conceptual do português on-line. In: XXI Encontro da Associação Portuguesa de Linguística, Porto, Portugal (2005)

11. Gonçalo Oliveira, H., Santos, D., Gomes, P., Seco, N.: PAPEL: A lexical ontology for Portuguese. In: Quot; Workshop on Language Resources for Teaching and Research Faculdade de Letras da Universidade do Porto, April 23 (2008)
12. Grassi, M., Piazza, F.: Towards an RDF encoding of conceptNet. In: Liu, D. (ed.) ISNN 2011, Part III. LNCS, vol. 6677, pp. 558–565. Springer, Heidelberg (2011)
13. Horrocks, I., Patel-Schneider, P.F., van Harmelen, F.: From $\mathcal{SHIQ}$ and RDF to OWL: The making of a web ontology language. J. of Web Semantics 1(1), 7–26 (2003)
14. Huang, X.X., Zhou, C.L.: An OWL-based WordNet Lexical Ontology. Journal of Zhejiang University - Science A 8(6), 864–870 (2007)
15. Liu, H., Singh, P.: ConceptNet: A Practical Commonsense Reasoning Toolkit. BT Technology Journal 22(4), 211–226 (2004)
16. Lloyd, J.W.: Foundations of Logic Programming, 2nd edn. Springer-Verlag New York, Inc., Secaucus (1993)
17. Maziero, E.G., Pardo, T.A.S., Di Felippo, A., Dias-da Silva, B.C.: A Base de Dados Lexical e a Interface Web do TeP 2.0: Thesaurus Eletrônico para o Português do Brasil. In: Companion Proceedings of the XIV Brazilian Symposium on Multimedia and the Web, pp. 390–392. ACM (2008)
18. McGuinness, D.L., Van Harmelen, F., et al.: OWL Web Ontology Language Overview. W3C Recommendation 10, 2004–03 (2004)
19. Miller, G.A.: WordNet: a Lexical Database for English. Communications of the ACM 38(11), 39–41 (1995)
20. Minsky, M.: A Framework for Representing Knowledge (1974)
21. Mitkov, R.: The Oxford Handbook of Computational Linguistics. Computational Linguistics 30(1) (2005)
22. Pinheiro, V.: SIM: Um Modelo Semântico Inferencialista para Expressão e Raciocínio em Sistemas de Linguagem Natural. Phd Thesis, Universidade Federal do Ceará (2010)
23. Pinheiro, V., Pequeno, T., Furtado, V., Assunção, T., Freitas, E.: SIM - Um Modelo Semântico-Inferencialista para Sistemas de Linguagem Natural. In: Companion Proceedings of the XIV Brazilian Symposium on Multimedia and the Web, pp. 353–358. ACM (2008)
24. Pinheiro, V., Pequeno, T., Furtado, V., Franco, W.: InferenceNet.Br: Expression of Inferentialist Semantic Content of the Portuguese Language. In: Pardo, T.A.S., Branco, A., Klautau, A., Vieira, R., de Lima, V.L.S. (eds.) PROPOR 2010. LNCS, vol. 6001, pp. 90–99. Springer, Heidelberg (2010)
25. Salomão, M.M.M.: FrameNet Brasil: Um Trabalho em Progresso. Calidoscópio (2009)
26. Santos, D., Barreiro, A., Freitas, C., Oliveira, H.G., Medeiros, J.C., Costa, L., Silva, R., Gomes, P.: Relações Semânticas em Português: comparando o TeP, o MWN.PT, o Port4NooJ e o PAPEL. In: Brito, A.M., Silva, F., Veloso, J., Fiéis, A. (eds.) Textos seleccionados. XXV Encontro Nacional da Associação Portuguesa de Linguística, APL 2010, pp. 681–700 (2010)

Real-Time MRI for Portuguese
Database, Methods and Applications

António Teixeira[1], Paula Martins[2], Catarina Oliveira[2], Carlos Ferreira[3], Augusto Silva[1], and Ryan Shosted[4]

[1] DETI/IEETA, University of Aveiro, 3810 193 Aveiro, Portugal
{ajst,augusto.silva}@ua.pt
[2] ESSUA/IEETA, University of Aveiro, 3810 193 Aveiro, Portugal
{pmartins,coliveira}@ua.pt
[3] Institute of Biomedical Research in Light and Image, Faculty of Medicine, University of Coimbra, 3030 190 Coimbra, Portugal
c_dferreira@ua.pt
[4] Dep. of Linguistics, University of Illinois at Urbana-Champaign, Illinois, USA
rshosted@illinois.edu

Abstract. In this paper, we present a database of synchronized audio and Real-Time Magnetic Resonance Imaging (RT-MRI) in order to study dynamic aspects of the production of European Portuguese (EP) sounds. Currently, data have been acquired from one native speaker of European Portuguese. The speech corpus was primarily designed to investigate nasal vowels in a wide range of phonological contexts, but also includes examples of other EP sounds. The RT-MRI protocol developed for the acquisition of the data is detailed. Midsagittal and oblique images were acquired with a frame rate of 14 frames/s, resulting in a temporal resolution of 72 ms. Different image processing tools (automatic and semi-automatic) applied for inspection and analysis of the data are described. We demonstrate the potential of this database and processing techniques with some illustrative examples of Portuguese nasal vowels, taps, trills and laterals.

Keywords: Real-Time MRI, European Portuguese, speech dynamics, nasal vowels, laterals, taps, trills.

1 Introduction

In the field of human speech production, real-time magnetic resonance imaging (RT-MRI) is one of the most promising techniques. It provides dynamic information of the entire vocal tract (instead of a few points across time such as in EMA or X-Ray microbeam) with reasonable spatial and temporal resolution in a non-invasive manner. These advantages make it suitable for the observation and quantification of articulatory movements, as well as visualization of the vocal tract shape.

H. Caseli et al. (Eds.): PROPOR 2012, LNAI 7243, pp. 306–317, 2012.

The acquisition of real-time images of the vocal tract has benefited greatly from recent improvements in MRI technology. Refinements include ultra fast sequences (gradient and spin eco), high performance gradients, high field scanners, parallel imaging, new k-space sampling schemes (e.g., radial and spiral filling techniques) and extraordinary improvements in coil technology [9,8].

However, analysis of high-dimensional image data and extraction of relevant articulatory information (i.e. location of constriction events) from a vast stream of MR images are complex tasks [21,3,11]. To deal with the image processing problem, many solutions have been devised [2,4,20,11,21].

Recently, several studies have been carried out, using the latest processing tools, to study syllable structure effects on velum-oral coordination [6]; to analyse vocal tract shape in English sibilant fricatives [5]; to study emotional speech production [12]; to automatically identify constriction location targets [21]; and to model articulatory data using a data-driven machine learning framework [3].

This paper focuses on the description of a database of synchronized audio and RT-MRI recently acquired by our team to study dynamic aspects of speech production in European Portuguese (EP), such as 1) velum lowering and coordination of oral-nasal gestures for EP nasal vowels; 2) tongue tip motion during the production of the alveolar tap and trill; and 3) timing patterns that distinguish initial and final /l/. Preliminary efforts toward the development of image processing tools for data analysis will also be outlined, together with some illustrative examples of relevant features that can be extracted.

One of the most important distinctive characteristics of Portuguese is the dynamic pattern of nasality [10], which is typically incremental over the vowel [22]. Previous articulatory studies based on EMA data [18,17] showed that the velum gesture is somewhat delayed with respect to the tongue body gesture, inducing an oral vowel onset. Furthermore, a final consonantal segment can appear, depending on the degree of overlap among adjacent gestures [23]. Teixeira, in [24], highlighted the important role of velar dynamics in the perception of synthesized nasal vowels. The novel RT-MRI database will be a rich new source of information for the study of velum movement, intergestural timing, lingual articulation, velopharyngeal opening, etc., which require data from dynamic productions.

Detailed RT-MRI is also of great interest to examine the mechanisms involved in the production of EP tap and trill, for which the available data are very scarce.

Furthermore, this RT-MRI database can actively contribute to clarify the controversial question of the effects of syllabic position on the articulatory properties of EP /l/ in a way complementary to other acoustic [1] and articulatory techniques already used, such as EMA [19] and static MRI [13,16].

The paper is organized as follows: section 2 details methods of MR image acquisition and describes the tools used for data analysis; section 3 provides some preliminary results; finally in section 4 we briefly discuss our results and summarize directions for future work.

2 Method

2.1 Corpus

The corpus was essentially designed to study nasal sounds, with the main purpose of characterizing gestural dynamics (e.g., velum, lips). In addition, this corpus was chosen to allow for comparisons with available EMA data [17].

The first stimulus set consisted of nonsense words containing the five EP nasal vowels ([6~], [e~], [i~], [o~], [u~], in SAMPA) in three prosodic conditions: word-initial, word-internal and word-final (e.g. [6~p6], [p6~p6], [p6~]). The nasal vowels were flanked by the bilabial stop or the labiodental fricative. A subset of these items were embedded in the carrier sentence, "Diz...após" (Say...after). The second stimulus set included the five nasal monophthongs and the seven oral vowels ([E], [e], [i], [a], [O], [o], [u]), to be pronounced in isolation by the speaker. Another set of stimuli included word-internal nasal consonants ([n], [m], [J]) next to the vowels [i, a, u] (e.g [ama], [ana], [aJa]. Finally, a few real words containing the lateral [l] and the EP rhotics were also acquired, in order to prepare future MRI acquisitions. [l] was produced in: (i) syllable-initial position (e.g. [lak6] "hairspray") (ii) syllable-final position (e.g. [sal] "salt"), and (iii) intervocalically (e.g. [sal6] "living-room"). The phonological context surrounding the tap and the trill was kept constant ([ka4u] "expensive" vs [kaR\u] "car"). The trill was made by the speaker either at the uvular [R\] place or with the tip of the tongue [r]. Each stimulus set was repeated 3-4 times, according with the pre-defined duration of the scan.

2.2 MRI Acquisition

The MRI experiment was carried out in a Magnetic Resonance Imaging Unit at Coimbra (Institute of Biomedical Research in Light and Image). The images were acquired on an unmodified 3.0 T MR scanner (Magneton Tim Trio, Siemens, Erlanger, Germany) equipped with high performance gradients (Gmax $= 45mT/m$, rise time $= 0.2s$, slew Rate $= 200\ T/m/s$, and FOV $= 50\ cm$). A custom 12-channel head and 4-channel neck phased-array coils were used for data acquisition. A body coil was used for radio frequency (RF) transmission. Parallel imaging (GRAPPA 2) together with magnetic field gradients operating at FAST mode were used to speed up the acquisition. An MRI screening form and informed consent was obtained before the study to comply with security and ethics rules. The subject lay supine in the MR scanner, while producing the stimuli and wore headphones to protect the ears from the noise. The imaging protocol was an evolution of a pilot study conducted by [14].

After localization images, a T1 W 2D-midsagittal MRI slice of the vocal tract was obtained, using an Ultra-Fast RF-spoiled Gradient Echo (GE) pulse sequence (Single-Shot TurboFLASH), with a slice thickness of 8 mm and the following parameters: TR/TE/FA $= 72ms/1.02ms/5^{\circ}$, Bandwidth $= 1395$ *Hz/pixel*, FOV(mm^2)$=$ 210 x 210, reconstruction matrix of (128 x 128) elements with 50 % phase resolution, in-plane resolution (mm ^{2}) $=$ 3,3 x 1,6, yielding a frame rate of 14 images/second. The acquisition of each item took about 5 seconds, resulting in 75 midsagittal images. An example of 5 frames is shown in Fig. 1.

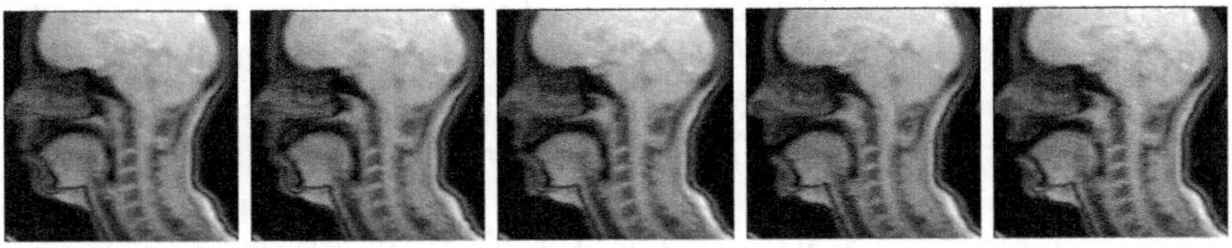

Fig. 1. Example of 5 midsagittal MRI images (frames 22 to 26) obtained during the beginning of the production of [6˜p6]

For selected items of the corpus, a coronal oblique slice was also taken at the velum, with the main purpose of obtaining information about the velum port area. (This allows for the calculation of a ratio between nasal port opening and oral cavity.) One of the sagittal slices was used to better determine the orientation of the oblique slice. Figure 2 (leftmost image) illustrates the orientation used to obtain the coronal-oblique slice.

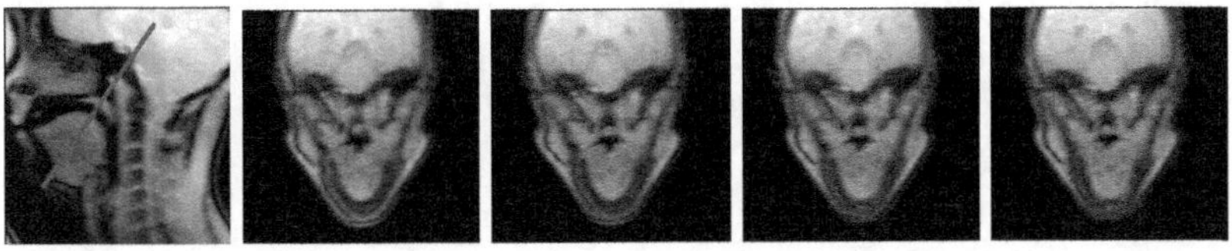

Fig. 2. Midsagittal image that provides a reference for proper orientation of the coronal oblique slice (at left). Sample of four coronal oblique frames obtained during the production of the nasal vowel [6˜], in the word [6˜p6] (at right).

In addition, two simultaneous coronal slices (one at the velum and another at lips) were also acquired. The acquisition of more than one slice was possible with a decrease of the temporal resolution. The compromise included a decrease to 7 frames/s, i.e. a temporal resolution of about 140 ms.

Because the subject is a member of the research team (the third author), she was already familiar with the stimuli. Nevertheless, before the acquisition, a small training session was undertaken to adjust the microphone and communication procedures. Each stimulus set was prompted orally by one of the experimenters over the intercom system (e.g., "Say ampa [6˜p6, p6˜p6, p6˜]"). The speaker was instructed to speak at a natural rate and volume. The recording time was about 90 minutes, including acquisition of all items of the corpus and some experiments in order to prepare for future acquisitions.

2.3 Audio Recordings

Audio was recorded simultaneously with the RT-images inside the MR scanner, at a sampling rate of 16000 Hz, using a fiberoptic microphone (Optoacoustics

FOMRI III Dual Channel MRI microphone, Or Yehuda, Israel). The microphone was fixed on the head coil, with the protective popscreen placed directly against the speaker's mouth, according to the manufacturer's instructions.

A computer running OptiMRI software (version 3.1), located in the adjacent MRI control room, recorded the dual-channel microphone outputs, the filtered speech processed by DSP and up to 3 TTL pulses, all of them synchronized with high accuracy (FOMRI III user manual).

In our experiment, a TTL pulse generated from the MRI scanner allowed the synchronization between MRI images and speech from the speaker.

To accomplish that, the sequence code of the TurboFLASH was customized to generate a TTL pulse every 72 ms. The sequence of the events was as follows: 1) the speaker was instructed by the operator about the sequence of words that might be produced, the experimenter started recording with the microphone, the sequence was launched (a TTL pulse was generated), the speaker waited 2 seconds after starting to hear the sequence noise and then started the production.

2.4 Speaker

The speaker was a 33-year-old female with no history of hearing or speech disorders. She is a native speaker of EP from the center of the country with phonetic and linguistic knowledge. She has participated in previous MR experiments.

2.5 Image Processing - Estimating Articulators Trajectory and Areas

Even using the most advanced techniques, processing RT-MRI data is still a challenging task. The quality of the images is poor due to constraints in acquisition time. To achieve high frame rates, while still capturing the main features of the moving vocal tract, requires a compromise between temporal and spatial resolution. As a consequence the images are of low resolution and too noisy (frame sizes of 64x64 or 128x128 pixels are typically used). Moreover, the high dimensionality of the acquired data further complicates the data analysis [11]. Several semi-automatic and automatic approaches have been used in recent years ([2,4,20]. More recently, a rapid automatic extraction method has been proposed, to determine constriction location targets and to estimate articulatory trajectories [21,11]. The method directly uses pixel intensity values obtained from selected regions of the vocal tract. Motivated by the results reported by [21,11], we replicated the method to explore our database. All the procedures have been performed in Matlab, using code written by the first author. The theoretical foundations of the method are detailed in [11,21].

To describe the various steps of the imaging processing technique we will refer to Fig. 3. The example uses the frames obtained from the sequence [6˜p6], [p6˜p6], [p6˜]. The 75 frames were interpolated (linear interpolation), resulting in 150 frames. To start the process, the operator manually selected one or more regions of interest (ROIs).

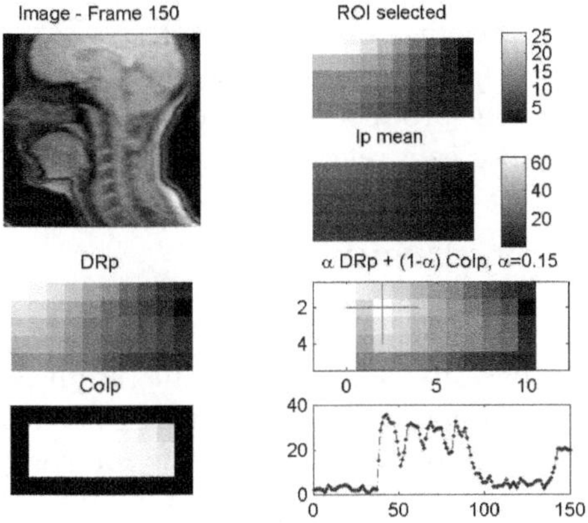

Fig. 3. Steps of the Pixel Intensity based method used to automatically estimate articulatory trajectories. The images correspond to the sequence of [6~p6], [p6~p6], [p6~].

In the example provided in Fig. 3 (top, leftmost image) the ROI was placed between the lips. After that, the algorithm calculated the mean Intensity value (Ip) of all the pixels, in the selected region, within a 4-pixel neighborhood. Dynamic range (DRp) was also calculated over the frames. Correlations (CoIp) between one pixel and the neighboring pixels were established within an 8-pixel neighborhood.

The point where the weighted DRp and CoIP are maximized was chosen (using $\alpha = 0.15$); Intensity variation observed in that point is used to determine constriction location. In the example, four peaks are displayed, corresponding to the bilabial closure for [p].

A semi-automatic segmentation technique (Live-wire [7]) was used in order to obtain the areas of the nasal and oral cavities, in the oblique views. The process was established using a customized, Live-wire-based pipeline, implemented in MevisLab [15].

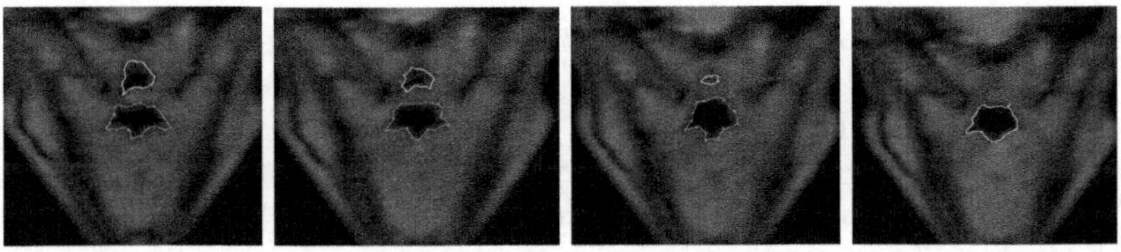

Fig. 4. Coronal slices corresponding to the frames 43 to 46, during the production of the vowel [i~] in the word [pi~p6]. Frame 43 (leftmost image) shows the frame where the highest nasal area is reached. In frame 46 (rightmost), the velum is already closed for the production of the plosive [p].

Fig. 4 presents an example of segmented images, from which the time-varying areas of the nasal and oral cavity were extracted.

The recorded speech (Optical Microphone output) and each series of MR images, obtained during the production of different items of the corpus, were aligned using the TTL pulse as a reference.

3 First Exploratory Results

In this section, we will present examples of the type of information that can be extracted from our database to highlight possible applications of our dynamic method to the production of nasal vowels, laterals and rhotics. A detailed articulatory description and a comprehensive phonetic interpretation of the findings are not within the scope of this paper and would, in any case, require data from a larger number of speakers for empirical validation.

3.1 Nasal Vowels

Comprehensive information about the nasal vowels has been difficult to obtain, as the analysis of velum movement is a very challenging task. Unlike other invasive techniques, such as EMA, RT-MRI provides a full midsagittal view of the moving vocal tract (including the velum), with reasonable temporal resolution, which make it suitable to examine: 1) temporal evolution of the velum and tongue body, 2) coordination of the nasal and oral gestures and 3) variation of the nasal/oral areas along the time.

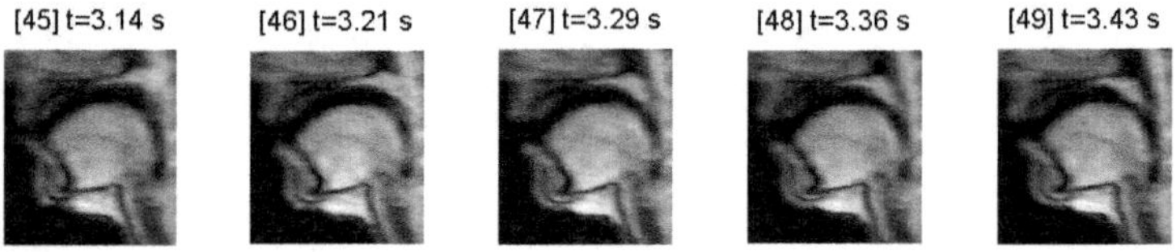

Fig. 5. Midsagittal images (frames 45 to 49) obtained during the production of the nasal vowel [6˜] in word-final position ([p6˜])

Fig. 5 shows the motion of the velum during the production of the EP nasal vowel [6˜]: at frame 45/46, it is raised, which is consistent with the oral vowel onset described in earlier studies [18,17]; after a gradual lowering, observed in frames 47 and 48, the velopharyngeal reaches its maximum aperture at frame 49. The total duration of the raising-lowering movement could be estimated from the images. In this particular case, the lowering movement of the velum is approximately 290 ms. Fig. 6 illustrates the pixel intensity changes in two ROIs (velum and lips). High intensity values are associated with lip closure and velum raising. In order to better estimate the gestural landmarks, tissue velocity at the constriction could be calculated [21].

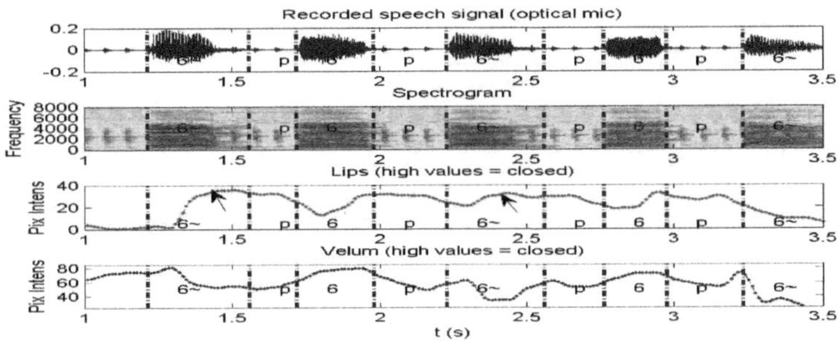

Fig. 6. Recorded speech signal from fiberoptic microphone (top), changes in intensity over time, in the selected ROI's: lips (3rd row) and velum (bottom). The speech segment presented corresponds to the production of [6˜p6], [p6˜p6], [p6˜](2.5 s of speech).

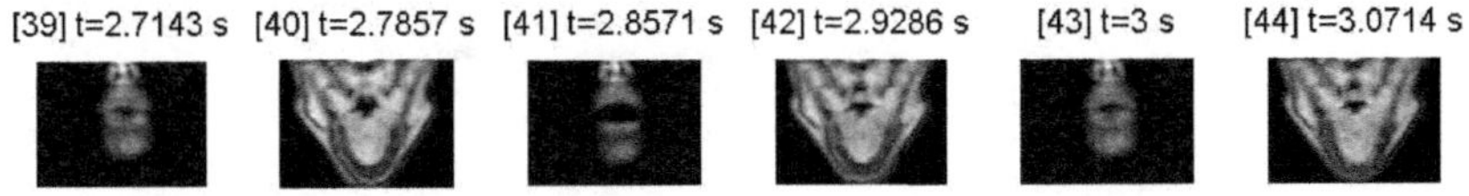

Fig. 7. Example of the acquisition of two coronal slices: one at lips and the other at velum. Frame rate decreases from 14 to 7, temporal resolution for each slice is 140 ms.

The coronal slices simultaneously obtained at the velum and lips (see Fig. 7) also provide information about the patterns of coordination between these articulators.

The semi-automatic segmentation performed over the coronal oblique slices of the velum (see Fig. 4) allow for the quantification of nasal and oral areas, from which the nasal/oral ratio can be computed, as illustrated in Fig. 8.

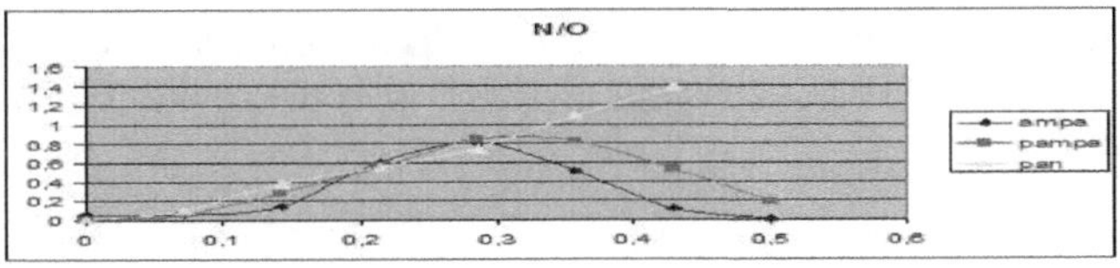

Fig. 8. Nasal-Oral areas ratio (N/O) over time during the production of [6˜p6] (blue line), [p6˜p6] (pink) and [p6˜] (yellow). The frames of interest were identified after acoustic signal/ image alignment.

The nasal-oral ratio pattern for [6˜] is similar in the three word contexts. The nasality increases over the vowel, moving gradually from low to high values. When the vowel is followed by the consonant, the velum quickly starts the raising gesture, which is reflected in a significant decrease in nasal area. These observations were consistent with the previous finding that in EP the opening movement is longer than the closing gesture [17]. Furthermore, in general the nasal areas are lower than the oral ones.

3.2 Alveolar Lateral

As the corpus only contemplates a few productions of the alveolar lateral, we were mainly interested in assessing the potential of the pixel intensity method for providing detailed information about the tongue tip closure gesture and the dorsal retraction gesture of /l/ (cf. [21]).

Changes in mean pixel intensity for [lak6] and [sal] are shown in Fig. 9. The constriction location and constriction degree for /l/ can be estimated and quantified using this method.

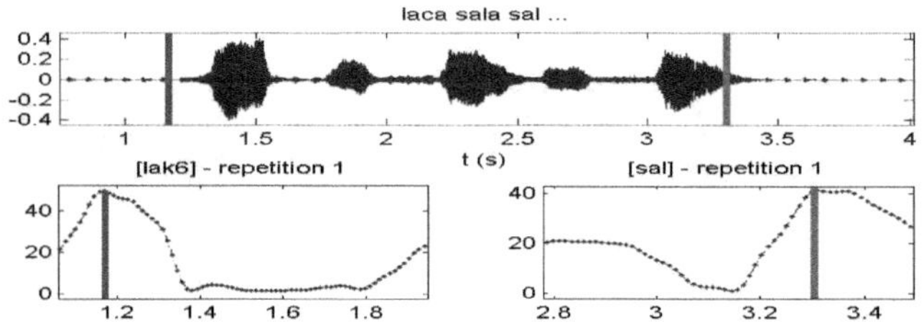

Fig. 9. Speech signal (top) and intensity function (bottom) for alveolar lateral in onset ([lak6]) and coda ([sal]). Maximum value (see vertical lines) corresponding to maximum contact (tongue tip/ alveolar contact).

The dorsal retraction gesture, typically associated with coda /l/, could also be automatically identified using this method. These new data will certainly contribute to a better understanding of the intergestural coordination of the tongue tip and tongue dorsum gestures in the production of EP /l/. It could also bring new insights into the effects of syllable structure on spatial magnitude, durational properties, and timing patterns of gestures associated with the production of /l/.

3.3 Taps and Trills

Fig. 10 presents midsagittal RT-MR images of a tap (first row) and trills (uvular, second row; alveolar, third row), both uttered in intervocalic position. The production of the tap involves a single brief closure, that can only be observed in frame 19 (row one). As shown in the image (second row), instead of a contact between the articulators during the production of [R\], there is only an approximation between them. The uvular fricative is the most common pronunciation of EP [R\]. The dynamics of the tongue tip during the alveolar trill is much more difficult to observe in the RT-MR images (third row), because of the insufficient temporal resolution.

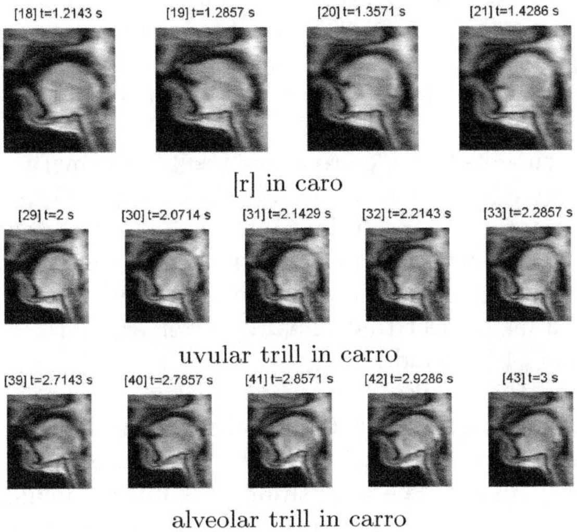

Fig. 10. Midsagittal images during the acquisition of EP tap/trills. Top row: EP tap (word [ka4u]); Middle and bottom rows: uvular and alveolar trills (word [kaR\u]).

As illustrated in Fig. 11, the pixel-based method is suitable for examining the duration of the apical gesture, albeit short, associated with a tap. On the contrary, the high-speed vibrating cycle associated with the trill was not captured, as the temporal resolution is too low.

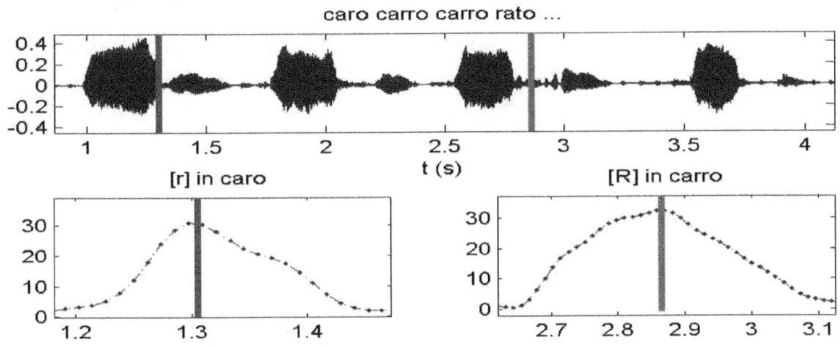

Fig. 11. Change in intensity at alveolar constriction: tap (left) and alveolar trill (right)

4 Conclusions

In this paper, the first EP database of Real-Time MRI with synchronized audio was presented. It represents a considerable step towards a better characterization of the dynamic aspects involved in the production of EP nasal vowels, laterals, taps and trills.

The temporal resolution achieved, 14 frames/s, although insufficient for detecting rapid articulatory movements (e.g., tongue tip vibration during trill production) is adequate to capture important features of other moving articulators during the production of different sounds (e.g., raising and lowering of the velum, lip closure).

While the experiments provide promising results, the method can of course be improved. The possibility of having two 2D-slices, acquired simultaneously, even with a decrease in temporal resolution (7 frames/s), offers new opportunities to address relevant topics in the field of articulatory phonology, including the temporal organization of gestures. The work presented here represents a preliminary step in a much larger project that aims to better explain the dynamic aspects in the production of EP sounds.

At this moment, the acquisition of more speakers (2 or 3) is already planned, in order to account for interspeaker variability. More than selectively sampling information that can be obtained from the images, a full exploration of the database is being planned. The acquisition of real-time imaging produces large amounts of images. The implementation of fully automatic segmentation techniques is therefore of utmost importance in the field. However, the task is challenging because the images still suffer from low spatial resolution and noise. Considerable efforts will be dedicated to image segmentation and image processing. Important technical developments in the field of MRI hold the promise of further advances in the field of speech production (e.g., higher temporal resolution).

Acknowledgments. This research was partially funded by FEDER through the Operational Program Competitiveness factors - COMPETE and by National Funds through FCT (Foundation for Science and Technology) in the context of the Project HERON II (FCT reference PTDC/EEA-PLP/098298/2008). The second author acknowledges FCT PhD grant-reference SFRH/BD/65183/2009. The authors also thank IBILI Director.

References

1. Andrade, A.: On /l/ velarization in European Portuguese. In: ICPhS. San Francisco (1999)
2. Bresch, E., Adams, J., Pouzet, A., Lee, S., Byrd, D., Narayanan, S.: Semi-automatic processing of real-time MR image sequences for speech production studies. In: 7th ISSP, Ubatuba (2006)
3. Bresch, E., Katsamanis, A., Goldstein, L., Narayanan, S.: Statistical multi-stream modeling of real-time MRI articulatory speech data. In: InterSpeech, Japan (2010)
4. Bresch, E., Narayanan, S.: Region segmentation in the frequency domain applied to upper airway real-time Magnetic Resonance Images. IEEE Transactions on Medical Imaging 28(3), 323–338 (2009)
5. Bresch, E., Riggs, D., Goldstein, L., Byrd, D., Lee, S., Narayanan, S.: An analysis of vocal tract shaping in English sibilant fricatives using real-time Magnetic Resonance Imaging. In: InterSpeech, Brisbane (2008)

6. Byrd, D., Tobin, S., Bresch, E., Narayanan, S.: Timing effects of syllable structure and stress on nasals: A real-time MRI examination. Journal of Phonetics, 97–110 (2009)
7. Falcão, A., Udupa, J., Samarasekera, S., Sharma, S., Hirsch, B., Lotufo, R.: User-steered image segmentation paradigms: Live wire and live lane. Graphical Models and Image Processing 60(4), 233–260 (1998)
8. Huettel, S., Song, A., McCarthy, G.: Functional Magnetic Resonance, vol. 1. Sinauer Associates Inc., Massachusetts (2004)
9. Kim, Y.C., Narayanan, S., Nayak, K.: Accelerated three-dimensional upper airway MRI using compressed sensing. Magnetic Resonance in Medicine 61, 1434–1440 (2009)
10. Lacerda, A., Head, B.: Análise de sons nasais e sons nasalizados do português. Revista do Laboratório de Fonética Experimental de Coimbra 6, 5–70 (1966)
11. Lammert, A., Proctor, M., Narayanan, S.: Data-driven analysis of real time vocal tract MRI using correlated image regions. In: InterSpeech, Japan (2010)
12. Lee, S., Bresch, E., Adams, J., Kazemzadeh, A., Narayanan, S.S.: A study of emotional speech articulation using a fast Magnetic Resonance imaging technique. In: InterSpeech, Pittsburgh (2006)
13. Martins, P., Oliveira, C., Silva, A., Teixeira, A.: Articulatory characteristics of European Portuguese laterals: An 2D and 3D MRI study. In: FALA, Vigo (2010)
14. Martins, P., Carbone, I., Pinto, A., Silva, A., Teixeira, A.: European Portuguese MRI based speech production studies. Speech Communication 50(11-12), 925–952 (2008)
15. MevisLab (2010), `http://www.mevislab.de`
16. Oliveira, C., Martins, P., Marques, I., Couto, P., Teixeira, A.: An articulatory and acoustic study of European Portuguese /l/. In: ICPhS, Hong Kong (2011)
17. Oliveira, C., Martins, P., Teixeira, A.: Speech rate effects on European Portuguese nasal vowels. In: InterSpeech, Brighton (2009)
18. Oliveira, C., Teixeira, A.: On gestures timing in European Portuguese nasals. In: ICPhS, Saarbrücken (2007)
19. Oliveira, C., Teixeira, A., Martins, P.: Towards an articulatory characterization of European Portuguese /l/. In: 3rd ISCA Workshop on Experimental Linguistics, Athens (2010)
20. Proctor, M., Bone, D., Katsamanis, A., Narayanan, S.: Rapid semi-automatic segmentation of real-time Magnetic Resonance images for parametric vocal tract analysis. In: InterSpeech, Japan (2010)
21. Proctor, M., Lammert, A., Katsamanis, A., Goldstein, L., Hagedorn, C., Narayanan, S.: Direct estimation of articulatory kinematics from real-time Magnetic Resonance Image sequences. In: InterSpeech, Florence (2011)
22. Sampson, R.: Nasal Vowel Evolution in Romance. Oxford University Press (1999)
23. Shosted, R.: Excrescent nasal codas in Brazilian Portuguese: an electropalatographic study. In: ICPhS, Hong Kong (2011)
24. Teixeira, A., Vaz, F., Príncipe, J.: Influence of dynamics in the perceived naturalness of Portuguese nasal vowels. In: Ohala, J.J., Hasegawa, Y., Ohala, M., Granville, D., Bailey, A.C. (eds.) ICPhS, University of California, Berkeley, San Francisco (1999)

Production and Modeling of the European Portuguese Palatal Lateral

António Teixeira[1], Paula Martins[2], Catarina Oliveira[2], and Augusto Silva[1]

[1] DETI/IEETA, University of Aveiro, 3810–193 Aveiro, Portugal
{ajst,augusto.silva}@ua.pt
[2] ESSUA/IEETA, University of Aveiro, 3810–193 Aveiro, Portugal
{pmartins,coliveira}@ua.pt

Abstract. In this study, an articulatory characterization of the palatal lateral is provided, using MRI images of the vocal tract acquired during the production of /L/ by several speakers of European Portuguese. The production of this sound involves: a complete linguo-alveolopalatal closure; inward lateral compression of the tongue and a convex shape of the posterior tongue body, allowing airflow around the sides of the tongue; large cross-sectional areas in the upper pharyngeal and velar regions. The lengths and area functions derived from MRI are analysed and used to model the articulatory-acoustic relations involved in the production of /L/. The results obtained in the first simulations show that the vocal-tract model (VTAR) is reasonably able to estimate the frequencies of the first formants and zeros. The lateral channels combined with the supralingual cavity create pole-zero clusters around and above F3, in the frequency range of 2-4.5 kHz.

Keywords: European Portuguese, magnetic ressonance, palatal lateral, acoustic modeling.

1 Introduction

The lateral consonants have much more complex and also more variable articulatory configurations than other sounds. The effects of this complexity on the sound spectrum are not clear.

European Portuguese (EP) has two lateral consonants: /l/ and /L/. The former is characterized by a linguo-alveolar contact together with inward lateral compression of the posterior tongue body that enables the formation of flow channels around the sides of the tongue [7]. Previous investigations also revealed a retraction (pharyngealization) and/ or raising (velarization) of the posterior tongue body [10], consistent with the low F2 frequencies obtained in the acoustic studies [6,10,1]. For the latter, the available descriptions are scarce, either for Portuguese or for other Romance languages (e.g. Catalan, Italian, Spanish). In general, the traditional descriptions consider /L/ as a dorso-palatal consonant [12], but articulatory data from Romance languages [11] and from Portuguese [7,8,3] revealed that this consonant is in general articulated at the alveolopalatal

H. Caseli et al. (Eds.): PROPOR 2012, LNAI 7243, pp. 318–328, 2012.

zone, presumably due to manner requirements [11]. Acoustically, little is known about EP /L/, but data from other languages and dialects [4,13] suggest that palatal laterals are long segments, with a low F1 frequency (300-400 Hz) and relatively high frequencies for the second and third formants.

Several computer vocal tract models [16,17,9,2] have been developed in order to better understand the production of liquids and to investigate the role of possible sources of pole-zero (e.g., supralingual side branch or lateral channels of different lengths). Most of the studies focused primarily on American English /l/, due to the amount of data available.

The first goal of this paper is to provide an accurate description of the vocal tract geometry of the /L/, using information obtained from MRI and data from Portuguese. Apart from the brief description of the EP /L/ outlined by [7], there are no MRI data for the palatal lateral available in the literature.

The second goal is to use the vocal tract areas derived from MRI to model the articulatory-acoustic relations involved in the production of the /L/. As there are no specific models for the palatal lateral, acoustic modeling experiments will be conducted using a simple tube vocal-tract model [15,16] specifically designed for /l/, but presumably suitable for lateral sounds in general. In order to test the validity of this last assumption, a set of simulations were performed to understand the contribution of the different vocal-tract cavities in the production of the palatal lateral.

The paper will be organized as follows: after a brief overview of the topic, section 2 describes image acquisition, segmentation techniques and modeling steps. The main articulatory properties of /L/ and the simulations conducted are presented in section 3. Finally, in section 4, the conclusions and ideas for future work are outlined.

2 Methods

2.1 MRI Set-up

Seven EP speakers [three females (CO, MC, ER) and four males (JPM, JH, LCR, AS)], aged between 21 and 39 years, were recruited for the study. They were all volunteers from the center of the country, with no history of hearing or speech disorders. An MRI screening form and informed consent were obtained before their participation in the study.

MRI corpus included the EP /l/ and /L/ next to the vowels /i/, /a/, /u/. The former was produced in different word positions: word-initial (e.g *litro* "liter" ([lit4u]), intervocalic (e.g *bilis* "bile" ([biliS]) and word-final (e.g *til* "tilde" [til]). The latter was acquired only in intervocalic context (e.g palha "straw" [paL6]), due to phonotactic constraints.

The MRI experiment was carried out at IBILI- Coimbra, using a 3.0 T scanner (Magneton Tim Trio, Siemens, Germany) equipped with high performance gradients (Gmax=45mT/m, rise time= 0.2s, slew rate= 200 T/m/s; and FOV =50 cm). A standard 12-channel head and neck phased-array coils and parallel imaging (Generalized Autocalibrating Partially Parallel Acquisition - GRAPPA) were used in all data acquisition sessions. The acquisition protocol was based on a previous MRI study conducted by our research team [8]. Subjects lay comfortably in a supine position in the MR machine using headphones.

After acquiring reference images, a T1 W 5 mm thickness midsagittal MRI slice of the vocal tract was obtained using a Turbo Spin Echo (TSE) sequence (TR/TE/FA=400 ms/7.8 ms, 120°), FOV=240 x 240 mm; matrix (256 x 256) resulting in a pixel size of (0.938, 0.938). The acquisition time was 6 seconds. After that, a volume covering the entire vocal tract was obtained in the sagittal plane with a 3D ultra fast spoiled GE sequence (Volume Interpolated Breath-Hold Examination - VIBE). The volume comprises 52 slices, from the right to the left temporo-mandibular joint. The parameters used are TR =4.32 ms, TE=1.37 ms, FA=10°, matrix (224 x 256), effective slice thickness = 2mm, voxel size (1.055, 1.055, 2) and acquisition time of 19 seconds. The speakers sustained the sound during the period of acquisition; the sequence was launched when the /L/ was produced. Finally, a 3D high-resolution sequence was obtained for each of the speakers, without phonation, so as to allow the extraction and co-registration of dental casts.

Speech sounds from each subject were also recorded in an anechoic chamber during a separate session, in conditions that simulated the MRI environment.

2.2 Image Processing

Segmentation of both the vocal tract and tongue was based on the general deformable models framework.

Image processing, data analysis and visualizations were performed in MevisLab (MeVis Medical Solutions), ITK-SNAP toolkit and Matlab. Segmentation of the vocal tract, in order to obtain area functions and information regarding the area and the extension of the lateral channels, was found to be more demanding than tongue segmentation. The overall process involved the following steps: 1) mandible/maxilla segmentation and extraction of 3D masks from an MRI volume acquired for each speaker (Fig.1, top); 2) corresponding 3D masks were resampled and co-registered (interactive affine transformation) with vocal tract data sets acquired during sound production (Fig.1, middle); 3) curved Multiplanar Reconstruction (MPR) of the vocal tract was performed to obtain areas perpendicular to the centerline of the tract and 4) segmentation of the re-sliced volume was performed, every 3 mm, from the glottis to the lips (Fig.1, bottom). This process was established using several pipelines, live-wire based [5], implemented in MevisLab. After that, contour lists (CSO) were exported to Matlab allowing the extraction of area functions and tract visualizations.

2.3 Acoustic Modeling

Acoustic modeling was attempted using the frequency-domain model for the vocal tract acoustic response (VTAR) [15,16]. This was chosen because it is publicly available and includes both parallel lateral channels and a supralingual

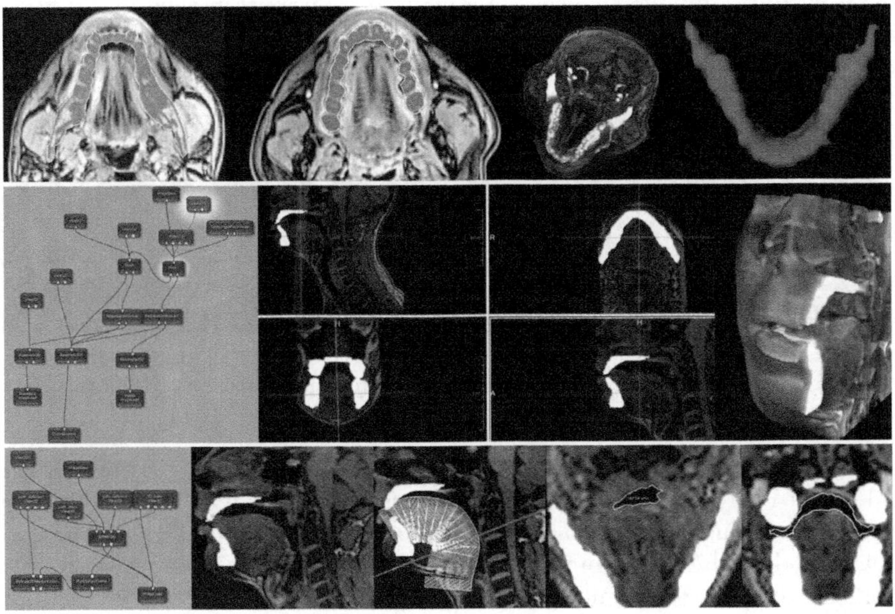

Fig. 1. Detailed pipeline used for image processing: 1) mandible/ maxilla segmentation and corresponding 3D masks (top); 2) resampling and coregistration of the masks with vocal tract data sets (middle); 3) Curved Multiplanar Reconstruction (MPR) of the vocal tract to obtain areas perpendicular to the centerline of the vocal tract and segmentation of the re-sliced volume (bottom)

side branch. Although it was originally developed for the acoustic modeling of /l/ (as produced in American English), these two last properties make it suitable to deal with other lateral sounds, including /L/.

Cross-sectional area functions obtained from MRI images of the vocal tract during a sustained production of /L/ by four speakers (CO, MC, JPM, LCR) were used in the simulations. These area functions are nonuniform along the length.

3 Results

In this section, an articulatory description of the palatal lateral is provided (3.1), followed by information about the length and areas of the different cavities (3.2). The results from the first simulations are presented in section 3.3.

3.1 Articulatory Characterization of /L/

As shown in midsagittal MRI images of Fig. 2, /L/ is produced with a complete contact of the tongue blade and/or pre-dorsum at the alveolopalatal zone.

The front of the tongue is always in a high position, but, in general, the tongue tip remains lowered, touching the lower incisors.

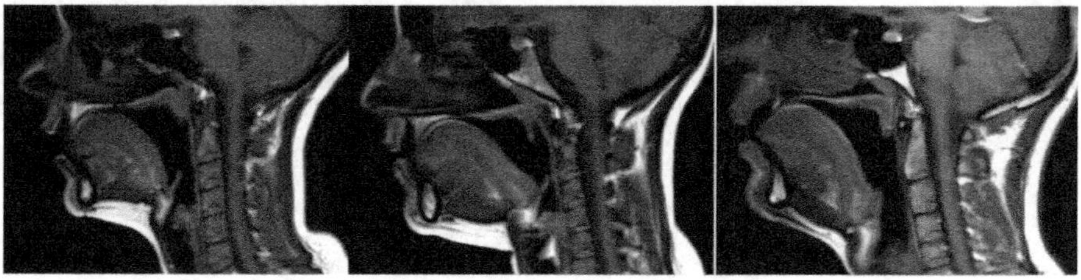

Fig. 2. Midsagittal MRI images of EP /L/ from 3 speakers (CO, MC and LCR)

The length of the lingual contact varies between 0.8-2.4 cm (for all speakers and vowel contexts acquired), which means that /L/, in EP, involves larger degrees of tongue contact than the alveolar /l/ [10]. The middle and posterior tongue body exhibits a convex shape and inward lateral compression of the tongue body toward the midsagittal plane (see Fig. 3), so that the air is allowed to flow along the sides of the tongue (and also centrally above the tongue) from the velar to the postpalatal region (leftmost image of Fig. 4).

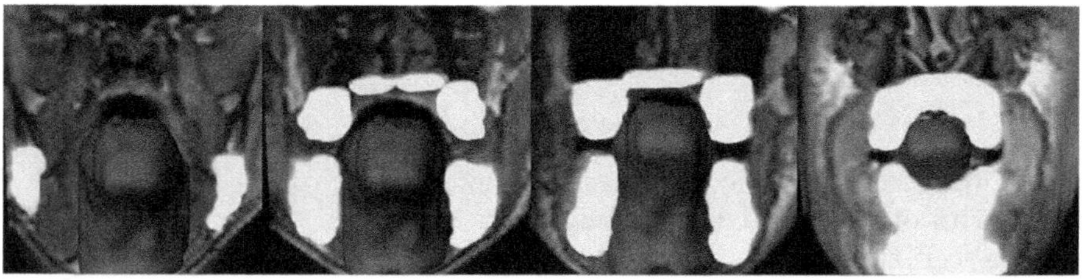

Fig. 3. 3-D volume rendering (speaker LCR) of /L/ (in the word *palha* "straw"). From left to right: coronal views from the velum to the closure.

As the sides of the tongue rise and make contact with the palate on both sides, preventing the airstream passing out over the borders of the tongue, the air continues to escape, above the tongue and between the teeth in the first place (two central images of Fig. 4), and only laterally once the alveolopalatal closure is established (rightmost image of Fig. 4).

For most subjects two slightly asymmetrical lateral channels were found. However, for one of the subjects (AS) only one lateral channel was observed in the coronal scans.

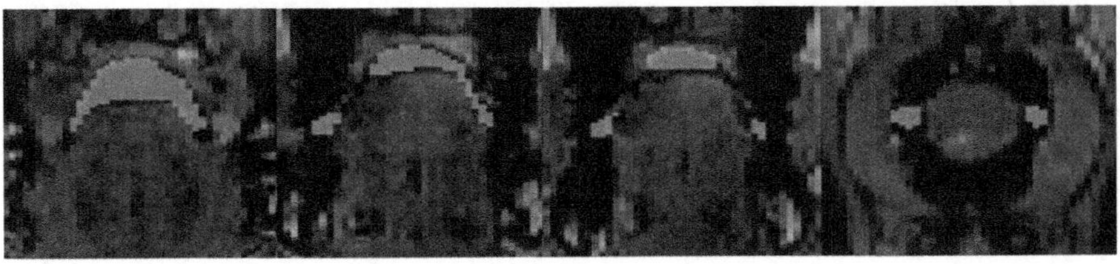

Fig. 4. Coronal MR slices with the airway passages highlighted (speaker LCR) (/L/ as in the word *palha* "straw"). From left to right: velar, palatal and contact areas.

3.2 The Area Functions of /L/

From the MRI segmentations, vocal-tract area functions were obtained. Fig. 5 shows the area functions for /L/ from a subset of speakers (2 male and 2 female). The numerical area functions values (length and average areas of the back, supralingual and front cavities and two lateral channels), obtained by averaging the area function over its length, are given in Table 1.

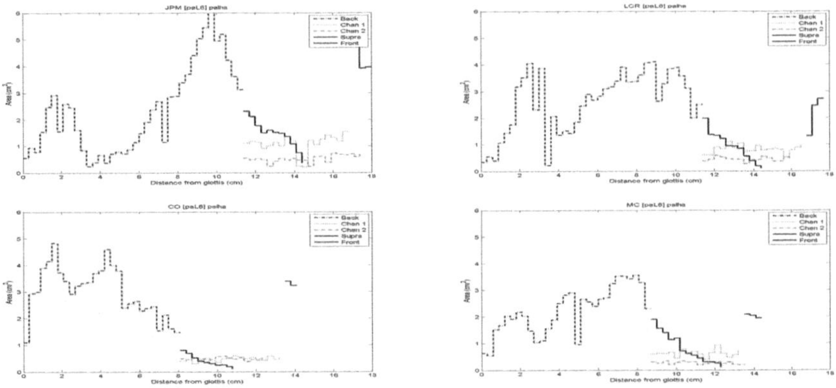

Fig. 5. MRI-derived area functions of the vocal tract (back, supralingual and front cavities and two lateral channels) from 2 male (top) and 2 female (bottom) speakers

The back cavity is longer for male than for female speakers, attaining values above 11 cm. The length of the supralingual cavity is always greater than 3 cm. Taking into account the differences in vocal-tract length between male and female, the supralingual cavity is proportionally smaller in male speakers.

The lateral channels are of different length and area, as can be observed in Table 1.

Table 1. Length and average area of the different cavities of /L/ (*palha* "straw") for 4 speakers; M=Male; F=Female; SD= standard deviation

Speaker		Length (cm)	Area (cm^2)
CO (F)	Back	8.10	3.00 (SD= 0.97)
	Front	0.60	3.30 (SD= 0.11)
	Channel 1	5.40	0.47 (SD= 0.08)
	Channel 2	4.50	0.49 (SD= 0.05)
	Supralingual	3.00	0.38 (SD= 0.22)
MC (F)	Back	8.70	2.22 (SD= 0.89)
	Front	0.90	2.02 (SD= 0.07)
	Channel 1	4.80	0.63 (SD= 0.13)
	Channel 2	5.10	0.26 (SD= 0.05)
	Supralingual	3.90	0.80 (SD= 0.57)
LCR (M)	Back	11.40	2.59 (SD= 1.15)
	Front	0.90	2.16 (SD= 0.74)
	Channel 1	5.40	0.82 (SD= 0.15)
	Channel 2	4.80	0.42 (SD= 0.13)
	Supralingual	3.30	0.89 (SD= 0.57)
JPM (M)	Back	11.40	2.33 (SD= 1.64)
	Front	0.90	4.32 (SD= 0.68)
	Channel 1	5.70	1.11 (SD= 0.35)
	Channel 2	6.30	0.52 (SD= 0.13)
	Supralingual	3.30	1.44 (SD= 0.56)

3.3 Simulations

Although "translating a 3D-geometry into a set of area functions is a simplification process" that "might have to be done based on a trial-and-error process" [17, p.94], the vocal tract configuration for /L/ can be modeled by a set of tubes representing the back cavity, the supralingual cavity, two lateral channels and a front cavity [15,16]. Inspired by the methods recently proposed by [17], the following approximations were also assumed: 1) the shortest lateral channel was (artificially) extended to start at the same position of the longest channel; this extension (less than 1cm) is attained including the corresponding area of the back cavity in the area of the lateral channel. With this option the length of the back cavity was slightly shortened.

Since there is no previous work in /L/ vocal tract modeling, as a preliminary step, data from /l/ were used in order to validate our approach. The resulting acoustic response (i.e., frequencies of formants and zeros) was close to that measured from natural acoustic productions of /l/ [10,6] and similar to the spectrum pattern described for the American English [15,16,17]. Fig. 6 shows the acoustic response of VTAR.

The vocal tract resonance frequency values of /L/ were estimated from the tube model and were compared to the values measured from the subjects natural

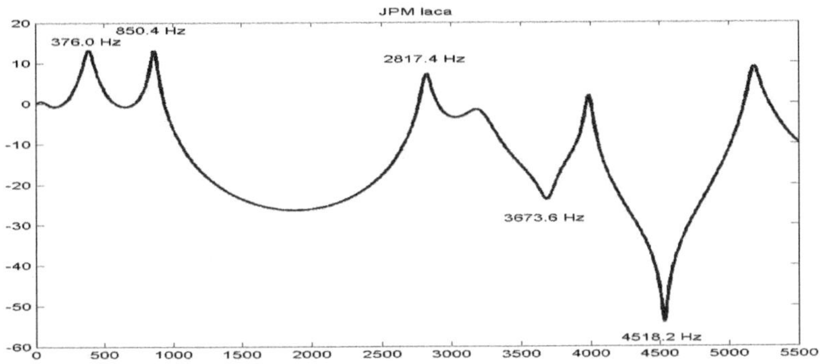

Fig. 6. Acoustic response function of /l/ for speaker JPM (producing the /l/ as in the word *laca* "hairspray")

speech (average values of sustained and normal productions of /L/). The results for four speakers are summarized in Table 2. Examples of the vocal tract acoustic response functions for one female and one male speaker are presented in Fig. 7.

Table 2. Formants and zeros of /L/ compared with calculated values from VTAR simulations (in Hz), for 4 speakers

Spk	Sim/Nat	F1	F2	F3	Zero1	Zero2
CO	Simu	468	2198	3135	3456	4033
	Nat	413	2442	3658	3165	4162
	Error (%)	13	-12	-14	9	-3
MC	Simul	446	1755	3102	3390	3623
	Nat	283	2229	3013	3024	4470
	Error (%)	58	-21	3	12	-19
LCR	Simul	440	1511	2564	3251	4055
	Nat	288	1842	3029	3453	4487
	Error (%)	53	-18	-15	-6	-10
JPM	Simul	463	1011	2653	3262	3451
	Nat	323	1739	2677	3132	4119
	Error (%)	43	-42	-1	4	-16

The frequencies obtained from simulations, when compared with natural productions, resulted in an acceptable percentage of error, especially for second and third formants. The location of the zeros is fairly accurate (particularly the first one). F1 simulated values are clearly above the frequencies measured from natural productions.

Although a somewhat oversimplified approximation, such a model can provide some insights into the origin of the pole-zero clusters observed in the spectrum

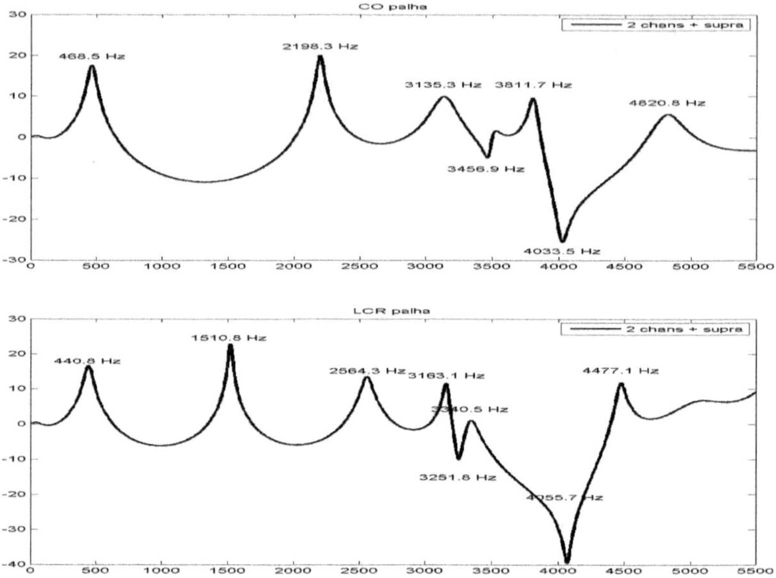

Fig. 7. Acoustic response functions from CO and LCR speakers (producing the /L/ as in the word *palha*"straw")

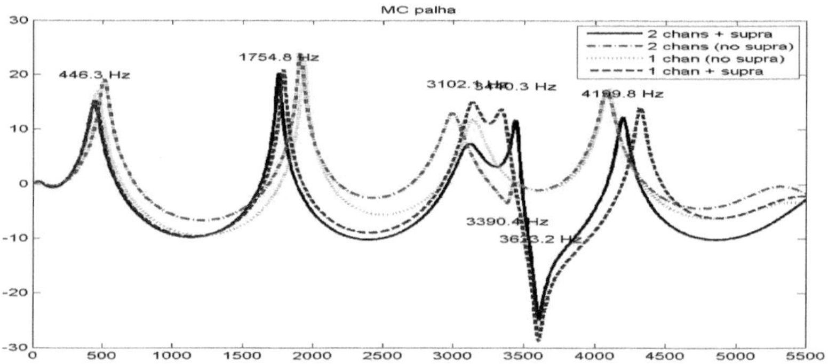

Fig. 8. Acoustic response functions obtained in several simulation conditions (speaker MC producing the word *palha* "straw")

of /L/. An example for one of the female speakers is presented in Fig. 8. The first zero is caused by the presence of the two lateral channels and the other is introduced by the supra-lingual cavity.

The estimated F2 value from JPM was considerably lower than that obtained from natural speech, resulting in a high error percentage. Additional experiments were conducted to attain a better matching. Instead of using nonuniform-areas for the back cavity, an approximation was performed: the back cavity was divided into 3 tube sections and the average area of each section was used. With this procedure the F2 rose from 1011 to 1536 Hz and the error percentage decreased from -42 to -12%.

4 Conclusions

This paper presents new 3D MRI articulatory data for the palatal lateral, from several speakers of EP. Qualitative data (e.g., tongue shape and position) and quantitative information (such as area functions, area and length of the lateral passages and of the contact) are provided, allowing for a better characterization of this lateral sound. The results regarding the /L/ closure are in agreement with the data reported for other languages [11] (e.g., Catalan) and support preliminary findings for EP [7] that pointed to a more anterior articulation of /L/, conditioned by aerodynamic factors. Based on these data, better acoustic models for /L/ can be developed.

The first simulations give fairly accurate predictions for the formant structure of /L/. Although promising, these results could be improved, specifically regarding F1 frequencies. Moreover, as stated by [17], the process of dividing the vocal tract into different cavities is not straightforward and the model is not able to perfectly handle the asymmetry (in terms of length) of the lateral channels. Acoustic analysis itself is also a challenging process, since there is no systematic method for detecting zeros. Nevertheless, these simulations allow us to better understand the effects of the different cavities in the sound spectrum.

As future work, the integration of this information into our articulatory synthesizer [14] is planned, but an adaptation of the implemented models is required in order to simulate lateral sounds.

Acknowledgements. This research was partially funded by FEDER through the Operational Program Competitiveness factors (COMPETE) and by National Funds through Foundation for Science and Technology (FCT) in the context of the Project HERON II (PTDC/EEA-PLP/098298/2008). The second author acknowledges the PhD grant from FCT (SFRH/BD/65183/2009). The authors thank Professor Miguel Castelo-Branco (IBILI Director) and the speakers involved in the study.

References

1. Andrade, A.: On /l/ velarization in European Portuguese. In: International Congress of Phonetic Sciences, San Francisco (1999)
2. Bangayan, P., Alwan, A., Narayanan, S.: From MRI and acoustic data to articulatory synthesis: A case study of the lateral approximants in American English. In: Proceedings of the 4th International Conference on Spoken Language Processing, Philadelphia, USA (1996)
3. Cagliari, L.C.: Airflow in palatal laterals. Work in Progress, Edinburgh University, Department of Linguistics 10, 113–115 (1993)

4. Colantoni, L.: Reinterpreting the CV Transition: Emergence of the glide as an allophone of the palatal lateral. In: Auger, J., Clements, J.C., Vance, B. (eds.) Contemporary Approaches to Romance Linguistics. Selected Papers from the 33rd Linguistic Symposium on Romance Languages, Bloomington, Indiana, pp. 83–102. John Benjamins (April 2003)
5. Falcão, A., Udupa, J., Samarasekera, S., Sharma, S., Hirsch, B., Lotufo, R.: User-steered image segmentation paradigms: Live wire and live lane. Graphical Models and Image Processing 60, 233–260 (1998)
6. Marques, I.: Variação Fonética da Lateral Alveolar no Português Europeu. Dissertação de mestrado, Universidade de Aveiro (2010)
7. Martins, P., Oliveira, C., Silva, A., Teixeira, A.: Articulatory characteristics of European Portuguese laterals: a 2D and 3D MRI study. In: FALA, Vigo, Spain (2010)
8. Martins, P., Carbone, I., Pinto, A., Silva, A., Teixeira, A.: European Portuguese MRI based speech production studies. Speech Communication 50(11-12), 925–952 (2008)
9. Narayanan, S., Bird, D., Kaun, A.: Geometry, kinematics and acoustics of Tamil liquid consonants. Journal of the Acoustical Society of America 106, 1993–2007 (1999)
10. Oliveira, C., Martins, P., Marques, I., Couto, P., Teixeira, A.: An articulatory and acoustic study of European Portuguese /l/. In: International Congress of Phonetic Sciences, Hong Kong (2011)
11. Recasens, D., Espinosa, A.: Articulatory, positional and contextual characteristics of palatal consonants: Evidence from Majorcan Catalan. Journal of Phonetics 34, 295–318 (2006)
12. Sá Nogueira, R.: Elementos para um tratado de Fonética Portuguesa. Imprensa Nacional de Lisboa, Lisboa (1938)
13. Silva, A.: Para a descrição fonético-acústica das líquidas no português brasileiro: dados de um informante paulistano. Dissertação de mestrado, Universidade Estadual de Campinas, Instituto de Estudos da Linguagem (1996)
14. Teixeira, A., Martinez, R., Silva, L.N., Jesus, L., Príncipe, J., Vaz, F.: Simulation of human speech production applied to the study and synthesis of European Portuguese. EURASIP Journal on Applied Signal Processing 9, 1435–1448 (2005)
15. Zhang, Z., Espy-Wilson, C., Tiede, M.: Acoustic modeling of American English lateral approximants. In: Proceedings of the 8th European Conference on Speech Communication and Technology (Eurospeech), Geneva, Switzerland (2003)
16. Zhang, Z., Espy-Wilson, C.: A vocal tract model of American English /l/. Journal of the Acoustical Society of America 115, 1274–1280 (2004)
17. Zhou, X.: An MRI-Based articulatory and acoustic study of American English liquid sounds /r/ and /l/. Ph.D. thesis, University of Maryland (2009)

A New Methodology for Comparing Speech Rhythm Structure between Utterances: Beyond Typological Approaches

Plínio A. Barbosa and Wellington da Silva

Instituto de Estudos da Linguagem, State University of Campinas, Brazil
pabarbosa.unicampbr@gmail.com, dablio_w@yahoo.com.br

Abstract. This paper proposes a new methodology for automatically comparing the speech rhythm structure of two utterances. Eleven parameters were automatically extracted from 44 pairs of audiofiles yielding 11-size difference vectors. The parameters include speech rate, duration-related stress group rate, prominence and prosodic boundary strength, f_0 peak rate, as well as the coupling strength between underlying syllable and stress group oscillators. The 11-parameter difference vectors were used to infer the perceptual differences identified by a group of 10 listeners who judged the same 44 pairs of audiofiles . The results indicate that duration-related prominence or prosodic boundary rate and speech rate, taken together, predict up to 71 % of the response variance. To a minor extent, prominence/boundary strength mean and non-prominent VV unit rate predict up to 60 % of the response variance when combined with prominence or prosodic boundary rate.

Keywords: speech rhythm, prominence, rhythm perception, speech rate.

1 Introduction

This paper explores a formal device for answering two related questions: what makes utterances sound prosodically distinct? Or what makes utterances differ as to the manner of speaking? We think the response to these questions concerns differences in speech rhythm structure. Speech rhythm is related to the variable interaction of a structuring component with a regularity component [1]. Since timing and prominence organisation are the main variables which define rhythm, a methodology to examine different aspects of timing and prominence organisation throughout an utterance is thought to be relevant.

Two main approaches on speech rhythm have been proposed by researchers. One group of researchers is interested in finding typological rhythmic differences among languages or language varieties. This group proposed several measures (nPVI, rPVI, VarCo, %V, ΔC, inter alia. See, for instance, [2,3,4]) for examining patterns of data with different clusters of data corresponding to different rhythm classes. The main problem with this approach is the lack of a universal principle of speech rhythm applicable to all languages, because its tenants presuppose an

H. Caseli et al. (Eds.): PROPOR 2012, LNAI 7243, pp. 329–337, 2012.

a priori classification of languages into two or three rhythm classes. Furthermore, events related to phonotactically-conditioned processes and hypoarticulation processes are usually invoked to explain why the data associated to a particular rhythm class occupy a particular region of the mathematical spaces formed by the proposed indexes. All proposals by this group of researchers reveal part of the consequences of speech production onto phoneme-sized variables. However, to advance the knowledge on speech rhythm, a key aspect of rhythm production (and perception) should be considered: the interplay between regularity and structuring constraints taking place between the syllable and the higher-level units (see [1] for emphasis on the importance of this interplay to all aspects of human movements). Structuring has to do with the hierarchical pattern of prominence or prosodic boundary levels. Regularity has to do with the (quasi-)regular recurrence of syllables and prominent syllables in time.

The other group of researchers has proposed hierarchical models such as [5,6], which propose the coupling between two or more levels of interacting oscillators, such as between the syllable and the stress-group oscillators. This coupling allows to explain both universal and language-specific properties of rhythm by means of general principles of production applicable to all languages, which are parameterised by a coupling strength variable (ω). These hierarchical models separate the contributions of segmental factors from prosodic factors in speech rhythm and this is one of their main strengths as compared with the typological approach. In fact, the hierarchical models capture the view of rhythm as the alternation of strong and weak beats as speech unfolds through time and fulfill the three properties proposed by [7] of an adequate model of speech rhythm: predictivity, expliciteness, and universality.

Unfortunately, the hierarchical models present either a level of abstraction that renders difficult the inference of relevant information from short excerpts of speech data [5,8,6] or evaluate speech rhythm in very particular experimental settings, such as speech synchrony and utterance repetition [9,10]. In this paper we present a methodology which takes prominence organisation and timing into account, while allowing a multiparametric comparison of any two excerpts of speech in terms of speech rhythm.

2 Methodology

For evaluating possible rhythmic differences across distinct speaking styles, reading vs storytelling styles were chosen. This choice is motivated by the fact that storytelling presents elements which can be found in spontaneous conversation, such as hesitations due to macro- and microplanning of the discourse. Though hesitations can occur in read speech, these are much less frequent than in the case of storytelling. This feature is important to be considered in developing an approach to describe speech rhythm in natural conditions and to investigate the possible differences between less and more controlled situations of utterance production.

2.1 Corpora

The corpus consisted of texts recorded by three speakers of Brazilian Portuguese (henceforth BP). Two native female and one native male speakers read a 1,500-word text on the origin of the Belém pastries (reading style, RE). After the reading, the three subjects told what the text was about (storytelling style, ST). The speakers were Linguistics students aged between 30 and 45 years at the time of recording. For the perception tests (see section 2.3), a subset of the corpus was used so that sessions last no more than 25 minutes. This subset was formed by the ST style of one of the female speakers, and the RE style of the other two speakers (one male and one female). Excerpts from 8.9 to 18.2 seconds were extracted from several parts of the material in order to make up the subsets for the discrimination tasks. The reason for choosing relatively long stretches of speech was guided by the high standard-deviations of the listeners' responses obtained in a previous study for excerpts of 1 to 2 seconds [11]. The long-duration excerpts allow the listeners to more accurately evaluate the manner of speaking than short-duration excerpts (see a similar extension for voice similariry judgement in [12]). The 44 excerpts were segmented and labelled in VV units.

2.2 Measuring Techniques and Parameters Extracted

According to a traditional approach in speech research [13,14,15], syllables were phonetically segmented by tracking two consecutive vowel onsets (VO). The segmentation was performed semi-automatically in Praat [16] in two stages: automatic VO detection by using the BeatExtractor Praat script available in [17], followed by manual correction, where applicable. Two consecutive VOs define a VV unit, which contains only a single vowel, starting at the first VO. The BeatExtractor script detects points in the speech signal where changes in the previously filtered energy envelope are relatively fast and positive (from low to high energy). According to [18], the speech signal energy in the region of the first and second formants simulates the spectral region our auditory system tracks for detecting syllables.

Duration-related stress groups were then delimitated by automatically detecting duration-related phrase stress positions throughout the utterances. The sequence of phrase stress positions was automatically tracked by serially applying two techniques for normalising the VV durations: a $z - score$ (z) transform (equation 1):

$$z = \frac{dur - \sum_i \mu_i}{\sqrt{\sum_i var_i}}, \tag{1}$$

where dur is the VV duration in ms, the pair (μ_i, var_i), the reference mean and variance in ms of the phones within the corresponding VV unit. These references are found in [17, pp. 489-490] for BP and Standard French, followed by a 5-point moving average filtering (equation 2):

$$z^i_{filt} = \frac{5.z^i + 3.z^{i-1} + 3.z^{i+1} + 1.z^{i-2} + 1.z^{i+2}}{13} \tag{2}$$

The normalisation technique and the detection of duration-related phrase stress positions (detection of z_{filt} maxima) were performed by the Praat script SGdetector. The computation of both the stress group duration and the number of VV units in the stress group is automatically performed by this script. This two-step normalisation technique aims at minimising the effects of phoneme-size intrinsic duration in the VV unit. This normalised duration maxima signal both prominence degree and prosodic boundary strength, indistinctly. This is not a drawback of this approach, since the salience of these two prosodic functions on a particular word, equally signals perceived rhythm. Listeners of Romance languages often attribute both functions to a prominent or a pre-boundary word when evaluating these functions in their own languages [19].

As presented in [6], the ratio between the intersect and the slope of the regression line designed to explain stress group duration from the number of VV units in the group (predictor variable) is related to a measure of the amount of stressing in a particular language, the coupling strength ω. The higher the coupling strength, the greater the influence of the structuring component onto VV duration regularity, that is, VV duration becomes less regular. The coupling strength defined as the aforementioned ratio is the first parameter extracted from the corresponding annotation file of each audio excerpt. Besides this parameter, ten other parameters were automatically computed by using a new RhythmParameterExtractor script running on Praat, which automatically extracts the 11 parameters from each excerpt by using a pair of audiofile and annotation file (TextGrid file in Praat).

The second parameter is speech rate in VV units per second, extracted from the corresponding TextGrid file. The third to fifth parameters are the mean, standard-deviation and skewness of the z_{filt} maxima, which reveal the structure of duration-related pooled prominence degree and boundary strength in the excerpt. The use of prominence/boundary distribution is crucial to produce an accurate description of speech rhythm, as recently claimed by [20,21]. The sixth parameter is the rate of the z_{filt} maxima in peaks per second, which is meant to stand for the prominence or prosodic boundary rate, for the reasons mentioned before. The seventh parameter is the rate of f_0 peaks in peaks per second. The sequence of f_0 peaks is obtained from the audiofile in five steps: (1) extracting the f_0 trace using Praat (limits between 75 and 600 Hz), (2) smoothing the obtained contour with a 5-Hz filter, (3) interpolating the gaps due to unvoiced segments, (4) automatically counting the number of peaks in the contour, and (5) dividing the number of peaks by the total duration of the excerpt. The next three parameters are the coefficients of variation (the ratio between standard-deviation and mean) for the following sequence of variables: the number of VV units per stress group, the duration of the stress group, and the VV duration. The last parameter is the rate of non-prominent VV units, which is close to the articulation rate, for non-prominent VV units do not contain silent pauses (as well as final-lengthened acoustic segments). Non-prominent VV units are those that do not have peaks of normalised duration for the respective VV units.

Their rate was computed by dividing the number of such units in a particular utterance by the total duration delimited between the first and the last VO of the utterance.

2.3 Perception Test: Discrimination Tasks

Two discrimination tasks were designed for comparing two randomly-combined audio excerpts. Each excerpt was also delexicalised using the technique developed by [22]. Delexicalisation is the method of suppressing the segmental information from the speech signal to render it unintelligible while preserving their prosodic characteristics. Vainio and colleagues' method combines inverse filtered glottal flow and all-pole modelling of the vocal tract with the advantage of preserving voice quality. Each pair of excerpt was combined in two orders of presentation (AB and BA), either in the delexicalised version or in the original version. This design allowed us to examining the degree of consistency when evaluating the same audiofiles in the two different orders. Consistency was defined as the absolute difference in judgment response for the AB and BA orders (perfect consistency has zero difference). Two subsets of 44 audio pairs were built, one with the delexicalised pairs (DS), and the other with the same pairs in their original version (OS). Each listener judged the DS first (task 1), and then the OS (task 2), instructed by this single sentence: "evaluate how different is the manner of speaking (*modo de falar* in Portuguese) of the excerpts in the pair in a scale of 1 (same manner of speaking) to 5 (very different manner of speaking)". In each subset, the excerpts in the pair were separated by a 1,000-Hz short tone to signal the boundary between the audio files being evaluated. A group of ten listeners, all Linguistics majors, participated in the experiment.

As regards their performance in the two tasks, we evaluated two hypotheses: (1) the task with the DS has higher and less variable consistency than the task with the OS (because the listeners would focus their attention to prosodic elements only in the case of the DS), and (2) one or more parameters among the 11 difference values for the two paired excerpts can satisfactorily predict the listeners' responses. This second hypothesis presupposes a link between perceptual and production differences, at least in terms of the 11 parameters proposed in this paper.

3 Results and Partial Discussion

The responses scale from 1 to 5 was linearly transformed into a scale from -1 to 1, with zero standing for a neutral response. As regards consistency, that is, the ability of the listener to choose the same response for the pair of excerpts, irrespective of the presentation order, the results indicated a higher and lesser variable degree of consistency for the OS, contradicting our first hypothesis (mean,standard-deviation): $(0.7, 0.6)$, for the DS, and $(0.4, 0.5)$ for the OS (significant difference with $t_{df=398} = 4.2$, $p < 10^{-4}$). Both means are close to the distance between two points in the transformed response scale (0.5). Apparently, the lexical and acoustic segmental information in the original subset of indexical (speaker recognition), lexical or semantic memory, helps maintaining the same judgment for the same pair of excerpts in different orders. The reason for the listeners not doing the same evaluation when exchanging the order is related to slight changes in the manner of speaking throughout the excerpts as well as to increased memory burden. Due to limits of post-recognition temporal buffer of up to 10 or 20 seconds [23, p. 56–66], probably only the final parts of the first excerpt is retained in the memory to compare with the second excerpt. As regards the responses themselves, there was no significant difference (paired-t-test, $t_{df=439} = 1.6$, $p < 0.2$) between the OS and DS.

In order to predict the responses from the 11-parameter difference vectors, multiple linear regression models were designed. The predicted variable was the mean transformed values related to the ten listners' responses to the original subset of excerpts. The reason for that is related to the fact that, although there was no difference in judgment between the OS and DS, the OS produced more consistent responses across presentation order. Only the pairs with response consistency equal or inferior to 0.5 were chosen for the models. From these pairs, only those with standard-deviation across listeners inferior to 0.5 were considered for analysis. The rationale behind this choice is the use of relatively homogeneous judgments. Each predicted variable for a selected pair is the mean response value for the two presentation orders. This produced a set of 15 paired excerpts comparing subjects and styles. The predictor variables were the 11-order absolute difference vector corresponding to each pair of excerpts. The best model explained 71 % of the variance of the listeners' responses (lr):

$$lr = -1.5 + 10.4pr + 2.65sr - 10.75pr * sr, \tag{3}$$

with $p - value$ of at least 0.009 for all coefficients ($F_{3,11} = 12.4$, $p < 0.0008$). This model predicts the listeners' responses from two production parameters: speech rate (sr) difference and z_{filt} maxima peak rate (pr) difference. Taken separately, these two parameters explain 40 % (pr) and 50 % (sr) of the responses' variance. Duration-related prominence/boundary strength mean (z_{filt} maxima mean) difference also explains 50 % of the responses' variance. Non-prominent VV rate (ur) difference, on the other hand, explains 31 % of the variance by itself. These four parameters are the best single predictors. Taken together, z_{filt} maxima mean and z_{filt} maxima peak rate differences explain 60 % of the variance, whereas combining z_{filt} maxima peak rate with non-prominent VV rate difference explains 56 % of the variance, all coefficients significant or marginally significant for all models with p-value from 0.08 to 0.003. It is not necessary to use logistic regression to restrict the predicted values to the $[-1, , 1]$ interval because values lesser than -1 can be interpreted as highly similar, whereas values greater than 1, as highly distinct in the manner of speaking.

As regards intra-speaker differences for different excerpts as well as differences between speaking styles and between speakers, response means (and standard-deviations) are: $-0.7(0.2)$ for excerpts from the same speakers, which indicates a similar manner of speaking; 0(0.4) for excerpts from the same style (reading) in different speakers, which indicates that the two speakers are reading in relatively different ways; and 0.5(0.6), for differents speaking styles (and speakers).

4 General Discussion

It can be inferred from the results that, at least for explaining what is perceived as differences in the manner of speaking, the listeners seem to rely on up to four parameters: speech rate, duration-related prominence or boundary rate (and not rate of f_0 peak, non significant in any combination), mean of prominence/boundary strength (estimated by $z-_{filt}$ maxima mean), and non-prominent VV rate. These results confirm hypothesis 2 above: at least two parameters related to rate (of syllable and of stress group) satisfactorily predict the listeners' responses, and explain 71 % of the response variance. None of the variability descriptors, i.e., coefficients of variation of VV duration, of number of VV units in the stress group and of stress group duration, made significant predictions for the responses. The successful predictors, speech rate in VV units per second, and stress group rate, as well as duration-related prominence/boundary, and non-prominent VV unit rate per second are closely related to the parameters which predict judgments of voice similarity [12]: pausing and articulation rate. These latter are, in some extent, included in speech rate and prominence degree/boundary strength rate. The articulation rate, on the other hand, is very close to the non-prominent VV unit rate. This result does not mean that the other descriptors are not useful for describing rhythm at all, but that the four descriptors presented above seem to be used for perceiving rhythm (or the influences of rhythm in the judgment of the manner of speaking).

As regards differences across styles and speakers, it seems from above that inter-style differences were well perceived, but also, in some extent, differences between readers (as the RE style was evaluated with two speakers). The methodology shown here seems quite robust and indicates that is probably better to evaluate rhythmic differences and their degree than to try to use typological approaches to classify speech rhythm. In speech technology, our approach can be used to automatically detect rhythmic differences between a pre-recorded utterance from one or more reference databases, and a new utterance whose rhythmic structure is unknown. The multiple regression equation 3 can be used to predict how apart a reference utterance and a new utterances are in terms of perceived rhythm. This figure can help taking decisions on prosodic differences for devices that automatically detect/recognise languages and dialects in person-machine dialogue systems.

According to the results presented here, the questions that open this paper (what makes utterances sound prosodically distinct? What makes utterances differ as to the manner of speaking?) have the following answer: speech rate and stress group (and not f_0 peak) rate as well as duration-related prominence/boundary strength mean and non-prominent VV rate (the latter is close to the articulation rate). All these measures are related to syllable and stress group rates, combined with a measure of prominence or boundary strength measure.

Acknowledgments. The first author acknowledges a grant from CNPq number 300371/2008-0, and Sandra Madureira for proof-reading. We thank our listeners and speakers, and Juva Batella for adapting the text from European Portuguese to BP. The original text is from the INESC-Lisboa.

References

1. Fraisse, P.: Les Rythmes. Journal Français d'Oto-Rhino-laryngologie Supplément 7, 23–33 (1968)
2. Dellwo, V.: The Role of Speech Rate in Perceiving Speech Rhythm. In: Proc. Speech Prosody 2008, Campinas, Brazil, pp. 375–378 (2008)
3. Low, E.L., Grabe, E., Nolan, F.: Quantitative Characterisations of Speech Rhythm: Syllable-Timing in Singapore English. Language and Speech 43, 377–401 (2000)
4. Ramus, F., Nespor, M., Mehler, J.: Correlates of Linguistic Rhythm in the Speech Signal. Cognition 73, 265–292 (1999)
5. Barbosa, P.A.: From Syntax to Acoustic Duration: a Dynamical Model of Speech Rhythm Production. Speech Communication 49, 725–742 (2007)
6. O'Dell, M.L., Nieminen, T.: Coupled Oscillator Model of Speech Rhythm. In: Proc. of ICPhS 1999, San Francisco, USA, pp. 1075–1078 (1999)
7. Bertinetto, P.M., Bertini, C.: Towards a Unified Predictive Model of Natural Language Rhythm. Quaderni Del Laboratorio Di Linguistica Della SNS 7 (2008)
8. Barbosa, P.A.: Measuring Speech Rhythm Variation in a Model-based Framework. In: Proc. of Interspeech 2009 - Speech and Intelligence, Brighton, UK, pp. 1527–1530 (2009)
9. Cummins, F., Port, R.: Rhythmic Constraints on "Stress-timing" in English. J. Phon. 26, 145–171 (1998)
10. Cummins, F.: Entraining Speech with Speech and Metronomes. Cadernos de Estudos Linguísticos 43, 55–70 (2002)
11. Silva, W., Barbosa, P.A.: Caracterização Semiautomática da Tipologia Rítmica do Português Brasileiro. Anais do Colóquio Brasileiro de Prosódia da Fala. ID [2432011] (2011), `http://www.experimentalprosodybrazil.org/III_CBPF_Anais.html`
12. Öhman, L., Eriksson, A., Granhag, P.A.: Mobile Phone Quality vs Direct Phone Quality: How the Presentation Format Affects Earwitness Identification Accuracy. The European Journal of Psychology Applied to Legal Context 2(2), 161–182 (2010)
13. Classe, A.: The Rhythm of English Prose. Blackwell, Oxford (1939)
14. Lehiste, I.: Suprasegmentals. MIT Press, Cambridge (1970)
15. Dogil, G., Braun, G.: The PIVOT Model of Speech Parsing. Verlag, Wien (1988)
16. Boersma, P., Weenink, D.: Praat: Doing Phonetics by Computer. Version 5.2.44, `http://www.praat.org`
17. Barbosa, P. A.: Incursões em torno do Ritmo da Fala. Pontes/FAPESP, Campinas (2006)
18. Scott, S.K.: Perceptual Centres in Speech: an Acoustic Analysis. PhD Thesis, University College London (1993)
19. Beckman, M.E.: Evidence for Speech Rhythms across Languages. In: Tohkura, Y., et al. (eds.) Speech Perception, Production and Linguistic Structure, pp. 457–463. IOS Press, New York (1992)

20. Kohler, K.J.: Rhythm in Speech and Language: A New Research Paradigm. Phonetica 66, 29–45 (2009)
21. Cumming, R.E.: The Language Specific Interdependence of Tonal and Durational Cues in Perceived Rhythmicality. Phonetica 68, 1–25 (2011)
22. Vainio, M., et al.: New Method for Delexicalization and its Application to Prosodic Tagging for Text-to-Speech Synthesis. In: Proc. of Interspeech 2009 - Speech and Intelligence, pp. 1703–1706 (2009)
23. Cowan, N.: Attention and Memory. An Integrated Framework. Oxford University Press, New York (1997)

Constructing Physically Realistic VCV Stimuli for the Perception of Stop Voicing in European Portuguese

Daniel Pape[1], Luis M.T. Jesus[1,2], and Pascal Perrier[3]

[1] Institute of Electronics and Telematics Engineering of Aveiro (IEETA), University of Aveiro, 3810-193 Aveiro, Portugal
[2] School of Health Sciences (ESSUA), University of Aveiro, 3810-193 Aveiro, Portugal
[3] DPC/Gipsa-Lab, UMR CNRS 5216, Grenoble INP, Grenoble, France
{danielpape,lmtj}@ua.pt,
{Pascal.Perrier}@gipsa-lab.grenoble-inp.fr

Abstract. In this book chapter we present the generation of physically realistic stimuli with a biomechanical speech production model, with the aim to produce perceptually appropriate VCV sets for the European Portuguese (EP) voicing distinction. The duration measures necessary for the biomechanical model were extracted from an extensive EP speech production database, recorded for this aim. The same database was used to generate realistic voicing extinction contours for the perceptual continuum. To assess the realistic accuracy of the biomechanically generated stimuli, we compared the biomechanical stimuli set to linear interpolation between articulatory targets, traditionally used for speech synthesis.

Keywords: biomechanical modelling, perceptual cues, cue weighting, European Portuguese, voicing perception.

1 Introduction

The work outlined in this book chapter consists of perceptual stimuli modelling as part of a research project on the importance of *voicing maintenance* in both speech production and perception in European Portuguese (EP) compared to other languages. For velar stop perception, we used and compared extracted voicing patterns and durational values from real speech productions (see Pape & Jesus 2011) in a matched cross-linguistic speech perception study, with the aim to examine the actual use and interaction (cue weighting) of the perceptual cues vowel duration, consonant duration and voicing maintenance. The speech material generated for the perceptual experiments consisted of biomechanically modelled stimuli acoustically synthesized with a three mass vocal fold model. The biomechanical modelling has the main advantage that all obtained tongue movements, trajectories and phoneme targets are comparable to natural speech, but with the additional possibility to manipulate all important temporal and glottal source parameters while maintaining articulatory realism. In sum, the use of biomechanical modelling is the best compromise to guarantee highly realistic perceptual stimuli, and to independently control parameters such as duration, transition and targets.

H. Caseli et al. (Eds.): PROPOR 2012, LNAI 7243, pp. 338–349, 2012.

1.1 Perceptual Cues for Stop Voicing

For speech production, phonological voicing distinction is defined as the presence or absence of vocal fold vibrations during consonant production (Jakobson et al. 1952). For speech perception, a stop voicing distinction is mainly based on Voice Onset Time (VOT) (Lisker & Abramson 1964, 1967). Cross-linguistic differences in voicing perception are captured by changes in the VOT boundaries, i.e., by the location of the identification boundaries between voicing categories on a VOT continuum (Hoonhorst et al. 2009). In languages with three voicing categories (voiced, voiceless and voiceless aspirated) the mean VOT boundaries are located around ±30ms. Two-category languages differ in the nature of their voicing categories. In languages with a voiceless aspirated contrast the boundary is at +30ms (Lisker & Abramson 1970), whereas for languages with voiced/voiceless contrast without aspiration the boundary is 0ms (for Spanish: Williams 1977; for French: Serniclaes 1987). Infants below six months of age raised in an English environment are sensitive to both VOT boundaries (±30ms, 0ms), although only the positive VOT boundary is phonological in English (Aslin et al. 1981) or the 0ms VOT in the other language (French: Hoonhorst et al. 2009; Spanish: Lasky et al. 1975).

VOT is one of the most dominant cues for characterising stops in a number of languages, but a number of additional perceptual cues are found to influence the perception of voicing: consonant and adjacent vowel duration (Luce & Charles-Luce, 1985; Jessen 1998; Cuartero 2002; Viana 1984), and loudness (Repp 1979), among others. These other cues distinguish voicing, in combination with VOT but also without VOT, i.e., when VOT is ambiguous or missing. Further, the literature shows (Morrison 2005; Escudero et al. 2009) that human perception does not rely only on a single perceptual cue, but rather on a combination of different cues to guarantee a stable and robust perceptual outcome. Taking into account the variety of perceptual cues for stop voicing, the question arises how different languages weight the available cue to achieve robust perception. However, few studies (Francis et al. 2000) attempted to study the simultaneous variation and cue weighting for stop voicing distinction, and (to our knowledge) no studies examined this cue weighting framework for stop voicing in a cross-linguistic context. The last point is of utmost importance when one takes into account speech production differences, for example, in voicing maintenance between different languages (see, e.g., Solé 2011 for Spanish vs. English, and Pape et al. (submitted) for EP vs. German and Italian). Given these cross-linguistic differences, the question arises whether these are also reflected in the perception, and if so, how these differences influence the cue weighting of the available cues.

1.2 Modelling Physically Realistic Perceptual Stimuli

One important point to take into account when designing a valid multidimensional cue weighting experiment are the transitions between the phoneme targets: For example, velar stops show a strong articulatory forward loop (see, e.g., Mooshammer et al. 1995), additionally the shape of the loops differs for voiceless and voiced consonantal targets (Brunner et al. 2011). Assuming that these loops could play a perceptual role, in the design of multidimensional perceptual stimuli one preferably aims for the most realistic synthesis model, with the aim to not disregard possible important influences on the perceptual system. However, it is not well understood how the temporal changes of articulatory pattern and thus the resulting articulatory trajectories are processed by the perceptual system. Thus, the particular model that is used to define and build a multidimensional stimuli space for perceptual experiments should be able to generate realistic articulatory movements, while simultaneously allowing for independent parametrical control of perceptual parameters, such as closure duration, vowel duration and transition duration. That is, in order to generate adequate perceptual stimuli for a cross-linguistic study on voicing distinction it is important to take into account the time-varying articulatory changes while producing a realistic phoneme chain with adequately realistic transitions. The most realistic perceptual stimuli would consist of naturally recorded speech, but for this condition one cannot control for the presence or lack of possible perceptual cues, therefore introducing an unwanted perceptual bias into the experiment.

There is no shortage of different models for articulatory motivated synthesis (Maeda 1990; Teixeira et al. 2005; Birkholz et al. 2010). However, very few of these models can claim to respect adequate articulatory movements. As soon as articulatory realistic transitions between the targets are required, only biomechanical models are currently capable of producing the required physically realistic articulatory trajectories. Thus, the use of a biomechanical modelling is the best compromise to guarantee highly realistic perceptual stimuli and independently control the varying parameters. Taking all these considerations into account, we used the advanced 2D tongue model described in Perrier et al. (2003) for the generation of suitable stimuli.

According to the biomechanical model described in Payan & Perrier (1997) and the improved version in Perrier et al. (2003), the tongue trajectories are obtained by activating (and thus deforming) different tongue muscles (Posterior and Anterior Genioglossus, Hyoglossus, Styloglossus, Verticalis, and Inferior longitudinalis) in a 2D Finite Element biomechanical model of the vocal tract, controlled on a target-to-target basis. The necessary vocal tract contours were extracted from X-ray data. The target for each phoneme is specified as a set of motor commands for each muscle. The movements between targets are achieved by a constant time shift of the motor commands between successive target values (as described in Perrier et al. 1996), resulting in time-varying vocal tract shapes (i.e., one complete 2D vocal tract shape each sampling period). Following, each of the obtained mid-sagittal vocal tract shapes is then converted to a time-dependent area function (Perrier et al. 1992). Then, the area functions were acoustically synthesized with a reflection-type line analog of the vocal tract (Story et al., 2000). Vocal folds oscillations are generated and controlled with a numerical implementation of the three mass model designed by Story & Titze (1995) based on lumped-elements (Titze & Story, 2002).

The biomechanical model was chosen because it has been shown to accurately account for articulatory trajectory shaping (Perrier et al. 2003) and velocity profiles (Payan & Perrier 1997) in the mid-sagittal plane, all possibly important for consonant perception (Sussman et al. 1998, Perrier & Fuchs 2008). Further, the model respects the relations between curvature and speed, which is found to be important in the correct perception of movements in general (Viviani & Stucchi 1992), and also for articulatory movements and thus speech (Perrier & Fuchs 2008). As can be seen in figure 3 for the movement of the different tongue nodes, the biomechanical model accurately accounts for the articulatory forward loops observed in natural velar stop productions. These articulatory loops – and thus the naturalness of the model – mark an important difference between simple kinematic modelling (i.e., interpolating between consecutive articulatory target positions) and realistic biomechanical models. As described, the obtained accuracy of the transitions in the biomechanical model is important for our aims of the perceptual experiments, requiring that the resulting stimuli being as realistic as possible, but with the simultaneous independent control of all important parameters.

For the generation of the tongue contours, the biomechanical model needs the phoneme sequence identity (e.g., /aka/), the holding phase of each phone (e.g., 100ms) and the transition time from one phone to the other. Since all values for the biomechanical model are defined at the muscle command level that means that, e.g., the required holding phase does not correspond to the acoustic duration of a phone, but rather to a combination of transition and holding time (see section 2.1).

2 Method

2.1 Parameterisation of the Biomechanical Tongue Model

The biomechanical model, as described in Perrier et al. (2003), can be controlled with two different approaches as the input to the generation of the tongue contours: Inverse Synthesis; directly by entering the corresponding *lambda* commands for the different muscles[1]. For the EP stimuli generation we used the Inverse Synthesis approach. This Inverse Synthesis is implemented as follows: (1) Convex target regions have been defined in the (F1,F2,F3) acoustic space to specify the spectral characteristics of the elementary sounds of EP; (2) relations between motor commands and formant values have been learned in the form of a radial basis functions model based on a large number of simulations with the biomechanical model; (3) for a given sequence of phonemes, the Inverse Synthesis Model finds the sequence of motor commands that minimizes the global change in motor commands while making sure that the successive target regions are reached at with the right timing. For more details about the procedure see Perrier et al. (2005). It is optimised by the inclusion of real speech formant and position data and their variance, thus Inverse Synthesis guarantees correct articulatory targets, which were extracted from cross-linguistic EMMA data. The biomechanical model's Inverse Synthesis approach accepts the input of phoneme identity, phoneme duration and transition duration to generate from this information the resulting appropriate motor commands and corresponding muscle forces.

[1] To obtain input of raw muscle data, the model accepts the input of *alpha* values for each target and muscle. These values contribute to specify muscle activation together with feedback information about current muscle length and length change rate. Muscle tissues stiffness is increased as a function of muscle activation. The more force is applied, the stiffer the muscle fiber remains. This possibility can be used to generate preferred articulatory targets and trajectories not already predefined, i.e., to force model to generate non-existing or impossible articulatory targets.

Figure 1 shows a block diagram illustrating the process of controlling the biomechanical model with the EP values extracted from the EP speech production database described in section 2.2. For our perceptual experiments in EP, the VCV stimuli were generated by the biomechanical model by defining the following parameters as the input:

- identity of the context vowels adjacent to the velar stop (e.g., /aka/ or /iki/);
- vowel duration of the preceding and following vowel (from 70ms to 130ms);
- stop closure duration (from 50ms to 150ms);
- voicing maintenance during closure (from fully voiced to fully devoiced).

We modelled the first three factors based on the durational values for EP as described in section 3.1. For this reason, we chose the vertices and step sizes of the parameters in accordance with the durational values shown in figure 2. Table 1 shows all parameters and the corresponding values used as orthogonal factors for the perceptual experiments.

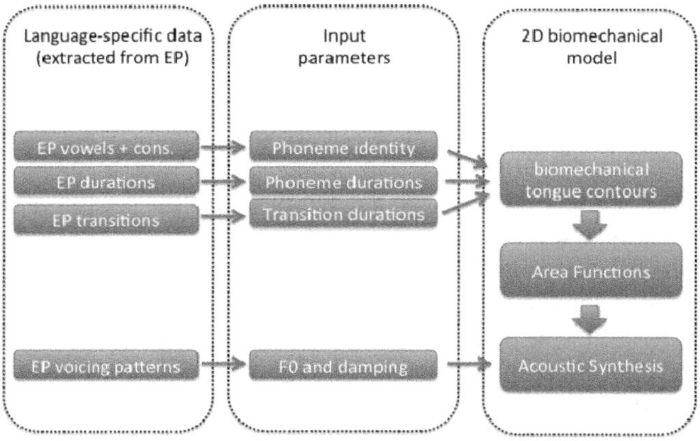

Fig. 1. Block diagram illustrating the process of integrating the EP production data into the biomechanical model

Table 1. Factors and step size for the generation of the perceptual stimuli. The steps are determined in accordance with the durational values given in section 3.1.

steps	vowel duration (ms)	consonant duration (ms)	consonant voicing (%)
1	70	50	0 (fully devoiced)
2	100	75	12.5
3	130	100	25
4		125	37.5
5		150	50
6			75
7			100 (fully voiced)

While the parameters *vowel duration* and *consonant duration* are controlled during the synthesis of the biomechanical tongue contours (by selecting appropriate *t_hold* and *t_transition* values for consonants and vowels), the parameter *consonant voicing* (or *voicing maintenance*) is generated during the following acoustic synthesis of the stimuli with the Story et al. (2000) model (see section 3.3).

In sum, for each *vowel-stop-vowel* item 15 fully factorized stimuli are generated (3 *vowel_durations* paired with 5 *consonant_durations*), which then are acoustically synthesized with 7 different *consonant_voicing* curves. The phoneme-to-phoneme transition time is set to a standard value of 60ms obtained from the EP database.

2.2 EP Real Speech Corpus

To obtain the holding and transition phase as the input into the biomechanical model it was necessary to measure durational values in the acoustic space from real EP speech data. For this aim, we computed durational measures from an extensive speech production database (see Pape & Jesus 2011) generated for this purpose. The database was recorded for 4 EP speakers, consisting of /CV_1CV_2/ items in the frame sentence *Diga CVCV outra vez*, with all EP stops and fricatives /p b t d k g f v s z ʃ ʒ/ with the (identical) preceding and following (V_1,V_2) four vowel contexts /i e o a/. Thus, the consonants can be compared in initial (**C**V_1CV_2) and medial (CV_1**C**V_2) position. Each item was randomly repeated 9 times. Vowels and consonant boundaries are defined on base of the onset and offset of stable formant structure. We concentrate here on velar stops, the consonant modelled later in different contexts.

3 Results

3.1 Durational Measures from the EP Real Speech Corpus

Figure 2 shows the durational measures for the preceding vowel and velar stop in medial position. As can be seen, the consonant duration is higher for voiceless stops. Further, the vowel preceding the consonant is nearly twice as long for the voiced stop when compared to its voiceless counterpart. Thus, our EP data show clear differences for both preceding vowel and consonant duration between voiceless and voiced velar stops. The consonant durations are in line with EP durations found by Veloso (1995a, 1995b) and Delgado Martins (1975). With respect to the expected intrinsic vowel duration differences (with low vowels being longer than high vowels (Lehiste 1970)) our database confirms this tendency for the preceding vowel duration.

We computed an ANOVA with the duration measures as dependent variable and the consonant identity /k g/ as factor. For both the initial and the medial consonant position, the consonant duration was highly significantly (initial: $p < 0.001$; medial: $p < 0.001$). Further, the preceding vowel duration was highly significant for the medial consonant position ($p < 0.001$) but not for the initial position ($p = 0.54$). The duration of the following vowel was not significant for the medial position ($p = 0.35$). For initial position, nothing can be concluded about the following vowel duration, since the following vowel for the initial position is the preceding vowel for the medial position, so it is not clear which of the consonants influences the durational cues.

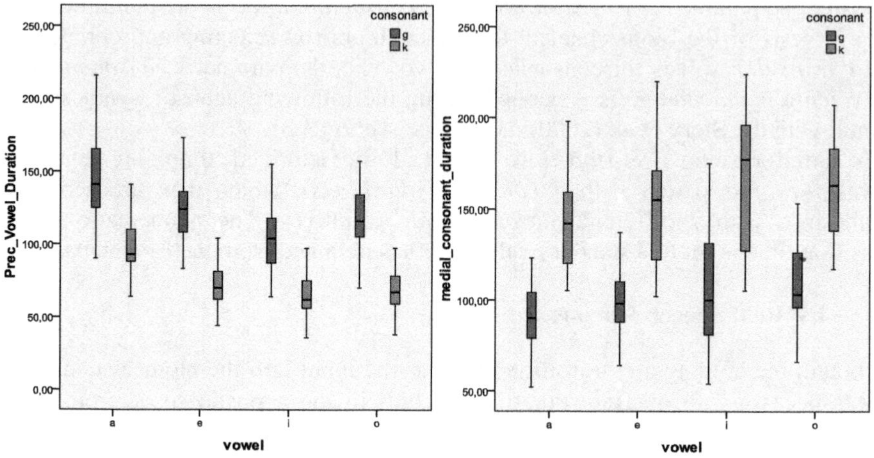

Fig. 2. Boxplots of the preceding vowel duration (left panel) and velar stop duration (right panel) for medial position (all values in milliseconds), with the neighbouring vowel context on the x-axis. The voiced stop is shown in darker shading, the voiceless stop in lighter shading. The voiced and voiceless character of each item is defined by its phonological status.

3.2 Tongue Contour Trajectories for the EP Stimuli

For the biomechanical EP VCV stimulus (velar consonant, 100ms vowel duration, 125ms consonant duration), figure 3 shows the trajectories of selected tongue nodes for the production of /aka eke iki oko/ stimuli. All four plots clearly confirm that the biomechanical model correctly reproduces the forward articulatory loop that can be observed for natural velar stop production (Mooshammer et al. 1995).

We compared our modelled biomechanical stimuli with target-to-target synthesis approaches. These approaches are the standard in articulatory synthesis. Examples can be found in the Maeda (1990) approach, the 3D articulatory synthesizer (Birkholz et al., 2010), the SpeechTrainer (Kröger 2003) and SAPWindows (Teixeira et al. 2005). For all these synthesis approaches, a number of articulatory targets are defined, and trajectories are achieved by a linear interpolation between targets. Figure 4 shows the illustration of the difference between several points on the tongue contour in the biomechanical model (in black) and targets for a target-to-target synthesis approach (dots in light colour). As can be seen in the right panel, the definition of one articulatory consonant target (here the highest point of the tongue contour) would results in very different articulatory tongue trajectories when comparing the biomechanical model (solid lines) and target-to-target approaches (dashed lines). It has to be noted that the differences between biomechanical and other synthesis approaches are less notorious when two consonantal targets are considered (e.g., the beginning and end of velar contact), however this approach is rarely used (Pape et al. 2011). Thus, the target-to-target approaches traditionally used for articulatory synthesis suffer from unrealistic modelling of the articulatory trajectories between targets, and are therefore unable to successfully model the articulatory loops seen in real speech articulatory data. Only the biomechanical model is able to reproduce these loops, as shown for all vowels and tongue nodes in the trajectories in figure 4.

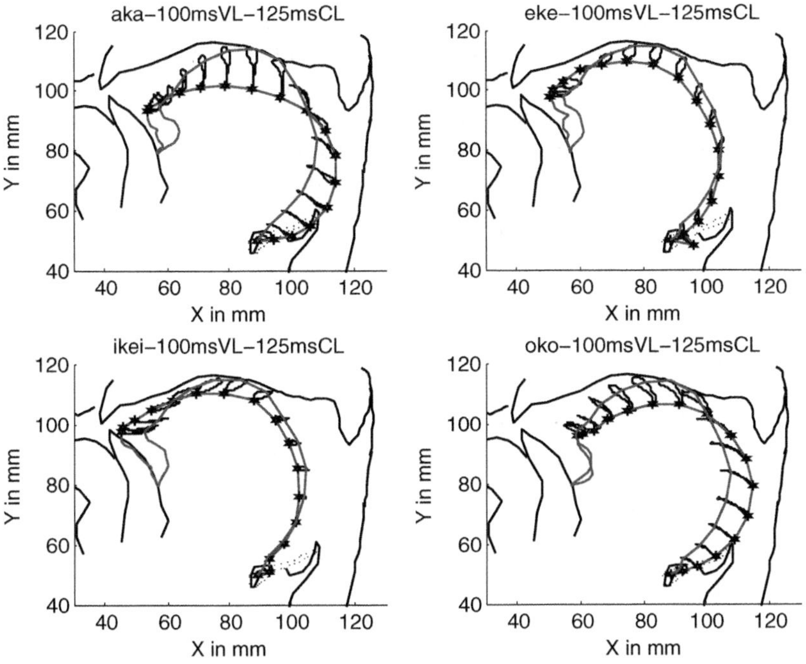

Fig. 3. Trajectories of selected tongue nodes for /VkV/ stimuli for the vowel contexts /a e i o/. The lines in light colour show the vowel and velar stop target positions.

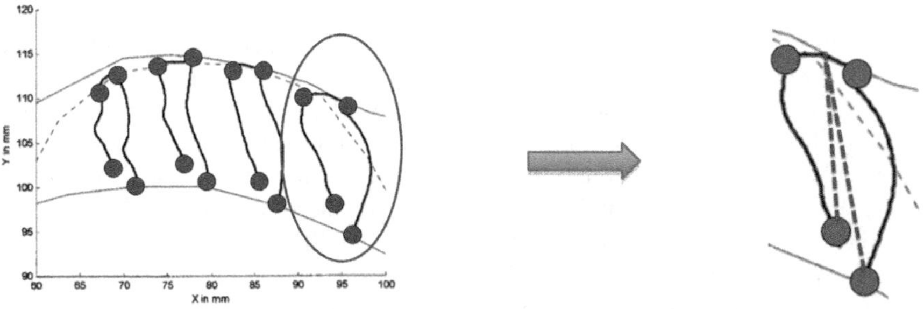

Fig. 4. The left panel shows four tongue surface trajectories for the /aka/ stimulus generated with the biomechanical model (in black). The dots in light colour exemplify the articulatory targets for a target-based synthesis. The right panel shows the differences in one tongue node trajectory between the biomechanical model (solid lines) and a given linear target-to-target trajectory (dashed lines, articulatory target at the highest point of the tongue contour).

3.3 Consonant Voicing Contours

To obtain the shapes of the velar stop consonant voicing curves we computed the mean curve over all repetitions and vowel contexts for each speaker of the EP database. Figure 5 shows the mean voicing probability estimate (9 repetitions, four vowels) computed for 10 consecutive equidistant landmarks throughout normalised consonant duration (see Pape & Jesus 2011). As can be seen, none of the voiced velar stops shows a linear decrease in voicing probability. Instead, the voicing decreases slower at the onset/offset of the curves, but faster during the mid.

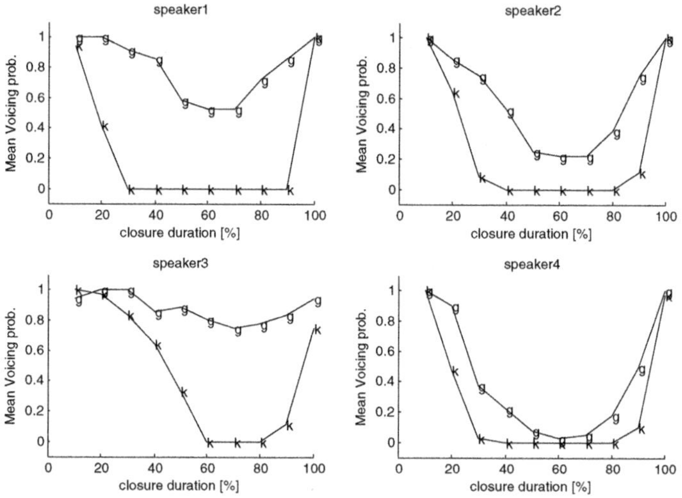

Fig. 5. Mean voicing probability curves (over all vowel contexts and repetitions) for the EP velar stops during the normalised stop closure. Each panel corresponds to one EP speaker of the database, each line represents the mean of 36 items (9 repetitions x 4 vowel contexts).

Due to this non-linear behaviour throughout the consonant closure, we chose to model our EP biomechanical stimuli by applying more realistic non-linear amplitude damping throughout the consonant closure. This was achieved by multiplying a weight to the consonant closure of the acoustically synthesized stimulus (i.e., the acoustic synthesis of the biomechanical tongue contours). In acoustic terms, this stimulus would correspond to the acoustic prototype of the /VgV/ biomechanical synthesis without amplitude damping due to the closure of the vocal tract and glottal damping (Pape et al. 2011). The obtained amplitude damping for the generation of the different consonant voicing contours was generated by applying a weighting function to the amplitude of the /VgV/ biomechanical stimulus. The shape of the weighting function was set as a cosine function ranging from 0 to $\pi/2$. The length of the cosine decrease was set to one quarter of the complete consonant duration. It was held identical for all *consonant_voicing* curves (steps given in table 1). The percentages of the *consonant_voicing* steps from table 1 define the landmarks where the cosine weighting reaches zero ($\pi/2$). Further, to avoid possible audible distortions due to jumps from full devoicing (zero) to full voicing amplitude we modelled the last 10ms of the consonant closure by weighting the fully voiced consonant with a sine function ranging from 0 to $\pi/2$. Figure 6 shows the modelled normalised six different weighting curves of the *consonant_voicing* steps (125ms stop duration, f_s=48kHz). The procedure is illustrated for the 50% *consonant_voicing* curve (step5): At the consonant onset, no damping is applied to the acoustic synthesis output. Ranging from the first quarter until the consonant duration midpoint the amplitude is damped with a cosine weighting. The fully devoiced status is maintained until 10ms prior to consonant offset, where the 10ms sine-shaped increase to the consonant offset begins. The fully devoiced condition (step1), was modelled by a 5ms cosine decrease at consonant onset and identical 10ms sine rise at consonant offset. The right panels in figure 6 show oscillograms of the acoustic synthesis for four different *consonant_voicing* steps with the overlaid weighting function envelopes.

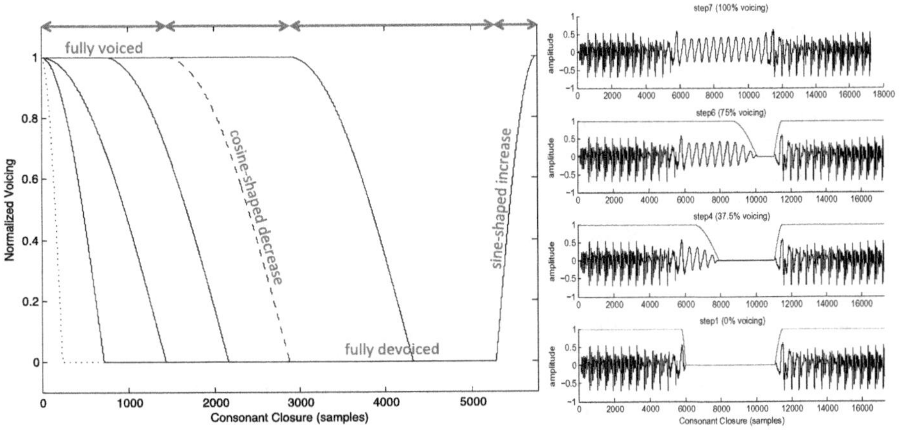

Fig. 6. The left panel shows modelled devoicing curves (step1 to step6 in table 1) for a 120ms stop closure. The voicing extinction curve for step5 (full devoicing at consonant mid) is shown as a dashed line, the fully devoiced condition (step1) is shown as a dotted line. The four different voicing statuses throughout the consonant closure for step5 are given in light colour, with arrows corresponding to the duration. The right panel shows the oscillograms for four steps of the consonant voicing continuum: top and bottom panel show the vertices (fully voiced and fully devoiced); second and third panel show the partly voiced conditions (second panel step6 = 75% voiced; third panel step4 = 37.5%voiced).

4 Conclusions

For a perceptual experiment examining European Portuguese (EP) stop voicing we modelled physically realistic stimuli by means of a biomechanical model. The input to the biomechanical model, i.e., phoneme identities, phoneme durations, transitions between phonemes and voicing extinction curves, were generated from an extensive EP real speech corpus. The resulting EP biomechanical tongue contours comply with articulatory and aerodynamic laws, e.g., articulatory loops observed for velar stops. The models (voicing curves) for the generation of different *devoicing* conditions were based on the voicing curves in the EP database. The extraction of the EP velar stops devoicing slopes showed a nonlinear behaviour in real speech.

Acknowledgements. This work was partially supported by the Portuguese Fundação para a Ciência e a Tecnologia (FCT), Portugal (grant SFRH/BPD/ 48002/2008).

References

Aslin, R., Pisoni, D., Hennessey, B., Perry, A.: Discrimination of voice onset time by human infants: New findings and implications for the effect of early experience. Child Development 52, 1135–1145 (1981)

Birkholz, P., Kröger, B., Neuscheafer-Rube, C.: Articulatory synthesis and perception of plosive-vowel syllables with virtual consonant targets. In: Proc. Interspeech 2010, pp. 1017–1020 (2010)

Brunner, J., Perrier, P., Fuchs, S.: Supralaryngeal control in Korean velar stops. Journal of Phonetics 39, 178–195 (2011)

Cuartero, N.: Voicing Assimilation in Catalan and English. Ph.D. Thesis, Universitat Autònoma de Barcelona, Barcelona, Spain (2002)

Delgado Martins, M.R.: Vogais e consoantes do Português: estatística de ocorrência, duração e intensidade. Bol. De Filologi 14, 1–11 (1975)

Escudero, P., Benders, T., Lipski, S.: Native, non-native and L2 perceptual cue weighting for Dutch vowels: The case of Dutch, German, and Spanish listeners. Journal of Phonetics 37, 452–466 (2009)

Francis, A., Baldwin, K., Nusbaum, C.: Effects of training on attention to acoustic cues. Attention, Perception & Psychophysics 62, 1668–1680 (2000)

Fuchs, S., Perrier, P.: On the complex nature of speech kinematics. ZAS Papers in Linguistics 42, 137–165 (2005)

Hoonhorst, I., Colin, C., Markessis, E., Radeau, M., Deltenre, P., Serniclaes, W.: French native speakers in the making: From language-general to language-specific voicing boundaries. Journal of Experimental Child Psychology 104, 353–366 (2009)

Jakobson, R., Fant, C., Halle, M.: Preliminaries to Speech Analysis. MIT Press, Cambridge (1952)

Jessen, M.: Phonetics and phonology of tense and lax obstruents in German. John Benjamins, Amsterdam (1998)

Kröger, B.: Einvisuelles Modell der Artikulation. Laryngo-Rhino-Otologie 82, 402–407 (2003)

Lasky, R., Syrdal-Lasky, A., Klein, R.: VOT discrimination by four to six and a half month old infants from Spanish environments. Journal of Experimental Child Psychology 20, 215–225 (1975)

Lehiste, I.: Suprasegmentals. MIT Press, Cambridge (1970)

Lisker, L., Abramson, A.: Cross-language study of voicing in initial stops. Word 20(3), 384–422 (1964)

Lisker, L., Abramson, A.: Some effects of context on voice onset time in English stops. Language and Speech 10, 1–28 (1967)

Lisker, L., Abramson, A.: The voicing dimension: Some experiments in comparative phonetics. In: Proc.ICPhS, pp. 563–567 (1970)

Luce, P., Charles-Luce, J.: Contextual effects on vowel duration, closure duration, and the consonant vowel ratio in speech production. JASA 78, 1949–1957 (1985)

Maeda, S.: Compensatory articulation during speech: Evidence from the analysis and synthesis of vocal-tract shapes using an articulatory model. In: Hardcastle, W., Marchal, A. (eds.) Speech Production and Speech Modeling, pp. 131–149 (1990)

Mooshammer, C., Hoole, P., Kühnert, B.: On loops. Journal of Phonetics 23, 3–21 (1995)

Morrison, G.: An appropriate metric for cue weighting in L2 speech perception: Response to Escudero&Boersma (2004). Studies in Second Language Acquisition 27(4), 597–606 (2005)

Pape, D., Jesus, L.: Devoicing of phonologically voiced obstruents: Is European Portuguese different from other Romance languages? In: Proc. ICPhS, pp. 1566–1569 (2011)
Pape, D., Jesus, L., Hall, A.: A cross-linguistic comparison of stop and fricative devoicing, Part I: Speech production. Journal of Phonetics (submitted)
Pape, D., Perrier, P., Fuchs, S., Kandel, S.: Les trajectoiresformantiquesrespectant les lois de la physique contribuent-elles à unemeilleure perception de la parole? In: Actes JEP (2010)
Pape, D., Perrier, P., Fuchs, S., Kandel, S.: Does physical realism of articulatory modeling improve the perception of synthetic speech? In: Proc. ISSP (2011)
Payan, Y., Perrier, P.: Synthesis of V-V Sequences with a 2D biomechanical tongue model controlled by the Equilibrium Point Hypothesis. Speech Communication 22, 185–205 (1997)
Perrier, P., Boë, L., Sock, R.: Vocal tract area function estimation from midsagittal dimensions with CT scans and a vocal tract cast: Modeling the transition with two sets of coefficients. Journal of Speech and Hearing Research 35, 53–67 (1992)
Perrier, P., Fuchs, S.: Speed-curvature relations in speech production challenge the 1/3 power jaw. Journal of Neurophysiology 100, 1171–1183 (2008)
Perrier, P., Ma, L., Payan, Y.: Modeling the production of VCV sequences via the inversion of a biomechanical model of the tongue. In: Proc.Interspeech, pp. 1041–1044 (2005)
Perrier, P., Payan, Y., Zandipour, M., Perkell, J.: Influences of tongue biomechanics on speech movements during the production of velar stop consonants: A modeling study. JASA 114(3), 1582–1599 (2003)
Repp, B.: Relative Amplitude of Aspiration Noise as a Voicing Cue for Syllable-Initial Stop Consonants. Language and Speech 22(29), 173–189 (1979)
Serniclaes, W.: Etude expérimentale de la perception du trait de voisement des occlusives du français. Ph.D. Thesis, Université Libre de Bruxelles, Brussels, Belgium (1987)
Solé, M.: Articulatory Adjustments in Initial Voiced Stops in Spanish, French and English. In: Proc. ICPhS, pp. 1878–1881 (2011)
Story, B., Laukkanen, A., Titze, I.: Acoustic impedance of an artificially lengthened and constricted vocal tract. Journal of Voice 14(4), 455–469 (2000)
Story, B., Titze, I.: Voice simulation with a body-cover model of the vocal folds. JASA 97, 1249–1260 (1995)
Sussman, H., Fruchter, D., Hilbert, J., Sirosch, J.: Linear correlates in the speech signal: The orderly output constraint. Behavioral and Brain Sciences 21, 241–299 (1998)
Teixeira, A., Martinez, L., Silva, O., Jesus, L., Principe, J., Vaz, F.: Simulation of human speech production applied to the study and synthesis of European Portuguese. EURASIP Journal on Applied Signal Processing 9, 1435–1448 (2005)
Titze, I., Story, B.: Rules for controlling low-dimensional vocal fold models with muscle activation. JASA 112, 1064–1076 (2002)
Veloso, J.: The role of consonantal duration and tenseness in the perception of voicing distinctions of Portuguese stops. In: Proc. ICPhS, pp. 266–269 (1995a)
Veloso, J.: Aspectos da percepção das "ocusivas fricatizadas" do português. Contributo para a compreensão do processamento de contrastes alofónicos. M.Sc. Thesis, Universidade do Porto, Porto, Portugal (1995b)
Viana, M.: Etude de deuxaspects de consonantismeduportugais: fricatisationetdevoisement. Ph.D. Thesis, LSHA Strasbourg, France (1984)
Viviani, P., Stucchi, N.: Biological movements look uniform: evidence of motor-perceptual interactions. Journal Experimental Psychological Human Perceptual Performance 18, 603–623 (1992)
Williams, L.: The voicing contrast in Spanish. Journal of Phonetic 5, 169–184 (1977)

Automatic Phonetic Transcription by Phonological Derivation

Marcos Garcia[1] and Isaac J. González[2]

[1] Center for Research in Information Technologies (CITIUS), University of Santiago de Compostela
[2] Cilenis Language Technology
marcos.garcia.gonzalez@usc.es, isaacjgonzalez@cilenis.com
http://gramatica.usc.es/pln
http://www.cilenis.com

Abstract. Automatic phonetic transcription tools usually perform phonetic transcriptions directly from orthographic representations. Although these approaches often achieve good results, theoretical studies suggest that including morphophonological knowledge allows those systems to improve their performance. Following this idea, we developed a tool which first obtains an underlying representation of each word, using small lexica and dedicated lemmatizers. For each representation, a phonological derivation generates the phonetic transcription by applying linguistically motivated rules. Since most of these rules are added as optional parameters, the system permits to generate dialect-specific transcriptions. This system is not only a grapheme-to-phone tool, but it also obtains phonological representations and evaluates several linguistic processes occurring during the derivation. Preliminary experiments emulating a phonological system of Galician (using as input words spelled in European Portuguese) show that the underlying representation of most words can be obtained using small lexica and also that the derivation produces high-quality phonetic transcriptions.

Keywords: grapheme-to-phoneme, phonetics, phonology, galician, portuguese.

1 Introduction

Automatic Phonetic Transcription (APT) is a crucial task for many applications of different areas. Besides text-to-speech systems, which need high quality transcriptions, APT tools are also used in theoretical and applied linguistics. These tools are useful in many areas (e.g., phonetics, phonology, dialectology, language learning, etc.) in order to obtain preliminary transcriptions of large corpora.

Rule-based APT systems often generate a phonetic transcription directly from the orthographic form. These approaches achieve good performance in languages whose spelling systems permit to *easily* infer their phonetic representation, such as Italian, Spanish or Portuguese. However, algorithms which do not take into

H. Caseli et al. (Eds.): PROPOR 2012, LNAI 7243, pp. 350–361, 2012.

account morphophonological information may produce some errors due to their lack of this linguistic knowledge [1]. In this respect, the transcription of Galician and Portuguese varieties presents some problems. For example, the orthographic vowels <e> and <o> could represent up to six phones (e.g., [ɛ, e, ɨ, o, ɔ, u]), some of them not being predictable from their context.

The transcription of these not predictable examples could be performed through rules and lists of exception words. Nevertheless, if there is no morphophonological knowledge, these lists may be very large, since they should include inflected forms: nominal forms (from 'festa') such as 'f[ɛ]sta', 'f[ɛ]stas', 'f[ɛ]stinha' (or 'f[e]stinha') and verbal forms (from 'levar') such as 'l[ɛ]vo', 'l[ɛ]vas', 'l[ɛ]va', etc. Thus, the creation of these lists would be time-consuming.

In this paper, a strategy to overcome these limitations is proposed. The behaviour of the phonemes in the root of the words is consistent and predictable among their inflected forms, so we use shorter lists of exceptions, including only those lemmas which have characters which do not follow the default conversion rules. This way, by applying a rule-based lemmatizer on the inflected words —as well as a set of specific rules for each linguistic variety— the system obtains the target lemma and verifies whether it is in an exception list. This process allows the system to generate an underlying (phonological) representation, which is an abstract form of the word before the application of the phonological rules.

Then, a phonological derivation produces phonetic transcriptions by applying rules on the underlying representations, as shown in Figure 1.

<festinha> / <festiña>	**Orthographic Representation**
<festa> (Exception: root vowel: /ɛ/)	*Lemma*
/fɛstiɲa/	**Underlying Representation**
/fɛs.ti.ɲa/	*Syllable Split*
/fɛsˈti.ɲa/	*Stress assignment*
/fɛsˈtĩ.ɲa/	*Nasalization*
/fɛsˈtĩ.ɲɐ/	*Unstressed vocalism*
[fɛsˈtĩ.ɲɐ]	**Phonetic Representation**

Fig. 1. Example of the conversion of the word 'festinha/festiña'

It is worth noting that different rules —belonging to several dialects— generate diverse transcriptions from the same underlying form. Thus, the selection of the rules involves different outputs (for dialects sharing the same underlying form). In our system, dialectal rules were added as optional parameters, so it is capable of automatically transcribing lexica for different dialects.

The system presented in this paper (released under GPL license,[1] and whose first version was presented in [12]) emulates the phonological system of Galician varieties by following the mentioned strategy. The input can be written in two different spellings: ILG/RAG [14] and European Portuguese (EP).

[1] `www.gnu.org/licenses/gpl.txt`

The main advantages of using the proposed method are that the user can (i) obtain phonological representations, (ii) analyze phonological processes such as syllabification, (iii) evaluate the phonological derivation as well as interact with it, and (iv) create phonetic representations for different linguistic varieties.

Experiments performed on a Portuguese corpus show that the system obtains underlying representations accurately. Furthermore, the results also indicate that the phonological derivation generates high quality phonetic transcriptions.

This paper is organized as follows: Section 2 shows some related work concerning automatic phonetic transcription of Galician and Portuguese. In Section 3, we briefly introduce the theoretical background of our method, whose architecture is presented in Section 4. Then, Section 5 contains the performed experiments and, finally, conclusions and further work are addressed in Section 6.

2 Related Work

Much of the work on automatic phonetic transcription has been done focused on text to speech synthesis. However, there is also other approaches related to our work that have to be taken into account.

Rule-based models were implemented in order to perform grapheme-to-phone conversion of Galician (and EP) varieties, achieving results of more than 98% precision [3,5]. Another work focused on the automatic transcription of Galician presents the main characteristics and difficulties of this task [13]. Besides the segmental features mentioned above (the transcription of <e>, <o> or <x>), this paper also introduces some other aspects, such as contraction forms or intonation issues. Moreover, the development of a corpus and a lexicon for Galician text-to-speech systems is presented in [8].

The disambiguation of heterophonic homographs is another important process of the automatic phonetic transcription task. For Galician, the main difficulties are presented in [19].

There are much work focused on the APT of Portuguese varieties, from rule-based to stochastic models [4,21,25,27]. For our goals, it is interesting to refer [26], which improves the precision of syllable splitting by applying phonological theories, namely the onset-rhyme theory. Another work which includes phonological processing is [24], which generates phonetic variants for speech recognition.

In [1], is presented a project whose aim is the population —taken into account morphological information— of a large lexicon of Portuguese with different phonetic transcriptions of several linguistic varieties.

Finally, another tool which uses linguistic knowledge is FreP [28]. FreP automatically extracts data about frequency and linguistic contexts of phonological entities, allowing a phonologist to easily obtain this kind of information.

3 Theoretical Background

The system presented in this paper was designed following some phonological theories, that we briefly describe in this section.

The main idea of our method is the use of several representation levels as proposed in classical generative phonology [9]. This theory describes the phonology of languages as derivational systems with different abstraction levels. In this view, the underlying form of a word —also known as *deep* or *phonological* representation— is transformed into a surface (or phonetic) form by the application of transformational rules in the phonological derivation process.

Lexical phonology [18], proposed as a refinement of classical generative phonology, introduced two main components in the description of the phonology of a language: lexical and post-lexical. In the first one, information of different levels (morphology, syntax, etc.) plays a role in the derivation, while the post-lexical component has no access to that knowledge. The application of post-lexical rules on the output of the lexical component generates the final surface form.

From this point of view, it can be postulated that the members of a linguistic community share the underlying forms of their language. So, different rules (or rule order) will generate several surface forms, corresponding to different variants of the same phonological word. For instance, we can assume that the underlying form of 'feira' for EP speakers is /feiɾa/. Thus, different realizations such as 'f[ej]ra', 'f[ɐj]ra' or 'f[e]ra' are generated due to differences in the derivation.

Classical generative phonology did not take into account syllables as units of analysis, so the explanation of many phonological processes within syllables were rarely simple. Different non-linear models of phonology emerged in order to introduce the syllable —and its internal structure— into phonological theory.

Onset-rhyme (syllable) theory defines syllables as phonological units which organize segmental melodies in terms of sonority [2]. Moreover, in this theory syllables have an internal structure, exemplified in Figure 2. The phonology of a particular language will define (i) if a syllabic constituent (Onset, Coda) is obligatory or optional, (ii) what segments could occupy each position, as well as (iii) the maximum number of segments which can be anchored to each constituent. This structure is useful to explain, for instance, some phonological processes which affect phonemes depending on the syllabic position in which they occur.

These (and other) theories were applied for describing the phonology of Portuguese varieties [17]. Taking into account the differences, much of this work is useful for analyzing Galician dialects. Specifically for Galician, some phonological and morphological aspects are described in works such as [7,22,10].

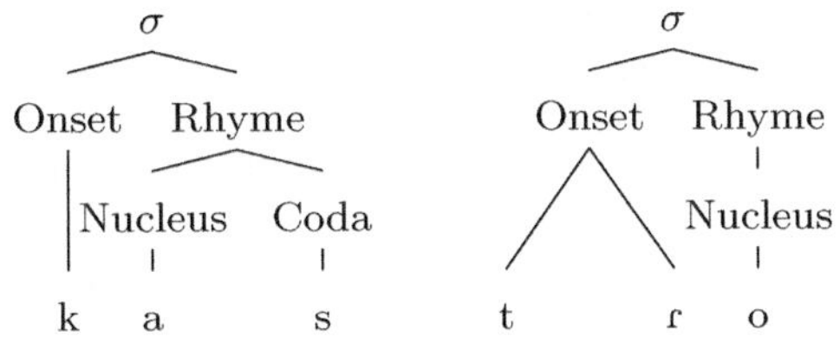

Fig. 2. Internal structure of the syllables for the word 'castro' (Onset-Rhyme Theory)

The phonological theories described in this section were taken into account in the development of our system. They were implemented with minor variations, in order to build an APT tool capable of generating different phonetic representations from a single underlying form. Furthermore, note that the system is also useful to obtain deep phonological representations of words as well as to automatically verify the behaviour of derivational rules on large corpora.

In this paper, we use the terms 'phoneme' and 'phone' to designate phonological and phonetic entities, respectively.

4 Architecture

Our system is composed of several modules applied in a pipeline. The different components are described in this section.

4.1 Phonological Conversion

The first step of our method is the conversion of an orthographic word into an underlying representation. Two different modules perform this process, depending on the spelling of the input (ILG/RAG or EP orthographies).

Both modules are compound of a set of transformational rules as well as lists of exceptions. The rules simply substitute orthographic characters by representations of phonological segments taking into account their context (<v>→ /b/, <ch>→ /t͡ʃ/, <rr>→ /r/, <c(aou)>→ /k(aou)/, etc.).

Those characters whose representation is not predictable by their context are included in the exception lists. We use the following, extracted from [23] (and verbal forms automatically extracted from large lexica):

- List of lemmas with <x> as /ks/ ('se/ks/o').
- Lists of lemmas with <e> as /ɛ/ and with <o> as /ɔ/ ('f/ɛ/sta', 'r/ɔ/da').[2]

By default, <x>, <e> and <o> are transformed into /ʃ/, /e/ and /o/. The words included in these lists are spelled in ILG/RAG, so we use a transliteration method in order to perform a look-up with words written in EP. For this spelling, we also use the following lists, semi-automatically extracted:

- List of lemmas with <qu/gu> followed by <e/i> as /kw/, /gw/ (fre/kw/ente).
- List of lemmas ending in <ão> as /an/ (p/an/).
- List of exceptions: *non-galician* forms ('fiz' → /fiʃen/, 'sim' → /si/).

We have to note that the transliteration strategy may produce some errors, since some words are not easily transformed with this method. However, its performance was successfully tested in several cases [15,16].

[2] To be more precise, lists of words with /ɛ, ɔ/ are not pure lemmas, but singular forms containing these vowels in the root. This way, the first list contains forms such as the adjective 'nova', but not the plural 'novas'.

Apart from the referred lists, two sets of Galician heterophonic homographs (one for each spelling system) were compiled, each one of about 600 pairs. The representation of these words depends on their PoS-tag (verb: '/ɔ/lho'; noun: '/o/lho'), semantic meaning (*weapon*: 'b/ɛ/sta'; *animal*: 'b/e/sta') or verbal mood (present: 'c/ɔ/me'; imperative: 'c/o/me'). The system does not perform disambiguation, so it generates both outputs of the heterophonic homographs.

Dedicated Lemmatizer: The use of large lists of exceptions (including all the inflected forms which behave as their lemmas) is avoided as follows:

An open-source rule-based lemmatizer which detects if the input word is an inflected form is applied. Then, the system verifies whether the input (or the obtained lemma) exists in the exception list. If it exists, the underlying representation is generated by applying the exception rules of each list (<e>→ /ɛ/, <x>→ /ks/, etc.). The lemmatizer is compound of two sub-modules: one for nominal forms (inspired in [6]), and a dedicated stemmer for verbs. Figure 3 shows some examples of this process.

Nominal metaphony in Galician dialects is usually different from Portuguese metaphony. Galician plurals maintain the vowel of the singular form ('[o]lho / [o]lhos', and not '[o]lho / [ɔ]lhos'), but the vowel changes in both singular and plural feminine forms ('n[ɔ]va / n[ɔ]vas': the masculine singular of these adjectives could be 'n[o]vo' or 'n[ɔ]vo' depending on the dialect). This way, and taking into account that the lists of these words also include singular feminine forms, the nominal lemmatizer only applies rules to infer the singular form.

Note that this implementation slightly differs from lexical phonology. In our system, some morphophonological processes (such as nominal and verbal inflection) are replaced by the use of lists and by lemmatization. Furthermore, the underlying forms could be different to others proposed in theoretical literature, which may include morpheme information or other phonological abstractions.

The output of the phonological conversion module is an underlying representation of the input word, obtained by including morphophonological processing. This output is the input of the next module, which performs syllabification.

4.2 Syllabification and Stress Assignment

The function of this module is to split the input word into syllables and to build the internal structure of each syllable, following the onset-rhyme theory. Then, the stressed syllable is detected and marked.

Input		**Lemma**		**Exception?**		**Underlying**
<levam>	$\xrightarrow{lemmatizer}$	levar	$\xrightarrow{look-up}$	yes	$\xrightarrow{Conversion\ rules}$	/lɛban/
<festas>	$\xrightarrow{lemmatizer}$	festa	$\xrightarrow{look-up}$	yes	$\xrightarrow{Conversion\ rules}$	/fɛstas/
<cestas>	$\xrightarrow{lemmatizer}$	cesta	$\xrightarrow{look-up}$	no	$\xrightarrow{Conversion\ rules}$	/sestas/

Fig. 3. Examples of the underlying form generation

A simplified variant of the algorithm proposed in [17], adapted to the Galician syllable structure, was implemented. The sonority hierarchy was removed, so we established which segments (and groups of segments) could occupy each syllabic position. Thus, in order to build the syllable structure, we follow these steps:

1. Falling diphthongs are marked as nucleus (/peiʃe/).
2. Vowels are marked as nucleus (/peiʃe/).
3. Complex onsets are marked (/pɾatos/).
4. Simple onsets followed by nucleus are marked (/pɾatos/).
5. Complex (and *non-patrimonial*) codas are marked (/abstɾaiɾ/).
6. Simple codas are marked (/abstɾaiɾ/).
7. Exception (uncommon codas, foreign words, etc.) are identified (/gangster/).

Then, syllable boundaries are marked (i) before each onset, (ii) before an initial nucleus, (iii) between two nucleus and (iv) between a coda and a nucleus.

Once syllable split is performed, the stressed syllable of each word has to be marked, in order to apply the phonological rules accurately. The system maintains the orthographical input during the conversion, so this process applies the accentuation rules inversely. It first verifies if the word has an accent, marking as stressed the syllable with it. If there is no accent, the inverse accentuation rules are applied until find the stressed syllable. There is also a small set of unstressed words, such as prepositions and pronouns ('com', 'me', 'te', etc.).

4.3 Derivational Rules

Phonological rules are applied after the syllabification process and the stress assignment, allowing them to incorporate syllabic knowledge. However, it is worth noting that some derivational rules may imply re-syllabification processes.

Derivational rules are applied sequentially, so the order of application is crucial for obtaining the desired output. Current version of our system contains both a set of *universal* rules (occurring in most of Galician dialects), and optional rules (corresponding to specific phonological processes which do not occur in every Galician variety), all of them present in the theoretical literature. The selection of some optional rules allows the user to obtain transcriptions of the standard variety as well as of non-standard Galician dialects [23].

- Universal rules:
  - Semi-vocalization: this rule transforms some adjacent vocalic phonemes into a diphthong (/so.si'al/ → [so'sjal]).
  - Unstressed vocalism: unstressed vowels are elevated and centralized in some contexts, namely in final word position (/'ko.mo/ → ['ko.mʊ]).
  - Voiced plosives: voiced plosives /b, d, g/ are pronounced as approximants ([β̞, ð̞, ɣ̞]) in some contexts (/a'bɾiɾ/ → [a'β̞ɾiɾ]).
  - Nasalization: this rule changes the place of articulation of implosive nasals. It could perform an assimilation (the nasal segment acquires the place of articulation of the following consonant: /'kan.pʊ/ → ['kam.pʊ])

or a velarization (implosive nasals are always velar: [ˈkaŋ.pʊ]). Furthermore, an additional rule nasalizes the vowels occurring in some contexts, such as between nasal segments (/ˈmaŋ.ta/ → [ˈmãŋ.ta]).

- Optional rules:
  - Thetacismo: this is a special rule to deal with a main characteristic of Galician dialects. Some of them have these two sibilants phones: [s, θ] ('caçar': [kaˈθaɾ]; 'casar': [kaˈsaɾ]), while others only have [s] ('caçar', 'casar': [kaˈsaɾ]). This exceptional rule is not derivational, and it also affects the underlying representation of the word, including both phonemes /s, θ/ or only /s/, depending on the application.
  - *Gheada*: another dialect-specific process. This rule changes every /g/ by [ħ], except when occurring after a nasal segment, allowing some other realizations ('água': [ˈa.ħwa]; 'bingo': [ˈbiŋ.gʊ], [ˈbiŋ.ħʊ] or [ˈbiŋ.kʊ]).
  - Complex codas: this rule simplifies complex codas as well as removes some *non-patrimonial* codas (/aβ̞sˈtrak̠.tʊ/ → [asˈtra.tʊ]).
  - Rhotacism: this rule affects the phoneme /s/ in some coda positions (depending on the features of the following consonant), changing its realization by [ɾ] (/ˈmes.mʊ/ → [ˈmeɾ.mʊ]).
  - Voicing: this rule will represent /s, θ/ in coda (when occurring before voiced onsets) by their voiced allophones (/ˈdes.dɪ/ → [ˈdez.dɪ]).

This module implements —with some differences— the behaviour of a phonological system as proposed by the theories described in Section 3. Rules are similar to those of classic generative phonology, but taking advantage of the knowledge provided by onset-rhyme theory as well as of regular expressions. This way, the addition, modification and deletion of the rules is a simple task that allows the user to test the behaviour of new phonological rules.

Note that the system performs phonological and phonetic conversion of individual words, so it does not include post-lexical rules applying across word boundaries. Moreover, current version of the system neither has a specific module for proper nouns, numeric expressions and acronyms.

5 Experiments

We carried out several evaluations in order to know the performance of the system transcribing into Galician a corpus written in European Portuguese.[3]

For this purpose, we randomly selected from CETEMPúblico a 10,000 token corpus.[4] Abbreviations, numbers and acronyms without vowels were removed.

First, the corpus was tokenized and PoS-tagged with FreeLing [20,11], allowing us to disambiguate several heterophonic homographs. Then, the system automatically transcribed each word. The chosen Galician variety —selected through the optional rules— was those proposed in [23] (with minor changes, due to the use of a different spelling). This output was manually revised by an

[3] A previous version of this system was evaluated on a ILG/RAG corpus in [12].

[4] `http://www.linguateca.pt/cetempublico/`

Table 1. Types of error of the orthography to phonology conversion (*Full* evaluation)

Type of Error	Errors
<x>	8
<e/o>	168
<qu/gu>	1
<ão/ões>	2
foreign words	63

expert, creating a gold standard corpus. For those words for which [23] offers more than one choice, the gold standard only includes the first transcription.[5]

The obtained corpus has about 8,500 tokens (without punctuation), 18,000 syllables and 40,000 phonemes. The internal structure of the syllables were not revised, so the evaluation of the syllabification process only takes into account the boundaries between these elements.[6]

Evaluations of two processes were performed: "orthography to phonology" and "phonology to phonetics". In order to carry out these evaluations, the errors produced during the conversion were manually revised, classifying them according to the two mentioned processes (note that PoS-tagging errors were computed as errors of the system). Furthermore, two different results were calculated: *Full*, taking into account the transcription of the whole corpus and *noForeign*, which does not evaluate the transcription of proper nouns and foreign words.

In the evaluation of the "orthography to phonology" module, we counted the errors produced during the grapheme-to-phoneme conversion, thus ignoring rising diphthongs (Table 1). Most of the errors concern the transcriptions of <e> and <o>, mainly due to errors in the transliteration of the input. Furthermore, eight of them were produced by PoS-tagging errors.

In order to evaluate the "phonology to phonetics" conversion, we computed the errors produced by (i) the syllabification, (ii) the stress assignment as well as (iii) the derivational modules.

First, the precision of the syllabification was calculated, also taking into account the derivational rules which involve re-syllabification. The system produced 97 errors (75 with a *noForeign* evaluation) on 14,155 syllable boundaries (the splits between words were not evaluated), achieving a precision of 99.31%.

In our system, stress assignment is highly influenced by the previous process (syllable splitting). Thus, 62 of the 102 errors produced in this stage were motivated by previous splitting errors. Nevertheless, these 102 errors point out that this module has a precision of 99.46%.

Table 2 contains the results of the two processes. The first two lines show that the orthography to phonology conversion achieved a precision of more than 99%.

---

[5] The software and the gold standard corpus will be available at the following website: `http://gramatica.usc.es/~marcos/software/`

[6] Syllable boundaries were annotated following the formal register described in [22]. In addition, rising diphthongs were considered as phonetic rather than phonological: <social>: /sɔθial/ → /sɔ.θi.al/ → [...] → [sɔ'θjal].

Table 2. Results of the two processes as well as the full phonetic conversion

Process	Evaluation	Phonemes	Errors	Precision
Orthography → Phonology	*Full*	39,394	242	99.39%
	noForeign	36,052	153	99.58%
Phonology → Phonetics	*Full*	39,394	199	99.48%
	noForeign	36,052	130	99.64%
Orthography → Phonetics	*Full*	39,394	441	98.88%
	noForeign	36,052	283	99.22%

Central lines indicate the precision of the phonology to phonetics conversion, which also scored more than 99%. Finally, the bottom lines of Table 2 show the results of the whole conversion (orthography to phonetics).

A deep evaluation of the errors shows that, as said above, many of them were produced by orthographic issues. The use of external resources (such as the exception lists and the lemmatizers) in a different spelling system necessarily involves some errors during the conversion. Other errors were generated by not listed exceptions (namely cases of hiatus/diphthong, and few errors produced by the lemmatization strategy), by foreign/uncommon words and by heterophonic homographs whose disambiguation needs semantic information. However, taking the above into account, the obtained results show that the performance of our method is comparable to other state-of-the-art systems for similar languages.

6 Conclusions and Further Work

In this paper, an open-source system which performs automatic phonetic transcription by phonological derivation was presented.

The performed experiments, although preliminary, show that the use of morphophonological processing permits to obtain accurate phonological representations (even from two different spelling systems). Moreover, the application of phonological theories as well as the use of linguistically motivated rules overcome some limitations of previous grapheme-to-phoneme approaches, allowing the system to automatically generate high-quality phonetic transcriptions of different dialects.

Further work will be focused on correcting and increasing the exception lists, on the inclusion of dedicated rules for guessing the vowel height of unknown words as well as on the improvement of the transliteration strategy. Moreover, the system also needs to include a preprocessing module capable of analyzing abbreviations, numbers and acronyms.

Finally, we plan to evaluate this method on the automatic transcription of European and Brazilian Portuguese dialects, by taking into account the morphophonological differences compared to Galician varieties.

References

1. Ashby, S., Ferreira, J.P.: The Role of Morphology in Generating High-Quality Pronunciation Lexica for Regional Variants of Portuguese. In: Pardo, T.A.S., Branco, A., Klautau, A., Vieira, R., de Lima, V.L.S. (eds.) PROPOR 2010. LNCS, vol. 6001, pp. 162–165. Springer, Heidelberg (2010)
2. Blevins, J.: The Syllable in Phonological Theory. In: Goldsmith, J.A. (ed.) The Handbook of Phonological Theory, pp. 206–244. Blackwell, Cambridge (1995)
3. Braga, D., Coelho, L.: Letter-to-sound conversion for Galician TTS systems. In: Actas de las IV Jornadas en Tecnologia del Habla, Zaragoza, pp. 171–176 (2006)
4. Braga, D., Coelho, L., Resende Jr., F.: A Rule-Based Grapheme-to-Phone Converter for TTS Systems in European Portuguese. In: Proceedings of the VI International Telecommunications Symposium (ITS 2006), Fortaleza, pp. 328–333 (2006)
5. Braga, D., Freixeiro, X.R.: Algoritmos de Conversão Grafema-Fone em Galego para Sistemas de Conversão Texto-Fala. In: Estudos galegos de Tradución & Paratradución no século XXI, Xerais, Vigo (2007)
6. Branco, A., Silva, J.: Very High Accuracy Rule-Based Nominal Lemmatization with a Minimal Lexicon. In: Actas do XXI Encontro Anual da Associação Portuguesa de Linguística (2007)
7. Castro, O.: Aproximación a la fonología y morfología gallegas. PhD Thesis, Georgetown University (1989)
8. Campillo, F., Braga, D., Mourín, A.B., García-Mateo, C., Silva, P., Sales Dias, M., Méndez, F.: Building High Quality Databases for Minority Languages such as Galician. In: Proceedings of the 7th International Conference on Language Resources and Evaluation (LREC 2010), ELRA, La Valleta (2010)
9. Chomsky, N., Halle, M.: The Sound Pattern of English. Harper and Row, New York (1968)
10. Dubert García, F.: Máis sobre o rotacismo de /s/ en galego. In: Álvarez, R., Vilavedra, D. (eds.), Cinguidos Por Unha Arela Común. Homenaxe ó Profesor Xesús Alonso Montero, pp. 367–387. Universidade de Santiago de Compostela (1999)
11. Garcia, M., Gamallo, P.: Análise Morfossintáctica para Português Europeu e Galego: Problemas, Soluções e Avaliação. Linguamática. Revista para o Processamento Automático das Línguas Ibéricas 2(2), 59–67 (2010)
12. Garcia, M., González, I.J.: Conversión Fonética Automática con Información Fonológica para el Gallego. Procesamiento del Lenguaje Natural 47, 283–291 (2011)
13. González González, M., Banga, E.R., Campillo, F., Méndez, F., Rodríguez Liñares, L., Iglesias, G.: Specific features of the Galician language and implications for speech technology development. Speech Communication 50, 874–887 (2008)
14. ILG/RAG: Normas Ortográficas e Morfolóxicas do Idioma Galego. Real Academia Galega and Instituto da Lingua Galega, Vigo (2005)
15. Malvar, P., Pichel, J.R., Senra, Ó., Gamallo, P., García, A.: Vencendo a escassez de recursos computacionais. Carvalho: Tradutor Automático Estatístico Inglês-Galego a partir do corpus paralelo Europarl Inglês-Português. Linguamática. Revista Para o Processamento Automático Das Línguas Ibéricas 2(2), 31–38 (2010)
16. Malvar, P., Pichel, J.R.: Generación semiautomática de recursos de Opinion Mining para el gallego a partir del portugués y el español. In: ICL: Workshop on Iberian Cross-Language NLP tasks. 27th Conference of the Spanish Society for Natural Language Processing. Huelva (2011)

17. Mira Mateus, M.H., Andrade, E.d.: The Phonology of Portuguese. Oxford University Press, Oxford (2000)
18. Mohanan, K.P.: The Theory of Lexical Phonology. Dordrecht, Reidel (1986)
19. Mourín, A., Braga, D., Coelho, L., García-Mateo, C., Campillo, F., Dias, M.: Homograph Disambiguation in Galician TTS Systems. In: IX Congreso Internacional da Asociación Internacional de Estudos Galegos. A Coruña - Santiago de Compostela - Vigo (2009)
20. Padró, L.: Analizadores Multilingües en FreeLing. Linguamática. Revista para o Processamento Automático das Línguas Ibéricas 3(2), 13–20 (2011)
21. Paulo, S., Oliveira, L.C., Mendes, C., Figueira, L., Cassaca, R., Viana, C., Moniz, H.: DIXI – A Generic Text-to-Speech System for European Portuguese. In: Teixeira, A., de Lima, V.L.S., de Oliveira, L.C., Quaresma, P. (eds.) PROPOR 2008. LNCS (LNAI), vol. 5190, pp. 91–100. Springer, Heidelberg (2008)
22. Regueira, X.L.: A sílaba en galego: lingua, estándar e ideoloxía. In: Lorenzo, R. (ed.), Homenaxe a Fernando R. Tato Plaza, pp. 235–254. Universidade de Santiago de Compostela, Santiago de Compostela (2002)
23. Regueira, X. L.: Dicionario de Pronuncia da Lingua Galega. Real Academia Galega and Instituto da Lingua Galega, A Coruña (2010)
24. Seara, I.C., Pacheco, F.S., Seara Júnior, R., Kafka, S.G., Klein, S., Seara, R.: Geração Automática de Variantes de Léxicos do Português Brasileiro para Sistemas de Reconhecimento de Fala. In: Actas do XX Simpósio Brasileiro de Telecomunicações. Rio de Janeiro (2003)
25. Siravenha, A.C., Neto, N., Macedo, V., Klautau, A.: Uso de Regras Fonológicas com Determinação de Vogal Tônica para Conversão Grafema-Fone em Português Brasileiro. In: Proceedings of the 7th International Information and Telecommunication Technologies Symposium (I2TS 2008), Foz do Iguaçu (2008)
26. Oliveira, C., Castro Moutinho, L., Teixeira, A.J.S.: On European Portuguese automatic syllabification. In: Proceedings of Interspeech 2005, pp. 2933–2936 (2005)
27. Veiga, A., Candeias, S., Perdigão, F.: Generating a Pronunciation Dictionary for European Portuguese Using a Joint-Sequence Model with Embedded Stress Assignment. In: Proceedings of the 8th Brazilian Symposium in Information and Human Language Technology (STIL 2011), pp. 144–153 (2011)
28. Vigário, M., Martins, F., Frota, S.: A ferramenta FreP e a frequência de tipos silábicos e classes de segmentos no Português. In: Selecção de Comunicações apresentadas no XX Encontro Nacional da Associação Portuguesa de Linguística, pp. 675–687 (2006)

The C-ORAL-BRASIL I: Reference Corpus for Informal Spoken Brazilian Portuguese

Tommaso Raso* and Heliana Mello*

Universidade Federal de Minas Gerais, Belo Horizonte, Brazil
{tommaso.raso,heliana.mello}@gmail.com

Abstract. The C-ORAL-BRASIL is a Brazilian Portuguese spontaneous speech corpus, representative of the state of Minas Gerais diatopy (primarily from the capital city, Belo Horizonte,metropolitan area). The corpus was compiled following the same architecture and segmentation criteria adopted by the C-ORAL-ROM [1] as well as its alignment software, the WinPitch [2]. The corpus comprises 139 informal speech texts, 208,130 words, 21:08:52 hours of recording (6.1 GB wav files). The mean word number per text is 1,500. The recordings were carried out with high resolution, non-invasive wireless equipment, generally with clip-on, monodirectional microphones, and a mixer whenever there were more than two interactants, in a few occasions omnidirectional microphones were used. The texts are transcribed following the CHAT format [3], implemented for prosodic annotation [4]. The main goals for the corpus architecture are the documentation of the diaphasic and diastratic variations in Brazilian Portuguese speech.

Keywords: Spontaneous speech, Brazilian Portuguese, Corpus compilation, PoS tagging, information structure.

1 Introduction

The C-ORAL-BRASIL is a Brazilian Portuguese spontaneous speech corpus compiled following the same architecture and segmentation criteria adopted by the C-ORAL-ROM [1] as well as its alignment software, the WinPitch [2]. The compilation project foresaw the documentation of actional speech in order to record a broad variety of illocutionary acts. In so doing, this corpus can be used not only for pragmatic studies, but for morphosyntactic, prosodic and semantic inquiries as well. In this paper we will only introduce the corpus informal half (C-ORAL-BRASIL I), which has been concluded and will be published in February 2012 [3]. The formal half of the corpus is still under construction. The C-ORAL-BRASIL brings the sound files, their transcriptions and sound-text alignments besides the morphosyntactic tagged files. Its recordings were carried out with high resolution, non-invasive wireless equipment, generally with clip-on, monodirectional microphones, and a mixer whenever there were more than two interactants, in a few occasions

* The authors would like to acknowledge their research grants from CNPq and FAPEMIG.

H. Caseli et al. (Eds.): PROPOR 2012, LNAI 7243, pp. 362–367, 2012.

omnidirectional microphones were used. This guarantees a high acoustic quality, which allows for several types of acoustic analysis and is always sufficient for F0 calculation, even though several recordings took place in rowdy contexts with background noise. The texts are transcribed following the CHAT format [4], implemented for prosodic annotation [5]. The main goals for the corpus architecture are the documentation of the diaphasic and diastratic variations in Brazilian Portuguese speech.

2 The Design

The corpus is entirely dedicated to informal spontaneous speech. By spontaneous speech it is understood the speech which is programmed while being executed and does not realize preexisting text in part or as a whole [6].

The corpus is made up of family/private context (159,364 words and 105 texts) and public context (48,766 words and 34 texts). In each of the two contexts the number of texts was equally divided among monologues, dialogues and conversations. Besides the social context and interaction typology division, great attention was paid to variety of situational contexts. This diaphasic variation is taken to be the major reason for the structural variation in speech.

In the monologues the structure of speech depends mainly on textual typology: life history, professional explanation, argumentative text, joke, recipe, fable, etc. In dialogues and conversations the variation is basically linked to the activity that the interlocutors are carrying out: a conversation among friends at home will be structured in a very different way from a row between a couple, etc. It is clear that the illocutions that should be performed through speech change radically. Each situation stimulates the emergence of different speech acts, different turn sizes, different utterance size and structure, larger or smaller silence periods, etc. Only 14 texts are chats or interviews.

The diastratic variation is amply represented and documented. There are 362 speakers in the corpus. Sex, age, origin, and schooling are registered for 68.23% of the total of speakers. The nearly 30% who are not documented for the above mentioned parameters represent speakers who entered the recording context unpredictably. Table 1 shows the cluster statistics with the relation between number of speakers and uttered words.

The female/male balance is very precise as far as number of uttered words is concerned: 50.36% of words are uttered by (203) females and 49.64% of words are uttered by (159) males.

There is a balance (Figure 1) as far as number of uttered words/age rate is concerned as well: 27.13% words for (76) speakers in the 18-25 years old stratum (level A), 30.28% words for (89) speakers in the 25-40 years old stratum (level B), 31,01% words for (64) speakers in the 40-60 years old stratum (level C), 8.05% words for (15) speakers in the above 60 years old stratum (level D) and 1.61% words for (11) speakers underage (level M). Level X shows the number of words uttered for a high number of people whose age in unknown.

Table 1. Cluster statistics for number of speakers and uttered words

CLUSTER	*speakers*
1 - 247 words	161
280 - 627 words	81
649 - 908 words	37
933 - 1016 words	16
1134 - 1400 words	26
1455 - 1663 words	17
1777 - 1994 words	7
2140 - 2455 words	10
2611 - 2901 words	2
3550 - 3738 words	2
4211 - 4327 words	2
6309 words	1
TOTAL	362

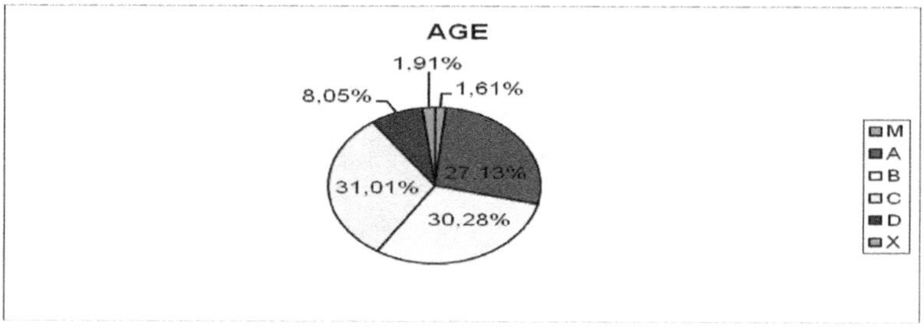

Fig. 1. Percentage of words uttered per age group

As far as schooling is concerned (Figure 2), 15.78% of words are uttered by level 1 speakers, i.e., speakers without formal schooling or that have up to 7 years of schooling (incomplete primary school level); 40.76% words for (102) level 2 speakers (up to undergraduate degree as long as not having a profession related to a university degree); 40.66% words for (10) level 3 speakers (those who work in professions dependent on a university degree).

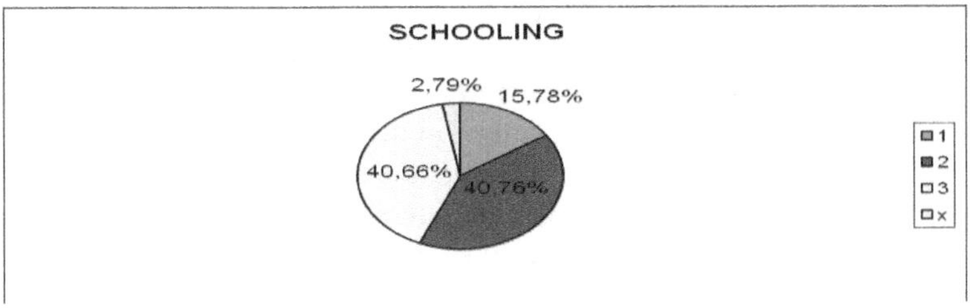

Fig. 2. Percentage of uttered words per school level

Diastraty is, therefore, very well balanced in all aspects, favoring speakers who belong to middle to middle-high schooling level, which allows for the corpus to be representative of a synchronous standard. However, lower school levels are also significantly represented. Speakers' occupations are also much diversified.

A corpus with these dimensions cannot represent diatopic variation. Therefore, the chosen diatopy was that of the state of Minas Gerais, particularly that of the metropolitan area of the capital city Belo Horizonte (this is the same procedure adopted for the C-ORAL-ROM in which the chosen cities were Florence, Aix-Marselle, Madrid and Lisbon).

Table 2. Speakers' origin

Origin	**Speakers**
Belo Horizonte	138
Other cities of Minas Gerais state	89
Other Brazilian states	19
Other countries	2
Unknown	114
Total	**362**

3 Transcription and Segmentation

The C-ORAL-BRASIL is transcribed through orthographic criteria implemented to capture speech phenomena in the process of grammaticalization and lexicalization which would be missed in a purely orthographic transcription. These are phenomena such as: nominative pronouns cliticization, verbal paradigm reductions, demonstrative reductions, absence of verb ser (to be) in interrogatives, clefts and other focus structures, apheretic forms, among others. These criteria were implemented for the morphosyntactic tagger/parser PALAVRAS [7] and allow for the quantitative/statistical study of phenomena known in the literature but never before approached with such potentiality.

The transcriptions were validated, with special attention paid to non-orthographic criteria, taking a 10% random sample from each text and, for some phenomena, a

15% from each text. The error frequency for the total number of words was 0.81%, 0.57% for phenomena involved in the non orthographic criteria. The highest error percentage for any individual phenomena was 3.25%.

In the C-ORAL-BRASIL, just as in the C-ORAL-ROM, speech flux is segmented according to prosodic criteria. The segmentation is based on the Language Into Act Theory – LACT [8], which designates the utterance as the reference unit to speech. The utterance constitutes the linguistic counterpart of a unit of action; therefore, for a locution there corresponds an illocution[9-10].

The utterance and tonal unit segmentation was carried out along with the transcription process, when transcribers (seven, divided into two groups), after a long training process, reached a Kappa [11] equal or superior to 0.80 for terminal breaks (considered to be an excellent score) and 0.60 or higher for non-terminal breaks (considered to be a good score). After transcriptions were completed and revised for a first time, the group which had reached the best agreement score was submitted to a new test, this time an overall Kappa of 0.86 was reached (0.87 for terminal and 0.78 for non-terminal breaks).

4 Morphosyntactic Annotation

The PoS, morphological and syntactic tagging for C-ORAL-BRASIL corpus has been carried out using the PALAVRAS parser [6]. PALAVRAS is a Constraint Grammar (CG) parser that is mostly used for the annotation of written data. With lexical adaptation and various filter programs, the parser can be adapted to non-standard language varieties, historical texts, spoken language, and other textual varieties.

Given the general architecture and the rule methodology of the parser, three main tasks were performed with regard to its application to oral data, affecting lexical recall on the one hand and contextual disambiguation on the other:

- the text flow normalization, which includes the treatment of corpus meta information from the non-grammatical annotation layers (speaker names, overlapping and disfluency phenomena);
- the treatment of non-standard word forms;
- the definition, in the absence of ordinary punctuation, of the reference units that can provide delimited windows for contextual disambiguation of both PoS and syntactic dependencies.

Exploiting prosodic break markers as reference units did improve performance at all levels. However, the effect was much more marked for syntax than for part of speech, lemmatization and morphology, reflecting the wider contextual scope of syntactic tags and the ensuing greater need for precise and correct segmentation. Interestingly, while syntactic performance can be further increased by pause/break disambiguation, this is not obvious for the more local tag categories. Thus, for inflexion tags (morphology), all-break performance was higher than for the pause/break run, and only for part of speech a slight improvement was observed.

5 Conclusions

The C-ORAL-BRASIL corpus is the first Brazilian Portuguese spontaneous speech aligned corpus. It is based on the C-ORAL-ROM, implementing several of its methodological aspects. The chosen diatopy is that of the metropolitan Belo Horizonte area, in the state of Minas Gerais. Its dimensions are about 35% bigger than those of each individual corpus in the C-ORAL-ROM. Its diaphasic variation is much larger than that of the Italian C-ORAL-ROM subcorpus, which is the most varied within that project. This was possible due to the careful planning of recording contexts, the selection of the best one third of the overall number of recordings made and to the very modern recording equipment used which allowed for excellent quality recordings in noisy natural environments as well as recordings with subjects in motion. Subjects are always carrying out some action other than speech in most recordings. The diastractic balance reached as far as sex, age and school level is excellent. An innovative transcription methodology for the study of language change in Brazilian Portuguese was implemented. Besides that, new validation methodologies for the transcription and segmentation validations were developed with a view to reaching optimal agreement levels before the actual segmentation process was started, so as to guarantee that after the revision process was concluded, the results would be highly reliable.

References

1. Cresti, E., Moneglia, M.: C-ORAL-ROM. Integrated Reference Corpora for Spoken Romance Languages. John Benjamins, Amsterdam/Philadelphia (2005)
2. Martin, P.: http://www.winpitch.com
3. Raso, T., Mello, H. (eds.): O Corpus C-ORAL-BRASIL. Editora UFMG, Belo Horizonte (2012)
4. MacWhinney, B.J.: The CHILDES Project. Tools for Analyzing Talk, vol. 2. Lawrence Erlbaum, Mahwah (2000)
5. Moneglia, M., Cresti, E.: Lĺntonazione e i criteri di trascrizione del parlato adulto e infantile. In: Bortolini U., Pizzuto E. (eds) Il Progetto CHILDES-Italia: Contributi di ricerca sulla lingua italiana. Del Cerro, Pisa (1997)
6. Nencioni, G.: Di scritto e di parlato: Discorsi Linguistici. Zanichelli, Bologna (1983)
7. Bick, E.: The Parsing System Palavras - Automatic Grammatical Analysis of Portuguese in a Constraint Grammar Framework. Aarhus University Press, Aarhus (2000)
8. Cresti, E.: Corpus di italiano parlato, vol. 2. Accademia della Crusca, Firenze (2000)
9. Austin, J.: How to do things with words. Oxford University Press, Oxford (1962)
10. Moneglia, M.: Spoken Corpora and Pragmatics. In: Revista Brasileira de Linguística Aplicada, pp. 479–520 (2011)
11. Fleiss, J.L.: Measuring nominal scale agreement among many raters. Psychological Bulletin 76, 378–382 (1971)

A European Portuguese Children Speech Database for Computer Aided Speech Therapy

Carla Lopes[1,2], Arlindo Veiga[1,3], and Fernando Perdigão[1,3]

[1] Instituto de Telecomunicações
[2] Instituto Politécnico de Leiria-ESTG
[3] Universidade de Coimbra – DEEC, Pólo II, P-3030-290 Coimbra, Portugal
{calopes,aveiga,fp}@co.it.pt

Abstract. This paper introduces a European Portuguese speech database containing spoken material recorded from children. The need for such database arose from the need of train phone models for the development of a computer aided speech therapy system. Articulatory disorders affect a significant number of children in pre-school age. We propose a system intended to assist and reinforce the conventional speech therapy programs. Through the systematic use of games, it learns the phones where the child has more difficulty to pronounce. The child is then taken to train the production of those phones by playing games. Another interest of a children speech database is that accurate children's phone recognition is only possible using training data that reflects the population of users. It is a difficult task due to the high pitch of children's speech.

Keywords: Phone recognition, computer aided speech therapy, children speech.

1 Introduction

We were not born speaking, but in a short time and effortlessly, we become experts in language acquisition – one of the most complex and sophisticated systems that is known. The process of language acquisition (in terms of speed and perfection) is often seen as one of the most amazing things humans can do. Language development is an iterative process. All children begin having problems in speaking some words, but as they grow, they develop abilities to improve their articulation and pronunciation. However, if some phones are easily learned, others are more difficult to learn and are often omitted or replaced. When this learning process runs away from typical time boundaries (due to several reasons, pathologic or not) the difficulties on verbal communication appear. These typically occur in pre-school ages and can occur in many different ways. The problem gets worse when the child begins its education, not only because of the difficulty on communication but also due to the social problems that appear. Speech therapy is the area of expertise related to the analysis and direct intervention in the recovery of these problems. The correction program consists of interactive sessions between the child and therapist. Sessions are usually individual, which makes it impractical to include an expanded program of speech therapy to the pre-school children population in general. Moreover, the success of a

H. Caseli et al. (Eds.): PROPOR 2012, LNAI 7243, pp. 368–374, 2012.

program of speech therapy is limited if it is only based on the sessions with the therapist (usually weekly). The family and kindergarten teachers play a key role on the stimulation, correction and training. Studies, conducted in several Portuguese regions, show that 20% to 30% of Portuguese children in pre-school and first cycle ages requires intervention at the level of speech therapy [1].

Using our know-how in speech processing and phone recognition, the aim of the present work is to create tools with application in games/simple interactive challenges for children to improve its language skills, especially phones in which they present more difficulty in uttering.

These kind of systems dates back to the 70s when the first training speech computer system appeared – the "Skill builder", [2]. The computer equipment gives the user visual feedback in order to support phono-audiology therapy. This work prompted several studies in the area. In 1988, Bernstein *et al* [3] make a state-of-the-art of the systems developed so far, emphasizing 6 systems. The importance of the problem led to, in 2002, a EU-funded research project (Ortho-Logo-Paedia) for speech therapy, [4] targeted to people with articulatory impairments. Tools supporting speech therapists were developed, as well as an automatic system that evaluates the speech production and web tools to make it available remotely using distance learning techniques. Also in this field, it is presented in [5] interactive tools designed to facilitate the acquisition of language skills in the areas of basic phonatory skills, phonetic articulation and language understanding primarily for children with neuromuscular disorders like dysarthria.

The works found in literature can be grouped into two major groups: those relating to the rehabilitation of individuals with articulation disorders [5],[6],[7] (rehabilitation involves the training of the voice activity, intensity, breathing, tone and vocalization) and deaf persons, [3],[8],[9]. The goal of the present work differs from the referred to systems since it intends to solve problems of dyslalia (non-neurological impairments associated with the external speech organs), [12], in children of 5-6 years old. As referred to in [3], although development considerations indicate that the most effective use of training aids must be made during the years of childhood, and especially during the pre-school period of language acquisition, most aids appear to have been designed without regard for the specific linguistic, cognitive and attention attributes of young children. The proposed system is entirely design to fit speech and features of this age group of children which made it obligatory to have a database with speech of children of this age.

In the literature there are systems of speech therapy in several languages: Romanian, [6], Arabic, [11], English, [2], Spanish [5], Brazilian Portuguese, [8], etc but as far as we know, no European Portuguese system has yet been developed. In [8], it is presented a very interesting work for Brazilian Portuguese. Several games were developed improving phonoarticulatory coordination of children and young deaf people (e.g., control of the airflow and fundamental frequency and positioning articulation of fricatives and vowels). Carvalho in [10] proposes an interactive application completely controlled by the utterance of 5 European Portuguese vowels. The goal of the present work is to extend the speech training to all Portuguese phones.

2 System Overview

In this section it is presented the outline of a system that attempts to fight against functional dyslalia - a functional articulation disorder [12].

Speech disorders can be broadly classified into two main classes: dysarthria and dyslalia. Dysarthria is an articulation disorder produced by peripheral or central nerve imperfect functioning. Dyslalia is a functional articulation disorder that refers to an immaturity on the speech pronunciation process or to some wrong speech habits. The system is intended to 5-6 years old children aiming to help them to overcome pronunciation problems until the beginning of the first cycle of education. The system is not yet fully developed but it works as described: it starts performing an evaluation on the phones that the child has more difficulties in pronouncing. This module (evaluation module) ask the user to say several words (showing pictures), which are subsequently transformed in a sequence of phones by a phone recognition system, which will be based on Hidden Markov Models or Artificial Neural Networks. Matching these sequences with the reference ones, the system is able to output information about the phones in which the child has difficulties and he/she must strengthen. The games run in a similar way, but when the system detects a mispronounced phone (e.g.: the child says "bona" instead of "bola", probably it has problems with the lateral phone [l]), it emits an alert.

The system synthesizes the mispronounced word (with the correct pronunciation) and asks the user to repeat the word. If the mispronunciation remains it shows (through a sequence of visemes) to the child the correct position of the lips and tongue. Figure 1 shows a block diagram of the system.

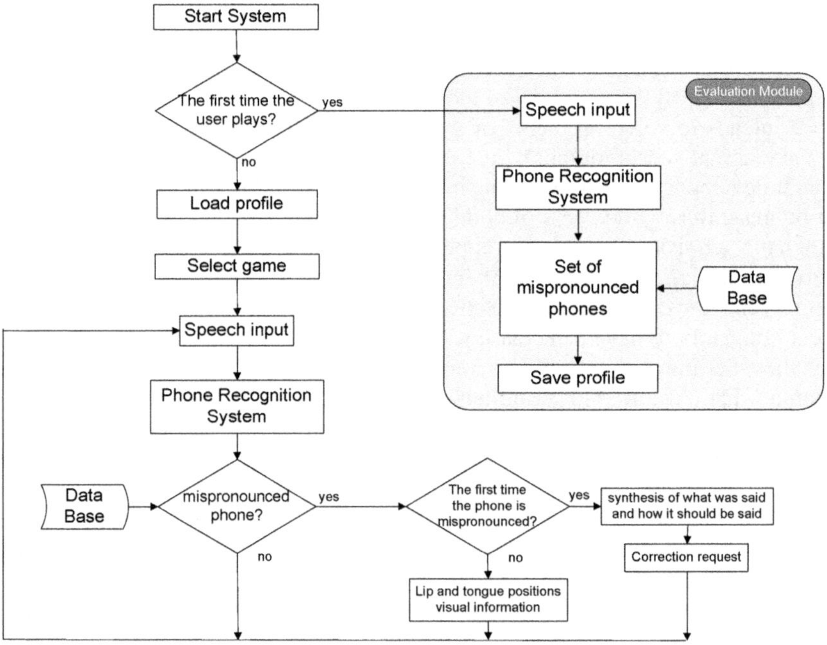

Fig. 1. Block diagram of the proposed dyslalia therapy system

To perform phone recognition, speech models must be firstly trained. This is a complex task since we are dealing with children speech and there is no database of

European Portuguese children speech. In [10], the only published work in European Portuguese in this field, a database has not been used: only a few utterances collected by the author were used for training the speech acoustic models. The cost of using models trained with speech from adults undermines the proposed system's success. There are several differences between speech from adults and children. Due to a shorter vocal tract, children have higher formants than adults and also high fundamental frequencies. High pitch values leads to severe sampling of the envelope of the speech spectrum and the usual speech analysis methods tend to lock to the pitch spectral lines instead of the envelope. This implies that speech parameters that are used in conventional speech recognition systems may be not adequate for children's acoustic models. In [13] experiments shows that using speech models, trained with adults' data, on the recognition of children speech make the recognition rates drop from 97% to 61%. This study reveals the requirement of using models trained on the same kind of the test data.

3 European Portuguese Children Speech Corpus

Data collection is always a complex task, but this complexity increases when dealing with young children, as is the present case (children with 5 and 6 years old). When recording data, the children reaction ranged from shyness (uttering the words very softly), fear (some reject to talk), kind of hysteria (screaming the words) and play (laugh while saying the words). Data collection was not done in laboratory (i.e., in a controlled environment, as desired), but in kindergartens, in rooms with poor acoustics, making it very difficult or even impossible to control the noise level.

3.1 Data Collection

For recording the speech signals it was developed a specific application, running in a notebook and using a headset microphone (Labtec stereo 242). The sampling frequency was 8000 samples per second. Each session have a registry with the number of the session, the initials of the child, its age and gender. The application includes a set of 200 pictures easily identified by children of 5 and 6 years. In each session 40 pictures are randomly selected for recording, resulting in a different file for each. The words to record, in addition to being part of the vocabulary of children, were chosen so that all phones occur frequently enough to allow an efficient training of its models.

The set of selected words may be found in [14] and the frequency of phone occurrences in all the 200 words are shown is Figure 2 (annotated in SAMPA).

The data was collected from 111 different speakers (55 male and 56 female) in different kindergartens of the center region of Portugal. Each speaker uttered 40 words but only 3726 files were considered valid.

Data collection was much more complex than expected. We had to control the recording application, the environment noise and the recording position and, in addition, we had to manage the children behavior. Some of them faced the task as a funny task (and laugh while uttering the words) and others get into stress because they did not recognize the pictures. All of them hesitate a lot. The data is full of hesitations, such as fillers, word cut-offs, repetitions and segmental extensions. Several utterances had to be discarded.

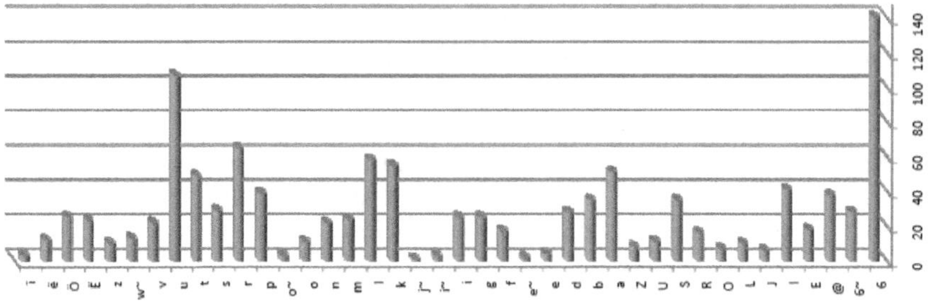

Fig. 2. Phone occurrences (in SAMPA) in the 200 words

3.2 Data Validation and Segmentation

All utterances were manually validated and annotated at the word level using software developed for speech annotation. In Figure 3 an example of annotation is given.

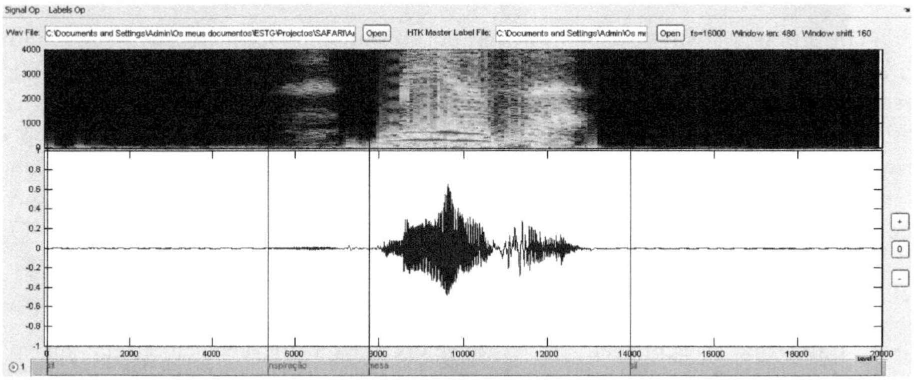

Fig. 3. Speech annotation example

Data processing led to more labels than those who were initially supposed (silence-word-silence): hesitations (e.g. vowel extensions); sometimes the spoken words are not perceptible; also noise and breathing in and out are present in the utterances. Six types of labels were then considered: *Hesitação* (hesitations), *Lixo* (rubbish), *Silêncio* (silence), *Respiração* (breath), *Ruído* (noise) and *Palavras proferidas* (words), with attributes corresponding to the spoken words. All labels were set manually and almost every utterance has a different situation. In the case were the child should say "*flor*" (flower) it says "*eu acho que é uma flor*" ("I think it is a flower"). In this case six words were annotated. The more difficult situation occurs when the words were mispronounced ("Foia" [foi6] instead of "Folha" [foL6], or "nanareijo [n6n6reiZu]" instead of "caranguejo" [k6ra~geiZu]). In these cases the mispronounced words were added to the dictionary. The 3726 recorded files resulted in 14099 labels. Figure 4 presents the labels distribution. As can be seen, noise is frequent in the utterances; nevertheless it characterizes the environment and allows the estimation of a noise model. As expected (in stress situations children tend to make inspirations), breath is also frequent in the database. Less frequent are hesitation and rubbish phenomena.

Words may be *connection words* or phonemes or *single words*, (e.g. "é uma", "de"). In the 4511 *words* there are 968 connection words and 3543 single words.

All files can be found in [14]. Annotated files followed HTK format of master label files (.mlf, [15]) using units of 100ns.

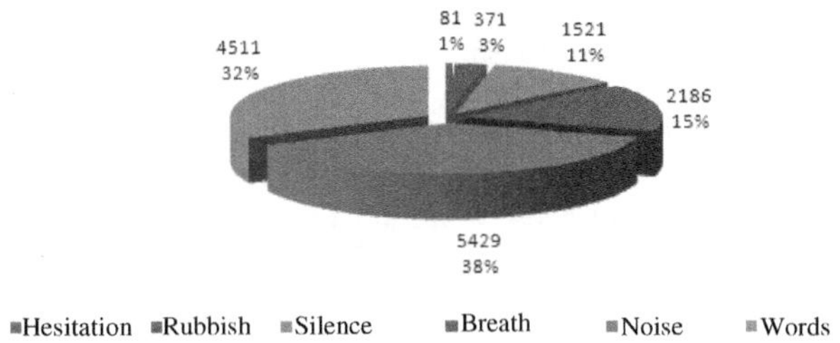

Fig. 4. Distribution of labels in each category in the complete set of utterances

3.3 Corpus Summary

The European Portuguese Children Speech Corpus that was collected contains a total of 3726 sentences (158 minutes), from 56 female and 55 male speakers. All sentences were manually segmented at word level and event level (silence, hesitations, breath, noise, rubbish). Table 1 resumes the Corpus information.

Table 1. Corpus main features

# recorded files	# speakers	#labels	#minutes	#words
3726	111	14099	158 minutes	4511

4 Conclusions

In view of the development of a computer aided speech therapy system to help young children to surpass dyslalia problems, it was necessary to collect a European Portuguese children speech database.

The data was collected with 111 different speakers with 5 and 6 years old in several kindergartens of the center region of Portugal. All data was manually annotated resulting in 4511 words, coming from 3726 files. This data will make possible to study children's speech and build accurate acoustic models to use in an automatic phone recognition system, which will be the key module of the computer aided speech therapy system. It is not intended that the system replaces a speech therapist, but that works as a complement, such that it can improve the therapeutic intervention, by allowing repetitive tasks that stimulate the child and improves its skills in the pronunciation of some phones.

References

1. Schreck, I.: Milhares de crianças têm de corrigir a fala. Jornal de Noticias, 15 de Março de (2009)
2. Nickerson, R.S., Kalikow, D.N., Stevens, K.N.: Computer-aided speech training for the deaf. Journal of Speech and Hearing Disorders 41, 120–132 (1976)
3. Bernstein, L., Goldstein, M., Mahshie, J.: Speech training aids for hearing-impaired individuals: Overview and aims. Journal of Rehabilitation Research and Development 25(4), 53–62 (1988)
4. Öster, A.-M., House, D., Protopapas, A., Hatzis, A.: Presentation of a new EU project for speech therapy: OLP (Ortho-Logo-Paedia). In: Fonetik 2002, Speech Music and Hearing Quarterly Progress and Status Report, TMH-QPSR, vol. 44, pp. 45–48 (2002)
5. Saz, O., Yin, S., Lleida, E., Rose, R., Vaquero, C., Rodríguez, W.R.: Tools and Technologies for Computer-Aided Speech and Language Therapy. Speech Communication 948–967 (2009)
6. Schipor, O.-A., Schipor, D.-M., Pentiuc, S.-G.: Improving Computer Based Speech Therapy Using a Fuzzy Expert System. Computing and Informatics (F:0.456), Slovakia 29(2), 303–318 (2010) ISSN: 1335-9150
7. Blter, O., Engwal, O., Ster, A.M., Kjellstrm, H.: Wizard-of-Oz Test of ARTUR - a Computer-Based Speech Training System with Articulation Correction. In: Seventh International ACM SIGACCESS Conference on Computers and Accessibility, Baltimore, MD, pp. 36–43 (2005)
8. de Lima Araújo, A.M.: Jogos Computacionais Fonoarticulatórios para Crianças com Deficiência Auditiva, PhD Thesis, de (2000)
9. Zahorian, S.A., Zimmer, A.M., Meng, F.: Vowel classification for computer-based visual feedback for speech training for the hearing impaired. In: International Conference on Spoken Language Processing, vol. 78, pp. 973–976 (2002)
10. Carvalho, M.: Interactive Game for the Training of Portuguese Vowels., Master Thesis, Faculdade de Engenharia da Universidade do Porto (2008)
11. Benselama, Z.A., Guerti, M., Bencherif, M.A.: Arabic speech pathology therapy computer aided system. Journal of Computer Science 3(9), 685–692 (2007)
12. Lima, R.: Alterações nos sons da fala: o domínio dos modelos fonéticos. Saber (e) Educar. Porto: ESE de Paula Frassinetti. N.° 13, pp. 149–157 (2008)
13. Elenius, D., Blomberg, M.: Comparing speech recognition for adults and children. In: Proc of Fonetik 2004, Stockholm, pp. 156–159 (2004)
14. European Portuguese Children Database, `http://www.co.it.pt/~calopes/child_database.htm`
15. Young, S.J., et al: The HTK book. Revised for HTK version 3.4, Cambridge University Engineering Department, Cambridge (December 2006)

Baseline Acoustic Models for Brazilian Portuguese Using CMU Sphinx Tools

Rafael Oliveira, Pedro Batista, Nelson Neto, and Aldebaro Klautau

Federal University of Pará, Signal Processing Laboratory,
Rua Augusto Correa 1, 660750110, Belém, PA, Brazil
{rafaelso,pedro,nelsonneto,aldebaro}@ufpa.br
http://www.laps.ufpa.br

Abstract. Advances in speech processing research rely on the availability of public resources such as corpora, statistical models and baseline systems. In contrast to languages such as English, there are few specific resources for Brazilian Portuguese. This work describes efforts aiming to decrease such gap. Baseline acoustic models for Brazilian Portuguese were built using the CMU Sphinx toolkit and public domain resources: speech corpora, phonetic dictionary and language model. Experiments were carried on for dictation and grammar tasks and the obtained results can be used to support further researches. Part of the trained acoustic models and a reference speech corpus were made publicly available.

Keywords: Speech recognition, Brazilian Portuguese, CMU Sphinx.

1 Introduction

Sphinx is a public domain software package maintained by the Carnegie Mellon University (CMU) for implementing automatic speech recognition (ASR) systems. Sphinx has been investigated in many works [1–3] and its current version has performance equivalent to other open source softwares widely used on the community, such as HTK [4]. But the HTK tools cannot be freely distributed, unlike Sphinx. In addition, free softwares compatible with the HTK acoustic models, like the Julius decoder, have not yet reached the same performance obtained with the HTK decoders [5].

Therefore, the development of specific resources for Sphinx has gained importance because there is no an eficiente HTK to Sphinx model converter and the PocketSphinx and the Sphinx3 decoders are interesting alternatives. Sphinx has been largely used for many languages [6–8]. However, to the best of the author's knowledge, there are no previous works using Sphinx for Brazilian Portuguese (BP). This work presents a first effort for developing baseline acoustic models for BP based on the Sphinx toolkit. In the sequel the resources used to build and evaluate this acoustic models are described.

H. Caseli et al. (Eds.): PROPOR 2012, LNAI 7243, pp. 375–380, 2012.

2 Resources Used

The phonetic dictionary used for developing the acoustic models is described in [9]. This dictionary has approximately 60 thousand words transcribed in the SAMPA alphabet. The speech corpora used to build the acoustic models were: (1) the West Point Brazilian Portuguese corpus suggested in [10]; (2) the LapsStory corpus [5] composed by audio files from audiobooks; and (3) a corpus collected by the Centro de Estudos em Telecomunicações (CETUC) [11]. The publicly available dictation test corpus LapsBenchmark [5] was adopted to evaluate the models. Table 1 summarizes the characteristics of these speech databases.

Table 1. Speech corpora used to build and evaluate the acoustic models

Database	Hours	Speakers	Words	Acoustic environment
West Point	8	128	484	not controlled
LapsStory	16.28	8	8257	controlled
CETUC	142.83	101	3528	not controlled
LapsBenchmark	0.96	35	2731	not controlled

The trigram language model (LM) was trained with 710,000 sentences extracted from CETENFolha [12] and LapsStory corpora as described in [5]. It has perplexity 170 measured with a test set of 10,000 sentences randomly selected from CETENFolha (unseen during the training stage).

In this work, the development of a free speech database named LapsMail was initiated. This corpus was designed to represent a basic set of commands to control a electronic mail application, aiming establish it as a reference benchmark corpus for this context. Actually, the LapsMail corpus consists of 86 BP sentences including 43 commands and 43 names spoken by 25 volunteers (21 male and 4 female) which corresponds to 84 minutes of audio. Its vocabulary has 95 different words. The following are three typical sentences in the corpus.

```
(1) <s> abrir caixa de entrada </s>
(2) <s> responder ao remetente </s>
(3) <s> ana carolina </s>
```

The LapsMail corpus was recorded using a high quality microphone (Shure PG30), sampled at 16 kHz and quantized with 16 bits. The acoustic environment was not controlled. The LapsMail corpus is publicly available [13].

3 Building Acoustic Models for BP

Both continuous and semi-continuous acoustic models were trained using the SphinxTrain package tools [14]. The acoustic waveforms from the training corpus were parametrized into 13-dimensional cepstrum. For continuous models, these features along with computed delta and delta-deltas were used (1s_c_d_dd). For semi-continuous models four feature streams were used (s2_4x). Cepstral mean normalization was performed using the current utterance during decoding.

The initial context-independent models for the 39 phonemes (38 monophones and 1 silence model) used the 3-state left-to-right with self-loops and no skip transitions topology for all the hidden Markov models (HMMs). When training semi-continuous models, 4 codebooks with 256 codewords entries were used. Codewords were found using k-means clustering. The mixture weights and transition probabilities in each state of each HMM were initialized with uniform distributions. For continuous models, the mean and variance of the Gaussian in each state of each HMM is set to the mean and variance of the training data. Mixture weights and transition probabilities are initialized equiprobable.

The context-dependent models were created for each triphone that occurred in the training data and reestimated until convergence. After that, the states of all triphones (including seen and unseen triphones in the training data) were tied based on a decision tree built from a set of linguistic questions automatically generated by a data-driven algorithm [15]. Finally, the tied-state context-dependent models were reestimated until convergence. The Baum-Welch algorithm was used to reestimate the models.

4 Baseline Results

Several continuous and semi-continuous acoustic models were trained following the procedure described in Section 3. During the train stage, the number of tied-states (senones) were varied between 500 and 8,000, and the number of Gaussians per mixture between 2 and 64. The built models are publicly available [13].

The acoustic models were evaluated in two experiments and measured in terms of word error rate (WER) and real time factor (xRT). The continuous and semi-continuous models were tested using the Sphinx3 and the PocketSphinx decoders, respectively. The decoding parameters were not tuned and kept constant during all the experiments. Table 2 shows the main parameters values. All the tests were executed on a computer with Intel Xeon processor (E5450 3.0 GHz) and 4 GB of RAM.

Table 2. Decoders parameters for Sphinx3 and PocketSphinx

Parameter	Sphinx3	PocketSphinx
Pruning beam width	1e-55	1e-48
Word beam width	1e-35	7e-29
Phone beam width	1e-50	1e-48
Word insertion penalty	0.7	0.65
Language model scale factor	9.5	6.5

4.1 Results for a Dictation Task

This experiment focus on comparing the trained acoustic models performance on a dictation task. The trigram LM and the LapsBenchmark corpus described in Section 2 were used here. The experiment results showed that increasing the number of tied-states reduced the WER for both continuous and semi-continuous models, but not for those models with more than 4,000 tied-states. Having more tied-states requires the decoder to compute more Gaussian likelihoods per observation, what justifies the increased xRT for the higher number of tied-states.

Increasing the number of Gaussians per mixture resulted in WER reduction, however values beyond 16 caused a significant increase in xRT. The WER with 2,000 tied-states and 8 Gaussians is 16.41%, being the best for a xRT around one. The WER and xRT graphs are shown in Fig. 1 and 2, respectively.

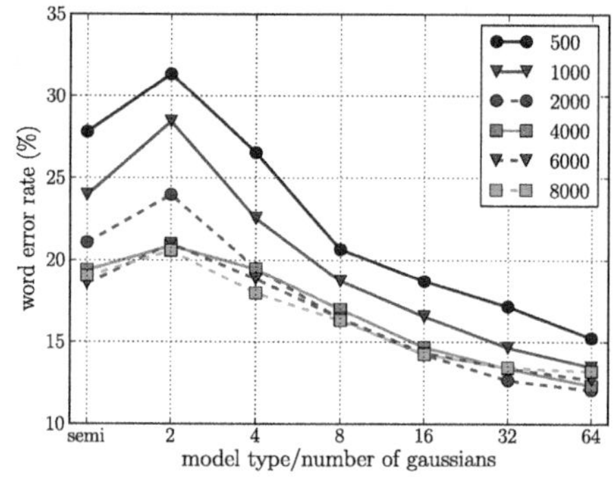

Fig. 1. WER (%) for dictation task using the LapsBenchmark corpus

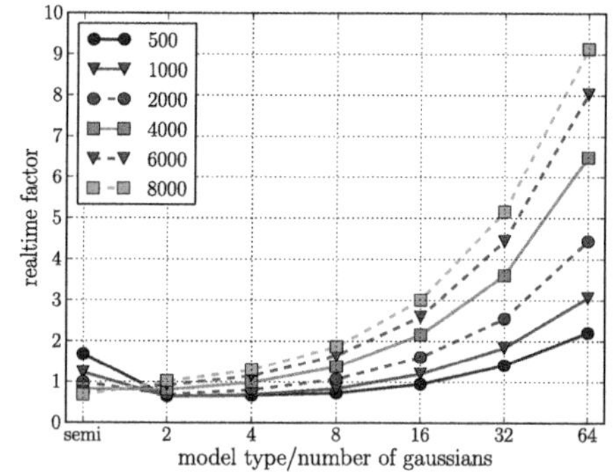

Fig. 2. xRT for dictation task using the LapsBenchmark corpus

4.2 Results for a Grammar Task

First, the acoustic model evaluated in Section 4.1 with 2,000 tied-states and 8 Gaussians per mixture was tested with the LapsMail corpus. The results showed a bad performance, with 6.21% of WER, worse than expected based on the small vocabulary length (only 95 words). Analyzing the results, we found that 165 of the 185 errors computed were caused by fails in recognition of names sentences. The complexity of the name recognition task was discussed in [16, 17], where approaches were proposed to increase the performance of ASR systems.

In another experiment the LapsMail names sentences were not included. The remaining 43 commands sentences with approximately 47 minutes of audio were used to evaluate the trained acoustic models. This experiment presented more reasonable results that can be seen in Fig. 3 (xRT was omitted for clarity). Similarly to the dictation task, increasing the number of Gaussians per mixture decreased the WER. In another hand, increasing the number of tied-states resulted in no WER reduction, possibly due the small vocabulary size. The WER with 500 tied-states and 64 Gaussians per mixture is 0.67%.

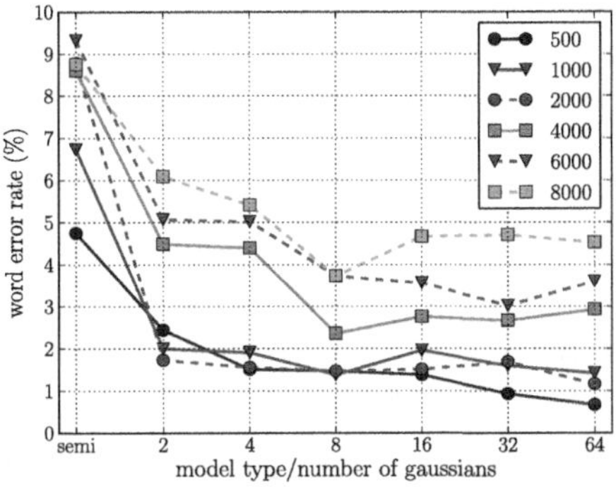

Fig. 3. WER (%) for grammar task using part of the LapsMail corpus

5 Conclusions

This paper presented recognition experiments in BP using the CMU Sphinx tools. The developed resources were made publicly available and allow for reproducing results across different sites. The final results are consistent with other developed large vocabulary ASR systems for BP [5] and the acoustic models distributed can be used in the development of sample applications with ASR support. In order to improve the models performance, the decoders parameters should be tuned and a more refined linguistic questions set will be investigated.

References

1. Vertanen, K.: Baseline WSJ acoustic models for HTK and Sphinx: Training recipes and recognition experiments. University of Cambridge, Tech. Rep. (2006)
2. Samudravijaya, K., Barot, M.: A comparison of public domain software tools for speech recognition. In: Proceedings of Workshop on Spoken Language Processing, pp. 125–131 (2003)
3. Ma, G., Zhou, W., Zheng, J., You, X., Ye, W.: A comparison between HTK and Sphinx on Chinese Mandarin. In: International Joint Conference on Artificial Intelligence (2009)
4. Young, S.E.: The HTK Book. Microsoft Corporation, Version 3.0 (2000)
5. Neto, N., Patrick, C., Klautau, A., Trancoso, I.: Free tools and resources for Brazilian Portuguese speech recognition. The Brazilian Computer Society 16, 53–68 (2010)
6. Varela, A., Cuayáhuitl, H., Nolazco-Flores, J.: Creating a Mexican Spanish version of the CMU Sphinx-III speech recognition system. In: Progress in Pattern Recognition, Speech and Image Analysis, pp. 251–258 (2003)
7. Satori, H., Hiyassat, H., Harti, M., Chenfour, N.: Investigation Arabic speech recognition using CMU Sphinx system. International Arab Journal of Information Technology 6 (2009)
8. Gulic, M., Lucanin, D., Simic, A.: A digit and spelling speech recognition system for the croatian language. In: 34th International Convention on Information and Communication Technology. Eletronics and Microeletronics, pp. 23–27 (2011)
9. Siravenha, A., Neto, N., Macedo, V., Klautau, A.: Uso de regras fonológicas com determinação de vogal tônica para conversão grafema-fone em Português Brasileiro. In: 7th International Information and Telecommunication Technologies Symposium (2008)
10. Santos, F., Barone, D., Adami, A.: A baseline system for continuous speech recognition of Brazilian Portuguese using the West Point Brazilian Portuguese speech corpus. In: International Conference on Computational Processing of the Portuguese Language (2010)
11. `http://www.cetuc.puc-rio.br/pos-novo.htm` (visited in November 2011)
12. `acdc.linguateca.pt/cetenfolha/` (visited in November 2011)
13. `http://www.laps.ufpa.br/falabrasil` (visited in January 2012)
14. `http://cmusphinx.sourceforge.net/wiki/tutorialam` (visited in November 2011)
15. Singh, R., Raj, B., Stern, R.M.: Automatic clustering and generation of contextual questions for tied states in hidden markov models. In: Proceedings of ICASSP, pp. 117–120 (1999)
16. Sethy, A., Narayanan, S., Parthasarthy, S.: A syllable based approach for improved recognition of spoken names. In: Proceedings of the ISCA Pronunciation Modeling Workshop (2002)
17. Maskey, S., Bacchiani, M., Roark, B., Sproat, R.: Improved name recognition with meta-data dependent name networks. In: Proceedings of ICASSP (2004)

A Fishervoice-SVM Language Identification System

Paula Lopez-Otero, Laura Docio-Fernandez, and Carmen Garcia-Mateo

Multimedia Technologies Group, Universidade de Vigo
E.E. Telecomunicación, 36310 Vigo, Spain
{plopez,ldocio,carmen}@gts.uvigo.es

Abstract. In this paper, a language identification system is described that implements the Fishervoice approach in order to reduce the dimensionality of the data. Fishervoice performs two-dimensional Principal Component Analysis (2D-PCA) and Linear Discriminant Analysis (LDA) to project the data into a discriminative subspace. After this transformation the speech utterances are transformed into supervectors and classified by means of a Support Vector Machine (SVM). Experiments performed on KALAKA-2 database, which includes speech in Spanish, Catalan, English, Basque, Galician and Portuguese, show that the Fishervoice-SVM system achieves good identification results while reducing dramatically the number of features needed to represent the speech utterances.

Keywords: Language identification, Fishervoice, Support Vector Machines.

1 Introduction

Language identification (LID) has a wide range of applications due to the emerging need for tools to process multimedia information. Automatic Speech Recognition (ASR) can obtain benefits from this task, because if a LID system can identify the language of the speech, a suitable language model can be chosen and recognition results will improve [22]. It is possible to classify audio streams, like broadcast news programs, by the language that is being spoken. LID can also be applied in multi-lingual conversational systems, as it makes it possible to have a human-machine interface where it is not necessary to select a language, it is automatically selected anytime a user speaks to the device [7]. Moreover, language identification can be used to ease the study of certain aspects of a spoken language, for example by classifying utterances of a language into its different varieties [20].

LID systems can be classified into two different approaches: those that recognize the language by means of its linguistic and phonotactic information [16], and those that use acoustic features [19]. These two approaches can be combined in order to construct a more complex system by fusing the results of different subsystems [4].

H. Caseli et al. (Eds.): PROPOR 2012, LNAI 7243, pp. 381–391, 2012.

In LID acoustic approaches, speech utterances can be modeled by means of Gaussian Mixture Models (GMM), and statistical models can be trained which represent the languages by means of low-level acoustic features [21]. To decide which language is spoken on a speech utterance, the likelihood of that utterance being generated by the different language models is computed, and the language whose model obtained the highest likelihood is selected. When using GMMs with a high number of Gaussian mixtures, speech utterances are represented by a large quantity of features, making the system more time and memory consuming, and the computation of the likelihoods will be too expensive. Dimensionality reduction strategies can be applied in order to deal with this problem.

In this paper, a continuation of the work described in [13] is proposed (from now on, this is the baseline system), where a dimensionality reduction strategy for language identification, namely the Fishervoice method, was proposed. The Fishervoice method first reduces the dimensionality of the data by applying two-dimensional Principal Component Analysis (2D-PCA), and then performs Linear Discriminant Analysis (LDA) to extract a discriminative subspace. This two-stage PCA+LDA strategy has been applied in face recognition [8] and in speaker identification [12]. This strategy not only reduces the dimensionality, but also projects the data into a more discriminative subspace. Once the speech utterances are represented by Fishervoice vectors, they have to be classified into one of the possible languages. In the baseline system this problem was solved by implementing an Euclidean distance-based classifier: first, the training vectors that are used to characterize the languages are transformed using the Fishervoice transformation, representing each language by means of a set of Fishervoice vectors corresponding to that language; then, to classify a Fishervoice vector into one of the candidate languages, the Euclidean distance between this vector and the vectors that model the different languages is computed, and the language of the training vector that achieved the minimum Euclidean distance is selected [13]. This classification method has the advantage of being faster than computing the likelihoods, but it does not lead to a very good performance.

In the system presented in this work, the speech utterances are classified into one of the candidate languages using a Support Vector Machine (SVM). An SVM is a supervised learning method that represents patterns in a high or infinite dimensional space in order to classify them into a given set of classes [3]. The language identification problem is treated as a pattern recognition one, as a set of patterns have to be classified into one of the available classes.

System performance is tested on the KALAKA-2 database [9]: this database was used on the Albayzin 2010 Language Recognition Evaluation (from now on, Albayzin-LRE) [17], so results obtained with the system presented in this work can be compared with the baseline system and with the other ones that participated in the evaluation.

The outline of the paper is as follows: in Section 2, the Fishervoice-SVM language identification system is described; in Section 3 the experimental framework and the database are summarized; in Section 4 experimental results are presented; and Section 5 explains some conclusions and future lines.

2 The LID System

The LID system presented in this work is depicted in Fig. 1. First of all the data representation must be described. In this work, utterances in waveform format are represented by means of the extracted Mel-frequency Cepstrum Coefficients (MFCC) with log-energy, delta and acceleration coefficients. These feature vectors are of length N. Some training data is employed to train a GMM-UBM with M Gaussians by expectation-maximization. Furthermore, there are two other datasets: the training one, composed by L_{train} utterances, and the test one, composed by L_{test} utterances. The GMM-UBM is Maximum a Posteriori (MAP)-adapted to each of these utterances; thus, each utterance will be represented by a set of M means of length N. Then, $A_{\text{train}} = \{A_{\text{train}_1}, \ldots, A_{\text{train}_{L_{\text{train}}}}\}$ is a set of L_{train} matrices of size $M \times N$, and in the same way, A_{test} is a set of L_{test} matrices of size $M \times N$, where each matrix represents a speech utterance. It has to be noticed that the language of each utterance A_{train_i} is known.

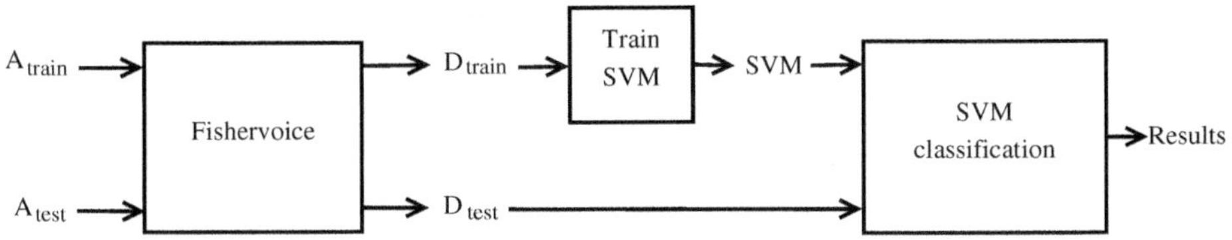

Fig. 1. Fishervoice-SVM LID System

A_{train} and A_{test} are transformed into $D_{\text{train}} = \{D_{\text{train}_1}, \ldots, D_{\text{train}_{L_{\text{train}}}}\}$ and D_{test} by applying the Fishervoice transformation. Then D_{train} is used to train an SVM and, once obtained, the recognition task is performed on D_{test}.

2.1 The Fishervoice Method

The dimensionality reduction strategy described in [12] is applied in this paper to perform language identification. The set A_{train} is used to compute two transformation matrices X and Y that will be employed to project the data into a low-dimensional subspace, thus reducing the dimensionality of the data.

Three scatter matrices (between-class D_{b}, within-class D_{w} and total D_{t}) are defined:

$$D_{\text{b}} = \sum_{i=1}^{L} P_i(\overline{A_{\text{train}_i}} - \overline{A_{\text{train}}})^{\text{T}}(\overline{A_{\text{train}_i}} - \overline{A_{\text{train}}}) \tag{1}$$

$$D_{\text{w}} = \sum_{i=1}^{L} \sum_{\substack{j=1 \\ j \neq i}}^{L} (A_{\text{train}_j} - \overline{A_{\text{train}_i}})^{\text{T}}(A_{\text{train}_j} - \overline{A_{\text{train}_i}}) \tag{2}$$

$$D_{\text{t}} = D_{\text{b}} + D_{\text{w}} \tag{3}$$

where L is the number of different languages (classes) in A_{train}, P_i is the a priori probability of the ith class, $\overline{A_{\text{train}_i}}$ is the mean matrix of the ith class ($i = 1, 2, \ldots, L$), $\overline{A_{\text{train}}}$ is the total mean matrix of A_{train}, and A_{train_j} is the $M \times N$ matrix of the jth segment in A_{train}. Thus, $\overline{A_{\text{train}}}$ represents the mean matrix of the whole set, and $\overline{A_{\text{train}_i}}$ is the mean matrix of language i. All the languages are assumed equiprobable, thus, $P_i = 1/L$.

The eigenvectors and eigenvalues of the total scatter matrix D_{t} are obtained by finding a matrix X that maximizes $J(X) = X^{\text{T}} D_{\text{t}} X$. To achieve the dimensionality reduction, some of the eigenvectors of X will be rejected. The proposed selection strategy keeps the e_{PCA} columns of X (eigenvectors) with the largest associated eigenvalues. Hence, a matrix X of dimension $N \times e_{\text{PCA}}$ is obtained, where $e_{\text{PCA}} \leq M$.

After obtaining X, the sample set A_{train} is transformed into a new space with a lower dimensionality by doing $B_{\text{train}} = A_{\text{train}} X$. Then, new between-class and within-class scatter matrices (R_{b} and R_{w}, respectively) are computed:

$$R_{\text{b}} = \sum_{i=1}^{L} P_i (\overline{B_{\text{train}_i}} - \overline{B_{\text{train}}})(\overline{B_{\text{train}_i}} - \overline{B_{\text{train}}})^{\text{T}} \tag{4}$$

$$R_{\text{w}} = \sum_{i=1}^{L} \sum_{\substack{j=1 \\ j \neq i}}^{L} (B_{\text{train}_j} - \overline{B_{\text{train}_i}})(B_{\text{train}_j} - \overline{B_{\text{train}_i}})^{\text{T}} \tag{5}$$

where $\overline{B_{\text{train}}}$ is the total mean matrix of the set B_{train}, and $\overline{B_{\text{train}_i}}$ is the mean matrix of the ith class in that set.

Applying the Fisher criterion, a matrix Y that maximizes $J(Y) = \frac{Y^{\text{T}} R_{\text{b}} Y}{Y^{\text{T}} R_{\text{w}} Y}$ is obtained, where the columns in Y are the corresponding eigenvectors. After this, Y becomes a $M \times e_{\text{LDA}}$ matrix by keeping the $e_{\text{LDA}} \leq N$ columns (eigenvectors) with the highest eigenvalues. Then, matrix Y is used to perform the transformation $C_{\text{train}} = Y^{\text{T}} B_{\text{train}}$, obtaining a dataset C_{train} composed by $e_{\text{LDA}} \times e_{\text{PCA}}$ matrices, which means that a new representation of A_{train} with lower dimensionality is obtained.

Matrices X and Y are employed to project A_{test} into the new low dimensionality subspace:

$$B_{\text{test}} = A_{\text{test}} X \tag{6}$$

$$C_{\text{test}} = Y^{\text{T}} B_{\text{test}} \tag{7}$$

It is necessary to transform the sets of matrices C_{train} and C_{test} into supervectors, in order to classify the speech utterances with the SVM. Thus, for each matrix C_{train_i} or C_{test_i} its rows are concatenated obtaining sets of supervectors D_{train} and D_{test} of dimension $e_{\text{LDA}} \cdot e_{\text{PCA}} \times L_{\text{train}}$ and $e_{\text{LDA}} \cdot e_{\text{PCA}} \times L_{\text{test}}$, respectively.

2.2 Classification

Once the data is transformed, it has to be classified into one of the candidate languages. This task can be seen as a pattern recognition problem: there are several languages (classes), and a speech utterance (pattern) has to be classified into one of those classes. In this system, this pattern recognition task is performed by an SVM.

An SVM separates patterns into two different classes by means of a kernel function. In this case, a radial basis function kernel is used [3]. In most cases, the set of patterns has to be divided into more than two classes, thus, a binary classifier is not enough to perform this task. There are two different strategies to overcome this multi-class problem: the one-against-one technique, which constructs one SVM for each pair of classes and classifies the pattern by maximum voting; and the one-against-all technique, which constructs SVMs to distinguish one class from all the others and classifies the pattern by selecting the SVM that achieved the maximum output [2].

In the experiments described in this work, a library for working with SVMs called LibSVM is used [5]. LibSVM includes functions to train SVMs, optimize the free parameters of the kernel function, scale the patterns and predict which class a pattern belongs to. This library has support for multi-class classification problems by means of the one-against-one strategy.

As can be seen in Fig. 1, the supervectors in D_{train}, where the language of each segment is known, are used to train the SVM. Then, once the SVM is obtained, the supervectors of D_{test} are classified by using this SVM, obtaining a set of identification results.

3 Experimental Framework

3.1 Description of the Database

The language identification system described in this work was assessed using the KALAKA-2 database [9], which was used for the Albayzin-LRE [17]. The database consists of speech stored in WAV files (PCM, 16 kHz, single channel, 16 bits/sample) extracted from different TV shows (news, documentaries, debates, interviews, etc.) spoken in Basque, Catalan, Galician, English, Portuguese and Spanish. There are three datasets (train, development and evaluation) featuring about ten hours of clean speech and two hours of noisy speech each. The development and evaluation datasets include some speech in unknown (out-of-set) languages. In these experiments, no noisy speech or out-of-set language will be considered. Table 1 contains the number of segments in each dataset, which were obtained in the following way: first, noisy speech, music, overlapped speech, etc. are removed from the audio files, obtaining segments of arbitrary duration with clean speech only; then, segments of 30 seconds are extracted from these segments.

There are some interesting issues about the languages in KALAKA-2. First of all, English is differenced from the other languages by the fact that it does not have its origins in Latin. The history of English starts with the arrival of three Germanic tribes to the British Islands in 500 BC: the Angles, the Saxons and the Jutes. These tribes crossed the North Sea from Jutland, which is the area located in the modern Denmark and the north of Germany. The Angles used to

Table 1. Number of segments of the datasets

	Train	Development	Test
Spanish	342	136	125
Catalan	687	120	149
English	249	133	135
Basque	406	146	130
Galician	464	137	121
Portuguese	387	164	146
Total	2189	836	806

name its own language as Englisc, word that gave the name to English. Before the arrival of these Germanic tribes, the inhabitants of Great Britain used to speak a language with Celtic origins [6].

Before 210 BC, the Iberian Peninsula was populated by Iberians on the South and by Indo-Europeans on the North (Celtic tribes coming from the South of Germany). In 210 BC Latin was brought to Iberia by the Romans, and it was spread quickly through the population, specially through the richer inhabitants that did not want to lose their privileges to the invaders. These richer inhabitants were mainly settled on the South of the Peninsula.

In III AC Vulgar Latin is spoken in all Iberia. Centuries later the Germans left an influence on Vulgar Latin, and then the Arabians arrived to Iberia in 711, taking the South under control. This is when the different languages in the Peninsula were born, as different dialects (which later became languages) appeared in the North of the Iberian Peninsula: the Galician-Portuguese, the Catalan, with an important Provençal influence, and the Castilian (also known as Spanish), that was more evolved than the others and was quickly spread to all the Iberian Peninsula with the expansion of the Kingdom of Castile. Galician-Portuguese was split into Galician, spoken on the North bank of River Miño, and Portuguese, spoken on the South bank and then spread to many parts of the world with the Portuguese Empire. The same happened with Spanish, it was extended to many regions of the world during the heyday of the Spanish Empire [11].

The origin of Basque is unknown, some theories point out that it is an Indo-European language, but this fact is not clear.

3.2 Description of the Experiments

The experiment performed to assess the Fishervoice-SVM system is the mandatory test condition of the Albayzin-LRE as described in [17]. It is a verification experiment: given a test segment and a target language, the system has to decide if the test segment is spoken in the target language. In this experiment six trials for each test segment of about 30 seconds are performed, one per target language. This is a closed-set experiment, which means that all the test segments are spoken in one of the six languages enumerated in Sect. 3.1. This verification experiment can be treated as an identification one, which makes it suitable to assess the Fishervoice-SVM system.

3.3 Metric

As stated in [17], the metric used to measure the performance of the language identification system is the average cost (C_{avg}):

$$\begin{aligned} C_{\text{avg}} = \tfrac{1}{L} \textstyle\sum_{i=1}^{L} \{ & C_{\text{miss}} \cdot P_{\text{target}} \cdot P_{\text{miss}}(i) \\ & + \textstyle\sum_{\substack{j=1 \\ j \neq i}}^{L} C_{\text{fa}} \cdot P_{\text{non-target}} \cdot P_{\text{fa}}(i,j) \\ & + C_{\text{fa}} \cdot P_{\text{OOS}} \cdot P_{\text{fa}}(i,0) \} \end{aligned} \tag{8}$$

$P_{\text{miss}}(i)$ is the miss rate computed on trials corresponding to target language i and $P_{\text{fa}}(i,j)$ is the false alarm rate computed on trials corresponding to other language j. The cost model is the same as in NIST Language Recognition Evaluations [14]:

$$\begin{aligned} C_{\text{miss}} &= C_{\text{fa}} = 1 \\ P_{\text{target}} &= 0.5 \\ P_{\text{non-target}} &= \tfrac{1 - P_{\text{target}} - P_{\text{OOS}}}{L-1} \end{aligned} \tag{9}$$

As the experiment is closed-set there are no out-of-set languages, thus, $P_{\text{OOS}} = 0$. Replacing these values in (8), the following C_{avg} is obtained:

$$C_{\text{avg}} = \frac{1}{2L} \sum_{i=1}^{L} \{ P_{\text{miss}}(i) + \frac{1}{L-1} \sum_{\substack{j=1 \\ j \neq i}}^{L} P_{\text{fa}}(i,j) \} \tag{10}$$

C_{avg} is a combination of two errors, the miss detection (the system decides that the test segment is not spoken in the target language when it is true) and the false alarm (the system decides that the test segment is spoken in the target language when it is false). The lower the C_{avg}, the better the performance of the system.

Some of the results in Sect. 4 are presented using the pairwise cost between languages i and j:

$$\begin{aligned} C(i,j) &= C_{\text{miss}} \cdot P_{\text{target}} \cdot P_{\text{miss}}(i) + C_{\text{fa}} \cdot (1 - P_{\text{target}}) \cdot P_{\text{fa}}(i,j) \\ &= 0.5 \cdot P_{\text{miss}}(i) + 0.5 \cdot P_{\text{fa}}(i,j) \end{aligned} \tag{11}$$

4 Experimental Results

4.1 Features

In this work, the acoustic features extracted from the speech utterances are 12 Mel-frequency Cepstral Coefficients (MFCC), extracted using a 25ms Hamming window at a rate of 10ms per frame, and augmented with the normalized log-energy and their delta and acceleration coefficients. Thus, the dimension of the feature space (N) used in these experiments is 39.

4.2 Results

KALAKA-2 database has three datasets: train, development and test. One half of the training data is used to train the GMM-UBM, and the other half is the one that composes the dataset A_{train}, as explained in Sect. 2. During the development stage, the dataset A_{test} is composed by the speech utterances in the development dataset, and during the test stage, it is composed by the speech utterances in the test dataset.

Development data is provided in Albayzin-LRE in order to tune the free parameters of the language recognition system. In this case there are three parameters that have to be tuned: the number of Gaussians of the GMM (M), the number of eigenvectors kept in the PCA stage (e_{PCA}) and the number of eigenvectors kept in the LDA stage (e_{LDA}). To achieve a dimensionality reduction, it can be concluded from Sect. 2.1 that $e_{\text{PCA}} < N$ and $e_{\text{LDA}} < M$. Several experiments were performed with $M = 32, 64, 128$ and different values of e_{PCA} and e_{LDA}, choosing the ones that achieved the lowest C_{avg}. In this case, $M = 64$, $e_{\text{PCA}} = 60$ and $e_{\text{LDA}} = 20$, which achieved a C_{avg} of 0.09. With these values, the dimensionality is reduced from $64 \times 39 = 2496$ to $60 \times 20 = 1200$.

Tables 2 and 3 show the C_{avg} achieved on the development set for different values of M, e_{PCA} and e_{LDA}. The value in bold is the lowest one, achieved with $M = 64$, $e_{\text{PCA}} = 60$ and $e_{\text{LDA}} = 20$ as stated before. As expected, C_{avg} is higher when the number of features is small, and performance improves as the number of features increases. Nevertheless, the best performance is not achieved when the number of features is the maximum, which shows that the Fishervoice transformation not only reduces the dimensionality but also improves the results by projecting the feature vectors into a more discriminative subspace.

Table 2. C_{avg} on the development set, $M = 32$

$e_{\text{LDA}}/e_{\text{PCA}}$	10	20	30
10	0.176	0.155	0.136
20	0.155	0.139	0.113
30	0.137	0.119	0.112

Performing the identification experiment on the test data with the values of M, e_{PCA}, e_{LDA} obtained in the development stage, the Fishervoice-SVM system achieved a C_{avg} equal to 0.084. When comparing to the results obtained in Albayzin-LRE, it can be seen that the Fishervoice-SVM system rises to the fifth position, which is a good result for a system that simple, while the baseline system was in the ninth position, as the C_{avg} was reduced in more than 50%. The other systems that participated in Albayzin-LRE are fusions of different phonotactic and acoustic subsystems [15] [1] [18], and their performance can be found in [17]. Numerical results are summarized in Table 4.

Table 3. C_{avg} on the development set, $M = 64$

e_{LDA}/e_{PCA}	10	20	30	40	50	60
10	0.153	0.134	0.104	0.117	0.137	0.127
20	0.156	0.117	0.109	0.1	0.101	**0.093**
30	0.138	0.119	0.105	0.096	0.094	0.1

Table 4. Experimental results on KALAKA-2 database

Experiment	C_{avg} Dev	C_{avg} Test	Dimensionality
GMM-SVM	0.102	0.091	2496
Fisher-Euclidean	0.206	0.192	1248
Fisher-SVM	**0.09**	**0.084**	**1200**

Table 5. $C(target\ language, segment\ language)$

		Target language					
		Spanish	Catalan	English	Basque	Galician	Portuguese
Segment language	Spanish	-	0.088	0.03	0.124	0.17	0.007
	Catalan	0.152	-	0.03	0.122	0.1	0.014
	English	0.132	0.099	-	0.115	0.057	0.014
	Basque	0.144	0.107	0.03	-	0.127	0.011
	Galician	0.169	0.096	0.03	0.112	-	0.011
	Portuguese	0.132	0.084	0.033	0.115	0.054	-
	C_{avg}	0.146	0.095	0.03	0.117	0.102	0.011

Table 5 represents the pairwise costs between each pair of languages and the C_{avg} of each individual language. It can be appreciated that the languages that are better distinguished from the others (the ones that achieved the lowest C_{avg}) are English and Portuguese: a possible explanation to this fact is that, while Spanish, Catalan, Basque and Galician are phonetically similar, Portuguese and English phonetics are easier to difference. Moreover, as Spanish coexists with Galician in Galicia, with Catalan in Catalonia and with Basque in Basque Country, it makes these four languages phonetically similar and thus, easier to be confused by an MFCC-based language identification system, as MFCCs reflect articulatory and hearing properties [10].

5 Conclusions and Future Work

A language identification system is presented in this paper, which first projects the feature vectors that represent the speech utterances into a low-dimensional discriminative subspace by using the Fishervoice technique, and then classifies these feature vectors by means of an SVM. Experimental results are compared to those achieved in the Albayzin-LRE evaluation by a previous Fishervoice-based system and by the other systems in the evaluation. This previous baseline system performed the classification by computing the Euclidean distance between a test utterance and each of the utterances used to model the different languages. Results on KALAKA-2 database, which includes speech in Spanish, Catalan, English, Basque, Galician and Portuguese, reflect that this relatively simple language identification system shows a good performance, specially when identifying languages that are phonetically different to the other target languages (in this case, the languages that were easier to identify were English and Portuguese).

The Albayzin-LRE evaluation establishes a protocol for performing language verification, but in this work this database was used to assess a language identification system. In future work, methods for performing language verification on Fishervoice vectors will be studied; thus, the SVM multiclass classifier will have to be replaced by a client-impostor classifier as, for example, a one-class SVM.

Acknowledgements. This work has been supported by the Galician Regional Government (CN2011/019, 2009/062), Spanish Government (TEC009-14094-C04-04 and FPI grant BES-2010-033358), and the European Regional Development Fund.

References

1. Abad, A., Koller, O., Trancoso, I.: The L2F Language Verification Systems for Albayzin-2010 Evaluation. In: Proceedings of FALA 2010 - VI Jornadas en Tecnología Del Habla and II Iberian SLTech Workshop, pp. 383–388 (2010)
2. Anthony, G., Gregg, H., Tshilidzi, M.: Image Classification Using SVMs: One-Against-One Vs One-Against-All. In: Proceedings of the 28th Asian Conference on Remote Sensing (2007)
3. Burges, C.J.C.: A Tutorial on Support Vector Machines for Pattern Recognition. Data Mining and Knowledge Discovery 2, 121–167 (1998)
4. Castaldo, F., Colibro, D., Cumani, S., Dalmasso, E., Laface, P., Vair, C.: Loquendo-Politecnico di Torino System for the 2009 NIST Language Recognition Evaluation. In: Proceedings of ICASSP, pp. 5002–5005 (2010)
5. Chang, C.-C., Lin, C.-J.: LIBSVM: a Library for Support Vector Machines. ACM Transactions on Intelligent Systems and Technology 2(3), article 27 (2011), `http://www.csie.ntu.edu.tw/~cjlin/libsvm`
6. Crystal, D.: The Cambridge Encyclopedia of the English Language, pp. 6–8. Cambridge University Press, Cambridge (2003)
7. Hazen, T.J., Hetherington, I.L., Park, A.: FST-Based Recognition Techniques for Multi-Lingual and Multi-Domain Spontaneous Speech. In: Proceedings of the European Conference on Speech Communication and Technology (2001)

8. Jing, X.Y., Wong, H.S., Zhang, D.: Face Recognition Based on 2D Fisherface Approach. Pattern Recognition 39(4), 707–710 (2006)
9. KALAKA-2. Speech database created for the Albayzin, Language Recognition Evaluation, organized by the Spanish Network on Speech Technology. Produced by the Software Technologies Working Group (GTTS), University of the Basque Country (2010), `http://gtts.ehu.es`
10. Kovács, G., Tóth, L.: Phone Recognition Experiments with 2D-DCT Spectro-Temporal Features. In: 6th IEEE International Symposium on Applied Computational Intelligence and Informatics, pp. 143–146 (2011)
11. Lapesa, R.: Historia de la Lengua Española. Escelicer, Gredos (1981)
12. Lopez-Otero, P., Docio-Fernandez, L., Garcia-Mateo, C.: A Fishervoice-based Speaker Identification System. In: Proceedings of FALA 2010 - VI Jornadas en Tecnología del Habla and II Iberian SLTech Workshop, pp. 139–142 (2010)
13. Lopez-Otero, P., Docio-Fernandez, L., Garcia-Mateo, C.: The UVigo-GTM Language Verification Systems for the Albayzin 2010 Evaluation. In: Proceedings of FALA 2010 - VI Jornadas en Tecnología del Habla and II Iberian SLTech Workshop, pp. 389–392 (2010)
14. Martin, A., Greenberg, C.: The 2009 NIST Language Recognition Evaluation. In: Odyssey 2010 - The Speaker and Language Recognition Workshop, paper 030 (2010)
15. Martínez, D., Villalba, J., Miguel, A., Ortega, A., Lleida, E.: ViVoLab UZ Language Recognition System for Albayzin 2010 LRE. In: Proceedings of FALA 2010 - VI Jornadas en Tecnología del Habla and II Iberian SLTech Workshop, pp. 375–376 (2010)
16. Matějka, P., Schwarz, P., Černocký, J., Chytil, P.: Phonotactic Language Identification using High Quality Phoneme Recognition. In: Proceedings of Interspeech, pp. 2237–2240 (2005)
17. Rodriguez-Fuentes, L.J., Penagarikano, M., Varona, A., Diez, M., Bordel, G.: Overview of the Albayzin 2010 Language Recognition Evaluation: Database Design, Evaluation Plan and Preliminary Analysis of Results. In: Proceedings of VI Jornadas en Technologa del Habla and II Iberian SLTech Workshop, pp. 309–316 (2010)
18. Saeidi, R., Soufifar, M., Kinnunen, T., Svendsen, T., Fränti, P.: UEF-NTNU System Description for Albayzin 2010 Language Recognition Evaluation. In: Proceedings of FALA 2010 - VI Jornadas en Tecnología del Habla and II Iberian SLTech Workshop, pp. 377–382 (2010)
19. Torres-Carrasquillo, P.A., Singer, E., Kohler, M.A., Green, R.J., Reynolds, D.A., Deller Jr, J.R.: Approaches to Language Identification Using Gaussian Mixture Models and Shifted Delta Cepstral Features. In: Proceedings of ICSLP, pp. 89–92 (2002)
20. Woehrling, C., de Mareil, P.B., Adda-Decker, M.: Linguistically-Motivated Automatic Classification of Regional French Varieties. In: Proceedings of Interspeech, pp. 2183–2186 (2009)
21. Wong, E., Pelenacos, J., Myers, S., Sridharan, S.: Language Identification Using Efficient Gaussian Mixture Model Analysis. In: Proceedings of Australian International Conference on Speech Science and Technology, pp. 300–305 (2000)
22. Zissman, M.A., Berkling, K.M.: Automatic Language Identification. In: Proceedings of the ESCA-NATO Workshop on Multi-Lingual Interoperability in Speech Technology, MIST (1999)

Summarizing Speech by Contextual Reinforcement of Important Passages

Ricardo Ribeiro[1,3] and David Martins de Matos[2,3]

[1] Instituto Universitário de Lisboa (ISCTE-IUL)
[2] Instituto Superior Técnico, Universidade Técnica de Lisboa
[3] L2F - INESC ID Lisboa
R. Alves Redol, 9, 1000-029 Lisboa, Portugal
{ricardo.ribeiro,david.matos}@inesc-id.pt

Abstract. We explore the use of contextual information of the same type, i.e., speech transcriptions, to assess the relevant content of a single information source. Our proposal consists in the use of topic-related additional information sources to contextualize the information of the main input source, improving the estimation of the most important passages. We analyse the impact of using as additional information both the full topic-related stories and just the passages from those stories that are closer to the passages of the input source to be summarized. A multi-document summarization framework, Latent Semantic Analysis (LSA), provides the means to assess the relevant content. To minimize the influence of speech-related problems, we explore several term weighting strategies. Evaluation is performed using an information-theoretic evaluation measure, the Jensen-Shannon divergence, that does not need reference summaries.

Keywords: Speech-to-text summarization, Latent semantic analysis, Term weighting.

1 Introduction

From a cognitive sciences perspective [6–8, 26], human summarization is characterized as a knowledge-based task. However, most of the work on speech summarization concentrates on two main research lines: (i) the analysis of new features, especially speech-related features (acoustic/prosodic) or discourse information [21, 24, 31], even though shallow text summarization approaches still seem to achieve a comparable performance [25]; (ii) the proposal of new summarization models [4, 17, 18]. In general, unsupervised approaches rely only on the information source to be summarized, and, in supervised approaches, summarization is cast as a classification problem, where the "additional information" consists of document/summaries pairs and is only used to train the summarization model.

We explore the use of additional related information to cope with the difficulties posed by speech-to-text summarization. Our proposal consists in the

H. Caseli et al. (Eds.): PROPOR 2012, LNAI 7243, pp. 392–402, 2012.

use of topic-related additional information sources to contextualize the information of the main input source, improving the estimation of the most important passages. The idea mimics the natural human behavior, in which information acquired from different sources is used to build a better understanding of a given topic. A multi-document summarization framework, LSA [11, 12], provides the means to assess the relevant content.

In the next section, we present an overview of the related work; Section 3 presents our summarization model; Section 4 describes the evaluation methodology and the obtained results; conclusions and future work close the document.

2 Related Work

As mentioned before, human summarization is a knowledge-based task. The models we here survey, although far from the descriptions of the human summarization process by Endres-Niggemeyer [6–8] or Pinto Molina [26], improve the detection of the relevant passages by using additional information sources. These models are closely related to ours, as we also try to combine different information sources to diminish the influence of the speech related problems in the assessment of the salient information.

In text summarization, CollabSum [30] explores multiple information sources to summarize a single information source. Given a collection of documents, the first step of CollabSum is to create clusters by grouping related documents. Following the clustering stage, the summarization stage starts by creating a weighted undirected graph that connects all the sentences in the cluster (weights correspond to sentence similarity). Additionally to this global graph, two other graphs are created: one only with intra-document links, and another only with cross-document links. The informativeness of each sentence is/may be computed using each graph in a similar way of LexRank [9] and TextRank [22]. The final informativeness can be estimated by combining the informativeness computed using the two non-global graphs. The last step consists in removing redundancy. Since the CollabSum model was not compared to other models, we are not able to analyse its performance.

Chatain et al. [2] base topic-adapted speech summarization on a sentence scoring function in which one of the components is an n-gram linguistic model that is computed from given data. Given a passage $P \triangleq w_1, w_2, ..., w_n$,

$$score(P) = \frac{1}{n} \sum_{i=1}^{n} \left[\alpha_C C(w_i) + \alpha_I I(w_i) + \alpha_L L(w_i)\right], \quad (1)$$

where $C(w_i)$ is the recognition confidence score of word w_i, $I(w_i)$ is the significance score (based on frequencies computed over corpora), and $L(w_i)$ is the linguistic score (computed using a language model); α_C, α_I, and α_L are the respective weighting values. However, in the two experiments performed, one using talks and the other using broadcast news, only the one using talks used a topic-adapted linguistic model (by interpolation with a baseline language model) and the data used for the adaptation consisted of the papers in the conference proceedings of the talk to be summarized. Results show that the adaptation of the language model improved the baseline.

Ribeiro and de Matos [27, 28] propose the use of additional related information sources in a summarization process based on LSA to overcome speech-related problems. Both data from different types of source [27], and the same type of source [28] are explored. The idea mimics the natural human behavior, in which information acquired from different sources is used to build a better understanding of a given topic. The mixed source type approach shows significant improvements over the baseline, both in terms of summary content and readability. We build on the same source type approach proposed by Ribeiro and de Matos [28], providing a better analysis of the influence of the global weights on the summarization process when using additional information sources, as well a new method for selecting the additional information.

Chen, Chen and Wang [4] propose a unified probabilistic generative framework for speech summarization,

$$P(S|D) = P(D|S)P(S)/P(D), \tag{2}$$

where S represents a sentence, and D represents a document. Sentence ranking is done by combining $P(D|S)$, relevance in context, and $P(S)$, relevance by itself. Several approaches are experimented with to compute $P(D|S)$: a language model using a general text news collection combined with the relevance model [14] based on the documents returned by an information retrieval (IR) system using the sentence as query; and, sentence [3] and word topical mixture models [5] based on text news documents. The computation of the sentence prior probability ($P(S)$) is also based on the set of documents returned by the IR system, to minimize the influence of recognition errors. Results show improvements over models like LSA, Maximal Marginal Relevance, and a classification-based approach using support vector machines.

3 Contextual Reinforcement of Important Passages

In the current work, we explore the use of additional information sources to assess relevance in the context of the summarization of broadcast news. SSNT [1] is a system for the selective dissemination of multimedia contents, working primarily with Portuguese broadcast news services. The system is based on an automatic speech recognition (ASR) module that generates the transcriptions later used by topic segmentation, topic indexing, and title&summarization modules. Our proposal consists in the use of news stories previously indexed by topic (by the topic indexing module) to identify the most relevant passages of the news stories to be summarized.

3.1 Selecting Background Information

The first step of our approach consists in the selection of background information to improve the detection of the most relevant passages of the information sources to be summarized. Given our case study (the SSNT system), the selection of background information sources is done by identifying and selecting the news stories from the previous n days that match the topics of the news story to be summarized. This procedure relies on the results produced by the topic indexing module, but a clustering approach could also be followed.

In addition to using the complete news stories of the same topic (T) of the news story to be summarized (IS), we also explore the use of the closest passages of the topic-related information sources in terms of content overlap (s and t are passages).

$$coverage(s) \triangleq \{t \in T : \frac{|s \cap t|}{|s|} > \epsilon\},\ s \in IS \tag{3}$$

The threshold (ϵ) was empirically set to the word error rate of the ASR module.

3.2 Summarization Process

The summarization process is characterized by the use of LSA. Originally, LSA was developed as a new theory of knowledge acquisition and representation [12, 13] that grounds its capabilities on the mathematical technique for factor analysis called Singular Value Decomposition (SVD) [10].

Pioneer work on LSA-based (text) summarization [11], as it happens in the general application of the theory, starts by representing the information source(s) in a terms by passages matrix similar to the one of Eq. 4 (T is the number of different terms; N is the number of passages; rows correspond to terms, and columns to passages).

$$A = \begin{bmatrix} a_{1,1} & \cdots & a_{1,N} \\ \cdots & & \\ a_{T,1} & \cdots & a_{T,N} \end{bmatrix} \tag{4}$$

Each element of the matrix A has two components: a local weight and global weight (Eq. 5). The local weight is a function of the occurrences of each term within each passage, and the global weight is a function of the number of passages in which each word occurs.

$$a_{ij} = L_{ij} \times G_i \tag{5}$$

Then, the SVD is applied to the matrix A, decomposing it into three matrices (Eq. 6): U is an $T \times N$ matrix of left singular vectors; Σ is the $N \times N$ diagonal matrix of singular values; and, V is the $N \times N$ matrix of right singular vectors (only possible if $T \geq N$).

$$A = U \Sigma V^T \tag{6}$$

The idea behind the method is that the decomposition captures the underlying topics (identifying the most salient ones) of the information source(s) by means of co-occurrence of terms (the latent semantic analysis), and identifies the best representative passages of each topic. Summary creation can be done by picking the best representatives of the most relevant topics according to a defined strategy. Our implementation follows the original ideas of Gong and Liu [11] and the ones of Murray, Renals and Carletta [23] for solving dimensionality problems. Prior to the application of the SVD, it is necessary to create the terms-by-passages matrix that represents the input source. It is in this step that the previously selected background information is taken into consideration: instead of using only the information source to be summarized to create the terms by sentences matrix, both the selected background information and the main information source are used. The matrix is defined as

$$\begin{bmatrix} a_{1d_1^1} \ldots a_{1d_s^1} & \ldots & a_{1d_1^D} \ldots a_{1d_s^D} & a_{1n_1} \ldots a_{1n_s} \\ & \ldots & & \\ a_{Td_1^1} \ldots a_{Td_s^1} & \ldots & a_{Td_1^D} \ldots a_{Td_s^D} & a_{Tn_1} \ldots a_{Tn_s} \end{bmatrix} \quad (7)$$

where $a_{id_j^k}$ represents the weight of term t_i, $1 \leq i \leq T$ (T is the number terms), in passage $s_{d_j^k}$, $1 \leq k \leq D$ (D is the number of selected documents to constitute the background knowledge) with $d_1^k \leq d_j^k \leq d_s^k$, of document d^k; and a_{in_l}, $1 \leq l \leq s$, are the elements associated with the news story to be summarized. To form the summary, only passages corresponding to the last columns of the matrix defined in Eq. 7 (which correspond to the sentence-like units of the news story to be summarized) are selected. As local weight we used the frequency of term in the passage. In order to assess the relevance of the background information, we explored different global weighting strategies. The global weights G_i were defined as follows:

- 1 (not using global weight);
- $-log(P(t_i))$ (self-information or surprisal: we use both terms interchangeably);
- *idf*, inverse document (passage) frequency.

$$P(t_i) = \frac{\sum_j C(t_i, s_j)}{\sum_i \sum_j C(t_i, s_j)}, \; s_j \text{ is a passage} \quad (8)$$

where

$$C(t_i, s_j) = \begin{cases} 1 & t_i \in s_j \\ 0 & t_i \notin s_j \end{cases} \quad (9)$$

4 Evaluation

As previously mentioned, our approach tries to diminish the influence of speech-related problems by using additional information sources of the same medium of the information source to summarize. Our goal is to understand if the content of the summaries produced using additional information is more informative than using a baseline LSA approach.

4.1 Corpora

To test our method, we selected a corpus composed by excerpts of broadcast news programs where it is possible to find topic related news stories in a chronological frame close to the information source to be summarized.

In this experiment, we extractively summarized the news stories of two episodes, 2008/06/09 and 2008/06/29, of a Portuguese news program (Table 1 top part) by selecting passages that are similar to the concept of sentence in written text: sentence-like units (SUs). According to Liu et al. [19], the concept of SU is different from the concept of sentence in written text, since, although semantically complete, SUs can be smaller than a sentence. As source of prior information, we used, for the 2008/06/09 show, the episodes from 2008/06/01 to 2008/06/08, and for the 2008/06/29 show, the episodes from 2008/06/20 to 2008/06/28 (Table 1 bottom part). All transcription were generated using the SSNT system.

Table 1. Properties of the news shows to be summarized and the corresponding background information

		2008/06/09	2008/06/29
Information sources to be summarized	# of stories	28	24
	# of SUs	443	476
	# of transcribed tokens	7733	6632
	# of different topics	45	47
	Avg. # of topics per story	3.54	4.92
Additional information sources	# of stories	205	268
	# of SUs	4127	4553
	# of transcribed tokens	62025	71470
	# of different topics	158	195
	Avg. # of topics per story	3.42	4.10

4.2 Evaluation Measure

In general, most evaluation methodologies for summarization need reference human summaries, or human annotation. Although not as expensive as a human evaluation (according to Lin [15], the human evaluation of a Document Understanding Conference[1], considering linguistic and content aspects, would require over 3000 hours of human labour), the cost of such resources is sufficiently high to motivate research in directions that minimize or do not require the use of this kind of resources. In that sense, Loius and Nenkova [20] and Saggion et al. [29] (Lin et al. [16] previously explored these measures, but maintained the use of reference human summaries) explore Information Theory-based methodologies that use the information sources to be summarized as references.

[1] `http://duc.nist.gov/`

As no reference summaries were available for the selected corpus, to evaluate the informativeness of the generated summaries, we used one of this recently studied information-theoretic measures [20, 29]: the Jensen-Shannon divergence (Eq. 10; P and Q are the probability distributions associated with the summary and the information source; t is a term), which was shown to have a strong correlation with human preferences [20].

$$D_{JS}(P||Q) = \frac{1}{2}\Big[\sum_t P(t)\log\big(\frac{2P(t)}{P(t)+Q(t)}\big) + \sum_t Q(t)\log\big(\frac{2Q(t)}{P(t)+Q(t)}\big)\Big] \quad (10)$$

The underlying idea is that the distribution of the terms in the information source to be summarized and in the summary should be similar.

4.3 Results

Figure 1 shows the number of times each method produced the most informative summary. In general, the methods using additional information sources, either the closest passages, or the full topic-related news stories, achieved the best results, especially when using global weights. In the subcorpus of 2008/06/09, the best results were obtained using the *idf* global weight, while in 2008/06/29 subcorpus, the best results were obtained by the *surprisal* global weight. In both corpus, the use of additional information without using global weights produced results comparable to the LSA baseline, which means that, as expected, global weights boost the influence of the additional information sources.

It is also interesting to observe that using full stories is better that using less information, although closer in terms of lexical content. In fact, it could be expected that more information could lead to a uniform reinforcement of all passages of the main input source, and consequently diminishing the influence of the additional information sources. Nevertheless, the results confirm our intuition: multiple information sources provide a better understanding of a single information source (even considering additional information that suffers from the same problems of the main information source).

Table 2 presents the average Jensen-Shannon divergence values for each summarization method. In general, the summarization approaches selected a higher number of times as best summary correspond to the ones with the best results in terms of average divergence. The only exception is tps_idf in the 2008/06/29 corpus, which has the second lowest divergence value and the worst result in terms of best summary analysis. However, this was mainly due to the good performance of the variants using the full stories.

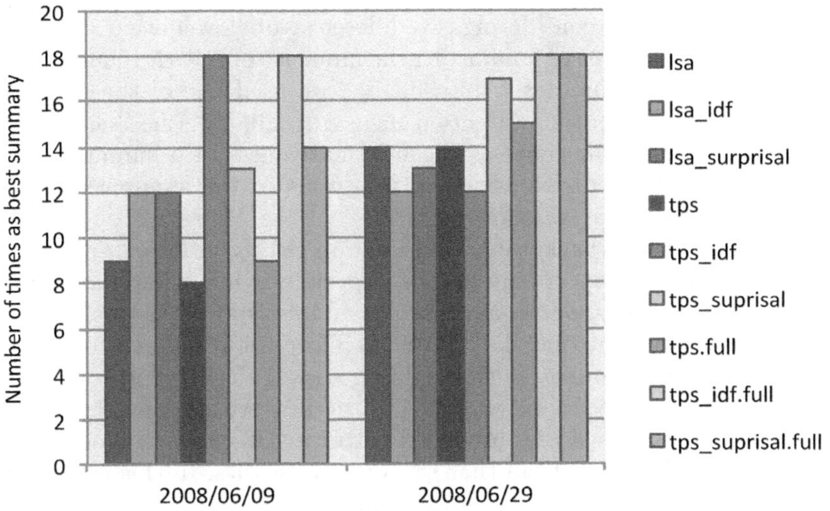

Fig. 1. Overall results for each summary creation method. We considered the best summaries, the ones with the lowest divergence values. **tps**: topic-directed approach (using additional information); **lsa**: LSA baseline; **idf**: using idf weights; **surprisal**: using surprisal weights; **full**: using the full news stories.

Table 2. Average Jensen-Shannon divergence values for each summary creation method (small numbers are better). **tps**: topic-directed approach (using additional information); **lsa**: LSA baseline; **idf**: using idf weights; **surprisal**: using surprisal weights; **full**: using the complete news stories.

	2008/06/09	2008/06/29
lsa	0.157	0.192
lsa_idf	0.159	0.198
lsa_surprisal	0.158	0.197
tps	0.163	0.194
tps_idf	0.150	0.191
tps_suprisal	0.154	**0.187**
tps.full	0.162	0.193
tps_idf.full	**0.149**	**0.187**
tps_suprisal.full	0.154	**0.187**

5 Conclusions and Future Work

We analysed the influence of the background information in the determination of the relevant content, considering two different global weighting strategies (idf and surprisal), as a means of minimizing the influence of speech-related problems. Since no reference summaries were available, we used one of the recently proposed information-theoretic evaluation measures [20, 29]: the Jensen-Shannon divergence. This measure assesses the informativeness of a summary by comparing the probability distributions of the terms in the summary and in the information source to be summarized.

Results showed that background information clearly influences and improves the content of the produced summaries, especially when using global weights, confirming the adequacy of our approach to address spoken language-related issues. Note, in particular, that the additional information suffered from the same problems of the input source. Both weighting strategies have better performance than the baseline, with idf achieving the best results. We also investigated if complete topic-related news stories produced better results that the closest passages in terms of lexical overlap from that sets of news stories. In this case, complete news stories achievied the best overall results.

As future work, there are two different research lines that should be addressed: new methods for combining the information; and, the analysis of the inclusion of speech-related features in this approach to summarization.

Acknowledgments. This work was partially supported by FCT (INESC-ID multiannual funding) through the PIDDAC Program funds. This work was also partly supported by FCT project CMU-PT/005/2007.

References

1. Amaral, R., Meinedo, H., Caseiro, D., Trancoso, I., Neto, J.P.: A Prototype System for Selective Dissemination of Broadcast News in European Portuguese. EURASIP Journal on Advances in Signal Processing 2007 (2007)
2. Chatain, P., Whittaker, E.W.D., Mrozinski, J.A., Furui, S.: Topic and Stylistic Adaptation for Speech Summarisation. In: Proc. of the 2006 IEEE International Conference on Acoustics, Speech and Signal Processing, pp. 977–980. IEEE (2006)
3. Chen, B., Yeh, Y.M., Yang, Y.M., Chen, Y.T.: Chinese Spoken Socument Summarization Using Probabilistic Latent Topical Information. In: Proc. of the 2006 IEEE International Conference on Acoustics, Speech and Signal Processing, pp. 969–972 (2006)
4. Chen, Y.T., Chen, B., Wang, H.M.: A Probabilistic Generative Framework for Extractive Broadcast News Speech Summarization. IEEE Transactions on Audio, Speech, and Language Processing 17(1), 95–106 (2009)
5. Chiu, H.S., Chen, B.: Word topical mixture models for dynamic language model adaptation. In: Proceedings of the 2007 IEEE International Conference on Acoustics, Speech and Signal Processing, pp. 169–172. IEEE (2007)
6. Endres-Niggemeyer, B.: Summarizing Information. Springer (1998)

7. Endres-Niggemeyer, B.: Human-style WWW summarization. Tech. rep., University for Applied Sciences, Department of Information and Communication (2000)
8. Endres-Niggemeyer, B.: SimSum: an empirically founded simulation of summarizing. Information Processing and Management 36(4), 659–682 (2000)
9. Erkan, G., Radev, D.R.: LexRank: Graph-based Centrality as Salience in Text Summarization. Journal of Artificial Intelligence Research 22, 457–479 (2004)
10. Golub, G.H., van Loan, C.F.: Matrix Analysis. In: Matrix Computations, 3rd edn., pp. 48–86. The Johns Hopkins University Press (1996)
11. Gong, Y., Liu, X.: Generic Text Summarization Using Relevance Measure and Latent Semantic Analysis. In: SIGIR 2001: Proc. of the 24st Annual Intl. ACM SIGIR Conf. on Research and Development in Information Retrieval, pp. 19–25 (2001)
12. Landauer, T.K., Dumais, S.T.: A Solution to Plato's Problem: The Latent Semantic Analysis Theory of Acquisition, Induction and Representation of Knowledge. Psychological Review 104(2), 211–240 (1997)
13. Landauer, T.K., Foltz, P.W., Laham, D.: An Introduction to Latent Semantic Analysis. Discourse Processes 25(2), 259–284 (1998)
14. Lavrenko, V., Croft, W.B.: Relevance Models in Information Retrieval. In: Language Modeling for Information Retrieval. The Information Retrieval Series, vol. 13. Springer (2003)
15. Lin, C.Y.: ROUGE: A Package for Automatic Evaluation of Summaries. In: Moens, M.F., Szpakowicz, S. (eds.) Text Summarization Branches Out: Proceedings of the ACL 2004 Workshop, pp. 74–81. Association for Computational Linguistics (2004)
16. Lin, C.Y., Cao, G., Gao, J., Nie, J.Y.: An Information-Theoretic Approach to Automatic Evaluation of Summaries. In: Proceedings of the Human Language Technology Conference of the North American Chapter of the ACL, pp. 463–470. Association for Computational Linguistics (2006)
17. Lin, S.H., Chen, B.: A Risk Minimization Framework for Extractive Speech Summarization. In: Proceedings of the 48th Annual Meeting of the Association for Computational Linguistics, pp. 79–87. Association for Computational Linguistics (2010)
18. Lin, S.H., Yeh, Y.M., Chen, B.: Extractive Speech Summarization – From the View of Decision Theory. In: Proc. of the 11th Annual Conference of the International Speech Communication Association (INTERSPEECH 2010), pp. 1684–1687 (2010)
19. Liu, Y., Shriberg, E., Stolcke, A., Hillard, D., Ostendorf, M., Harper, M.: Enriching Speech Recognition with Automatic Detection of Sentence Boundaries and Disfluencies. IEEE Transactions on Speech and Audio Processing 14(5), 1526–1540 (2006)
20. Louis, A., Nenkova, A.: Automatically Evaluating Content Selection in Summarization without Human Models. In: Proc. of the 2009 Conference on EMNLP, pp. 306–314. ACL (2009)
21. Maskey, S.R., Hirschberg, J.: Comparing Lexical, Acoustic/Prosodic, Strucural and Discourse Features for Speech Summarization. In: Proc. of the 9th EUROSPEECH-INTERSPEECH (2005)
22. Mihalcea, R., Tarau, P.: TextRank: Bringing Order into Texts. In: Proceedings of the 2004 Conference on Empirical Methods in Natural Language Processing, pp. 404–411. ACL (2004)
23. Murray, G., Renals, S., Carletta, J.: Extractive Summarization of Meeting Records. In: Proceedings of the 9th EUROSPEECH - INTERSPEECH (2005)

24. Murray, G., Renals, S., Carletta, J., Moore, J.: Incorporating Speaker and Discourse Features into Speech Summarization. In: Proc. of the Human Language Technology Conference of the North American Chapter of the ACL, pp. 367–374. ACL (2006)
25. Penn, G., Zhu, X.: A Critical Reassessment of Evaluation Baselines for Speech Summarization. In: Proceeding of ACL 2008: HLT, pp. 470–478. ACL (2008)
26. Pinto Molina, M.: Documentary Abstracting: Toward a Methodological Model. Journal of the American Society for Information Science 46(3), 225–234 (1995)
27. Ribeiro, R., de Matos, D.M.: Mixed-Source Multi-Document Speech-to-Text Summarization. In: Coling 2008: Proceedings of the 2nd Workshop on Multi-source Multilingual Information Extraction and Summarization, pp. 33–40 (2008)
28. Ribeiro, R., de Matos, D.M.: Using Prior Knowledge to Assess Relevance in Speech Summarization. In: 2008 IEEE Workshop on Spoken Language Technology, pp. 169–172. IEEE (2008)
29. Saggion, H., Torres-Moreno, J.M., da Cunha, I., SanJuan, E., Velázquez-Morales, P.: Multilingual Summarization Evaluation without Human Models. In: Coling 2010: Tthe 23rd Intl. Conf. on Computational Linguistics, vol. Posters, pp. 1059–1067 (2010)
30. Wan, X., Yang, J., Xiao, J.: CollabSum: Exploiting Multiple Document Clustering for Collaborative Single Document Summarizations. In: SIGIR 2007: Proc. of the 30th Annual Intl. ACM SIGIR Conference on Research and Development in Information Retrieval, pp. 143–150 (2007)
31. Zhang, J.J., Chan, R.H.Y., Fung, P.: Extractive Speech Summarization Using Shallow Rhetorical Structure Modeling. IEEE Transactions on Audio, Speech, and Language Processing 18(6), 1147–1157 (2010)

Incorporating ASR Information in Spoken Dialog System Confidence Score

José Lopes[1,2], Maxine Eskenazi[3], and Isabel Trancoso[1,2]

[1] INESC-ID Lisboa, Portugal
[2] Instituto Superior Técnico, Lisboa, Portugal
[3] Language Technologies Institute, Carnegie Mellon University, Pittsburgh, PA, USA
jose.david.lopes@l2f.inesc-id.pt

Abstract. The reliability of the confidence score is very important in Spoken Dialog System performance. This paper describes a set of experiments with previously collected off-line data, regarding the set of features that should be used in the computation of the confidence score. Three different regression methods to weight the features were used and the results show that the incorporation of the confidence score given by the speech recognizer improves the confidence measure.

Keywords: Spoken Dialog Systems, Confidence measures.

1 Introduction

Spoken Dialog Systems (SDS) have to deal with many sources of uncertainty before performing an action based on the user's input. First, there are errors introduced by the speech recognizer. Parsing may not be sufficient to resolve ambiguity, meaning that the same recognized word could represent different actions for the SDS. To reduce the effect of uncertainty it is very important to select a set features could help improve the accuracy of the confidence scores and consequently improve the quality of the interaction.

Since speech recognition is the major source of uncertainty, previous studies tried to add semantic and context information that could reduce the unpredictability of the speech recognition output.

The different approaches to the problem are very dependent on the quality of the specific modules used in the system. The study presented used a dialog system developed with the Olympus architecture [1], however to have speech recognition in European Portuguese, the in-house Automatic Speech Recognition (ASR) module [2] was plugged in. This introduced a new challenge to compute confidence scores, since in the previous architecture with PocketSphinx [3] the acoustic confidence score was not taken into account in the computation of the overall confidence score. The confidence annotator module already computes a set of features to help the Dialog Manager make a better decision, however only the ratio of uncovered words at each stage of the dialog has been used.

The introduction of confidence scores coming from this ASR module and other features was studied in order to observe their impact on the confidence measure

H. Caseli et al. (Eds.): PROPOR 2012, LNAI 7243, pp. 403–408, 2012.

computed by the system. Three different methods for computing regression were compared, simultaneously with various rejection thresholds.

In addition, since the previous studies with the integration of the new ASR module target the improvement of ASR performance by choosing the lexical primes that the system has proposed [4], it is very important to have reliable information from the ASR module in order to decide whether or not retain a prime.

2 Related Work

Many studies have been carried out that approach to this problem from the point of view of error recover. The intuitive idea here is that for SDS, a misunderstanding costs more than a rejection. Bohus et al [5] used two features to build two different regression models, the number of correctly and incorrectly transfered concepts for each dialog state, since they believed that rejection thresholds may vary along dialog states. Then, a logistic regression optimization was trained for task success and a Poisson optimization model was created for dialog duration.

Sarikaya et al in [6] state that for domain constraint systems, semantic information can be helpful. Two different features were used in the studies, the first relying only on the parsing result is based on the intuition that a grammatically correct sentence is easier to parse. The second uses the language model posterior to incorporate information from the recognition process into the confidence measure calculation.

Litman and Pan [7] designed strategies for adaptive behavior in an SDS. The system should change its behavior according to the confidence measure computed by the system. A corpus was labeled with semantic accuracy, that is the percentage of concepts affected by speech recognition errors in each user turn. A threshold was set to classify a dialog as good or bad according to the ASR performance. Together with this, a set of 23 features divided into five different categories was used: acoustic confidence, dialog efficiency and quality, experimental parameters and lexical. Since the used thresholds often requires tuning and the system may recover from an ASR error by adding context information when binding the ASR output to the concepts. In order to deal with this, a new feature was computed that tries to predict the percentage of misrecognitions. This feature uses the log-likelihood score from the ASR to predict that a turn is going to be misrecognized. If the log-likehood falls below a threshold, the turn is predicted to be a misrecognition.

Raymond et al. [8] tried to improve belief confirmation using confidence measures. They combine linguistic information, using a ratio between the intersection of the observed trigrams in the training corpus and all the trigrams occuring in a determined utterance, with the acoustic information given by the log-likelihood score given by the speech recognizer.

3 Data

The data set used for this study was collected from a set of tests done with an SDS that gives bus schedule information which is described in more detail in [4].

The study was carried out for two weeks where the users were supposed to complete three different usage scenarios in each week. The system received 256 calls during that period of time. This corresponded to 1592 user turns. These turns were labeled as correct if the system concepts were correctly acquired from the user turn, or incorrect if they were not. The data set was further divided into a training and a test set. 1144 turns were used for training and the remaining 448 were used for testing.

4 Experimental Procedure

This study aims at selecting a set of features that could help to improve the performance of the Helios confidence annotator, giving more accurate confidence scores to help in the dialog manager decision process. First, the data was analyzed selecting numeric features that could be used to compute the confidence score. The set of features selected included acoustic features that come from the ASR (average word confidence and average word confidence greater than 50% averaged within the turn), dialog performance feature (last turn was marked non-understood) and parsing features (the number of words in the turn and the number of words not covered by the parser).

Since this is a binary problem, logistic regression seemed to be adequate. There are several algorithms to run a logistic regression for feature selection. For the maximum entropy (MaxEnt) method we have used MegaM[9], for prior-weighted Logistic Regression we have used FoCal [10] with 0.5 prior and Stepwise regression functions available in MATLAB's statistical toolbox to compute stepwise regression. Then the results are computed for several rejections thresholds.

The performance of these new weighted feature selections was compared with the baseline system which confidence score was given by:

$$logit(confidence) = 1.6886 - 5.5482 \cdot Ratio\ of\ Uncovered\ Words \tag{1}$$

5 Results

On Table 1 we see the average word confidence is indeed the most weighted feature in every method. Non-understanding in the last turn, the number of uncovered words and the ratio of uncovered words are good predictors for misunderstandood turns. Some of the values are quite unexpected, such as the fact that the average word confidences greater that 0.5 give a negative contribution. In fact, the threshold of 0.5 could mean that it is not adjusted to the new ASR module. The fragment ratio gives an interesting clue: the more words, the less likely the concept is to be correctly bound.

The graphics in this section show Accuracy, Precision and Recall simultaneously, computed with the different algorithms mentioned before, Helios (the baseline), MegaM, FoCal and Stepwise regression computed with MATLAB.

Table 1. Weights for the features used in the confidence annotator training procedure

	MegaM	FoCal	Stepwise
Intercept	-0.61	-0.94	0.30
Word Confidence	2.95	4.09	0.71
Word Confidence > 0.5	-0.3	-0.97	-0.17
Fragment Ratio	-0.11	-0.36	0.00
Fragment Ratio > mean	0.38	0.57	0.00
Non-understanding in last turn	-0.57	-0.60	-0.10
Ratio of the number of parses	-0.62	-0.98	0.00
Number of uncovered words	-0.88	-1.80	-0.12
Number of uncovered words > 0	-0.38	-0.38	-0.31
Number of uncovered words > 1	-0.71	-0.03	0.00
Normalized number of uncovered words	0.97	2.13	0.00
Ratio of uncovered words	-0.62	-1.34	-0.20
Ratio of uncovered words > mean	-0.38	-0.39	0.00

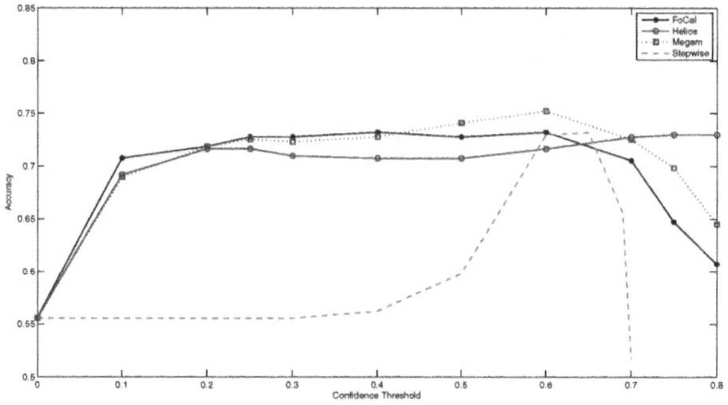

Fig. 1. Accuracy results for the compared methods

The best performance in terms of accuracy is achieved with the weights computed with MegaM setting the rejection threshold at 0.6. All the methods, except Helios, have their best performance at 0.6 threshold.

In a dialog system it is very important to minimize the number of misrecognitions. Precision gives an indication of the vulnerability of the system to misrecognitions. Figure 2 shows that FoCal shows clearly the best performance for all the thresholds. Both MegaM and FoCal outperform the Helios baseline.

Although rejections are not as problematic as misunderstandings in SDS, they can reveal if the system is minimizing them. In Figure 3 the Stepwise method has the best performance between 0.4 and 0.6 confidence thresholds. After that, Helios baseline has the best performance.

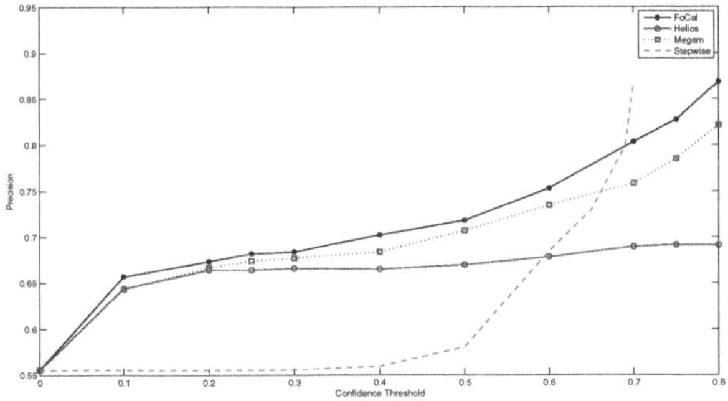

Fig. 2. Precision results for the compared methods

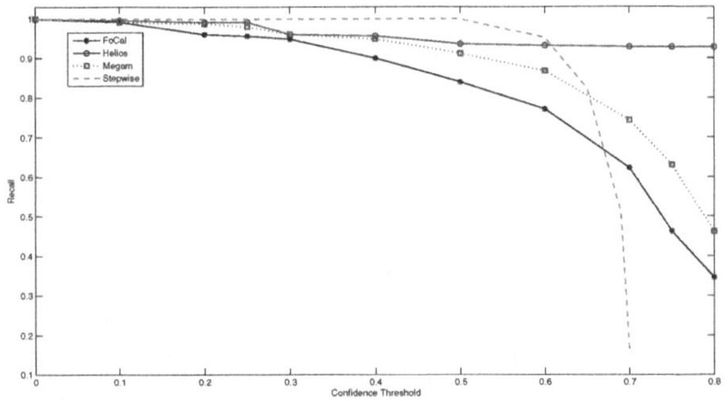

Fig. 3. Recall results for the compared methods

Among the methods presented the MegaM has a more balanced performance and it has best accuracy. We see that the rejection threshold of 0.6 seems to be the ideal point. The performance of all the methods that include the ASR confidence measure outperform the Helios baseline at this point.

6 Conclusions and Future Work

This paper describes a study of the improvement of the confidence score in an SDS, namely by including ASR confidence scores. A set of features was selected from the features computed by Helios. Three methods have been compared to baseline Helios confidence annotator: MaxEnt, prior-weighted Logistic Regression and stepwise regression. All the new approaches have outperformed the baseline if 0.6 confidence threshold is considered. The best performance was obtained using MaxEnt to compute the feature weights.

In the future, this study could be extended to other feature selection approaches. Semantic accuracy used in [7] and Word Error Rate could also be used to label the data set, instead of the binary classification that was used. It would also be interesting to see the impact of the results of this study with off-line data in an on-line system.

Acknowledgments. This work was partly supported by FCT project CMU-PT/005/2007.

References

1. Bohus, D., Raux, A., Harris, T.K., Eskenazi, M., Rudnicky, A.I.: Olympus: an open-source framework for conversational spoken language interface research. In: Bridging the Gap: Academic and Industrial Research in Dialog Technology Workshop at HLT/NAACL (2007)
2. Neto, J., Meinedo, H., Viveiros, M., Cassaca, R., Martins, C., Caseiro, D.: Broadcast News Subtitling System in Portuguese. In: Proc. ICASSP 2008, Las Vegas (2008)
3. Huggins-Daines, D., Kumar, M., Chan, A., Black, A.W., Ravishankar, M., Rudnicky, A.I.: PocketShpinx: A Free, Real-Time Continuous Speech Recognition System for Hand-Held Devices. In: Proc. ICASSP 2006, Toulouse, France (2006)
4. Lopes, J., Eskenazi, M., Trancoso, I.: Towards Choosing Better Primes for Spoken Dialog Systems. In: Proc. ASRU 2011, Hawaii, USA (2011)
5. Bohus, D., Rudnicky, A.I.: A principled approach for rejection threshold optimization in spoken dialog systems. In: Proc. Interspeech 2005, Lisbon, Portugal (2005)
6. Sarikaya, R., Gao, Y., Picheny, M., Erdogan, H.: Semantic Confidence Measurement for Spoken Dialog Systems. IEEE Trans. on Speech and Audio Processing 13 (July 2005)
7. Litman, D., Pan, S.: Designing and evaluating an adaptive spoken dialogue system. User Modeling and User-Adapted Interaction 12, 111–137 (2002)
8. Raymond, C., Estève, Y., Béchet, F., De Mori, R., Damnati, G.: Belief confirmation in Spoken Dialogue Systems using confidence measures. In: Proc. ASRU 2003, St. Thomas, US Virgin Islands (2003)
9. Daumé III, H.: Notes on CG and LM-BFGS Optimization of Logistic Regression (August 2004), `http://pub.hal3.name#daume04cg-bfgs`, `http://hal3.name/megam/`
10. Brummer, N.: Focal: Tools for Fusion and Calibration of automatic speaker detection systems, `http://www.dsp.sun.ac.za/nbrummer/focal/`

Transcription of Multi-variety Portuguese Media Contents

Alberto Abad[1], Hugo Meinedo[1], Isabel Trancoso[1,2], and João Neto[1,2]

[1] INESC-ID Lisboa, Portugal
[2] Instituto Superior Técnico, Lisboa, Portugal
alberto.abad,hugo.meinedo,
isabel.trancoso,joao.neto@l2f.inesc-id.pt
http://www.l2f.inesc-id.pt/

Abstract. Current automatic transcription technology applied to media contents is an important medium that not only allows generating subtitles, but also enables data search and retrieval capabilities over multimedia streams. Among others, one of the most important challenges that transcription systems have to deal with is speaker accent variability. In this work we study the importance of accent variability for three broad varieties of Portuguese: African Portuguese, Brazilian Portuguese and European Portuguese. Then, we propose a multi-variety transcription system based on the combination of variety identification followed by specific variety-dependent transcription systems.

Keywords: automatic speech recognition, speaker accent variability, accent identification, broadcast news transcription.

1 Introduction

The presence of speech of different accents of the same language and the need for robustly dealing with them poses a significant challenge in automatic speech recognition (ASR) systems [1]. Thus, the influence of accent as a speaker variability factor in speech recognition has been the focus of intense research both for dialectal and foreign/non-native speech. In general, approaches to deal with accented speech can be coarsely classified into two groups [2]. First, those based on the adaptation of a general non-accented speech recognizer to the particular characteristics of one speaker or a small group of speakers. Typically, in these approaches acoustic models [3] and pronunciation lexicons [4] are adapted with a small amount of data. A second possible approach when a large amount of training data for each accent is available consists of building complete accent or variety specific speech recognition systems [2]. The key point in this type of approach is the need for an automatic method to identify and select the appropriate system for each accent.

Besides applications such as voice operated interactive systems, the need for robust recognition of several language varieties can be also crucial in automatic

H. Caseli et al. (Eds.): PROPOR 2012, LNAI 7243, pp. 409–420, 2012.

media contents transcription. Let us take, for instance, the example of a news subscription service that crawls for different media sources to automatically transcribe and then provide results to the users according to their preferences. A certain user may be interested only in news in a specific language, but he/she may be indifferent to the specific variety, the variety information may not be available or even the media content may present a mixture of speech of different varieties. In any of these cases, it is necessary that the media transcriber system reacts robustly to this type of variability.

At the INESC-ID's Spoken Language Systems Laboratory (L^2F) we have developed in the last years a media content processing system that integrates several speech and language technologies for European Portuguese. Among other applications, this system has been used as the core of the fully automatic speech recognition subtitling system that is running on the main news shows of the public TV channel in Portugal (RTP), since March 2008 [5]. In fact, although Portuguese is the seventh most spoken language in the world with around 178 Million L1-speakers [6], only about five percent speak Portuguese with an European Portuguese (EP) accent. Consequently, one can expect that a considerable amount of media content is generated everyday in other Portuguese varieties. This fact motivated the development of variety dependent recognition modules for two other broad varieties of Portuguese: Brazilian Portuguese (BP) [7] and African Portuguese (AP) [8]. BP is the variety spoken in South America and it is the one with the largest number of speakers. AP encompasses the varieties spoken in one of the five PALOP countries (African Countries with Portuguese as Official Language): Angola, Cape Verde, Guinea-Bissau, Mozambique and São Tomé and Príncipe. The motivation to consider AP as a broad geographical variety is related to the difficulty to obtain data from each individual country, and also to the fact that a human benchmark [9] revealed that identifying African varieties in Broadcast News (our main source of data) is much harder than identifying accents of everyday's people on the street.

In this work, we take advantage of the existence of Portuguese variety-dependent systems to propose an automatic transcription system of media contents in different varieties of Portuguese based on automatic variety identification. The next section describes the main characteristics of the three variety-dependent ASR systems: AP, BP and EP. The variety identification system is described in section 3. It is composed by the fusion of 4 sub-systems (one acoustic and three phonotactic based) and is particularly designed for improved performance in close language/variety detection tasks. In section 4, cross-variety experiments show the importance of Portuguese accent as a speaker variability factor in speech recognition with a multi-variety corpus. Then, we evaluate the proposed multi-variety transcription system in contrast to an oracle system that uses the correct variety-dependent system for each test segment. Finally, this document finishes with the conclusions in Section 5.

2 Portuguese Variety-Dependent Transcription

2.1 The AUDIMUS Speech Recognizer

Our in house speech recognition engine [10] is a hybrid automatic speech recognizer that combines the temporal modeling capabilities of Hidden Markov Models (HMMs) with the pattern discriminative classification capabilities of Multi-Layer Perceptrons (MLPs). A block diagram is shown in Figure 1.

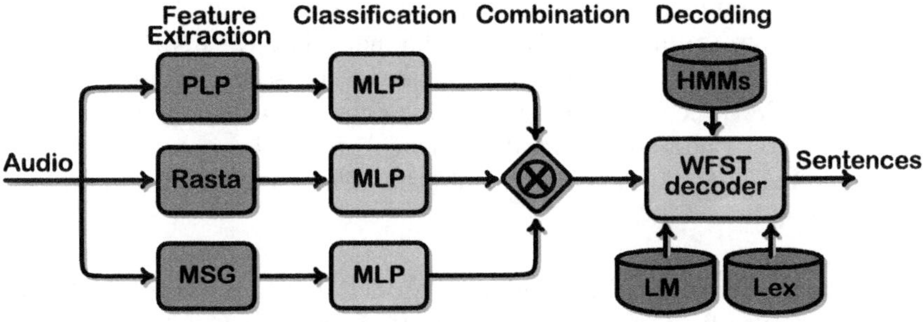

Fig. 1. Block diagram of AUDIMUS speech transcription system

Feature Extraction. The MLP/HMM acoustic model combines posterior phone probabilities generated by three phonetic classification branches. Different feature extraction and classification branches effectively perform a better modeling of the acoustic diversity, in terms of speakers and environments, present in data. The first branch extracts 26 PLP (Perceptual Linear Prediction) features, the second 26 Log-RASTA (log-RelAtive SpecTrAl) features and the 3rd uses 28 MSG (Modulation Spectrogram) coefficients for each audio frame.

MLP Classifiers and HMM Topology. Each MLP classifier incorporates local acoustic temporal context via an input window of several frames (between 7 and 15 frames) and is composed of two fully connected non-linear hidden layers and an output layer. Usually, the hidden layer size depends on the amount of training data available, while the number of softmax outputs of the output layer depends on the characteristic phonetic set of each language and the HMM topology. In the first versions of our recognizer single state context independent phoneme HMMs were trained, while in more recent versions a combination of multiple sate context independent units and intra-word context-dependent phone transition units are modeled [11].

Decoding Process. The decoder is based on the Weighted Finite-State Transducer (WFST) approach, where the search space is a large WFST that results from the integration of the HMM/MLP topology transducer, the lexicon transducer and the language model one [12]. This decoder uses a specialized WFST composition algorithm of the lexicon and language model components in a single step. Furthermore it supports lazy implementations, where only the fragment of the search space required in runtime is computed.

2.2 EP Transcription System

The EP transcription system was the first to be developed. Additional details can be found in [5].

EP Acoustic Model. The initial EP acoustic model was trained with 46 hours of manually annotated BN data collected from the public Portuguese TV. Currently, automatically collected and transcribed data is being reused to perform unsupervised training. The current iteration uses a total of 1000 hours of data mostly news shows from several EP TV channels. The EP MLPs are formed by 2 hidden layers with 2000 units each and have 500 softmax output units that correspond to 38 three state context independent phonemes of the EP language plus a single-state non-speech model (silence) and 385 phone transition units which were chosen to cover a significant part of all the transition units present in the training data.

EP Language Model. The Language Model (LM) is a statistically 4-gram model and results from the interpolation of three specific LMs. The first is a backoff 4-gram LM, trained on a 700M word corpus of newspaper texts, collected from the Web from 1991 to 2005. The second LM is a backoff 3-gram LM estimated on a 531k word corpus of broadcast news transcripts. The third model is a backoff 4-gram LM estimated on recent EP web newspapers texts, which are daily updated. The final interpolated LM was smoothed using Kneser-Ney modified discounting.

EP Vocabulary and Pronunciation Lexicon. The EP engine uses a 100k word vocabulary, which is also updated on a daily basis. The pronunciation lexicon is built automatically by classifying the words into "known" ones, for which the system retrieves a correct pronunciation from an in-house lexicon, and "unknown" ones. For the latter, a further split is made, which automatically detects spelled acronyms and foreign words. For "unknown" words which do not fit into these categories, the pronunciation is generated by our rule-based grapheme-to-phone (GtoP) conversion module [13]. For spelled acronyms, rule-based pronunciations are also generated. For foreign words, a further subdivision is made, in order to identify the ones that exist in the public domain lexicon provided by CMU[1], for which a nativized version is produced. For the words not included in the CMU lexicon, grapheme nativization rules are applied prior to using the GtoP module. The final multiple pronunciation EP lexicon includes 114k entries.

EP Results with BN Data. In one of our BN evaluation test sets (RTP07), which is composed by six one hour long news shows from 2007, our current EP BN transcription system achieves a word error rate (WER) performance of 18.4%.

---

[1] `http://www.speech.cs.cmu.edu/cgi-bin/cmudict`

2.3 BP Transcription System

The BP transcription system is the result of porting some of the key modules of the EP recognizer to cover the BP variety. Particularly, it was necessary to train new acoustic models based on BP data, to build new language models, and to develop a new GtoP module. Details of this process can be found in [7].

BP Acoustic Model. The initial set of acoustic models for BP was trained with about 15 hours of manually transcribed BN data recorded from the Record channel transmitted by cable TV in Portugal. Due to the reduced amount of data available compared to EP, the size of the two nonlinear hidden layers of the MLPs was reduced to 600 units and only monophone units were modeled (39 phonemes + silence). Afterwards we increased the training data with more 33 hours of automatically transcribed material, which allowed increasing the size of the hidden layers to 1000 units and modeling multiple-state and phone transition units (up to a total of 320 softmax outputs).

BP Language Model. The BP language model is a 4-gram backoff model created by interpolating three individual LMs built from three different sources: the CETENFolha corpus which has around 24M words, a recent newspapers corpora automatically obtained from the Internet which amounts to 62M words and the manual transcriptions of the training set. The language models were smoothed using Kneser-Ney discounting and entropy pruning.

BP Vocabulary and Pronunciation Lexicon. BP recognizer also uses a 100K word vocabulary that includes all the words of the transcriptions of the training corpus, completed with the most frequent words of the newspaper corpus. The pronunciation lexicon is generated similarly as for the EP, but with a specific GtoP module for the BP variety with new rules. In this module, the grapheme set was augmented with the symbol ü, covering words written before the recent orthographic convention, and the phoneme set was augmented with three symbols: the two affricate symbols [tʃ] and [dʒ] and an additional SAMPA symbol to take into account the different realization of "r's" in coda position most commonly found in BP. The symbol for [ə] was not included, neither was [l].

BP Results with BN Data. The current version of the BP transcription system achieves a WER of 21.6% in our BP test set formed by 13 short news shows, which amount to almost 2 hours of speech.

2.4 AP Transcription System

The AP transcription system was the last to be developed. In contrast to the BP, the GtoP module has not been ported to AP and our efforts were restricted to the acoustic and the language models. More details can be found in [8].

AP Acoustic Model. In terms of acoustic model training two distinct approaches were followed. The first one consisted of training new acoustic models using only the small amount of AP training data available which is around 7.5 hours and using around 17 hours of automatically detected and transcribed AP data [8]. The second approach consisted of adapting EP acoustic models using only the manually transcribed AP data. In this last case, MLPs of two hidden layers with 1500 units modeling single-state phonemes were considered. In practice, this last approach resulted the most convenient.

AP Language Model. The language model for AP is a 4-gram model created by interpolating three individual language models. The first is the same 4-gram LM from the EP, the second LM was built from recent AP newspapers automatically obtained from the Internet with 1.6M words and the third LM was built from the manual AP transcriptions of the training set which amount to 86k words. The language models were smoothed using Kneser-Ney discounting and entropy pruning.

AP Vocabulary and Pronunciation Lexicon. The same 100K word vocabulary and pronunciation lexicon of the EP system is used for AP.

AP Results with BN Data. The AP test set consists of 3 short BN shows of among 25 and 30 minutes each one. In this test set, the AP system achieves a WER performance of 23.7%. It is worth noting that the AP BN test set is quite challenging due to the high correlation between noisy spontaneous speech conditions and AP accented speech found in BN shows obtained from the RTP Africa channel. The reason is that in these BN shows the anchor usually speaks with an EP accent, while AP accented speech is most often found in out of studio interviews and reports.

3 Portuguese Variety Identification

The Portuguese variety identification system in this work follows the principals of the one in [8]. It combines a state of the art acoustic sub-system based on Gaussian supervectors with three variations of a recently proposed phonotactic approach that makes use of "specialized" tokenizers, known as mono-phonetic Phone Recognition followed by Language Modeling (PRLM) [14] approach.

For each identification sub-system, target variety models are trained with the Portuguese variety training data-set described in [15], which consists of a total of 4141 BN speech segments of different lengths from the three target varieties. The complete variety identification system is calibrated and assessed with the variety development data-set that consists of 1484 segments from speakers that are not in the variety training set: 610 AP, 462 BP and 412 EP segments [15].

3.1 Mono-Phonetic PRLM Sub-Systems

PRLM systems make use of phonetic classifiers to tokenize speech data into phonetic sequences. Then, for each target language/variety a different phonotactic n-gram language model is trained using all the phonetic sequences extracted from the training data of that particular language/variety. During test, the phonetic sequence of a given speech signal is obtained and the likelihood of each target n-gram model is evaluated to obtain target language/variety scores.

Our PRLM approach focuses on the phonetic classifier, trying to build a system with a highly specialized tokenizer that incorporates the differences between language/variety pairs at this level. To better characterize these differences, we divide all occurring phones in our varieties into the following two groups [16]:

1. mono-phones: phones in one language/variety, that overlap little or not at all with those in another language/variety.
2. poly-phones: phones that are similar enough across the languages/varieties to be equated.

Determining the set of phones which best characterize a certain variety, given its neighboring varieties, is not straightforward. Linguistic knowledge about the varieties' phonetic and phonological characteristics is crucial, but often not available, not sufficiently detailed or controversial. We use a computational method instead to find variety dependent unique phones. Binary multi-layer perceptrons (MLPs) are trained to discriminate between the same pairs of aligned phone classes. In this work, we have used 3 phonotactic sub-systems, one per each Portuguese variety pair: AP-EP, AP-BP and BP-EP. More details about the determination of the mono-phonetic units and the training of the mono-phonetic classifiers can be found in [15].

3.2 Gaussian Supervector Based Sub-System

Combining Gaussian mixture models (GMM) with Support Vector Machines (SVM) as a discriminative classifier [17], the so-called Gaussian supervector (GSV) approach, is a well-known state-of-the-art technique in the Language Identification field. In this work, we have built a GSV system based on mean supervectors: Maximum-a-Posteriori (MAP) adaptation of Gaussian means. The extracted features are Shifted Delta Cepstra of PLP-RASTA features [18]. The universal background model (UBM) of 1024 mixtures was trained with all Portuguese variety data. In this implementation, we have used an alternative scoring approach [19]. In contrast to the conventional GSV, each language SVM model is *pushed back* to a *positive* and a *negative* variety-dependent GMM model, which are then used to calculate log-likelihood ratio scores. In certain situations, especially on short utterances, this approach has shown improved accuracy. In fact, this is the typical situation in BN data where long speech segments are rare.

3.3 Back-End Calibration and Identification Results

A linear logistic regression back-end for simultaneous fusion and calibration of the detection sub-systems has been developed using the FoCal Multiclass Toolkit[2]. Five-fold cross-calibration strategy was applied for back-end parameter estimation.

[2] http://niko.brummer.googlepages.com/focalmulticlass

The complete system generates three variety-dependent detection calibrated scores for every test segment. Since we are interested in variety identification rather than in verification, we select as the identified variety the one with the largest variety-dependent detection score. The miss probability and false alarm averaged results for the three varieties in the development set are 7.53% and 3.74% respectively, which corresponds to an average variety identification cost of 5.63%. Notice, however, that these might be optimistic performance indicators, since the same data set was used for both calibration and evaluation.

4 Multi-variety Transcription

4.1 Multi-variety Evaluation Corpus

The multi-variety evaluation corpus is formed by three subsets extracted from each one of the previously described BN speech recognition evaluation sets of the three Portuguese varieties. Concretely, for each variety we have randomly selected speech segments with a total approximate duration of around 45 minutes, totaling 131 minutes of useful speech. Table 1 summarizes the most relevant characteristics of the multi-variety corpus.

Table 1. Multi-variety corpus

Test Data	AP	BP	EP	$\sum$
duration [min.]	45.0	44.9	41.8	131.7
segments	282	505	360	1147
words	6948	7641	7148	21737
∅ dur./segm. [s]	9.6	5.3	6.9	6.9

4.2 Cross-Variety Speech Recognition Tests

Table 2 shows the WER performance results obtained by each variety-dependent transcription system, including results for each separate variety data subset and for the whole multi-variety corpus.

Table 2. Variety matched and cross-variety WER results

	AP	BP	EP	all
AP-ASR	**24.5**	49.0	22.4	32.4
BP-ASR	52.2	**22.1**	62.1	44.8
EP-ASR	27.2	57.0	**16.7**	34.2

Attending to the partial results, it is clear that the best performance is always obtained in matched variety conditions with a considerably robust performance in the three Portuguese varieties. In the ideal case of knowing the variety of each test segment (oracle system), the WER achieved in the overall multi-variety test set is 21.1%.

With respect to the cross-variety figures, it can be observed that AP and EP are much closer among them than the BP variety in terms of ASR performance. Thus, we can even observe in the AP-ASR row a better WER performance in the EP subset than in the AP subset. First, it is worth recalling that the AP ASR system makes use of acoustic models adapted from EP, language models that include EP information and that the same pronunciation lexicon is used. Second, as noted previously in Section 2.4, this figure also reveals that the AP subset is more challenging in terms of ASR than the EP one. With respect to the BP variety, although it is the "most distant" one in terms of cross-variety ASR results, these figures also seem to show a certain closer proximity to AP than to EP variety.

Regarding the overall results, the best variability-dependent transcription system is the AP one with a WER of 32.4%. However, the performance of the best individual system is far of the oracle system results (21.1% WER).

4.3 Multi-variety Recognition Based on Variety Identification

In our approach to the transcription of multi-variety Portuguese media contents, we first apply the variety identification system in a per segment basis in order to select the appropriate variety-dependent transcription system. Then, the selected transcriber is applied for that particular test segment.

Variety Identification. Table 3 shows the variety identification results obtained in the multi-variety test corpus with the system described in previous Section 3. In this case, the average variety miss and false alarm rates are 22.4% and 10.0% respectively, which corresponds to an average identification cost of 16.2%. These results represent a strong degradation with respect to the reference identification results reported in Section 3.3. This difference may be partially explained by the optimistic method used to calibrate and measure the performance of the variety ID system. Particularly, there is a very significant increase in the number of EP segments that are misclassified as AP, which results in a considerably boost of EP miss rate and AP false alarm rate. This fact seems to indicate the existence of a strong mismatch between the multi-variety EP sub-set and the characteristics of the EP data used for variety ID training and calibration. On the other hand, the average identification cost for the BP variety is of 6.8%, which is reasonably low. In spite of the performance drop with respect to the reference, these results are still quite encouraging since misclassification happens most often among varieties that are closer between them according to the cross-variety WER results reported in Table 2. Consequently, we expect a low impact in terms of multi-variety performance transcription.

ASR Results. Table 4 shows the results of the oracle system and of the proposed multi-variety system that makes use of automatic variety identification to select the variety-dependent recognizer. Attending to the individual variety subsets, an almost equivalent performance can be observed for AP and BP with respect to the oracle. In the case of AP, this occurs in spite of the fact that the

Table 3. Portuguese variety identification results in terms of false alarm rates *Pfa(Lt,Ln)*, where *Lt* is the target variety and *Ln* de actual variety, and miss probability rate *Pmiss(Lt)* and average false alarm rate *avg Pfa(Lt)* for each target variety

100 x *Pfa(Lt,Ln)*		*Lt*	
Ln	AP	BP	EP
AP	—	7.09	4.61
BP	5.74	—	0.20
EP	41.39	8.06	—
100 x *Pmiss(Lt)*	11.7	5.94	49.4
100 x *avg Pfa(Lt)*	20.58	7.63	1.8

miss rate for AP is 11.7%. It is likely that the segments erroneously classified as EP correspond to slight accented AP speech, which may be recognized equivalently or even better with the EP transcriber. The low miss rate for BP explains the low degradation of the multi-variety transcription system for this variety. On the other hand, the largest performance drop is observed in the EP subset and it is related to large misclassification of EP segments as AP segments. Anyway, the impact is not dramatic since, as already noted, the AP transcriber shares several components from the EP system. With respect to the overall result of the multi-variety transcriber, a performance of 22.7% WER is achieved, which is very close to the oracle system and much better than any of the individual variety-dependent recognizers of Table 2.

Table 4. WER results of the oracle and of the proposed multi-variety transcriber

	AP	BP	EP	all
oracle ASR	24.5	22.1	16.7	**21.1**
multi-variety ASR	24.5	22.6	21.0	**22.7**

5 Conclusions

In this work we have addressed the speaker accent variability issue for the transcription of media contents in broad varieties of the Portuguese language. First, we have demonstrated the strong impact of Portuguese varieties into speech recognition performance through a set of cross-variety experiments. Then, we have proposed a multi-variety transcription system based on the combination of variety identification and variety-dependent automatic transcribers. The proposed system showed excellent results when compared to the oracle system that uses true variety identity information. In the future we expect to improve the system, reducing the misclassification rate for the African/European Portuguese varieties pair and improving African Portuguese speech recognition. For this purpose, some possibilities include augmenting the amount of transcribed training data, namely for AP, and building a specific pronunciation lexicon or developing separate complete speech recognizers for each PALOP country variety.

Acknowledgments. This work was partially supported by FCT (INESC-ID multi-annual funding) through the PIDDAC Program funds, and also through the EU-funded project *EUTV Adaptive Media Channels*.

References

1. Huang, C., Chen, T., Li, S., Chang, E., Zhou, J.L.: Analysis of speaker variability. In: Proc. European Conference on Speech Communication and Technology, Denmark, vol. 2, pp. 1377–1380 (2001)
2. Huang, C., Chang, E., Chen, T.: Accent Issues in Large Vocabulary Continuous Speech Recognition. Microsoft Research China Technical Report, MSR-TR-2001-69 (2001)
3. Wang, Z., Schultz, T., Waibel, A.: Comparison of acoustic model adaptation techniques on non-native speech. In: Proc. ICASSP 2003, pp. 540–543 (2003)
4. Humphries, J.J., Woodland, P.C., Pearce, D.: Using accent-specific pronunciation modelling for robust speech recognition. In: Proc, Fourth International Conference on Spoken Language, ICSLP, vol. 4, pp. 2324–2327 (1996)
5. Neto, J., Meinedo, H., Viveiros, M., Cassaca, R., Martins, C., Caseiro, D.: Broadcast news subtitling system in Portuguese. In: Proc. ICASSP 2008, Las Vegas, USA (2008)
6. Lewis, M.P.: Ethnologue: Languages of the World, 16th edn., SIL International, (May 2009), `http://www.ethnologue.com/`
7. Abad, A., Trancoso, I., Neto, N., Viana, M.C.: Porting an European Portuguese broadcast news recognition system to Brazilian Portuguese. In: Proc. Interspeech 2009, Brighton, UK (2009)
8. Koller, O., Abad, A., Trancoso, I., Viana, C.: Exploiting variety-dependent phones in portuguese variety identification applied to broadcast news transcription. In: Proc. Interspeech 2010, Makuhari, Japan (2010)
9. Rouas, J., Trancoso, I., Viana, C., Abreu, M.: Language and variety verification on broadcast news for Portuguese. Speech Communnication 50(11-12), 965–979 (2008)
10. Meinedo, H., Abad, A., Pellegrini, T., Trancoso, I., Neto, J.: The L2F Broadcast News Speech Recognition System. In: Proc. Fala 2010, Vigo, Spain (2010)
11. Abad, A., Neto, J.: Incorporating acoustical modeling of phone transitions in an hybrid ANN/HMM speech recognizer. In: Proc. Interspeech 2008, Brisbane, Australia, pp. 2394–2397 (2008)
12. Caseiro, D., Trancoso, I.: A specialized on-the-fly algorithm for lexicon and language model composition. IEEE Transactions on Audio, Speech and Lang. Proc. 14(4) (2005)
13. Caseiro, D., Trancoso, I., Oliveira, L., Viana, C.: Grapheme-to-phone using finite state transducers. In: Proc. 2002 IEEE Workshop on Speech Synthesis, Santa Monica, CA, USA (2002)
14. Zissman, M.A.: Comparison of Four Approaches to Automatic Language Identification of Telephone Speech. IEEE Transactions on Speech and Audio Processing 4(1) (1996)
15. Koller, O., Abad, A., Trancoso, I.: Exploiting variety-dependent phones in Portuguese variety identification. In: Odyssey 2010: The Speaker and Language Recognition Workshop (2010)

16. Berkling, K., Arai, T., Barnard, E.: Analysis of Phoneme-Based features for language identification. In: Proc. ICASSP, vol. 1, pp. 289–292 (1994)
17. Campbell, W.M., Campbell, J.P., Reynolds, D.A., Singer, E., Torres-Carrasquillo, P.A.: Support vector machines for speaker and language recognition. Computer Speech and Language 20(2-3), 210–229 (2006)
18. Torres-Carrasquillo, P.A., Singer, E., Kohler, M.A., Greene, R.J., Reynolds, D.A., Deller Jr., J.R.: Approaches to Language Identification using Gaussian Mixture Models and Shifted Delta Cepstral Features. In: Proc. ICSLP 2002, Denver, Colorado, pp. 89–92 (2002)
19. Campbell, W.M.: A covariance kernel for svm language recognition. In: Proc. ICASSP 2008, pp. 4141–4144 (2008)

Towards Automatic Classification of Speech Styles

Arlindo Veiga[1], Sara Candeias[1], Dirce Celorico[1],
Jorge Proença[1], and Fernando Perdigão[1,2]

[1] Instituto de Telecomunicações, pole of Coimbra, DEEC, Coimbra, Portugal
[2] Universidade de Coimbra, DEEC/FCTUC, Coimbra, Portugal
{aveiga,saracandeias,dircelorico,jproenca,fp}@co.it.pt

Abstract. In this paper we present results from a study seeking to distinguish "unprepared" from "prepared" speech in broadcast news media. The idea is to explore the results from a previous experiment concerning the characterization of filled pauses and extensions, extending the analysis of such hesitation phenomena to large audio corpus. Daily news broadcasts of Portuguese television were segmented and labeled manually in terms of several speech styles, over a range of background environments. An automatic detection of filled pauses and extensions in this audio data allowed us to correlate the presence of hesitation events with segments of unprepared speech. Distinguishing unprepared speech from prepared speech is of considerable practical interest for audio segmentation, speech processing and linguistic research. The long-term objective of this work is to automatically segment all audio genres and speaking styles as well as identify prosodic and linguistic features of the speech segments.

Keywords: unprepared and spontaneous speech, hesitations, broadcast news audio.

1 Introduction

1.1 Motivation

It is a fact that "unprepared speech" and "prepared speech", both under the head of "speech styles" are significantly different acoustically and linguistically in every language [6], [5]. Speech styles classification and segmentation is an interdisciplinary research field concerned with the identification and the automatic recognition of the properties of the speech variety. The present study explores the results of a previous experiment concerning the characterization of the hesitation events [14], such as filled pauses (FP) and vocalic extensions (EX), extending the analysis of such phenomena to an enlarged corpus, in order to classify fragments of unprepared- and prepared speech. In particular, we aim to explore the correlation between the detection of hesitation events and the presence of unprepared speech, obtaining a more accurate comprehension of the speech profile as well as a better understanding of the communication process.

H. Caseli et al. (Eds.): PROPOR 2012, LNAI 7243, pp. 421–426, 2012.

1.2 Background

The concept of speech style depends on a sense of what a speech style is. In fact, expressions such as "prepared", "planned", "spontaneous" or "read" speech are rather vague and susceptible to discussion. Often the term "read speech" is synonymous of "planned-" or "prepared speech". Also, "spontaneous" speech is used to refer to "unplanned speech" or "unprepared speech". It is a common sense that speaking without planning leads to the appearance of disfluencies. Such occurrences in the speech signal are a strong indicator of the "unprepared" speech style.

In general, speakers express their spontaneity in speech through morphologic-syntactic and acoustic-phonetic features, such as self-repairs including repetitions and restarts, interjections, unknown or mispronounced words, ellipsis, ungrammatical constructions or unusual orders, partial words, fillers including filled pauses and discourse markers, extensions, silence pauses, pitch variation, speaking-rate fluctuation, etc. (see, for instance [12], [13]). With respect to the morphological and syntactical structures, unprepared speech includes ill-formed phenomena. The characterization of these phenomena with a reliable feature set could be itself other topic for research, since it has not been well explained by conventional linguistic theory so far. Related to the acoustic-phonetic features, some cues have been indicated in the literature as being relevant for the characterization of the connected or continuous speech as a subset of unprepared speech. For Portuguese, we can mention works that had attempted to provide evidence of the hesitation events [14], [9] or the degree of relative hypo articulation of the surface forms [2], [4], in the continuous speech. Another study, [1], compares European and Brazilian Portuguese speaking styles, in order to characterize the rhythm of the speech. In [7], an overview is given of the phonetic and phonological correlates of speaking styles. In the context of the automatic speech recognition area, [10] presents a study describing the differences between acoustic characteristics of spontaneous and read speech.

Studies of the broadcast news audio segmentation and speech transcription in various languages [11], including the Portuguese, have not addressed the problem of speech style identification. The present work intends to fill this gap by using hesitation cues to automatically classify speech segments in terms of speech styles.

2 Corpus Description

Audio from broadcast news of a Portuguese television channel was used for training, test and evaluation purposes. The audio data is stored at a 16kHz sampling rate and the video information is discarded. A total of 30 daily news programs were considered for this study, with duration of about 25 hours. The sound material contains studio and out of studio recordings, as well as sessions recorded from the telephone. Utterances by anchors and professional speakers, commentators, reporters, interviews and interviewees are present in the audio. Prepared speaking style is dominant but, most of the times, speech is over background speech, noise or music. There are also non-speech events like jingles, laughter, coughing or clapping.

All the audio has been carefully examined and annotated manually, using the *Transcriber* software tool [3]. Four levels of annotation were considered, as shown in Table 1. At the signal level (first level), the labels are 'speech', 'music', 'jingle', 'noise', 'cough', 'laugh', 'claps', etc. The acoustical environment is the second level with the following labels: 'clean', 'music', 'stationary noise', 'speech overlapping', 'crowd noise', 'mixed-' or 'indistinct' background. In this level speech over a telephone system was also annotated. The speaking style is identified in the third level, labeled as 'Lombard', 'prepared' and 'unprepared' speech. Unprepared speech is still differentiated into with low-, average- and high-spontaneity, taking into consideration the occurrence of the hesitation events. The speaker information corresponds to the fourth level, in which the speaker is identified whenever possible. The occurrence of foreign languages in the signal was also included in the labels. To the work presented here, only the annotations with prepared- and unprepared speech styles were explored.

Table 1. Corpus annotation levels

1 - Signal Level	2 – Acoustical Env.	3 – Speech Style	4 - Speaker
Speech*	Clean (no noise)	Prepared	Anchor1
Silence	Music overlap	Lombard	Anchor2
Music	Speech overlap	Unprepared (High)	Journalist(M/F)
Jingle	Stationary noise	Unprepared (Average)	Male
Noise	Crowd noise	Unprepared (Low)	Female
Clapping, etc.	Mixed background		VIP(1,2,3...)
	Indistinct background		
	Telephone		

*Speech annotations have levels 2, 3 and 4 compulsorily. Other first level annotations are not classified further.

3 Hesitation Detection Procedure

In this paper we consider filled pauses (FP) and vowel extensions (EX) as hesitation events that could be used to characterize unprepared speech. These events are detected through a description of the speech at a phonetic level. For this purpose, the audio signal was analyzed according to a standard front-end for speech recognition: 12 Mel-frequency cepstral coefficients (MFCC), plus log energy, and their first and second order regression coefficients, at a frame-rate of 100Hz. The phone acoustic models are defined in terms of Hidden Markov Models (HMM), which were previously built using HTK 3.4 [15]. Initially, an automatic segmentation of the audio signal into acoustically consistent segments was done by means of the distBIC algorithm [16], [8]. Then, a HMM decoder is applied to those speech segments, providing an optimal phone sequence. The phone sequence is explored in order to identify a set of recognized vowels that remain for longer than a predefined period. The segments that correspond to this condition are assigned as a FP/EX event. We have found that a period of 350ms was adequate [14]. Several phone insertion penalties were considered to reach a good compromise between false positives and false negatives rates in the FP/EX detection.

Given the hypothesis that the occurrence of FP/EX is indicative of unprepared-speech, the detected events were matched to the tagged speech segments in order to validate the classification. A segment, given its endpoints, is classified as prepared or unprepared speech as a function of the number of FP/EX events present in that segment.

4 Results and Discussion

The underlying hypothesis is that FP and EX are prevalent in unprepared speech and that each speaking style segment is characterized by having or not these hesitation events. Thus, a true positive/negative corresponds to a correctly predicted unprepared/prepared speaking style; a false positive (or false alarm) indicates a wrong decision of unprepared speech due to the presence of hesitation events and a false negative (miss) indicates a wrong decision as prepared speech since no hesitation events were detected in the segment.

The decision of the speech style is based on the number of FP/EX events detected in the segment. For a threshold of 2 events (only a segment with 2 or more events is predicted as unprepared) we obtain a false positive rate of 23% and a false negative rate of 33%. Another threshold could be based on the FP/EX rate, (hesitation events per second), giving, however, similar results. These thresholds can be adjusted to tradeoff false-alarms and misses.

Fig. 1 shows the histograms of the number of FP/EX events, conditioned to the prepared or unprepared speech segments. As it can be seen, most of the segments of prepared speech have none or one hesitation and a larger number of events are usually detected in unprepared segments. However, there is a considerable overlap between the distributions, which accounts for the indicated error rates.

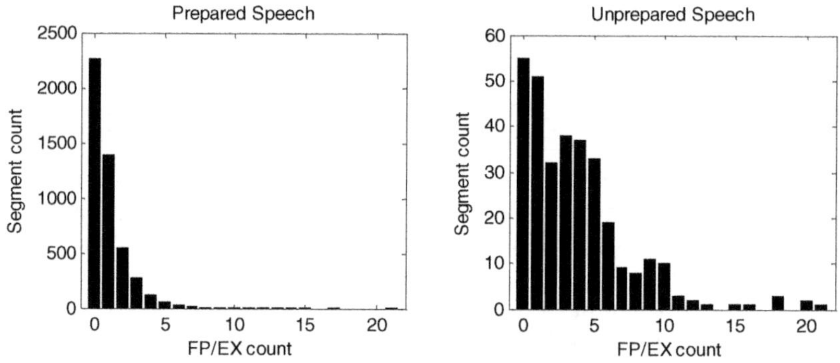

Fig. 1. Histograms of the number of FP/EX events detected in the reference segments of prepared speech (left) and unprepared speech (right)

Although not being a perfect indicator on its own to successfully separate these two styles of speech, it seems to be a very important parameter, possibly to consider simultaneously with other disfluency parameters for an automatic classification of speech style. We could explain some of the misses in unprepared speech by the presence of other events (besides FP and EX), which are not considered in our current method, such as truncated words, self-repairs and repetitions. Furthermore, our audio contains some acoustic environments that provide a variety of background noises or music during the considered speech. It is possible that some hesitations were extrapolated from vowels detected from the background during prepared speech, giving an incorrect classification. There are also other occurrences on prepared speech that can lead to erroneous classifications, such as the prominent words.

5 Conclusion

In this paper we presented results from a study seeking to distinguish "unprepared" from "prepared" speech in Portuguese broadcast news audio. The idea was to explore the results from a previous experiment concerning the characterization of FP and EX, extending the analysis of such hesitation phenomena to an enlarged audio corpus. Daily news broadcasts of Portuguese television were segmented and labeled manually in terms of several speech styles, over a range of background environments. We had hypothesized that unprepared speech segments could be classified knowing the presence of hesitation events. An automatic detection of such hesitation events was implemented in the audio data, allowing us to correlate the presence of FP/EX with segments of unprepared speech. In the presented study, only FP/EX were taken into consideration; however we verified that other acoustic and lexical events should be considered in order to improve the classification.

The long-term objective is to segment automatically all audio genres and speaking styles. Regarding this objective, we have already implemented several important features, such as audio segmentation using BIC, aspiration detection using word spotting, speaker identification using GMM [17], jingle detection based on audio fingerprint [18], and so on. The identification of both prosodic and linguistic features in the speech segments is another aim of this study.

Acknowledgments. This work is funded by Instituto de Telecomunicações, FCT (SFRH/BPD/36584/2007 and PTDC/CLE-LIN/11 2411/2009) and by QREN project TICE.Healty (13842).

References

1. Barbosa, P., Viana, M., Trancoso, I.: Cross-variety Rhythm Typology in Portuguese. In: Interspeech 2009, ISCA Brighton, UK (2009)
2. Braga, D., Freitas, D., Teixeira, J.P., Barros, M.J., Latsh, V.: Back Close Non-Syllabic Vowel [u] Behavior in European Portuguese: Reduction or Suppression. In: ICSP2001 (International Conference in Speech Processing), Taejon, Korea, August 22-24 (2001)
3. Barras, C., Geoffrois, E., Wu, Z., Liberman, M.: Transcriber: a Free Tool for Segmenting, Labeling and Transcribing Speech. In: First International Conference on Language Resources and Evaluation (LREC), pp. 1373–1376 (1998)

4. Candeias, S., Perdigão, F.: A realização do schwa no Português Europeu. In: 8th Symposium in Information and Human Language Technology (STIL 2011), II Workshop on Portuguese Description-JDP, Cuiabá, UFMG – Brazil (October 2011)
5. Furui, S.: Recent advances in spontaneous speech recognition and understanding. In: ISCA & IEEE Workshop on Spontaneous Speech Processing and Recognition (SSPR), Tokyo, pp. 1–6. IEEE Press, New York (2003)
6. Levelt, W.: Speaking. MIT Press, Cambridge (1989)
7. Llisterri, J.: Speaking styles in speech research. In: ELSNET/ESCA/SALT Workshop on Integrating Speech and Natural Language, Dublin, Ireland (July 1992)
8. Meinedo, H., Neto, J.: Audio Segmentation, Classification and Clustering in a Broadcast News Task. In: IEEE Transactions on Audio, Speech, and Language Processing Archive, vol. 18 (1). IEEE Press, Piscataway (2010)
9. Moniz, H., Trancoso, I., Mata, A.: Classification of disfluent phenomena as fluent communicative devices in specific prosodic contexts. In: Interspeech 2009, ISCA Brighton, UK, pp. 1719–1722 (2009)
10. Nakamura, M., Iwano, K., Furui, S.: Differences between acoustic characteristics of spontaneous and read speech and their effects on speech recognition performance. Computer Speech and Language 22, 171–184 (2008)
11. Rosenberg, A., Hirschberg, J.: Story Segmentation of Broadcast News in English, Mandarin and Arabic. In: HLT-NAACL 2006, New York (2006)
12. Shriberg, E., Stolcke, A., Hakkani-Tuor, D., Tur, G.: Prosody-based automatic segmentation of speech into sentences and topics. Speech Communication 32, 127–154 (2000)
13. Shriberg, E.: Spontaneous speech: How people really talk, and why engineers should care. In: Interspeech 2005, Lisbon, Portugal (2005)
14. Veiga, A., Candeias, S., Lopes, C., Perdigão, F.: Characterization of hesitations using acoustic models. In: 17th International Congress of Phonetic Sciences (ICPhS XVII), Hong Kong, August 17-21, pp. 2054–2057 (2011)
15. Young, S., Evermann, G., Hain, T., Kershaw, D., Moore, G., Odell, J., Ollason, D., Amnd Povey, D., Valtchev, V., Woodland, P.: The HTK Book (for HTK Version 3.4). Microsoft Corp. and Cambridge University Engineering Department, Cambridge (2006)
16. Delacourt, P., Welleken, C.J.: DISTBIC: A Speaker-Based Segmentation for Audio Data Indexing. Speech Communication 32, 111–126 (2000)
17. Reynold, D.A., Quatieri, T.F., Dunn, R.B.: Speaker Verification Using Adapted Gaussian Mixture Models. Digital Signal Processing 10, 19–41 (2000)
18. Lopes, C., Veiga, A., Perdigão, F.: Using Fingerprinting to Aid Audio Segmentation. In: VI Jornadas en Tecnología del Habla and II Iberian SLTech Workshop - FALA 2010, Vigo, Spain (2010)

Author Index

GPSR Compliance

The European Union's (EU) General Product Safety Regulation (GPSR) is a set of rules that requires consumer products to be safe and our obligations to ensure this.

If you have any concerns about our products, you can contact us on ProductSafety@springernature.com

In case Publisher is established outside the EU, the EU authorized representative is:

Springer Nature Customer Service Center GmbH
Europaplatz 3
69115 Heidelberg, Germany

Batch number: 10844619

Printed by Printforce, the Netherlands